Oxidative Stress in Applied Basic Research and Clinical Practice

Editor-in-Chief
Donald Armstrong

More information about this series at http://www.springer.com/series/8145

Note from the Editor-in-Chief

All books in this series illustrate point-of-care testing and critically evaluate the potential of antioxidant supplementation in various medical disorders associated with oxidative stress. Future volumes will be updated as warranted by emerging new technology, or from studies reporting clinical trials.

Donald Armstrong
Editor-in-Chief

Stephen C. Bondy • Arezoo Campbell
Editors

Inflammation, Aging, and Oxidative Stress

Springer

Editors
Stephen C. Bondy
Department of Medicine
Center for Occupational and Environmental Health
University of California
Irvine, CA, USA

Arezoo Campbell
Department of Pharmaceutical Sciences
College of Pharmacy
Western University of Health Sciences
Pomona, CA, USA

ISSN 2197-7224 ISSN 2197-7232 (electronic)
Oxidative Stress in Applied Basic Research and Clinical Practice
ISBN 978-3-319-33484-4 ISBN 978-3-319-33486-8 (eBook)
DOI 10.1007/978-3-319-33486-8

Library of Congress Control Number: 2016949388

Printed on acid-free paper

This Springer imprint is published by Springer Nature
The registered company is Springer International Publishing AG Switzerland

Preface

Aging carries inherent risk upon which distinct diseases are superimposed. Normal physiological processes tend to develop pathological aspects, which further disrupt declining tissue function. These age-related diseases often exaggerate the typical cellular changes accumulating with time, which are associated with senescence. For this reason, there has been a gradual merging of concepts regarding normal aging and specific disorders.

The aim of this book is to describe this interface, emphasizing two components (tissue damage by oxidant free radicals and chronic inflammation) which are associated with both aging and, more pronouncedly, age-related disease. It is remarkable how these processes appear to represent a commonality between apparently unrelated diseases. Oxidant and inflammatory phenomena often represent aberrations of the common effective functioning of reactive oxygen species and the immune system, in signaling and bactericidal events and in dispersal of materials of exogenous origin. In this book, we have emphasized cardiovascular and neurodegenerative disorders. The chapters that have been assembled are focused on oxidative stress and inflammatory events in both normal and aberrant aging, and how these may interact. Several chapters describe the underlying mechanistic basis of oxidative stress and inflammation. The promotion of these events by environmental factors is considered in other chapters.

The oxidant and inflammatory phenomena discussed appear to be inescapable components of the aging process and lay a foundation for many diseases which are present only in the elderly. The last section of this book provides several chapters that discuss the alleviation of these inevitable changes. These include lifestyle changes such as dietary, behavioral, and exercise-based strategies. Such measures can improve the length and quality of life for the elderly and are also likely to impact the incidence of specific age-linked diseases.

The last two of Shakespeare's "Seven Ages of Man" read as follows: "The sixth age shifts into the lean and slippered pantaloons, with spectacles on nose and pouch on side… and his big manly voice, turning again toward childish treble, pipes and whistles in his sound," followed by "Last scene of all, that ends this strange eventful history, is second childishness and mere oblivion, sans teeth, sans eyes, sans taste, sans everything."

These stages, while undesirable, are obviously inescapable. By appropriate environmental and dietary means, it is hoped that the earlier stages can be prolonged while these last phases can be collapsed into a relatively short time.

Irvine, CA, USA Stephen C. Bondy
Pomona, CA, USA Arezoo Campbell

Contents

Contributors

Syed F. Ali Division of Neurotoxicology, National Center for Toxicological Research/US FDA, Jefferson, AR, USA

Vellareddy Anantharam Parkinson's Disorder Research Laboratory, Iowa Center for Advanced Neurotoxicology, Department of Biomedical Sciences, Iowa State University, Ames, IA, USA

Jesus A. Araujo Division of Cardiology, Department of Medicine, David Geffen School of Medicine at UCLA, Los Angeles, CA, USA

Department of Environmental Health Sciences, Fielding School of Public Health, UCLA, Los Angeles, CA, USA

Francesc Baixauli Signaling and Inflammation Program, Spanish National Center for Cardiovascular Research Carlos III (CNIC), Madrid, Spain

Immunology, La Princesa Health Research Institute, Madrid, Spain

Steven W. Barger Geriatric Research Education and Clinical Center, Central Arkansas Veterans Healthcare System, Little Rock, AR, USA

Department of Geriatrics, University of Arkansas for Medical Sciences, Little Rock, AR, USA

Nancy Bartolotti Department of Anatomy and Cell Biology, College of Medicine, The University of Illinois at Chicago, Chicago, IL, USA

Michel Baudry Graduate College of Biomedical Sciences, Western University of Health Sciences, Pomona, CA, USA

Stephen C. Bondy Department of Medicine, Center for Occupational and Environmental Health, University of California, Irvine, CA, USA

Arezoo Campbell Department of Pharmaceutical Sciences, College of Pharmacy, Western University of Health Sciences, Pomona, CA, USA

Rudy Castellani Department of Pathology, School of Medicine, University of Maryland, Baltimore, MD, USA

Frej Y. Fyhrquist Minerva Institute for Medical Research, Helsinki, Finland

Anamitra Ghosh Parkinson's Disorder Research Laboratory, Iowa Center for Advanced Neurotoxicology, Department of Biomedical Sciences, Iowa State University, Ames, IA, USA

Mohammed Haddad National Institute of Scientific Research- Armand Frappier Institute, Laval, QC, Canada

Institute of Nutrition and Functional Foods, Laval University, Laval, QC, Canada

Rüdiger Hardeland Johann Friedrich Blumenbach Institute of Zoology and Anthropology, University of Göttingen, Göttingen, Germany

Dilshan S. Harischandra Parkinson's Disorder Research Laboratory, Iowa Center for Advanced Neurotoxicology, Department of Biomedical Sciences, Iowa State University, Ames, IA, USA

G. Jean Harry National Toxicology Program Laboratory, National Institute of Environmental Health Sciences, National Institutes of Health, Neurotoxicology Group, Research Triangle Park, NC, USA

Jau-Shyong Hong Neuropharmacology Section, Neurobiology Laboratory, National Institute of Environmental Health Sciences, National Institutes of Health, Research Triangle Park, NC, USA

Syed Z. Imam Division of Neurotoxicology, National Center for Toxicological Research/US FDA, Jefferson, AR, USA

Huajun Jin Parkinson's Disorder Research Laboratory, Iowa Center for Advanced Neurotoxicology, Department of Biomedical Sciences, Iowa State University, Ames, IA, USA

Arthi Kanthasamy Parkinson's Disorder Research Laboratory, Iowa Center for Advanced Neurotoxicology, Department of Biomedical Sciences, Iowa State University, Ames, IA, USA

Anumantha G. Kanthasamy Parkinson's Disorder Research Laboratory, Iowa Center for Advanced Neurotoxicology, Department of Biomedical Sciences, Iowa State University, Ames, IA, USA

Ismail Laher Department of Pharmacology and Therapeutics, Faculty of Medicine, The University of British Columbia, Vancouver, BC, Canada

Orly Lazarov Department of Anatomy and Cell Biology, College of Medicine, The University of Illinois at Chicago, Chicago, IL, USA

Ulrike Luderer Division of Occupational and Environmental Medicine, Department of Medicine, University of California Irvine, Irvine, CA, USA

Department of Developmental and Cell Biology, University of California Irvine, Irvine, CA, USA

Program in Public Health, University of California Irvine, Irvine, CA, USA

Bernard Malfroy-Camine MindSet Rx, Inc., Arlington, MA, USA

Christopher A. McPherson National Toxicology Program Laboratory, National Institute of Environmental Health Sciences, National Institutes of Health, Neurotoxicology Group, Research Triangle Park, NC, USA

Syed Jafar Mehdi Department of Geriatrics, University of Arkansas for Medical Sciences, Little Rock, AR, USA

Mark R. Miller Queens Medical Research Institute, University/BHF Centre of Cardiovascular Science, University of Edinburgh, Edinburgh, UK

Maria Mittelbrunn Instituto de Investigaci6n Sanitaria Hospital 12 de Octubre (i+12), Madrid, Spain

Esteban A. Oyarzabal Neuropharmacology Section, Neurobiology Laboratory, National Institute of Environmental Health Sciences, National Institutes of Health, Research Triangle Park, NC, USA

Department of Neurology, University of North Carolina at Chapel Hill, Chapel Hill, NC, USA

Department of Biomedical Research Imaging Center, University of North Carolina at Chapel Hill, Chapel Hill, NC, USA

Merle G. Paule Division of Neurotoxicology, National Center for Toxicological Research/US FDA, Jefferson, AR, USA

Morgane Perrotte National Institute of Scientific Research- Armand Frappier Institute, Laval, QC, Canada

Institute of Nutrition and Functional Foods, Laval University, Laval, QC, Canada

George Perry College of Sciences, The University of Texas at San Antonio, San Antonio, TX, USA

Joel Pincemail Liège CHU Department of Cardiovascular Surgery and Antioxidant Nutrition and Health Platform, Liège, Belgium

Kedar N. Prasad Antioxidant Research Institute, Premier Micronutrient Corporation, Novato, CA, USA

Gajalakshmi Ramanathan Division of Cardiology, Department of Medicine, David Geffen School of Medicine at UCLA, Los Angeles, CA, USA

Charles Ramassamy National Institute of Scientific Research- Armand Frappier Institute, Laval, QC, Canada

Institute of Nutrition and Functional Foods, Laval University, Laval, QC, Canada

Outi J. Saijonmaa Minerva Institute for Medical Research, Helsinki, Finland

Nada Sallam Department of Pharmacology and Toxicology, Faculty of Pharmacy, Cairo University, Cairo, Egypt

Anoop S.V. Shah Queens Medical Research Institute, University/BHF Centre of Cardiovascular Science, University of Edinburgh, Edinburgh, UK

Edward H. Sharman Department of Neurology, University of California, Irvine, Irvine, CA, USA

Yen-Yu Ian Shih Department of Neurology, University of North Carolina at Chapel Hill, Chapel Hill, NC, USA

Department of Biomedical Research Imaging Center, University of North Carolina at Chapel Hill, Chapel Hill, NC, USA

Department of Biomedical Engineering, University of North Carolina at Chapel Hill, Chapel Hill, NC, USA

Sandeep Kumar Singh Department of Biochemistry, Faculty of Science, Banaras Hindu University, Varanasi, UP, India

Parrisa Solaimani Department of Pharmaceutical Sciences, College of Pharmacy, Western University of Health Sciences, Pomona, CA, USA

About the Authors

Stephen C. Bondy received an M.A. in biochemistry from Cambridge University, in Cambridge, England, and his Ph.D. in neurochemistry from Birmingham University, in Birmingham, England. He subsequently held appointments at Columbia University, University of California Los Angeles, University of Colorado, and the National Institutes of Environmental Health Sciences. He is currently a Professor in the Center for Occupational and Environmental Medicine at the University of California, Irvine. His research focuses on identifying environmental factors which may accelerate brain aging, with emphasis on understanding the mechanisms that may underlie these events. In addition, Dr. Bondy is pursuing the development of therapeutic approaches with the potential to slow neurodegenerative processes.

Arezoo Campbell received a B.A. in philosophy, a B.S. in biological sciences, and a Ph.D. in environmental toxicology from the University of California, Irvine (UCI) in Irvine, California. She is currently an Associate Professor in the Department of Pharmaceutical Sciences at Western University of Health Sciences. Dr. Campbell's research emphasizes the means by which environmental exposures induce aberrant innate immune responses, which may accelerate the appearance of age-related neurodegenerative disease. Dr. Campbell also investigates the importance of inflammatory and oxidative stress in contributing to adverse changes within the brain.

Part I
Age-Related Cellular Events

Metal Toxicity, Inflammation and Oxidative Stress

Stephen C. Bondy

1 Introduction

This review aims to show that many unrelated adverse neurological health conditions are caused by neurotoxic metals or by inappropriate disposition of biologically essential metals. The toxicity of these metals is commonly associated with excess levels of reactive oxygen and nitrogen radicals, together with heightened indices of inflammation. This is especially true in the case of age-related neurodegenerative disease.

Both oxidative stress and excess inflammatory activity characterize cells that are in not in an optimal health state. They may be regarded as non-specific indices of pressures that cause an organism to depart from an ideal state of well-being. A wide range of challenges to organismic homeostasis can lead to such a common outcome. This reflection of an unhealthy cell is in itself injurious and can lead to further departure from normal cellular vigor. Excessive levels of oxidant radicals and inflammation together can thus lead to a progressively damaging intracellular milieu. Two major potential triggers of this cascade will be considered in relation to nervous function; namely the process of cellular senescence and the impact of harmful environmental agents. When these factors act in concert, an especially challenging situation is presented to the organism.

S.C. Bondy (✉)
Department of Medicine, Center for Occupational and Environmental Health,
University of California, Irvine, CA 92697-1830, USA
e-mail: scbondy@uci.edu

S.C. Bondy, A. Campbell (eds.), *Inflammation, Aging, and Oxidative Stress*,
Oxidative Stress in Applied Basic Research and Clinical Practice,
DOI 10.1007/978-3-319-33486-8_1

2 Age: Related Events

2.1 *Inflammation*

Brain aging is characterized by relatively high levels of inflammatory activity which cannot be attributed to provocation by obvious responses to bacterial infection or other exogenous agents [1–3]. This unprovoked elevated basal activity increase with aging. While the underlying cause of this is unclear, it has been attributed to a persistence of responses to earlier immune challenges. This inability of the brain to down-regulate reactivity following immune activation is distinctive and not found in other tissues [4]. It may reflect either some memorial function of the CNS or its limited ability to interact with the peripheral immune system. Another possibility is that it represents a continuous yet futile response to intracellular proteinaceous materials found to acccumulate in specific neurodegenerative diseases, such as amyloid peptide aggregates, neurofibrillary tangles in the case of Alzheimer's disease, AD, or Lewy bodies in the case of Parkinson's disease, PD and Lewy body dementia. These cannot be resolved by normal chaperone-directed protein folding and proteolytic events [5]. This is further suggested by the finding that aged brain of AD and PD victims have an even greater elevation of inflammatory activity than corresponding aged controls not suffering from neurological disease [6].

2.2 *Oxidative Stress*

In addition to evidence of apparently non-productive inflammation within the aging brain, there is also indication of a parallel elevation of the rate of free radical production with senescence. This can be detected by the increasing levels of oxidatively modified macromolecules with age. Both nucleic acids and proteins show such progressive changes [7–9]. This may represent the gradual accumulation of free radical-induced damage with time. In addition, the dynamic production of short-lived reactive oxidizing radicals can is elevated with senescence [10]. The reasons underlying these changes are uncertain but they may in part reflect the growing number of mitochondria that are less effective in the older animal [11]. Mitochondrial DNA is not subject to the repair process found with nuclear DNA. Those mitochondria which have a growing number of DNA deletions with time, may have a less efficient electron transport chain. However, they may be able to divide more rapidly than healthy mitochondria and thus the overall mitochondrial population can gradually diminish in effectiveness in handling the electron transport chain without outflow of noxious reactive species. Such mitochondria are more likely to leak oxidant radicals whose presence is minimal in younger organisms [12]. As is discussed in the next section, the tendency to an elevatedthe rate of production of reactive oxygen species (ROS) with age can also be due to the gradual onset of non-productive inflammation.

2.3 The Relation Between Oxidative Stress and Inflammation

There is considerable overlap between oxidant and inflammatory processes. Immune-related cells that respond to bacterial infection often possess the ability to approach or phagocytose microorganisms and kill them with a burst of oxidative activity. Thus generation of free radicals constitutes part of the normal inflammatory response. When such responses are excessively prolonged, the result can be self-inflicted oxidative damage to an organism. Since the relationship between oxidative stress and inflammation in nervous tissue is very close, the issue of causality can be difficult to address. While inflammatory events can lead to production of oxidative and nitrosylative species, conversely, the presence of such reactive moieties, some of which serve as signaling molecules, can also lead to microglial and astroglial activation, and thus initiate immune and further oxidant responses within the brain [13, 14]. This occurs largely by expression of glial NADPH oxidase and nitric oxide synthase (NOX-1), leading to elevated levels of oxidative and nitrosylative species [15]. NADPH oxidase is the major driver of ROS-production, leading both to direct neuronal damage and also to expression of inflammatory genes within microglia [10].

Induction of the Nrf-2/ARE system, which initiates the formation of several enzymes capable in inactivating oxidant radicals [16] is protective against neurotoxicity [17], but can also lead to suppression of inflammatory events [18]. Thus there is a reciprocal relation between inflammatory and oxidant processes and this interaction can potentiate either beneficial or adverse events. A large number of neurotoxic agents are known to lead to oxidative and inflammatory damage within the nervous system. What is relevant to this review are those low level yet continuing exposures that can gradually enhance the slow onset of age-related alterations in the redox status and immune responses of the cell.

2.4 Neurotoxic and Essential Metals That Increase Pro-Oxidant Conditions Within the Cell

Rather than exhaustively listing metals known to promote elevated free-radical production or impair anti-oxidant defense processes, this review seeks to identify common events within the cell that are targeted by various metal-containing compounds. Metal neurotoxicity falls into a few major categories in which a specific cell process is interfered with. There is much overlap between groupings as some metals may be active in more than one of these. However each of these classes illustrates a distinctive toxic route by which the milieu of the cell is shifted to a harmful pro-oxidant equilibrium.

A. VALENCE FLUX UNDER NORMAL PHYSIOLOGICAL CONDITIONS PROMOTING REDOX CYCLING OF PHENOLS/QUINONES AND OTHER ORGANIC SPECIES

Several essential metals constitute a key constituent of enzymes involved in redox regulation. These are largely the transition metals, Fe, Cu, Cr and Mn. Both Fe and Cu are found in the cytochromes of the respiratory chain, Fe is present in cytochrome P450 mixed function oxidases, while Cu and Mn are present in superoxide dismutases, permitting catabolism and ultimately detoxification of the superoxide anion. Peroxidases and catalases, which are involved in detoxification of reactive oxygen species, also contain Fe in the form of a heme group. The redox reactions effected by these systems are generally well-controlled with minimal leakage of oxidatively damaging molecules into the cytosol.

Iron, copper and manganese are transition metals whose atom has an incomplete inner valence d-sub-shell and which can adopt multiple valence states. The electrons of the inner 3d orbital can be available for chemical bonding. This allows all of these metals to form more than one valence state under physiological conditions and the flux between these valences constitutes much of the basis for their biological importance.

3 Iron

For Fe and Cu, the lower valence state (Fe^{2+}, Cu^{+}) has greater reactivity and readiness to catalyze free radical-generating reactions [19–21]. The low molecular weight ionic forms of Fe and Cu exist in very low cytosolic levels ([22]). Any elevation can rapidly become a basis for initiation of free radical production by the cycling of the Fenton reaction together with the Haber-Weiss reaction (Fig. 1). The essential outcome of this redox flux is the conversion of relatively stable species with little direct oxidizing potential such as hydrogen peroxide, superoxide and nitric oxide, to more reactive peroxynitrite anion and the very short-lived and highly reactive hydroxyl radical (Fig. 1). However, hydrogen peroxide and nitric oxide, precursors to highly oxidant species, while in themselves relatively inert, their very stability and lack of charge allows diffusion across cell membranes. In this way the focus of oxidation damage can be broadened beyond an original lesion.

The need for maintenance of the unchelated forms of these metals at very low levels is underscored by the existence of specific proteins, ferritin and ceruloplasmin, to sequester their ions. The presence of these protective proteins, together with regulation of uptake, makes the neurotoxicity of transition metals only able to manifest itself under unusual conditions. Iron neurotoxicity can be seen in cerebral hemorrhage where the hemoglobin from extravasated blood can be subjected to proteolysis with consequent liberation of Fe [23].

4 Copper

Copper uptake and disposition is also firmly regulated and copper neurotoxicity is confined to Wilson's disease, a genetic disorder involving copper toxicosis due to inability to effectively link copper to ceruloplasmin and to excrete copper [24].

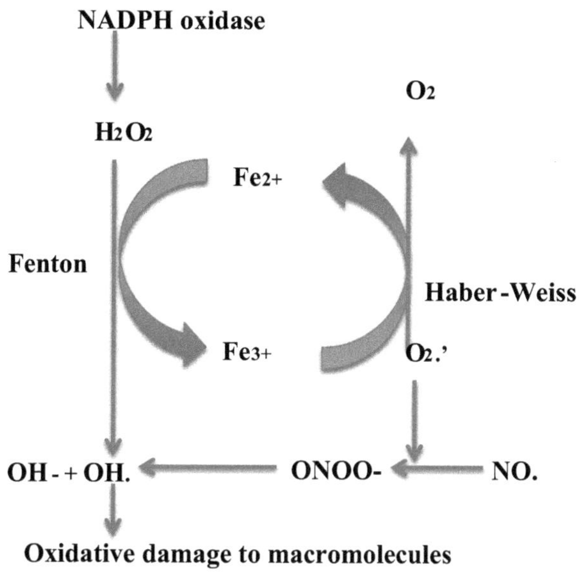

Fig. 1 Generation of reactive hydroxyl radicals ($OH^{\bullet}$). Iron (Fe^{2+}) in the form of low molecular weight salts, which can result from degradation of haem and other iron-containing species, reacts with hydrogen peroxide. This leads to the generation of very short-lived and powerfully oxidant hydroxyl radicals ($OH^{\bullet}$) while iron is simultaneously oxidized to Fe^{3+}. Superoxide ($O_2^{\bullet-}$) can then react with Fe^{3+} in the Haber-Weiss reaction leading to reduction to Fe^{2+}. This forms the basis of catalytic redox cycling. Hydroxyl radicals can damage macromolecules ultimately disrupting neuronal functioning. Superoxide and nitric oxide ($NO^{\bullet}$) form peroxynitrite (ONOO-), which also takes up a proton and forms to $OH^{\bullet}$ and NO_2

Aceruloplasminemia is another genetic disorder where ceruloplasmin is present in an abnormal form [71]. This leads to iron accumulation in the brain, as ceruloplasmin is critical for the complexing of iron by ferritin where it is present in a relatively harmless ferric form. There is conflicting evidence concerning whether ceruloplasmin may stimulate various inflammatory and oxidant processes in microglia [25] or be protective against cerebral iron toxicity [26].

5 Manganese

Manganese is the only transition metal of this family where excess absorption from an exogenous source can lead to oxidative damage in the brain. The damage incurred is focused on the dopaminergic system but not confined to it and there are major clinical differences seen between Parkinson's disease and manganism. Dopaminergergic damage may in part due to the manganese –catalyzed ability catecholamines to undergo redox cycling between quinone and aromatic ring structures leading to a catalytic production of superoxide [27]. However manganese can facilitate oxidation of thiol groups of cysteine-containing enzymes [28].

Unlike Fe or Cu, Mn is most stable in its lower valence form (Mn^{2+}) and this may account for its ability to act either in a pro- or anti-oxidant manner. Mn^{3+} is likely to be present in the cytosol as a n insoluble colloidal material. The presence of traces of Fe or Cu on surfaces of such nanoparticles may powerfully catalyze free radical production (see Sect. 4). There are indications that he neurotoxicity of manganic salts may be due to their insolubility and resulting colloidal nature, providing a surface on which low molecular weight iron can bind to and thus become redox active [20, 21].

6 Transition Metals and Protein Configuration

Imbalance of redox-active metals within the brain and the resulting oxidative stress is an important basis of neurotoxic damage in several neurodegenerative conditions. In prion disorders, the early evidence of oxidative stress suggests a role in the pathogenesis of these disorders [29]. The main cause of pathology in prion disorders is PrP-scrapie (PrPSc), a β -sheet–rich the alteration of the configuration of the normal the prion protein (PrPC) from an α-helix to a β-pleated sheet. This has been attributed to disruption of manganese sequestration processes [30] and normal prion protein has been reported as protective of manganese toxicity while the aberrant prion form leads to manganese accumulation [31]. Evidence for the ability of prion protein to regulate copper and iron uptake has also been presented [29]. The production of reactive oxygen species by beta-amyloid peptide is much enhanced in the presence of traces of copper or iron [32].

B. ATTRACTION TO SULFHYDRYL GROUPS OF AMINO ACID RESIDUES IN PROTEINS AND OF SOLUBLE ANTIOXIDANTS LIKE GLUTATHIONE

Another group of metals and metalloids can derange the redox balance of the cell primarily by their ability to react with sulfhydryl groups of amino acids within proteins, and gluthathione and lipoic acid ([33]). Toxic metals in this class include Hg, Pb, Cd and As, all of which can have an adverse effect on the functioning of the brain [34].

In addition to their redox-disrupting properties, many of this class of metals are able to mimic essential metal elements. By this means they can enter the cell through ion-selective channels and disrupt metabolic events by taking the place of physiological metals within key binding sites of enzymes. This entry and attachment to distinct proteins allows the concentration and expression of toxicity of xenobiotic metals.

7 Mercury

Not only is the valence of several metals an important variable but also the nature of the anion of metals salts can be an important determinant of their toxicity Inorganic mercury is much less neurotoxic than organic mercurials such as monomethyl mercury. This is largely due to the inability of ionic compounds to traverse the blood brain barrier. This distinction is less pronounced in the fetus since it lacks a developed blood brain barrier [72]. However, the long latency period between ingestion of methyl or dimethyl mercury before the onset of symptoms of toxicity suggests that redistribution and/or conversion to a more toxic metabolite takes place [35]. There is evidence that this metabolite is in fact inorganic Hg [36, 37].

The potential neurotoxicity of the organic mercury in fish following their human consumption is an active issue and fish consumption has frequently been discouraged. However, many fish containing organomercurials also have a high selenide content generally in molar excess with respect to mercury [38]. Selenium can bind and thereby detoxify mercury [39] and thus the hazard posed by these fish may be overestimated [40]. Furthermore, any cost-benefit evaluation of fish consumption should take into account the known highly beneficial qualities of omega fatty acids, whose content is rich in salmon, herring and trout [41].

Glutathione peroxidase is a key enzyme in the maintenance of adequate levels of glutathione, a major antioxidant species present in the cell in millimolar concentration. It is one of the few selenium-dependent enzymes and in view of the great affinity of mercury for selenium, especially vulnerable to mercurials. Lipoic acid, a co-factor for several enzymes of the citric acid cycle, is an important lipophilic organosulfur compound with potent anti-oxidant properties. Such binding of metals to sulfur or selenium atoms can lead to inactivation of enzymes and also diminish the cell's ability to inactivate free radicals.

8 Arsenic

Arsenic exhibits a wide range of toxic effects, notably as a carcinogen. Much of its toxicity is likely to be underlain by its ability to complex sulfydryl group, thus initiating oxidative stress [42].

In contrast to mercurials, ingested inorganic arsenic is more toxic than organic arsenic. Within the cell it is methylated to toxic derivatives of arsenous acid where the As is trivalent [43]. Trivalent inorganic As is much more toxic than the than pentavalent form. Glutathione is vital for preventing chronic arsenic toxicity. However extracellular glutathione can also facilitate the reduction of non-toxic pentavalent arsenic to its highly toxic trivalent form [44]. The extent to which As3+ compounds form a stable conjugate with glutathione may be an important determinant of the toxicity of arsenicals [45]. The neurotoxicity of most of the metals in this class, including arsenic, can be attenuated by zinc or calcium or by chelating agents [46] or by anti-oxidants [47] suggesting that their toxicity resides in the combination of competition with essential metals together with in their pro-oxidant properties.

9 Lead

Lead (Pb) competes with calcium and thus is deposited in bones with a long residence period [48]. Pb can also mimic Zn and displacement of Zn from enzymes of the heme synthetic pathway, especially d-aminolevulinic acid dehydratase (ALAD) accounts for the anemia characterizing lead poisoning [49]. The capacity of lead salts to enhance free radical production lies largely in its ability to bind to –SH groups leading to pronounced depletion of brain GSH and superoxide dismutase [50]. These multiple attributes make lead an especially potent developmental neurotoxin.

Part of the ability of lead to promote conditions of oxidative stress may result from its inhibition of δ -amino-levulinic acid dehydratase which can lead to elevated levels of the pro-oxidant δ-amino levulinic acid [51].

The toxicity of many xenobiotic metals often involves several aspects, not all of which relate to pro-oxidant events. This is especially true of Pb, which has a large range of individual toxic effects. The superimposing of oxidative stress upon other adverse biochemical changes can enhance the overall toxicity of such metals.

C. COMPETITION WITH ESSENTIAL MONOVALENT METALS

Thallium and lithium salts gain entry into the cell by mimicking potassium and sodium respectively and entering through specific channels. Once within the cell, the toxicity of these metals can be manifested and this injuriousness often includes a significant ability to promote pro-oxidant events.

10 Thallium

Although it can exist in the monovalent or a trivalent form, thallium, Tl has an ionic radius resembling that of K+ and can enter the cell through K+ channels. Once within the cell, Tl3+ can form an insoluble complex with riboflavin. This causes the metal to be retained within the cell. As little as 1.7 g can be a lethal dose in humans. The inactivation of riboflavin leads to its deficiency and causes some aspects of Tl poisoning to resemble beri-beri. Tl^{3+} also has an affinity for sulfhydryl groups and thus reduces antioxidant protection by lipoate and glutathione [52, 53]. This is illustrated by the finding that selenium can be protective against Tl toxicity. Tl salts have been used as a rodenticide, fungicide (ringworm treatment), bacteriocide (tuberculosis treatment), and in the manufacture of corrosion-resistant alloys, fireworks, cement and dyes. Radioactive Tl may be emitted during nuclear reactor mishaps.

11 Lithium

Lithium, Li is in current use as a neuroleptic as it is especially useful in therapy of manic-depressive disorder. Li enters the cell through sodium channels. However it is not well-recognized by the sodium pump and can thus accumulate within the cell.

Undesirable iatrogenic effects of Li include Parkinsonian tremor, muscle weakness, slurred speech and blurred vision. The therapeutic and toxic mechanisms are unclear but Li certainly enhances lipid peroxidative mechanisms [54].

D. PRESENCE OF COLLOIDAL NANOPARTICLES WHICH CAN PROVOKE IMMUNE RESPONSES AND SERVES AS A SURFACE FOR IRON-BASED GENERATION OF FREE RADICALS

12 Aluminum

Some metals and metalloids are relatively insoluble under physiological conditions.

Aluminum (Al) is largely in colloidal form at biological pH. Mineral micelles have a very large surface area and there is evidence that ionic iron and copper can be loosely bound on the surface of such colloids [55, 56]. This appears to promote the redox flux potential of these transition metal. The Fenton reaction can catalyze free radical formation especially rapidly on such surfaces [57].

13 Other Particulates

Similar promotion of iron-based free radical formation has been reported for colloidal manganic salts, as these are very insoluble in biological media. The ability of colloidal surfaces to bind redox-active metals has been clearly shown for silica nanoparticles. When washed with deferoxamine in order to remove superficial iron, such particles lose all ability to promote redox cycling [73]. Since a variety of inhaled or ingested nanoparticulate and colloidal species have been shown able to enter the brain, such findings, often in isolated systems have direct biological relevance. Inhaled or intranasally instilled silica nanoparticles have been found able to enter the central nervous system [58]. The ability of these insoluble materials to promote oxidant events is enhanced by their activation of glia and the induction of an inflammatory and phagocytic response [59]. Particles within the brain that are of biological origin such as melanin can also interact with transition metals to increase oxidant events [60, 61] but the sequestration of iron by melanin has also been reported to be protective against iron-induced oxidative damage [62]. It has been suggested that the increased tissue iron found in the melanin of the parkinsonian substantia nigra may saturate iron-chelating sites. This may result in a looser association between iron and melanin leading to increased, rather than decreased, production of free radical species [63]. Thus the distinction between effective and ineffective metal chelation can be very critical.

The relationships between the different means by which metals can act in a neurotoxic manner are summarized in a Venn diagram (Fig. 2). This suggests a breakdown into two major groups, namely inappropriate disposition of essential metals with redox potential, and xenobiotic metals with attraction to –SH groups or ability to compete with physiological metals.

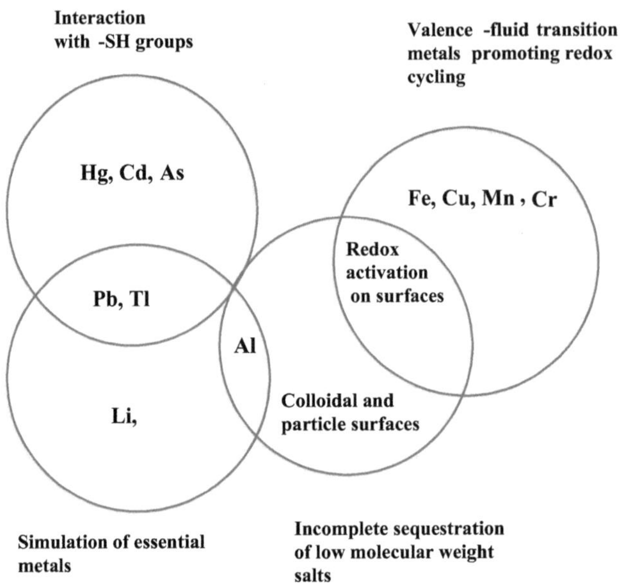

Fig. 2 Classification of means by which metals can induce production of reactive oxygen species. A Venn diagram illustrates the multiple means by which several metals promote ROS generation

14 Remediation of Metal-Induced Oxidative Stress

The interaction with low levels of metals enhancing oxidant conditions within the nervous system interacts with the increasing propensity of the aging brain to express an ever-increasing inflammatory and oxidative milieu. This suggests that a degree of protection against neurotoxic effects can be achieved by use of antioxidant or anti-inflammatory agents. Antioxidant administration can substantially reduce the development of learning deficits in aged animals [64] and anti-inflammatory agents such as COX-2 inhibitors, melatonin and various plant extracts, also slow down rates of neural aging [65–67]. However, whether increased ROS production is causal to aging (the "mitochondrial oxidative stress theory of aging") is not established. There is recent evidence that the slow increase in ROS production with age may act as a protective signal that can trigger changes of gene expression and thus attenuate the effects of aging [68]. This may account for the general failure to extend lifespan with dietary supplements, and apparent potential of several antioxidants, notably α-tocopherol, and anti-inflammatory agents (poly-unsaturated fatty acids), to be harmful [69, 70].

References

1. Bondy SC, Sharman EH. Melatonin and the aging brain. Neurochem Int. 2007;50:571–80.
2. Norden DM, Godbout JP. Review: microglia of the aged brain: primed to be activated and resistant to regulation. Neuropathol Appl Neurobiol. 2013;39:19–34.
3. Singhal G, Jaehne EJ, Corrigan F, Toben C, Baune BT. Inflammasomes in neuroinflammation and changes in brain function: a focused review. Front Neurosci. 2014;8:315.
4. Qin L, Wu X, Block ML, Liu Y, Breese GR, Hong JS, Knapp DJ, Crews FT. Systemic LPS causes chronic neuroinflammation and progressive neurodegeneration. Glia. 2007;55:453–62.
5. Bolshette NB, Thakur KK, Bidkar AP, Trandafir C, Kumar P, Gogoi R. Protein folding and misfolding in the neurodegenerative disorders: a review. Rev Neurol (Paris). 2014;170:151–61.
6. Sardi F, Fassina L, Venturini L, Inguscio M, Guerriero F, Rolfo E, Ricevuti G. Alzheimer's disease, autoimmunity and inflammation. The good, the bad and the ugly. Autoimmun Rev. 2011;11:149–53.
7. Nunomura A, Moreira PI, Castellani RJ, Lee HG, Zhu X, Smith MA, Perry G. Oxidative damage to RNA in aging and neurodegenerative disorders. Neurotox Res. 2012;22:231–48.
8. Santos RX, Correia SC, Zhu X, Smith MA, Moreira PI, Castellani RJ, Nunomura A, Perry G. Mitochondrial DNA oxidative damage and repair in aging and Alzheimer's disease. Antioxid Redox Signal. 2013;18:2444–57.
9. Sultana R, Butterfield DA. Oxidative modification of brain proteins in Alzheimer's disease: perspective on future studies based on results of redox proteomics studies. J Alzheimers Dis. 2013;33 Suppl 1:S243–51.
10. Qin L, Liu Y, Hong JS, Crews FT. NADPH oxidase and aging drive microglial activation, oxidative stress, and dopaminergic neurodegeneration following systemic LPS administration. Glia. 2013;61:855–68.
11. Vanitallie TB. Parkinson disease: primacy of age as a risk factor for mitochondrial dysfunction. Metabolism. 2008;57 Suppl 2:S50–5.
12. Holt IJ, Speijer D, Kirkwood TB. The road to rack and ruin: selecting deleterious mitochondrial DNA variants. Philos Trans R Soc Lond B Biol Sci. 2014;369(1646):20130451.
13. Niranjan R. The role of inflammatory and oxidative stress mechanisms in the pathogenesis of Parkinson's disease: focus on astrocytes. Mol Neurobiol. 2014;49:28–38.
14. Ortiz GG, Pacheco-Moisés FP, Bitzer-Quintero OK, Ramírez-Anguiano AC, Flores-Alvarado LJ, Ramírez-Ramírez V, Macias-Islas MA, Torres-Sánchez ED. Immunology and oxidative stress in multiple sclerosis: clinical and basic approach. Clin Dev Immunol. 2013;2013:708659.
15. Chéret C, Gervais A, Lelli A, Colin C, Amar L, Ravassard P, Mallet J, Cumano A, Krause KH, Mallat M. Neurotoxic activation of microglia is promoted by a nox1-dependent NADPH oxidase. J Neurosci. 2008;28:12039–51.
16. Sandberg M, Patil J, D'Angelo B, Weber SG, Mallard C. NRF2-regulation in brain health and disease: implication of cerebral inflammation. Neuropharmacology. 2014;79:298-306. Toxicol Appl Pharmacol. 2004;195:129–41.
17. Miller DM, Singh IN, Wang JA, Hall ED. Administration of the Nrf2-ARE activators sulforaphane and carnosic acid attenuates 4-hydroxy-2-nonenal-induced mitochondrial dysfunction ex vivo. Free Radic Biol Med. 2013;57:1–9.
18. Ye M, Wang Q, Zhang W, Li Z, Wang Y, Hu R. Oroxylin A exerts anti-inflammatory activity on lipopolysaccharide-induced mouse macrophage via Nrf2/ARE activation. Biochem Cell Biol. 2014;92:337–48.
19. Greenough MA, Camakaris J, Bush AI. Metal dyshomeostasis and oxidative stress in Alzheimer's disease. Neurochem Int. 2013;62:540–55.
20. Hamai D, Bondy SC, Becaria A, Campbell A. The chemistry of transition metals in relation to their potential role in neurodegenerative processes. Curr Top Med Chem. 2001;1:541–51.
21. HaMai D, Campbell A, Bondy SC. Modulation of oxidative events by multivalent manganese complexes in brain tissue. Free Radicol Biol Med. 2001;31:763–8.
22. Gutteridge JM. Iron and oxygen radicals in brain. Ann Neurol. 1992;32(Suppl):S16–21.

23. Xiong XY, Wang J, Qian ZM, Yang QW. Iron and intracerebral hemorrhage: from mechanism to translation. Transl Stroke Res. 2014;5:429–41.
24. Dusek P, Roos PM, Litwin T, Schneider SA, Flaten TP, Aaseth J. The neurotoxicity of iron, copper and manganese in Parkinson's and Wilson's diseases. J Trace Elem Med Biol. 2015;31:193–203.
25. Lee KH, Yun SJ, Nam KN, Gho YS, Lee EH. Activation of microglial cells by ceruloplasmin. Brain Res. 2007;1171:1–18.
26. Ayton S, Lei P, Adlard PA, Volitakis I, Cherny RA, Bush AI, Finkelstein DI. Iron accumulation confers neurotoxicity to a vulnerable population of nigral neurons: implications for Parkinson's disease. Mol Neurodegener. 2014;9:27.
27. Bisaglia M, Filograna R, Beltramini M, Bubacco L. Are dopamine derivatives implicated in the pathogenesis of Parkinson's disease? Ageing Res Rev. 2014;13:107–14.
28. Herszage J, dos Santos AM, Luther 3rd GW. Oxidation of cysteine and glutathione by soluble polymeric MnO2.v. Environ Sci Technol. 2003;37:3332–8.
29. Singh N, Singh A, Das D, Mohan ML. Redox control of prion and disease pathogenesis. Antioxid Redox Signal. 2010;12:1271–94.
30. Kralovicova S, Fontaine SN, Alderton A, Alderman J, Ragnarsdottir KV, Collins SJ, Brown DR. The effects of prion protein expression on metal metabolism. Mol Cell Neurosci. 2009;41:135–47.
31. Choi CJ, Anantharam V, Saetveit NJ, Houk RS, Kanthasamy A, Kanthasamy AG. Normal cellular prion protein protects against manganese-induced oxidative stress and apoptotic cell death. Toxicol Sci. 2007;98:495–509.
32. Choi DY, Lee YJ, Hong JT, Lee HJ. Antioxidant properties of natural polyphenols and their therapeutic potentials for Alzheimer's disease. Brain Res Bull. 2012;87:144–53.
33. Shanker G, Aschner M. Identification and characterization of uptake systems for cystine and cysteine in cultured astrocytes and neurons: evidence for methylmercury-targeted disruption of astrocyte transport. J Neurosci Res. 2001;66:998–1002.
34. Järup L. Hazards of heavy metal contamination. Br Med Bull. 2003;68:167–82.
35. Weiss B, Clarkson TW, Simon W. Silent latency periods in methylmercury poisoning and in neurodegenerative disease. Environ Health Perspect. 2002;110 Suppl 5:851–4.
36. Friberg L, Mottet NK. Accumulation of methylmercury and inorganic mercury in the brain. Biol Trace Elem Res. 1989;21:201–6.
37. Vahter ME, Mottet NK, Friberg LT, Lind SB, Charleston JS, Burbacher TM. Demethylation of methyl mercury in different brain sites of Macaca fascicularis monkeys during long-term subclinical methyl mercury exposure. Toxicol Appl Pharmacol. 1995;134:273–84.
38. Copat C, Vinceti M, D'Agati MG, Arena G, Mauceri V, Grasso A, Fallico R, Sciacca S, Ferrante M. Mercury and selenium intake by seafood from the Ionian Sea: A risk evaluation. Ecotoxicol Environ Saf. 2014;100:87–92.
39. Heath JC, Banna KM, Reed MN, Pesek EF, Cole N, Li J, Newland MC. Dietary selenium protects against selected signs of aging and methylmercury exposure. Neurotoxicology. 2010;31:169–79.
40. Khan MAK, Wang F. Mercury–selenium compounds and their toxicological significance: toward a molecular understanding of the mercury–selenium antagonism. Environ Toxicol Chem. 2009;28:1567–77.
41. Xun P, Hou N, Daviglus M, Liu K, Morris JS, Shikany JM, Sidney S, Jacobs DR, He K. Fish oil, selenium and mercury in relation to incidence of hypertension: a 20-year follow-up study. J Intern Med. 2011;270:175–86.
42. Lantz RC, Hays AM. Role of oxidative stress in arsenic-induced toxicity. Drug Metab Rev. 2006;38:791–804.
43. Xi S, Jin Y, Lv X, Sun G. Distribution and speciation of arsenic by transplacental and early life exposure to inorganic arsenic in offspring rats. Biol Trace Elem Res. 2010;134:84–97.
44. Sakurai T, Kojima C, Ochiai M, Ohta T, Sakurai MH, Waalkes MP, Fujiwara K. Cellular glutathione prevents cytolethality of monomethylarsonic acid. Toxicol Appl Pharmacol. 2004;195(2):129–41.

45. Watanabe T, Hirano S. Metabolism of arsenic and its toxicological relevance. Arch Toxicol. 2013;87:969-979. Environ Health Perspect. 1994;102(3):215–9.
46. Kadeyala PK, Sannadi S, Gottipolu RR. Alterations in apoptotic caspases and antioxidant enzymes in arsenic exposed rat brain regions: reversal effect of essential metals and a chelating agent. Environ Toxicol Pharmacol. 2013;36:1150–66.
47. Flora SJ, Bhadauria S, Kannan GM, Singh N. Arsenic induced oxidative stress and the role of antioxidant supplementation during chelation: a review. J Environ Biol. 2007;28 (2 Suppl):333–47.
48. Rabinowitz MB. Toxicokinetics of bone lead. Environ Health Perspect. 1991;91:33–7.
49. Wetmur JG. Influence of the common human delta-aminolevulinate dehydratase polymorphism on lead body burden. Environ Health Perspect. 1994;102(3):215–9.
50. Baranowska-Bosiacka I, Gutowska I, Marchlewicz M, Marchetti C, Kurzawski M, Dziedziejko V, Kolasa A, Olszewska M, Rybicka M, Safranow K, Nowacki P, Wiszniewska B, Chlubek D. Disrupted pro- and antioxidative balance as a mechanism of neurotoxicity induced by perinatal exposure to lead. Brain Res. 2012;1435:56–71.
51. Whittaker MH, Wang G, Chen XQ, Lipsky M, Smith D, Gwiazda R, Fowler BA. Exposure to Pb, Cd, and As mixtures potentiates the production of oxidative stress precursors: 30-day, 90-day, and 180-day drinking water studies in rats. Toxicol Appl Pharmacol. 2011;254:154–66.
52. Galván-Arzate S, Pedraza-Chaverrí J, Medina-Campos ON, Maldonado PD, Vázquez-Román B, Ríos C, Santamaría A. Delayed effects of thallium in the rat brain: regional changes in lipid peroxidation and behavioral markers, but moderate alterations in antioxidants, after a single administration. Food Chem Toxicol. 2005;43:1037–45.
53. Hanzel CE, Verstraeten SV. Thallium induces hydrogen peroxide generation by impairing mitochondrial function. Toxicol Appl Pharmacol. 2006;216:485–92.
54. Bhalla P, Chadha VD, Dhar R, Dhawan DK. Neuroprotective effects of zinc on antioxidant defense system in lithium treated rat brain. Indian J Exp Biol. 2007;45:954–8.
55. Bondy SC, Kirstein S. The promotion of iron-induced generation of reactive oxygen species in nerve tissue by aluminum. Mol Chem Neuropathol. 1996;27:185–94.
56. Wu Z, Du Y, Xue H, Wu Y, Zhou B. Aluminum induces neurodegeneration and its toxicity arises from increased iron accumulation and reactive oxygen species (ROS) production. Neurobiol Aging. 2012;33:199. e1–12.
57. Bondy SC. Nanoparticles and colloids as contributing factors in neurodegenerative disease. Int J Env Res Public Health. 2012;8:2200–11.
58. Wu J, Wang C, Sun J, Xue Y. Neurotoxicity of silica nanoparticles: brain localization and dopaminergic neurons damage pathways. ACS Nano. 2011;5:4476–89.
59. Xue Y, Wu J, Sun J. Four types of inorganic nanoparticles stimulate the inflammatory reaction in brain microglia and damage neurons in vitro. Toxicol Lett. 2012;214:91–8.
60. Double KL, Halliday GM. New face of neuromelanin. J Neural Transm Suppl. 2006;70:119–23.
61. Faucheux BA, Martin ME, Beaumont C, Hauw JJ, Agid Y, Hirsch EC. Neuromelanin associated redox-active iron is increased in the substantia nigra of patients with Parkinson's disease. J Neurochem. 2003;86:1142–8.
62. Zecca L, Casella L, Albertini A, Bellei C, Zucca FA, Engelen M, Zadlo A, Szewczyk G, Zareba M, Sarna T. Neuromelanin can protect against iron-mediated oxidative damage in system modelingiron overload of brain aging and Parkinson's disease. J Neurochem. 2008;106:1866–75.
63. Gerlach M, Double KL, Ben-Shachar D, Zecca L, Youdim MB, Riederer P. Neuromelanin and its interaction with iron as a potential risk factor for dopaminergic neurodegeneration underlying Parkinson's disease. Neurotox Res. 2003;5:35–44.
64. Thakurta IG, Banerjee P, Bagh MB, Ghosh A, Sahoo A, Chattopadhyay S, Chakrabarti S. Combination of N-acetylcysteine, α-lipoic acid and α-tocopherol substantially prevents the brain synaptosomal alterations and memory and learning deficits of aged rats. Exp Gerontol. 2014;50:19–25.

65. Miller MG, Shukitt-Hale B. Berry fruit enhances beneficial signaling in the brain. J Agric Food Chem. 2012;60:5709–15.
66. Strauss KI. Antiinflammatory and neuroprotective actions of COX2 inhibitors in the injured brain. Brain Behav Immun. 2008;22:285–98.
67. Zhou J, Yang F, Zhou L, Wang JG, Wen P, Luo H, Li W, Song Z, Sharman EH, Bondy SC. Dietary melatonin attenuates age-related changes in morphology and in levels of key proteins in globus pallidus of mouse brain. Brain Res. 2014;1546:1–8.
68. Yang W, Hekimi S. A mitochondrial superoxide signal triggers increased longevity in Caenorhabditis elegans. PLoS Biol. 2010;8, e1000556.
69. Bjelakovic G, Nikolova D, Gluud C. Meta-regression analyses, meta-analyses, and trial sequential analyses of the effects of supplementation with beta-carotene, vitamin A, and vitamin E singly or in different combinations on all-cause mortality: do we have evidence for lack of harm? PLoS One. 2013;8(9), e74558.
70. Mendelsohn AR, Larrick JW. Trade-offs between anti-aging dietary supplementation and exercise. Rejuvenation Res. 2013;16:419–26.
71. Xu X, Pin S, Gathinji M, Fuchs R, Harris ZL. Aceruloplasminemia: an inherited neurodegenerative disease with impairment of iron homeostasis. Ann N Y Acad Sci. 2004;1012: 299–305.
72. Feng W, Wang M, Li B, Liu J, Chai Z, Zhao J, Deng G. Mercury and trace element distribution in organic tissues and regional brain of fetal rat after in utero and weaning exposure to low dose of inorganic mercury. Toxicol Lett. 2004;152:223–34.
73. Fubini B, Fenoglio I, Elias Z, Poirot O. Variability of biological responses to silicas: effect of origin, crystallinity, and state of surface on generation of reactive oxygen species and morphological transformation of mammalian cells. J Environ Pathol Toxicol Oncol. 2001;20 Suppl 1:95–108.

Exosomes in the Preservation of Cellular Homeostasis

Francesc Baixauli and Maria Mittelbrunn

1 Introduction: Integration of Quality-Control Mechanisms in the Endolysosomal System

Cells are exposed to a variety of stress conditions throughout their lifetime. Surveillance systems inside cells detect altered proteins and coordinate their repair or elimination in order to maintain intracellular homeostasis. Defective functioning of the cellular surveillance mechanisms disrupts cellular homeostasis and survival, and may be the main cause of cumulative cellular damage during disease and aging.

Preserving protein homeostasis (proteostasis), involves an exquisite network of mechanisms that serve either to refold, degrade, sequester misfolded polypeptides or to secrete toxic protein products to the extracellular environment [1, 2]. A network of molecular chaperones recognizes misfolded proteins and promotes their refolding [3] or assists in their degradation via the ubiquitin-proteasome (UPS) or autophagy-lysosomal pathways. An additional cellular quality-control strategy to cope with misfolded proteins is sequestration into specialized compartments [4, 5]. When load of proteins destined for degradation saturates the capacity of the proteolytic systems, cells can defend themselves against proteotoxicity through secretion of toxic protein products to the extracellular environment in small vesicles called exosomes [6–11].

F. Baixauli (✉)
Signaling and Inflammation Program, Spanish National Center for Cardiovascular Research Carlos III, Madrid 28029, Spain

Immunology, La Princesa Health Research Institute, Madrid 28006, Spain
e-mail: fbaixauli@externo.cnic.es

M. Mittelbrunn
Instituto de Investigación Sanitaria Hospital 12 de Octubre (i+12), Madrid 28041, Spain
e-mail: mmittelbrunn.imas12@h12o.es

S.C. Bondy, A. Campbell (eds.), *Inflammation, Aging, and Oxidative Stress*, Oxidative Stress in Applied Basic Research and Clinical Practice,
DOI 10.1007/978-3-319-33486-8_2

These cellular clearance mechanisms share essential steps and components that are required for their proper function: cargo selection and tagging, recognition and delivery to the proteolytic core, and recycling of the constituents. Although cargo selection was thought to be required only for the UPS, growing evidence supports the existence of highly specific mechanisms regulating cargo selection and sorting in each degradative compartment. The UPS and autophagy-lysosome systems involve catalytic degradation of cargo in a confined compartment, the proteasomal catalytic chamber or the lysosomal lumen, followed by recycling of the constituents. In contrast, secretion of components in exosomes clears intracellular components without recycling. Recent studies illustrate the intricate mechanisms of sorting and clearance mediated through exosomes, and the consequences of the transmission of their content to receptor cells. Exosome secretion adds an additional layer of complexity to the understanding of how quality control mechanisms contribute to overall organismal fitness.

1.1 The Endolysosomal System

Two quality-control mechanisms for dealing with damaged, unwanted or toxic intracellular components converge at the endolysosomal compartment: degradation and recycling through the autophagy-lysosome catabolic and recycling pathway, and sequestration and secretion through exosomes. The endolysosomal system is a highly dynamic compartment in which membrane rearrangements modulate protein and lipid trafficking and degradation and also mediate the internalization of nutrients and growth factors to ensure cell survival, growth and differentiation. The formation and maturation of the endocytic and autophagic compartments involves a gradual structural and molecular remodelling of intermediate stages through continuous exchange of membranes and proteins through specific vesicular carriers, tubular connections or kiss-and-run fusion events [12].

The endocytic pathway starts at the plasma membrane and terminates in lysosomes, after moving through a series of endosomal intermediates that are distinguished by their content, morphology or pH. Vesicles formed from the plasma membrane by clathrin-dependent or –independent mechanisms fuse and deliver their membrane and protein content to Rab5- and EEA1-positive early endosomes, which undergo conversion from Rab5- to Rab7-positive endosomes [13]. During this conversion, a significant amount of the internalized content is recycled back to the plasma membrane through Rab11-positive recycling endosomes, while the remaining material is moved from the early endosome to the Golgi complex by the retromer complex or is transported and sequestered in intraluminal vesicles (ILVs) in late endosomes, also known as multivesicular bodies (MVBs) [14]. Fusion of MVBs with the plasma membrane releases the ILVs, eliminating the incorporated components through secretion to the extracellular environment. These extracellular vesicles, now termed exosomes, can be taken up by receptor cells and participate in cell-to-cell communication [15]. Alternatively, late endosomes and MVBs can degrade their cargo by fusing with lysosomes, the main catalytic compartment in the cell [16]. The three types of autophagy described to date—microautophagy, macroautophagy and chaperone-mediated autophagy—deliver intracellular material destined for degradation to lysosomes; this material can include portions of cytosol, proteins, aggregates and damaged organelles or invading pathogens. Autophagosomes must undergo a series of maturation steps in part by fusing with endocytic vesicles, including early and late endosomes and MVBs [17, 18].

Here, we discuss current knowledge on the quality-control mechanisms arising from the endolysosomal compartment. We outline the evidence that selective incorporation and release of cellular compounds in exosomes is a quality-control strategy to alleviate intracellular stress. Finally, we describe the molecular and functional crosstalk between exosomes and the lysosome-autophagy degradative compartment, and the impact of dysregulation of these processes in human disease and aging.

2 Exosomes: Clearance Through Secretion

Exosomes are small vesicles of endosomal origin, ranging from 50 to 150 nm, that are secreted to the extracellular environment by almost every cell type. They were first described in the 1980s as a cellular mechanism to eliminate unwanted, toxic or damaged material. Studying the maturation of reticulocytes into erythrocytes, the groups of Stahl and Johnstone described a new mechanism by which reticulocytes get rid of the transferrin receptor (TfR). These authors observed the presence of TfR molecules in vesicles inside endosomal compartments, which were released to the extracellular medium upon exocytosis of these endosomes [19, 20].

Since that time, molecular characterization of exosomes from different cell types and body fluids has revealed that these vesicles contain specific set of lipids, proteins and genetic material in the form of RNA and DNA [21]. The membranes of exosomes are enriched in ceramides, sphingolipids, cholesterol and glicerophospholipids, which support transmembrane proteins such as the tetraspanins CD63 and CD81, as well as associated proteins such as integrins and immunoglobulins. The luminal protein cargo of exosomes consists mostly of cytoskeletal proteins (tubulin, actin), heat-shock proteins (hsp70, hsp90), and molecules involved in exosome biogenesis (Alix, Tsg101) or vesicle trafficking (Rab GTPases, annexins and flotillins). Exosomes also contain RNA species, including mRNAs, miRNAs, and other regulatory non-coding RNAs [22]. The enrichment of a specific set of molecules in exosomes indicates the existence of specialized mechanisms that regulate their sorting, and reveals exosome biology to be much more complex than originally suspected [23]. Exosomes and their cargo can modulate the activity of receptor cells that take them up, and they are therefore increasingly recognized as important vehicles for cell-to- cell communication [15]. This novel role of exosomes seems to be involved in many physiological and pathological situations, including immune response regulation, cancer progression and metastasis, neuronal survival and communication, and the progression of neurodegenerative diseases [7, 24–28]. Deciphering the regulation of exosome biogenesis, composition, cargo sorting and secretion is thus critical to understanding the roles of these vesicles in health and disease.

2.1 Exosome Biogenesis and Sorting Mechanisms

The biogenesis of exosomes is coordinated with the sorting of specific molecules during the formation of ILVs in the MVB compartment [29]. A single MVB contains distinct ILV subpopulations that can be distinguished by their size and mechanisms of formation [30]. ILV formation, including the sorting of ubiquitinated membrane proteins, requires the ESCRT (endosomal sorting complexes required for transport) machinery [31]. ESCRT is composed of four multimeric complexes, ESCRT-0 to III, and the Vps4 ATPase, which function sequentially to mediate cargo recognition, ILV formation, and final ESCRT disassembly and ILV budding. ESCRT-0 is recruited to MVB by one of its subunits, Vps27, which contains a FYVE domain that specifically binds phosphatidyl inositol 3-phosphate (PtdIns(3)P), produced exclusively in MVB by endosomal class II PtdIns 3-kinase (PI3K) [32, 33]. Another ESCRT-0 subunit, HRS (hepatocyte growth factor–regulated tyrosine kinase substrate), recognizes and sequesters ubiquitinated transmembrane proteins in the endosomal membrane [14] and recruits and activates the ESCRT-I machinery through interaction with Tsg101 (tumor susceptibility gene 101). ESCRT-I then recruits ESCRT-II and mediates vesicle budding and recruitment of the adaptor protein Alix (ALG-2 interacting protein X), which stabilizes ESCRT-III assembly and promotes membrane curvature and vesicle abscission. Finally, dissociation and recycling of the ESCRT machinery require interaction with the ATPase Vps4 (vacuolar protein sorting 4), which drives ATP-dependent vesicle abscission. ESCRT-dependent exosome biogenesis involves interaction of syndecans with syntenin, which in turn interact with CD63 and Alix [34]. Silencing of syndecan and syntenin decreases the number of exosomes and impairs the exosomal accumulation of Alix, hsp70 and CD63, whereas it has no effect on flotillin-positive extracellular vesicles. Recently, systematic silencing of ESCRT machinery components by small interfering RNA in MHC-II expressing Hela cells has shed light on the heterogeneity and complexity of the regulation of exosome biogenesis and composition [35]. For instance, silencing of HRS induces the formation of uniformly small ILVs that contain and require CD63 [30], suggesting that ILV size is influenced by the cargo and mechanism of formation and pointing to the existence of ESCRT-dependent and -independent mechanisms of ILV formation within single MVBs.

The protein SIMPLE (small integral membrane protein of lysosomes and late endosomes), was recently shown to be secreted in association with exosomes. Fibroblasts expressing the mutant form found in Charcot-Marie-Tooth disease, CMT1C, showed altered MVB formation and secreted less CD63- and Alix-containing exosomes, although flotillin secretion was unaffected [36]. Mutation of SIMPLE in the PPxY motif, which mediates binding to E3 ubiquitin ligases, enhances its secretion in exosomes [36], possibly by impairing lysosomal targeting of these vesicles. Interaction of secreted proteins with E3 ubiquitin ligases is potentially relevant to their loading into exosomes. For instance, the export of the tumor suppressor PTEN into exosomes requires its binding to Ndfip1 [37], an adaptor protein for members of the Nedd4 family of E3 ubiquitin ligases.

Depletion of components of all four ESCRT complexes does not abolish MVB formation in mammalian cells [38], and an ESCRT-independent mechanism of ILV formation depending on lipids and tetraspanins has been reported. Sorting of premelanosomal protein requires the tetraspanin CD63 [39], which accumulates in ILVs even in the absence of ESCRT function [30, 38]. Consistently, CD63 was recently shown to be essential in the formation of small (<40 nm) ILVs, independently of HRS, in MVBs of HeLa cells [30]. CD81-enriched domains have been proposed recently as sorting platforms for exosomal proteins [40] and certainly might account for ESCRT-independent sorting of some cargoes and the formation of a population of ILVs: proteins that are known to interact with certain tetraspanins were found in exosomes via mass spectrometry, and in CD81-deficient animals, exosomes were found to be devoid of CD81-interacting molecules, which are normally loaded onto exosomes.

The lipid composition of the limiting membrane of MVBs could induce the inward budding and the formation of ILVs in an ESCRT-independent manner. Lipid metabolism enzymes such as neutral sphingomyelinase (nSMase), that control the hydrolysis of sphingomyelin into ceramide [41], and phospholipase D2, that hydrolyses phosphatidylcholine into phosphatidic acid [42] are involved in ILVs formation.

The mechanisms of inclusion of soluble cytosolic proteins into ILVs are still not very well understood, but a role for heat shock cognate chaperone of 70 kDa (Hsc70) has been proposed recently [43]. Hsc70 binds to soluble cytosolic proteins containing a KFERQ sequence and to phosphatidylserine (PS) on the MVB outer membrane, thus entering ILVs formed in a Tsg101- and VPS4- dependent manner. In addition, binding of Hsc70 to the cytosolic tail of the TfR has been shown to allow targeting of this transmembrane protein to exosomes [44].

Posttranslational modifications in proteins such as ubiquitination, sumoylation, phosphorylation and acetylation have been detected in exosomal preparations, which suggests that these modifications can in part regulate the sorting of specific molecules into exosomes [45]. Poly- and mono-ubiquitinated proteins have been detected in exosomes by mass spectrometry [46, 47]. Myeloid-derived suppressor cells release ubiquitinated proteins in exosomes associated with endosomal trafficking, along with pro-inflammatory high mobility group protein B1 and pro-inflammatory histones. Phosphorylated proteins have been also detected by mass spectrometry in exosomes. Phosphorylation and ubiquitination appear to co-regulate the sorting of Fas ligand into secretory lysosomes by controlling its entry into MVBs [48]. Phosphorylation is involved in incorporation of Annexin A2 into exosomal membranes [49]. In addition, aberrant phosphorylation of the protein tau promotes its incorporation into exosomes, resulting in the spreading of this abnormally processed protein in Alzheimer's disease patients [50]. Oxidation of gamma-synuclein confers prion-like properties and causes the formation of toxic aggregates that can be exported in exosomes [51].

Exosomes contain RNAs that can be incorporated into recipient cells. The RNA profile of exosomes is enriched in small RNA species, including miRNAs and mRNAs, but also small ribosomal RNA, tRNAs and structural RNAs such as vRNA, Y-RNA and SRP-RNA [22]. Some miRNAs are enriched in exosomes while others are barely present, which suggests the existence of mechanisms that regulate the active sorting of specific sets of miRNAs into exosomes. It was recently demonstrated that the loading of miRNAs into exosomes depends on the heterogeneous ribonucleoprotein hnRNPA2B1, which binds to specific sequences found in exosomal miRNAs (EXOmotifs) that are not found among cellular miRNAs [52]. Interestingly, hnRNPA2B1 is sumoylated in exosomes and this modification is required for efficient loading of miRNAs into exosomes [52]. In addition, microRNAs found in exosomes are frequently posttranslationally modified by the addition of uridines [53].

2.2 Signals That Promote Exosome Release

Depending on the cell type or cell status, specific sets of protein and genetic material can be secreted in exosomes, reflecting the adaptability of this system to respond to changing circumstances. Cells secrete exosomes in a constitutive manner, but environmental challenges, stress signals or activation stimuli increase exosome release and affect exosome composition [23].

Cellular activation stimuli, such as dendritic-cell activation by Toll-like receptor stimulation, T- cell activation by T-cell receptor engagement- or B-cell activation by α-IgM, induce exosome secretion [54, 55]. In macrophages, activation of P2X7R purinergic receptors induces the release of IL-1β in exosomes [56]. Calcium signaling also triggers exosome release in many cell types, including lymphocytes, neurons and oligodendrocytes [57]. Calcium entry through NMDA and AMPA receptors triggers oligodendroglial exosome secretion, which improves viability of the recipient neuronal cells upon exposure to conditions of cell stress [24].

Exosome release is also increased in stress situations, which are frequent in the tumor microenvironment and which cancer cells need to overcome to enable tumor progression. A number of studies indicate that hypoxia promotes exosome release in different tumor types [58]. Hypoxia enhances the secretion of exosomes containing proteins that facilitate angiogenesis and metastasis of cancer cells, including those associated with cell migration, extracellular matrix (ECM) remodeling, growth hormone signaling, clathrin-mediated endocytosis, and VGEF signaling [58, 59]. Hypoxia induces the HIF pathway, which in turn promotes RAB22A expression and microvesicle formation [60]. Hypoxic exosomes contain hypoxia-related microRNAs that can induce angiogenesis upon uptake by endothelial cells [59, 61]. Acidic conditions are also characteristic of tumor microenvironments, and increase exosome release and change the membranes lipid composition of exosomes modulating their fusion proprieties [62].

Thermal and oxidative stress can enhance the release and modify the composition of immunosuppressive exosomes from T and B cells [63], thus influencing the response to stress in receptor cells. Irradiation therapy can induce a senescent phenotype and increase exosome secretion in a p53-dependent manner [64]. Ischemic preconditioning of the heart is a process whereby repeated short episodes of ischemia/ reperfusion (I/R) protect the myocardium against subsequent ischemic insult. This protection can result from ischemia in a remote region, and a very recent report described that exosome secretion during preconditioning might induce cardioprotection [65, 66]. In this regard, heat stress induces the release of exosomes bearing heat shock proteins that improve proteostasis responses in receptor cells ([67]). Thus, molecular chaperones maintain proteostasis not only in a cell-autonomous manner but also in a non-cell-autonomous manner through exosomes, supporting the physiological role of exosomes in the regulation of organismal proteostasis.

3 Lysosomes: Clearance Through Degradation

3.1 Form and Function

Lysosomes are the main degradative compartment of the cell. First described by the Nobel laureate Christian de Duve in the 1950s, lysosomes are responsible for proteolytic degradation and recycling of functional and damaged proteins and organelles, and ensure continuous renewal and recycling of cellular constituents, thus avoiding accumulation of harmful components. In addition to their catabolic function, lysosomes participate in a range of physiological processes such as plasma membrane repair, bone and tissue remodeling, cell signaling, energy metabolism, immune responses, and cell death [68].

Lysosomes contain a single lipid bilayer membrane containing proteins involved in the acidification of the lysosomal lumen, fusion of the lysosome with other cellular structures such as late endosomes, autophagosomes and the plasma membrane, and transport of degradation products into and out of lysosomes [16]. The lysosomal membrane segregates and protects the cell from the acidic environment of the lumen by means of an internal thick glycocalyx that prevents lysosomal membrane degradation by luminal acid hydrolases. About 50 different acid hydrolases have been identified, and include members of the sulphatase, glycosidase, peptidase, phosphatase, lipase and nuclease protein families; this repertoire allows hydrolysis of a vast repertoire of biological substrates including proteins, nucleic acids, carbohydrates and lipids [69].

The most important biochemical feature of the lysosome is its acidic lumen (pH 4.5–5.0), which allows higher enzymatic activity of the resident acidic hydrolases, and facilitates partial unfolding of the substrate proteins, allowing endoproteases to access internal peptide bonds. The lysosomal pH gradient is maintained by the lysosomal membrane [70, 71], which contains more than 20 lysosomal membrane proteins, including lysosome-associated membrane protein 1 (LAMP1) and LAMP2, and more importantly, the vacuolar-type H(+)-ATPases (V-ATPases). The proton-pumping V-ATPase uses the energy of ATP hydrolysis to pump protons from the cytosol to the lumen, creating a H^+ concentration in the lysosome about

100-fold higher than that of cytosol. Lysosomal acidification also involves additional lysosomal membrane channels, such as the anion transporter chloride channel 7 (ClC7) [72–74], the cation transporter mucolipin 1 (MCOLN1; also known as TRPML1), and two-pore calcium channel 1 (TPC1) and TCP2, which mediate Ca^{2+} and Na^{+} release from the lysosome [75–78].

3.2 *Lysosome Biogenesis*

Lysosome formation involves an intersection between the endocytic and biosynthetic pathways of the cell. The complex events involved in the formation of a mature lysosome depend on the coordinated action of adaptor molecules such as clathrin, membrane-associated RAB GTPases, SNARE proteins, and components of the retromer complex SNX1, SNX2 or Vps26. These molecules mediate tethering and docking processes important for endolysosomal membrane trafficking, lysosome–endosome fusion, or homotypic fusion between endosomes [16].

Newly synthesized proteins can be targeted to lysosomes directly from the trans-Golgi network via the endosomal system, or indirectly, through transport to the plasma membrane and subsequent endocytosis. Resident lysosomal proteins are synthesized at the rough ER and targeted to the lysosome by specific sorting mechanisms [79]. Most lysosomal hydrolases are modified by mannose 6-phosphate (M6P) residues within their oligosaccharide chains. These modifications are recognized by two different mannose- 6-phosphate receptors (M6PRs) that cycle between the trans-Golgi network (TGN) and endosomes. Dissociation of bound lysosomal hydrolases from M6PRs occurs in the acidic interior of endosomes. The absence of M6PRs in lysosomes distinguishes these organelles from endosomal compartments. Other luminal lysosomal proteins utilize alternative sorting mechanisms [80]. Sortilin was identified as a sorting receptor of sphingolipid activator proteins (prosaposin and GM2 activator protein), acid sphingomyelinase and cathepsins D and H, whereas lysosome-integral membrane protein 2 (LIMP2) was found to mediate the lysosomal targeting of β-glucocerebrosidase [81–84].

Lysosomal biogenesis and function are subject to transcriptional regulation of essential genes by activation of the transcription factor EB [85]. TFEB maintains cellular homeostasis by coordinating the cellular responses to various stresses, including nutrient sensing starvations, metabolic stress, and lysosomal stress [68]. The activity of TFEB is regulated via protein phosphorylation and subcellular localization. TFEB is sequestered in the cytoplasm in resting cells, but upon activation, it translocates to the nucleus and binds to the CLEAR consensus sequence to activate de novo gene transcription. Of the more than 400 direct TFEB target genes identified, most are related to the lysosome and autophagy, including lysosomal hydrolases and accessory proteins, lysosomal V-ATPase pumps, and regulators of lysosomal biogenesis and autophagy. TFEB is negatively regulated by MTOR (mechanistic target of rapamycin complex), which forms a critical signaling axis linking TFEB, the lysosome, and control of autophagy [86, 87]. In the presence of amino acids, MTORC1 localizes to the lysosomal cytoplasmic surface via Rag

GTPases and interacts with a protein complex at the lysosomal surface called Ragulator, which is required for MTORC1 activation [88]. TFEB binds Ragulator at the lysosomal membrane, which allows MTORC1-mediated phosphorylation of TFEB at both Ser142 and Ser211 to control TFEB cytoplasmic and nuclear localization [89, 90]. The regulation of TFEB was recently shown to depend on the activity of the phosphatase calcineurin. Upon activation by calcium released from the lysosome via mucolipin 1 (MCOLN1), calcineurin binds and dephosphorylates TFEB, promoting its nuclear translocation and induction of autophagy and lysosomal biogenesis [91]. These data indicate that the lysosome is a hub for signaling pathways that regulate cellular homeostasis by linking lysosomal calcium signaling to both calcineurin regulation and autophagy induction.

3.3 Lysosomal Degradation

Lysosomes receive extracellular and intracellular material to be degraded via several pathways. Extracellular material and integral membrane proteins destined for degradation reach the lysosome mainly via the endocytic pathway. Depending on the nature of the cargo, the capture of extracellular material and integral membrane proteins and their incorporation in endosomes is regulated by specific endocytic mechanisms, such as phagocytosis, macropinocytosis, clathrin-mediated endocytosis, caveolin-mediated endocytosis and clathrin- and caveolin-independent endocytosis. The incorporated material can then be recycled back to the plasma membrane or targeted for degradation by fusion of the endosomes with lysosomes. Intracellular components are transported to the lysosome by the process of autophagy, a 'self-eating' catabolic pathway that is used by cells to capture their own cytoplasmic components destined for degradation and recycling.

4 Autophagy

4.1 Types of Autophagy

To date, three types of autophagy have been identified in mammalian cells: microautophagy, chaperone-mediated autophagy (CMA), and macroautophagy [92–94]. The role of all three types of autophagy in degradation and recycling processes is strictly dependent on lysosomal function.

In **macroautophagy**, here referred to as autophagy, a whole portion of the cytosol, that can include organelles and soluble proteins, is sequestered inside double-membraned organelles called autophagosomes that subsequently fuse with lysosomes to promote the bulk degradation of their cargo [93, 94]. Autophagy initiates when double-membraned structures called phagophores engulf portions of cytoplasm. After fusion of the isolation membranes, these autophagosomes distribute randomly throughout the cytoplasm and then traffic along microtubules towards the microtubule-organizing centre, where lysosomes are concentrated, enabling fusion and degradation of the contents by lysosomal hydrolases. In yeast, autophagosomes directly fuse with a single large acidic vacuole for degradation. In contrast, in higher eukaryotes, nascent autophagosomes must undergo a series of maturation steps in part by fusing with endocytic vesicles, including early and late endosomes and MVBs [17, 18]. The resulting hybrid organelles, called amphisomes, are more acidic and fuse with lysosomes to form degradative autolysosomes. Proper maturation of the autophagosome requires an intact endocytic trafficking pathway, components of the ESCRT pathway, and components involved in endocytic vesicle fusion [95, 96].

Chaperone-mediated autophagy (CMA) is a selective degradation process in which cytosolic proteins are directly translocated across the lysosomal membrane through chaperone- and receptor- mediated internalization, unfolding, and translocation via lysosome-associated membrane protein 2A (LAMP-2A) [92, 97]. All CMA substrates contain a specific KFERQ-like pentapeptide sequence [98]. This motif is selectively recognized in the cytoplasm by Hsc70 [99], which targets the substrate-chaperone complex to the lysosomal surface where it interacts with the receptor protein LAMP-2A [97]. Once unfolded, the substrate translocates across the membrane and is then rapidly degraded. Although basal CMA can be detected in almost all cell types, CMA is maximally activated in response to stress (starvation, oxidative stress and conditions that cause protein damage) [100, 101]. Alterations in CMA underlie the pathogenesis of various neurodegenerative disorders such as Parkinson's disease and certain tauopathies [102, 103]. A decrease in CMA activity, induced by the pathogenic protein that accumulates in the affected neurons, renders these cells susceptible to numerous stressors and often precipitates cell death. In addition, CMA contributes to the removal of some pathogenic forms of huntingtin, the protein that accumulates in Huntington's disease [104].

In **microautophagy**, cytosolic proteins or whole regions of the cytosol are engulfed in the lysosome through the direct invagination of lysosomal or endosomal membranes [105]. Recently, endosomal microautophagy has been found to transport cytosolic proteins into ILVs through a process dependent on the ESCRT machinery and electrostatic binding of Hsc70 to endosomal acidic phospholipids, and independent of substrate unfolding or LAMP-2A [43].

4.2 Insights into Autophagosome Biogenesis

The main protein components involved in autophagosome formation and autophagy degradation were initially described in the 1990s by yeast genetics, which provided powerful genetic and molecular tools to investigate autophagy [106]. These conserved autophagy-related genes are collectively known as *ATG* [107] and function sequentially in phagophore formation, extension, cargo recognition, and autophagosome fusion with the lysosome.

The core autophagy machinery includes two ubiquitin-like conjugation systems, for protein-to-protein and protein-to-lipid conjugation. The combined action of these conjugation cascades and a kinase nucleation complex induces the formation of a limiting membrane or phagophore, which elongates and fuses to form the autophagosome [108].

The ubiquitin-like molecule Atg12 is conjugated to Atg5 by Atg7 and Atg10, forming a complex that interacts with Atg16L1 and associates with phagophores, but that dissociates from completed autophagosomes. The formation of phagophore requires the class III phosphoinositide 3-kinase (PI3K) Vps34, which acts in a large macromolecular complex, together with Beclin-1 (mammalian atg6), atg14 and Vps15, to form PtIns(3)P [109, 110]. The activity of this complex is dependent on upstream autophagy regulators, including the mammalian Atg1 orthologs ULK1 and ULK2, atg13, and FIP2000 (focal adhesion kinase (FAK)-family interacting protein of 200 KDa), which functions as the scaffold for the recruitment of downstream Atg proteins. Under fed conditions, the ULK complex is bound to MTORC1 and is thus inactive. Upon amino acid starvation, MTORC1 is inactivated and dissociates from the ULK complex, which results in increased ULK1 and ULK2 kinase activity. The carboxy-terminal domain of ULK1 and ULK2 binds to membranes, and this property might mediate the recruitment of the complex to the site of autophagosome initiation [111]. Once activated and targeted to the site of autophagosome initiation, the ULK and PI3K complexes phosphorylate a specific set of proteins and produce a specific pool of PdIns(3)P that drives the nucleation of the isolation membrane and the recruitment of additional ATG proteins and autophagy-specific PtdIns(3)P effectors, such as DFCP1 and WIPI.

Following nucleation, the Atg5-Atg12-ATg16L complex is recruited to the membrane, where it functions as an E3-like ligase to mediate the second of the ubiquitin-like reactions and promote the lipid conjugation of phosphatidylethanolamine (PE) to LC3 and its family members GATE16 and GARABAP. This modification enables GATE16 and GARABAP to associate with the autophagosomal membrane while the isolation membrane is expanding [112]. Before closure of the isolation membrane, which creates the autophagosome, the ATG proteins bound to the membrane dissociate, whereas LC3 and its family members found inside autolysosomes remain attached.

Autophagy culminates in the maturation step, when nascent autophagosomes fuse with the endolysosomal system to create a fully functional degradative compartment, the autolysosome. The Atg16L complex was found to be required for specific recruitment of LC3 to sites of autophagosome formation [113]. However, fusion of autophagosomes with endosomes appears to require loss of Atg16L complex and LC3-II from the surface of the autophagosome.

The origin of the membrane that forms the autophagosome is diverse and may include multiple sources, including the endoplasmic reticulum, the ER-Golgi intermediate compartment (ERGIC), the Golgi apparatus, recycling endosomes, the plasma membrane, and mitochondria. Although incompletely understood, the inclusion of membranes from distinct locations may contribute essential lipids and proteins to the forming autophagosome [18]. Several autophagic proteins localize to the ER, suggesting that it might play a role in autophagosome formation [114]. The double FYVE domain containing protein 1 (DFCP1), Atg14L, and ULK1 localize to the ER in resting conditions, but upon starvation they form punctate structures on the ER and recruit a subset of class III PI3-kinase that promotes the formation of PtIns(3)P in the ER, turning this organelle into the platform for autophagosome formation [115]. The outer mitochondrial membrane has been proposed to contribute to autophagosome biogenesis in starved mammalian cells [116]. During starvation, components of the core autophagy machinery relocate to contact sites between mitochondria and the ER and promote autophagosome biogenesis [117]. In addition, autophagosome expansion includes a contribution from membrane-vesicles trafficked from the Golgi and endosomes by coat protein complex II (COPII) [118–120]. The plasma membrane is another important source of membrane involved in the formation of early autophagosomal structures under basal and induced conditions [121]. The ability of the plasma membrane to contribute to autophagosome formation is associated with the localization of Atg16L1 at the plasma membrane via Atg16L1-AP-2-clathrin heavy chain interactions and homotypic and heterotypic fusion between these vesicles [122–126].

4.3 Regulation of Autophagy

Autophagy is critical to the maintenance of proteostasis and to the preservation of proper organelle function by selective removal of damaged organelles, particularly of mitochondria. Basal and inducible macroautophagy are essential for cellular homeostasis in many different tissues [127–130]. A broad range of cell-stress-inducing conditions activate autophagy-mediated degradation of protein aggregates, oxidized lipids, damaged organelles and intracellular pathogens [131]. Autophagy is also essential for defense against exogenous and endogenous aggressors and in circumstances requiring extensive cellular remodeling [132]. The breakdown products generated during autophagy degradation are used to generate new cellular components and energy in response to the nutritional needs of the cell. The importance of this pathway is reflected in the several human diseases in which macroautophagy malfunction is observed, which include cancer, neurodegenerative and metabolic diseases, cardiac hypertrophy, diabetes, and pathogen infection, as well as aberrant differentiation of many cell lineages [93, 94].

The signals that regulate autophagy are diverse. Induction of autophagy in response to starvation is mediated in part via inactivation of MTOR and activation of Jun N-terminal kinase (JNK), while energy loss induces autophagy by activation of AMP kinase (AMPK) [133, 134]. Following amino-acid starvation or MTOR inhibition, activated ULK1 phosphorylates Beclin-1 on Ser 14, thereby enhancing the activity of the ATG14L-containing VPS34 complexes for full autophagic induction [135]. The metabolic sensor AMPK plays a key role in regulating different Vps34 complexes. AMPK directly phosphorylates Vps34 and Beclin-1 to regulate non- and pro- autophagic Vps34 complexes, which are involved in intracellular vesicle trafficking and autophagy induction, respectively [136].

4.4 Selective Autophagy

Several types of autophagy have been described, with the type depending on the nature of the substrate eliminated: protein aggregate-prone or misfolded proteins (aggregophagy), damaged mitochondria (mitophagy), excessive peroxisomes (pexophagy), lipid droplets (lipophagy), and bacteria and virus (xeophagy), among others.

The molecular mechanisms underlying cargo selection and regulation of selective types of autophagy are still largely unknown. These pathways appear to rely on specific cargo-recognizing autophagy receptors and adaptors, which connect the cargo to the autophagic membrane to allow specific sequestration of the substrate [137]. The autophagy receptors p62 and NBR1 (neighbour of BRCA1 gene 1) bind ubiquitinated protein aggregates through a ubiquitin-associated (UBA) domain and LC3 via their LIR (LC3-interacting region) motifs and thereby promote the specific degradation of ubiquitinated proteins [138–140]. p62 has also been implicated in autophagic degradation of other ubiquitinated substrates such as intracellular bacteria, viral capsid proteins, the midbody after cytokinesis, peroxisomes, damaged mitochondria, and bacteriocidal precursor proteins. Defects in these pathways disrupt tissue homeostasis, accompanied by extensive accumulation of p62-containing aggregates, which deregulates its function as a scaffold protein in several signaling cascades such as NF-kB signaling, apoptosis and Nrf2 activation.

Posttranslational protein modifications contribute to the regulation of the autophagy pathway. Protein phosphorylation regulates the activity of proteins involved in autophagy, whereas ubiquination regulates proteins involved in the degradation of the autophagic cargo. Phosphorylation of autophagy receptors might be a general mechanism for the regulation of selective autophagy. For instance, phosphorylation close to the LIR motif of the autophagy receptor optineurin enhances the LC3 binding affinity of optineurin and promotes selective autophagy of ubiquitinated cytosolic *Salmonella enterica* [141]. In relation to autophagy, ubiquitin has so far been ascribed as a signal for cargo degradation. Ubiquitination of aggregated prone proteins, as well as bacteria and mitochondria, serves as a signal for recognition by autophagy receptors like p62 and NRB1. However, the role in autophagy of the ubiquitin-like proteins SUMO and Nedd is so far unexplored. A role in selective autophagy has recently been reported for protein acetylation. Histone deacetylase 6 (HDAC6), initially identified as a mediator of the transport of misfolded proteins to the aggresome, is also implicated in the maturation of ubiquitin-positive autophagosomes [142] and the Parkin-mediated clearance of damaged mitochondria [143]. Furthermore, the acetylation of an aggregophagy cargo protein, mutant huntingtin, is important for its degradation by autophagy [144].

Analysis of autophagy-deficient mice reveals marked accumulation of aberrant organelles, including mitochondria, which appear swollen and deformed, and increased numbers of peroxisomes and lipid droplets. In addition, these mice accumulate polyubiquitinated proteins in almost all tissues, forming inclusion bodies whose size and number increase with aging [129].

Macroautophagy maintains mitochondrial quality control by selectively degrading dysfunctional mitochondria via a process known as mitophagy [145]. The autophagy receptor Nix/bnip3L is an outer mitochondrial membrane protein that interacts with GABARAP and plays an important role during erythroid differentiation by controlling mitochondrial elimination [146, 147]. PINK1, a mitochondrial kinase, and Parkin, an E3 ubiquitin ligase, have been genetically linked to Parkinson's disease (PD) and a pathway that prevents progressive mitochondrial damage and dysfunction [148]. Upon mitochondrial membrane depolarization, PINK becomes stabilized and recruits Parkin to the damaged mitochondria, where it ubiquitinates various outer mitochondrial membrane substrates and induces mitochondrial degradation. Although p62 has been implicated in the recognition of ubiquitinated mitochondria, p62-deficient cells do not show alteration in mitochondrial degradation [149–151].

When specific bacteria invade host cells, they are engulfed by a selective type of autophagy called xenophagy, restricting their growth. Invading bacteria such as *Salmonella enterica* and *Listeria monocytogenesis* become positive for ubiquitin when they access the cytosol after rupturing the limiting membrane of the endosome/phagosome. p62, NDP52, and optineurin have been proposed to be autophagy receptors that link ubiquitinated bacteria to LC3 [141]. Additionally, bacterial clearance through autophagy can lead to the production of antimicrobial peptides that kill invading bacteria [152].

Lipophagy has been implicated in the degradation of lipid droplets, where esterified triglycerides are stored, providing a mechanism to generate endogenous free fatty acids for energy production through beta-oxidation. Liver-specific Atg7-deficient mice display massive accumulation of triglycerides and cholesterol in the form of lipid droplets [130].

5 Crosstalk Between Endolysosomal Compartments in the Regulation of Cellular Proteostasis

To ensure efficient function in the regulation of intracellular homeostasis, the various proteolytic pathways need to maintain a constant communication. In recent years, it has become evident that the different autophagy routes act in coordination. Cells respond to blockade of CMA by activating macroautophagy and to blockade of macroautophagy by activating CMA. Despite this compensatory activation, the maintenance of cellular proteostasis involves interactions between the autophagy and endocytic degradation pathways. Induction of autophagy promotes autophagic flux by increased convergence with the endosomal pathway [153]. Both pathways intersect when autophagosomes fuse with MVBs, forming an amphisome. A functional endocytic pathway is critical for maintaining an efficient degradation process, and loss of trafficking complexes that control endosomes promote accumulation of autophagosomes and amphisomes.

Early endosomes are required for functional autophagy. In fact, loss of COPI function impairs transport into and through early endosomes and blocks autophagosome maturation and autophagic flux [154, 155]. Phagophore formation and expansion requires vesicles derived from Rab11-positive recycling endosomes and from the plasma membrane, coated with the clathrin adaptor protein AP2 and Atg16L1 [123].

Small GTP binding proteins are essential for endocytic trafficking and autophagy. Rab7 is involved in transport to late endosomes and in the biogenesis of the perinuclear lysosome compartment. In addition, Rab7 participates in the final maturation of late autophagic vacuoles [156] through the fusion of late endosomes and autophagosomes with the lysosome [157]. Similarly, the Rab5-related yeast endocytic module Vps21 has been implicated in early-to late-endosome transport and autophagy [158], whereas Rab11 facilitates crosstalk between autophagy and the endosomal pathway by regulating Hook localization [153].

This crosstalk is also displayed by molecular complexes common to the autophagy and endocytic pathways. For example, the UVRAG-containing class III PI3K complex regulates not only autophagosome formation and maturation but also endosomal fusion [159], and favors transport of autophagic and endocytic cargo to the degradative compartments. Under nutrient-enriched conditions, the UVRAG negative regulator Rubicon blocks the maturation of both the autophagosomal and endocytic pathways [160, 161]. Nutrient deprivation induces the release of UVRAG from Rubicon and promotes its association with the HOPS complex, enhancing autophagosome and endosome maturation [162].

The relationship between autophagy and MVB biogenesis modulates the secretion of exosomes [6]. The loss or mutation of several components of the ESCRT machinery, involved in the formation of ILVs in late endosomes, results in abnormal autophagic maturation and neurodegeneration due to impaired autophagosome fusion with the endolysosomal system and accumulation of ubiquitinated intracellular aggregates [163–166]. Induction of autophagy or overexpression of LC3 blocks exosome secretion, indicating that under conditions promoting autophagy, such as starvation, MVBs fuse with autophagosomes, with consequent inhibition of exosome release [167]. This finding supports the notion that the balance between autophagy and exosome release might be regulated by the cellular metabolic state. Recently, a functional interconnection has been described between the core autophagy machinery and ESCRT components that facilitates basal autophagic flux and Alix-associated activities in late endosomes [168]. The study shows that interaction between ATG12–ATG3 and the ESCRT-associated protein Alix regulates MVB distribution, exosome biogenesis, viral budding and trafficking of the late endosome to lysosomes. Additionally, Alix silencing impairs basal autophagy, similarly to ATG12–ATG3 deficiency. Interaction of Alix with the actin-binding protein cortactin is proposed to be required for autophagosome maturation through regulation of actin remodeling [168]. These studies further support the existence of molecular crosstalk between these different quality control mechanisms arising from the endolysosomal system in the control of intracellular proteostasis.

6 Endolysosomal Dysfunction and Aging

Aging is characterized by progressive deterioration of cells and organs due to accumulation of macromolecular and organelle damage [169]. Recycling of damaged components and replacement with newly synthesized ones ensures cellular homeostasis and delays the aging process. However, overall proteolytic activity of cellular clearance mechanisms declines with aging, and their dysfunction is associated with neurodegenerative disease.

Accumulating evidence implicates dysfunction of the autophagic-lysosomal pathway as a key feature of aging [170, 171]. Decreased macroautophagic activity with age has been described in different mammalian tissues [92, 172]. In addition, incomplete digestion of engulfed cargo inside lysosomes leads to the accumulation of undegraded products that through further modification and cross-linking give rise to the autofluorescent pigment known as lipofuscin, a marker of aging. Accumulation of lipofuscin reduces the degradative capacity of lysosomes and their ability to fuse with autophagic structures, which in turn results in further accumulation of waste material inside the lysosome [173]. Interestingly, leupeptin infusion in a transgenic mouse model of Alzheimer's disease (AD) results in the onset of neurodegenerative defects [174], providing evidence that lysosome dysfunction is involved in age-related diseases.

The activity of CMA decreases in almost all tissues of aged rodents and in senescent human fibroblasts in culture due to a decrease in the lysosomal levels of LAMP-2A, thus reducing binding and lysosomal translocation of substrate [175]. Restoring the levels of LAMP-2A in aged mice not only increased CMA but also upregulated the macroautophagic and proteasomal degradative pathways, increased the levels of intracellular ATP, and reduced the levels of cytosolic waste (oxidized and polyubiquitinated aggregated proteins) [176]. Interestingly, these changes coincided with a sharp increase in the number of autophagosomes, indicating the importance of autophagy in the aging process. Thus, preventing the decline in autophagic activity slows down cellular aging and preserves organ function, linking the integrity of the autophagosomal-lysosomal network to the progression of aging [176].

6.1 *Exosome Secretion During Aging*

Although the study of exosomes in aging is still in its infancy, there is evidence that senescent cells undergo specific changes in exosome composition and trafficking [177]. Insufficient digestion of damaged molecules might promote the release of partially digested or undigested materials through exosomes. This is important, for example, during age-related macular degeneration (AMD), where increased autophagy and the release of exosomes can contribute to the formation of drusen [178]. Senescence induced by DNA damage is characterized by an increase in exosome secretion dependent on the activation of p53. In addition, exosome release also increases during proliferative senescence in normal human diploid fibroblasts. These data support the hypothesis that senescence, initiated either by telomere attrition (e.g., aging) or DNA damage (e.g., radiotherapy), can induce a p53-dependent increase in the biogenesis of exosome-like vesicles [179]. p53 regulates the transcription of a set of genes encoding proteins that regulate the endosomal compartment and enhance exosome production, for example TSAP6 and CHMP4C [180]. Senescent cells secrete a plethora of soluble proteins, including various cytokines, chemokines, growth factors, and proteases, thereby generating a strong proinflammatory environment. This feature of senescent cells has been termed the senescence associated-secretory phenotype (SASP). How the autophagic-lysosomal pathway and exosome secretion are coordinated during organismal aging and how they modulate SASP are a largely unexplored questions.

Exosomes isolated from elderly people contain higher amounts of cytokine mRNAs than exosomes from younger people [181]. Also, the abundance of 5´tRNA halves changes in mouse serum during aging [182]. Endothelial cells secrete miR-214–enriched exosomes, which promote endothelial cell migration and angiogenesis. These exosomes repress the expression of ataxia telangiectasia mutated (ATM) in recipient cells in a miR-214–dependent manner, thereby avoiding senescence and stimulating blood vessel formation [183]. Parabiotic exposure of aged animals to a youthful systemic milieu can promote oligodendrocyte differentiation and improve remyelination [184]. This promyelinating effect is mediated by serum exosomes, and may involve peripheral exosome-mediated delivery of miRNAs. Moreover, in the brains of APP transgenic mice, intracerebral infusion of neuronal exosomes results in decreases in Aβ levels and amyloid deposition [185].

6.2 Endolysosomal Dysfunction in Neurodegenerative Diseases

The connection between these clearance mechanisms and their impact on cellular proteostasis is mostly reflected in aged-associated brain diseases. Endolysosomal dysfunction is increasingly recognized as a major pathogenic event in several neurodegenerative disorders, including Parkinson's disease (PD), Alzheimer's disease (AD) and related synucleinopathies characterized by accumulations of α-synuclein in Lewy bodies [186, 187]. AD and PD are characterized by the expression of several effector genes involved in the regulation of early endosomes (rab4 and rab5), late endosomes (rab7), exosome secretion (rab27), autophagy (Pink1 or Parkin), endosomal vesicle recycling (VPS35, LRRK2), and lysosome-related genes (GBA, ATP13A2), suggesting a mechanistic link between endolysosomal impairment and neurodegenerative disorders [188]. Functional MVBs are required to prevent accumulation of abnormal proteins that can lead to neurodegeneration. Depletion of ESCRT subunits (Tsg101, Vps24) or overexpression of a CHMP2B mutant associated with frontotemporal dementia inhibits autophagic degradation and leads to accumulation of ubiquitin-positive aggregates containing aggregate-prone proteins associated with neurodegenerative diseases [163, 164].

Neurological disorders include a group of inherited metabolic diseases collectively known as lysosomal storage disorders (LSDs), which show perturbations in lysosomal homeostasis caused by defects in enzymes and proteins required for lysosomal catabolism [189]. A common feature of all LSDs is the accumulation of macromolecules inside organelles of the endolysosomal system, which perturbs the homeostasis of other organelles and intracellular regulatory systems, highlighting the complexity of the maintenance of cellular clearance mechanisms and its consequences for pathogenesis. Abnormal storage of lipids promotes redistribution within cells via membrane trafficking, fusion, or altered trafficking pathways characteristic of these diseases. Endolysosomal molecules can also be disseminated via membrane contact sites between endolysosomes and the ER [190, 191], and by extracellular secretion of endolysosomal content, including exosome release. For example, Niemann-Pick type C1 (NPC1) disease is an autosomal-recessive lysosomal storage disorder that leads to the abnormal accumulation of free cholesterol and sphingolipids within the late endosomal and lysosomal compartments, resulting in progressive neurodegeneration and demyelination. NPC1 patients show upregulation of exosomal cholesterol release, which may be a compensatory mechanism to maintain cellular cholesterol homeostasis by bypassing the traffic block in this disease that results in toxic lysosomal cholesterol accumulation [192, 193]. In addition, primary kidney cells from arylsulfatase A-deficient mice secrete the accumulating lipids into the culture medium [194].

PD has been linked to mutations in lysosomal-related genes, such as glucocerebrosidase (GBA) and lysosomal type 5 P-type ATPase (ATP13A2). PD-related GBA deficiency and mutations reduce lysosomal function and lead to α-synuclein accumulation. PD-related mutations/deficiency in the ATP13A2 gene lead to a general lysosomal impairment characterized by lysosomal membrane instability, impaired lysosomal acidification, decreased processing of lysosomal enzymes, reduced degradation of lysosomal substrates, and diminished clearance of autophagosomes, collectively contributing to α-synuclein accumulation and cell death [195]. ATP13A2 localizes in MVBs, where it regulates zinc homeostasis and promotes α-synuclein externalization via exosomes [196, 197]. Reduced ATP13A2 function impairs delivery of endocytosed proteins/autophagy cargo to the lysosome, and decreases the release of exosomes and α-synuclein; in contrast, elevated ATP13A2 expression increases α-synuclein clearance through exosomes. These data suggest that ATP13A2 is a PD-associated gene involved in exosome biogenesis, and suggest a potential neuroprotective role of exosomes in PD. The contribution of disruption of the endolysosomal pathways to disease pathogenesis underscores the importance of regulating lysosomal function and exosome secretion as a therapeutic strategy for these disorders [198].

Because exosomes remove unwanted or damaged material to the extracellular environment, their release can be beneficial for the regulation of intracellular proteostasis. However, exosomes can be transferred from one cell to another, and their secretion can have a beneficial or pathological impact on neighbouring cells. Exosomes are involved in the spread of toxic proteins in neurodegenerative diseases such as AD, PD, Huntington's, and prion diseases [199]. For example, beta-amyloid peptides are released in association with exosomes [200]. Exosomes might provide catalytic environments that favor nucleation and aggregation of α-synuclein and beta-amyloid peptides [201]. For instance, exosomes can stimulate aggregation of beta-amyloid peptides, and preventing their secretion reduces amyloid plaque formation in vivo [202]. In addition, exosomes have been found containing misfolded proteins involved in ALS, such as SOD, TDP-43, and hnRNPA2B1 [52, 203–205].

Autophagy prevents aggregation of proteins associated with neurodegenerative diseases, such as hungtintin, tau, and alpha-synuclein [121, 206]. Impairment of autophagy is accompanied by accumulation of p62 (sequestosome 1/SQSTM1), which leads to the formation of large protein aggregates containing p62 and polyubiquitinated proteins in neurodegenerative pathologies [207]. It is thus possible that protein aggregates, which are not properly cleared by autophagy, may spread to neighboring neurons in a prion-like manner. In this regard, there is accumulating evidence that extracellular protein aggregates can be taken up by cells [208], thus promoting the formation of new protein aggregates in the recipient cell. Recent studies on α-synuclein oligomers indicate that these oligomers can be secreted either directly or associated with exosomes, with the secretion route being strongly influenced by autophagic activity [209]. These findings suggest that exosome-mediated release of α-synuclein oligomers is a mechanism through which cells clear toxic α-synuclein oligomers when autophagic mechanisms fail. Pharmacological activation of autophagy can attenuate accumulation of aggregated pathological proteins and prevent cell death, indicating that autophagy can attenuate proteotoxicity [210].

7 Concluding Remarks

The research findings discussed here support the complex contribution of exosome secretion and autophagy to the control of proteostasis during cellular fitness and aging. The numerous intersections between the diverse quality-control mechanisms originating at the endolysosomal system range from autophagy to exosome secretion, and expand the array of physiological and pathological roles of these processes. It is becoming clear that the autophagic machinery affects the secretion of exosomes, and that modulating autophagy and endolysosomal function, including the loading and secretion of exosomes, represents an attractive avenue for future therapeutic intervention in age-associated diseases, including neurodegenerative diseases and inflammation. It is important now to understand how these mechanisms are coordinated and regulated in order to identify the optimum therapeutic targets for treating disease.

Acknowledgements We thank F. Sanchez-Madrid for his continuous support and advice and S. Bartlett for English editing. F.B. is supported by ERC-2011-AdG 294340-GENTRIS and COST-Action BN1202. This work is supported by the research grant CP14/0019 (Fondo de Investigación Sanitaria del Instituto de Salud Carlos III), and co-funding by Fondo Europeo de Desarrollo Regional (FEDER). The Centro Nacional de Investigaciones Cardiovasculares (CNIC, Spain) is supported by the Spanish Ministry of Science and Innovation and the Pro-CNIC Foundation.

References

1. Powers ET, Morimoto RI, Dillin A, Kelly JW, Balch WE. Biological and chemical approaches to diseases of proteostasis deficiency. Annu Rev Biochem. 2009;78:959–91.
2. Taylor RC, Berendzen KM, Dillin A. Systemic stress signalling: understanding the cell non-autonomous control of proteostasis. Nat Rev Mol Cell Biol. 2014;15:211–7.
3. McClellan AJ, Tam S, Kaganovich D, Frydman J. Protein quality control: chaperones culling corrupt conformations. Nat Cell Biol. 2005;7:736–41.
4. Bagola K, Sommer T. Protein quality control: on IPODs and other JUNQ. Curr Biol. 2008;18:R1019–21.
5. Kaganovich D, Kopito R, Frydman J. Misfolded proteins partition between two distinct quality control compartments. Nature. 2008;454:1088–95.
6. Baixauli F, Lopez-Otin C, Mittelbrunn M. Exosomes and autophagy: coordinated mechanisms for the maintenance of cellular fitness. Front Immunol. 2014;5:403.
7. Bellingham SA, Guo BB, Coleman BM, Hill AF. Exosomes: vehicles for the transfer of toxic proteins associated with neurodegenerative diseases? Front Physiol. 2012;3:124.
8. Chen B, Retzlaff M, Roos T, Frydman J. Cellular strategies of protein quality control. Cold Spring Harb Perspect Biol. 2011;3:a004374.
9. Tyedmers J, Mogk A, Bukau B. Cellular strategies for controlling protein aggregation. Nat Rev Mol Cell Biol. 2010;11:777–88.
10. Vilchez D, Saez I, Dillin A. The role of protein clearance mechanisms in organismal ageing and age-related diseases. Nat Commun. 2014;5:5659.
11. Wolff S, Weissman JS, Dillin A. Differential scales of protein quality control. Cell. 2014;157:52–64.
12. Luzio JP, Pryor PR, Bright NA. Lysosomes: fusion and function. Nat Rev Mol Cell Biol. 2007;8:622–32.
13. Rink J, Ghigo E, Kalaidzidis Y, Zerial M. Rab conversion as a mechanism of progression from early to late endosomes. Cell. 2005;122:735–49.
14. Hanson PI, Cashikar A. Multivesicular body morphogenesis. Annu Rev Cell Dev Biol. 2012;28:337–62.
15. Mittelbrunn M, Sanchez-Madrid F. Intercellular communication: diverse structures for exchange of genetic information. Nat Rev Mol Cell Biol. 2012;13:328–35.
16. Saftig P, Klumperman J. Lysosome biogenesis and lysosomal membrane proteins: trafficking meets function. Nat Rev Mol Cell Biol. 2009;10:623–35.
17. Lamb CA, Dooley HC, Tooze SA. Endocytosis and autophagy: shared machinery for degradation. Bioessays. 2013;35:34–45.
18. Lamb CA, Yoshimori T, Tooze SA. The autophagosome: origins unknown, biogenesis complex. Nat Rev Mol Cell Biol. 2013;14:759–74.
19. Harding C, Heuser J, Stahl P. Endocytosis and intracellular processing of transferrin and colloidal gold-transferrin in rat reticulocytes: demonstration of a pathway for receptor shedding. Eur J Cell Biol. 1984;35:256–63.
20. Johnstone RM, Adam M, Hammond JR, Orr L, Turbide C. Vesicle formation during reticulocyte maturation. Association of plasma membrane activities with released vesicles (exosomes). J Biol Chem. 1987;262:9412–20.

21. Raposo G, Stoorvogel W. Extracellular vesicles: exosomes, microvesicles, and friends. J Cell Biol. 2013;200:373–83.
22. Nolte-'t Hoen EN, Buermans HP, Waasdorp M, Stoorvogel W, Wauben MH, t Hoen PA. Deep sequencing of RNA from immune cell-derived vesicles uncovers the selective incorporation of small non-coding RNA biotypes with potential regulatory functions. Nucleic Acids Res. 2012;40:9272–85.
23. Villarroya-Beltri C, Baixauli F, Gutierrez-Vazquez C, Sanchez-Madrid F, Mittelbrunn M. Sorting it out: regulation of exosome loading. 2014. Semin Cancer Biol.
24. Fruhbeis C, Frohlich D, Kuo WP, Amphornrat J, Thilemann S, Saab AS, Kirchhoff F, Mobius W, Goebbels S, Nave KA, et al. Neurotransmitter-triggered transfer of exosomes mediates oligodendrocyte-neuron communication. PLoS Biol. 2013;11, e1001604.
25. Gutierrez-Vazquez C, Villarroya-Beltri C, Mittelbrunn M, Sanchez-Madrid F. Transfer of extracellular vesicles during immune cell-cell interactions. Immunol Rev. 2013;251:125–42.
26. Mittelbrunn M, Gutierrez-Vazquez C, Villarroya-Beltri C, Gonzalez S, Sanchez-Cabo F, Gonzalez MA, Bernad A, Sanchez-Madrid F. Unidirectional transfer of microRNA-loaded exosomes from T cells to antigen-presenting cells. Nat Commun. 2011;2:282.
27. Peinado H, Aleckovic M, Lavotshkin S, Matei I, Costa-Silva B, Moreno-Bueno G, Hergueta-Redondo M, Williams C, Garcia-Santos G, Ghajar C, et al. Melanoma exosomes educate bone marrow progenitor cells toward a pro-metastatic phenotype through MET. Nat Med. 2012;18:883–91.
28. Robbins PD, Morelli AE. Regulation of immune responses by extracellular vesicles. Nat Rev Immunol. 2014;14:195–208.
29. Colombo M, Raposo G, Thery C. Biogenesis, secretion, and intercellular interactions of exosomes and other extracellular vesicles. Annu Rev Cell Dev Biol. 2014;30:255–89.
30. Edgar JR, Eden ER, Futter CE. Hrs- and CD63-dependent competing mechanisms make different sized endosomal intraluminal vesicles. Traffic. 2014;15:197–211.
31. Henne WM, Buchkovich NJ, Emr SD. The ESCRT pathway. Dev Cell. 2011;21:77–91.
32. Bilodeau PS, Winistorfer SC, Kearney WR, Robertson AD, Piper RC. Vps27-Hse1 and ESCRT-I complexes cooperate to increase efficiency of sorting ubiquitinated proteins at the endosome. J Cell Biol. 2003;163:237–43.
33. Katzmann DJ, Stefan CJ, Babst M, Emr SD. Vps27 recruits ESCRT machinery to endosomes during MVB sorting. J Cell Biol. 2003;162:413–23.
34. Baietti MF, Zhang Z, Mortier E, Melchior A, Degeest G, Geeraerts A, Ivarsson Y, Depoortere F, Coomans C, Vermeiren E, et al. Syndecan-syntenin-ALIX regulates the biogenesis of exosomes. Nat Cell Biol. 2012;14:677–85.
35. Colombo M, Moita C, van Niel G, Kowal J, Vigneron J, Benaroch P, Manel N, Moita LF, Thery C, Raposo G. Analysis of ESCRT functions in exosome biogenesis, composition and secretion highlights the heterogeneity of extracellular vesicles. J Cell Sci. 2013;126:5553–65.
36. Zhu H, Guariglia S, Yu RY, Li W, Brancho D, Peinado H, Lyden D, Salzer J, Bennett C, Chow CW. Mutation of SIMPLE in Charcot-Marie-Tooth 1C alters production of exosomes. Mol Bioi Cell. 2013;24(1619–1637):S1611–3.
37. Putz U, Howitt J, Doan A, Goh CP, Low LH, Silke J, Tan SS. The tumor suppressor PTEN is exported in exosomes and has phosphatase activity in recipient cells. Sci Signal. 2012;5:70.
38. Stuffers S, Sem Wegner C, Stenmark H, Brech A. Multivesicular endosome biogenesis in the absence of ESCRTs. Traffic. 2009;10:925–37.
39. van Niel G, Charrin S, Simoes S, Romao M, Rochin L, Saftig P, Marks MS, Rubinstein E, Raposo G. The tetraspanin CD63 regulates ESCRT-independent and -dependent endosomal sorting during melanogenesis. Dev Cell. 2011;21:708–21.
40. Perez-Hernandez D, Gutierrez-Vazquez C, Jorge I, Lopez-Martin S, Ursa A, Sanchez-Madrid F, Vazquez J, Yanez-Mo M. The intracellular interactome of tetraspanin-enriched microdomains reveals their function as sorting machineries toward exosomes. J Biol Chem. 2013;288:11649–61.
41. Trajkovic K, Hsu C, Chiantia S, Rajendran L, Wenzel D, Wieland F, Schwille P, Brugger B, Simons M. Ceramide triggers budding of exosome vesicles into multivesicular endosomes. Science. 2008;319:1244–7.

42. Ghossoub R, Lembo F, Rubio A, Gaillard CB, Bouchet J, Vitale N, Slavik J, Machala M, Zimmermann P. Syntenin-ALIX exosome biogenesis and budding into multivesicular bodies are controlled by ARF6 and PLD2. Nat Commun. 2014;5:3477.
43. Sahu R, Kaushik S, Clement CC, Cannizzo ES, Scharf B, Follenzi A, Potolicchio I, Nieves E, Cuervo AM, Santambrogio L. Microautophagy of cytosolic proteins by late endosomes. Dev Cell. 2011;20:131–9.
44. Geminard C, De Gassart A, Blanc L, Vidal M. Degradation of AP2 during reticulocyte maturation enhances binding of hsc70 and Alix to a common site on TFR for sorting into exosomes. Traffic. 2004;5:181–93.
45. Moreno-Gonzalo O, Villarroya-Beltri C, Sanchez-Madrid F. Post-translational modifications of exosomal proteins. Front Immunol. 2014;5:383.
46. Burke MC, Oei MS, Edwards NJ, Ostrand-Rosenberg S, Fenselau C. Ubiquitinated proteins in exosomes secreted by myeloid-derived suppressor cells. J Proteome Res. 2014;13:5965–72.
47. Lee HS, Jeong J, Lee KJ. Characterization of vesicles secreted from insulinoma NIT-1 cells. J Proteome Res. 2009;8:2851–62.
48. Zuccato E, Blatt EJ, Holt O, Sigismund S, Shaw M, Bossi G, Griffiths GM. Sorting of Fas ligand to secretory lysosomes is regulated by mono-ubiquitylation and phosphorylation. J Cell Sci. 2007;120:191–9.
49. Valapala M, Vishwanatha JK. Lipid raft endocytosis and exosomal transport facilitate extracellular trafficking of annexin A2. J Biol Chem. 2011;286:30911–25.
50. Saman S, Kim W, Raya M, Visnick Y, Miro S, Saman S, Jackson B, McKee AC, Alvarez VE, Lee NC, et al. Exosome-associated tau is secreted in tauopathy models and is selectively phosphorylated in cerebrospinal fluid in early Alzheimer disease. J Biol Chem. 2012;287:3842–9.
51. Surgucheva I, Sharov VS, Surguchov A. gamma-Synuclein: seeding of alpha-synuclein aggregation and transmission between cells. Biochemistry. 2012;51:4743–54.
52. Villarroya-Beltri C, Gutierrez-Vazquez C, Sanchez-Cabo F, Perez-Hernandez D, Vazquez J, Martin-Cofreces N, Martinez-Herrera DJ, Pascual-Montano A, Mittelbrunn M, Sanchez-Madrid F. Sumoylated hnRNPA2B1 controls the sorting of miRNAs into exosomes through binding to specific motifs. Nat Commun. 2013;4:2980.
53. Koppers-Lalic D, Hackenberg M, Bijnsdorp IV, van Eijndhoven MA, Sadek P, Sie D, Zini N, Middeldorp JM, Ylstra B, de Menezes RX, et al. Nontemplated nucleotide additions distinguish the small RNA composition in cells from exosomes. Cell Rep. 2014;8:1649–58.
54. Saunderson SC, Schuberth PC, Dunn AC, Miller L, Hock BD, MacKay PA, Koch N, Jack RW, McLellan AD. Induction of exosome release in primary B cells stimulated via CD40 and the IL-4 receptor. J Immunol. 2008;180:8146–52.
55. Yeh YY, Ozer HG, Lehman AM, Maddocks K, Yu L, Johnson AJ, Byrd JC. Characterization of CLL exosomes reveals a distinct microRNA signature and enhanced secretion by activation of B-cell receptor signaling. 2015. Blood.
56. Qu Y, Franchi L, Nunez G, Dubyak GR. Nonclassical IL-1 beta secretion stimulated by P2X7 receptors is dependent on inflammasome activation and correlated with exosome release in murine macrophages. J Immunol. 2007;179:1913–25.
57. Savina A, Furlan M, Vidal M, Colombo MI. Exosome release is regulated by a calcium-dependent mechanism in K562 cells. J Biol Chem. 2003;278:20083–90.
58. Azmi AS, Bao B, Sarkar FH. Exosomes in cancer development, metastasis, and drug resistance: a comprehensive review. Cancer Metastasis Rev. 2013;32:623–42.
59. King HW, Michael MZ, Gleadle JM. Hypoxic enhancement of exosome release by breast cancer cells. BMC Cancer. 2012;12:421.
60. Wang T, Gilkes DM, Takano N, Xiang L, Luo W, Bishop CJ, Chaturvedi P, Green JJ, Semenza GL. Hypoxia-inducible factors and RAB22A mediate formation of microvesicles that stimulate breast cancer invasion and metastasis. Proc Natl Acad Sci U S A. 2014;111:E3234–42.
61. Umezu T, Tadokoro H, Azuma K, Yoshizawa S, Ohyashiki K, Ohyashiki JH. Exosomal miR-135b shed from hypoxic multiple myeloma cells enhances angiogenesis by targeting factor-inhibiting HIF-1. Blood. 2014;124:3748–57.

62. Parolini I, Federici C, Raggi C, Lugini L, Palleschi S, De Milito A, Coscia C, Iessi E, Logozzi M, Molinari A, et al. Microenvironmental pH is a key factor for exosome traffic in tumor cells. J Biol Chem. 2009;284:34211–22.
63. Hedlund M, Nagaeva O, Kargl D, Baranov V, Mincheva-Nilsson L. Thermal- and oxidative stress causes enhanced release of NKG2D ligand-bearing immunosuppressive exosomes in leukemia/lymphoma T and B cells. PLoS One. 2011;6, e16899.
64. Amzallag N, Passer BJ, Allanic D, Segura E, Thery C, Goud B, Amson R, Telerman A. TSAP6 facilitates the secretion of translationally controlled tumor protein/histamine-releasing factor via a nonclassical pathway. J Biol Chem. 2004;279:46104–12.
65. Feng Y, Huang W, Wani M, Yu X, Ashraf M. Ischemic preconditioning potentiates the protective effect of stem cells through secretion of exosomes by targeting Mecp2 via miR-22. PLoS One. 2014;9, e88685.
66. Giricz Z, Varga ZV, Baranyai T, Sipos P, Paloczi K, Kittel A, Buzas EI, Ferdinandy P. Cardioprotection by remote ischemic preconditioning of the rat heart is mediated by extracellular vesicles. J Mol Cell Cardiol. 2014;68:75–8.
67. Takeuchi T, Suzuki M, Fujikake N, Popiel HA, Kikuchi H, Futaki S, Wada K, Nagai Y. Intercellular chaperone transmission via exosomes contributes to maintenance of protein homeostasis at the organismal level. 2015. Proceedings of the National Academy of Sciences of the United States of America.
68. Settembre C, Fraldi A, Medina DL, Ballabio A. Signals from the lysosome: a control centre for cellular clearance and energy metabolism. Nat Rev Mol Cell Biol. 2013;14:283–96.
69. Lubke T, Lobel P, Sleat DE. Proteomics of the lysosome. Biochim Biophys Acta. 2009;1793:625–35.
70. Forgac M. Vacuolar ATPases: rotary proton pumps in physiology and pathophysiology. Nat Rev Mol Cell Biol. 2007;8:917–29.
71. Mindell JA. Lysosomal acidification mechanisms. Annu Rev Physiol. 2012;74:69–86.
72. Graves AR, Curran PK, Smith CL, Mindell JA. The Cl-/H+ antiporter ClC-7 is the primary chloride permeation pathway in lysosomes. Nature. 2008;453:788–92.
73. Wartosch L, Fuhrmann JC, Schweizer M, Stauber T, Jentsch TJ. Lysosomal degradation of endocytosed proteins depends on the chloride transport protein ClC-7. FASEB J. 2009;23:4056–68.
74. Wartosch L, Stauber T. A role for chloride transport in lysosomal protein degradation. Autophagy. 2010;6:158–9.
75. Calcraft PJ, Ruas M, Pan Z, Cheng X, Arredouani A, Hao X, Tang J, Rietdorf K, Teboul L, Chuang KT, et al. NAADP mobilizes calcium from acidic organelles through two-pore channels. Nature. 2009;459:596–600.
76. Kim HJ, Soyombo AA, Tjon-Kon-Sang S, So I, Muallem S. The Ca(2+) channel TRPML3 regulates membrane trafficking and autophagy. Traffic. 2009;10:1157–67.
77. Soyombo AA, Tjon-Kon-Sang S, Rbaibi Y, Bashllari E, Bisceglia J, Muallem S, Kiselyov K. TRP-ML1 regulates lysosomal pH and acidic lysosomal lipid hydrolytic activity. J Biol Chem. 2006;281:7294–301.
78. Wang X, Zhang X, Dong XP, Samie M, Li X, Cheng X, Goschka A, Shen D, Zhou Y, Harlow J, et al. TPC proteins are phosphoinositide- activated sodium-selective ion channels in endosomes and lysosomes. Cell. 2012;151:372–83.
79. Braulke T, Bonifacino JS. Sorting of lysosomal proteins. Biochim Biophys Acta. 2009;1793:605–14.
80. Coutinho MF, Prata MJ, Alves S. A shortcut to the lysosome: the mannose-6-phosphate-independent pathway. Mol Genet Metab. 2012;107:257–66.
81. Canuel M, Korkidakis A, Konnyu K, Morales CR. Sortilin mediates the lysosomal targeting of cathepsins D and H. Biochem Biophys Res Commun. 2008;373:292–7.
82. Lefrancois S, Zeng J, Hassan AJ, Canuel M, Morales CR. The lysosomal trafficking of sphingolipid activator proteins (SAPs) is mediated by sortilin. EMBO J. 2003;22:6430–7.
83. Ni X, Morales CR. The lysosomal trafficking of acid sphingomyelinase is mediated by sortilin and mannose 6-phosphate receptor. Traffic. 2006;7:889–902.

84. Zhao Y, Ren J, Padilla-Parra S, Fry EE, Stuart DI. Lysosome sorting of beta-glucocerebrosidase by LIMP-2 is targeted by the mannose 6-phosphate receptor. Nat Commun. 2014;5:4321.
85. Sardiello M, Palmieri M, di Ronza A, Medina DL, Valenza M, Gennarino VA, Di Malta C, Donaudy F, Embrione V, Polishchuk RS, et al. A gene network regulating lysosomal biogenesis and function. Science. 2009;325:473–7.
86. Settembre C, De Cegli R, Mansueto G, Saha PK, Vetrini F, Visvikis O, Huynh T, Carissimo A, Palmer D, Klisch TJ, et al. TFEB controls cellular lipid metabolism through a starvation-induced autoregulatory loop. Nat Cell Biol. 2013;15:647–58.
87. Settembre C, Di Malta C, Polito VA, Garcia Arencibia M, Vetrini F, Erdin S, Erdin SU, Huynh T, Medina D, Colella P, et al. TFEB links autophagy to lysosomal biogenesis. Science. 2011;332:1429–33.
88. Sancak Y, Bar-Peled L, Zoncu R, Markhard AL, Nada S, Sabatini DM. Ragulator-Rag complex targets mTORC1 to the lysosomal surface and is necessary for its activation by amino acids. Cell. 2010;141:290–303.
89. Roczniak-Ferguson A, Petit CS, Froehlich F, Qian S, Ky J, Angarola B, Walther TC, Ferguson SM. The transcription factor TFEB links mTORC1 signaling to transcriptional control of lysosome homeostasis. Sci Signal. 2012;5:42.
90. Settembre C, Zoncu R, Medina DL, Vetrini F, Erdin S, Erdin S, Huynh T, Ferron M, Karsenty G, Vellard MC, et al. A lysosome-to-nucleus signalling mechanism senses and regulates the lysosome via mTOR and TFEB. EMBO J. 2012;31:1095–108.
91. Medina DL, Di Paola S, Peluso I, Armani A, De Stefani D, Venditti R, Montefusco S, Scotto-Rosato A, Prezioso C, Forrester A, et al. Lysosomal calcium signalling regulates autophagy through calcineurin and TFEB. Nat Cell Biol. 2015;17:288–99.
92. Cuervo AM, Wong E. Chaperone-mediated autophagy: roles in disease and aging. Cell Res. 2014;24:92–104.
93. He C, Klionsky DJ. Regulation mechanisms and signaling pathways of autophagy. Annu Rev Genet. 2009;43:67–93.
94. Mizushima N, Levine B, Cuervo AM, Klionsky DJ. Autophagy fights disease through cellular self- digestion. Nature. 2008;451:1069–75.
95. Itakura E, Kishi-Itakura C, Mizushima N. The hairpin-type tail-anchored SNARE syntaxin 17 targets to autophagosomes for fusion with endosomes/lysosomes. Cell. 2012;151:1256–69.
96. Takats S, Nagy P, Varga A, Pircs K, Karpati M, Varga K, Kovacs AL, Hegedus K, Juhasz G. Autophagosomal Syntaxin17-dependent lysosomal degradation maintains neuronal function in Drosophila. J Cell Biol. 2013;201:531–9.
97. Cuervo AM, Dice JF. A receptor for the selective uptake and degradation of proteins by lysosomes. Science. 1996;273:501–3.
98. Dice JF, Terlecky SR, Chiang HL, Olson TS, Isenman LD, Short-Russell SR, Freundlieb S, Terlecky LJ. A selective pathway for degradation of cytosolic proteins by lysosomes. Semin Cell Biol. 1990;1:449–55.
99. Chiang HL, Terlecky SR, Plant CP, Dice JF. A role for a 70-kilodalton heat shock protein in lysosomal degradation of intracellular proteins. Science. 1989;246:382–5.
100. Kiffin R, Christian C, Knecht E, Cuervo AM. Activation of chaperone-mediated autophagy during oxidative stress. Mol Biol Cell. 2004;15:4829–40.
101. Massey A, Kiffin R, Cuervo AM. Pathophysiology of chaperone-mediated autophagy. Int J Biochem Cell Biol. 2004;36:2420–34.
102. Cuervo AM, Stefanis L, Fredenburg R, Lansbury PT, Sulzer D. Impaired degradation of mutant alpha-synuclein by chaperone-mediated autophagy. Science. 2004;305:1292–5.
103. Martinez-Vicente M, Talloczy Z, Kaushik S, Massey AC, Mazzulli J, Mosharov EV, Hodara R, Fredenburg R, Wu DC, Follenzi A, et al. Dopamine-modified alpha-synuclein blocks chaperone-mediated autophagy. J Clin Invest. 2008;118:777–88.
104. Koga H, Martinez-Vicente M, Arias E, Kaushik S, Sulzer D, Cuervo AM. Constitutive upregulation of chaperone-mediated autophagy in Huntington's disease. J Neurosci. 2011;31:18492–505.
105. Mijaljica D, Prescott M, Devenish RJ. Microautophagy in mammalian cells: revisiting a 40-year-old conundrum. Autophagy. 2011;7:673–82.

106. Yang Z, Klionsky DJ. Eaten alive: a history of macroautophagy. Nat Cell Biol. 2010;12:814–22.
107. Klionsky DJ, Cregg JM, Dunn Jr WA, Emr SD, Sakai Y, Sandoval IV, Sibirny A, Subramani S, Thumm M, Veenhuis M, et al. A unified nomenclature for yeast autophagy-related genes. Dev Cell. 2003;5:539–45.
108. Ohsumi Y, Mizushima N. Two ubiquitin-like conjugation systems essential for autophagy. Semin Cell Dev Biol. 2004;15:231–6.
109. Itakura E, Kishi C, Inoue K, Mizushima N. Beclin 1 forms two distinct phosphatidylinositol 3-kinase complexes with mammalian Atg14 and UVRAG. Mol Biol Cell. 2008;19:5360–72.
110. Liang XH, Jackson S, Seaman M, Brown K, Kempkes B, Hibshoosh H, Levine B. Induction of autophagy and inhibition of tumorigenesis by beclin 1. Nature. 1999;402:672–6.
111. Chan EY, Longatti A, McKnight NC, Tooze SA. Kinase-inactivated ULK proteins inhibit autophagy via their conserved C-terminal domains using an Atg13-independent mechanism. Mol Cell Biol. 2009;29:157–71.
112. Romanov J, Walczak M, Ibiricu I, Schuchner S, Ogris E, Kraft C, Martens S. Mechanism and functions of membrane binding by the Atg5-Atg12/Atg16 complex during autophagosome formation. EMBO J. 2012;31:4304–17.
113. Fujita N, Itoh T, Omori H, Fukuda M, Noda T, Yoshimori T. The Atg16L complex specifies the site of LC3 lipidation for membrane biogenesis in autophagy. Mol Biol Cell. 2008;19:2092–100.
114. Axe EL, Walker SA, Manifava M, Chandra P, Roderick HL, Habermann A, Griffiths G, Ktistakis NT. Autophagosome formation from membrane compartments enriched in phosphatidylinositol 3-phosphate and dynamically connected to the endoplasmic reticulum. J Cell Biol. 2008;182:685–701.
115. Matsunaga K, Morita E, Saitoh T, Akira S, Ktistakis NT, Izumi T, Noda T, Yoshimori T. Autophagy requires endoplasmic reticulum targeting of the PI3-kinase complex via Atg14L. J Cell Biol. 2010;190:511–21.
116. Hailey DW, Rambold AS, Satpute-Krishnan P, Mitra K, Sougrat R, Kim PK, Lippincott-Schwartz J. Mitochondria supply membranes for autophagosome biogenesis during starvation. Cell. 2010;141:656–67.
117. Hamasaki M, Furuta N, Matsuda A, Nezu A, Yamamoto A, Fujita N, Oomori H, Noda T, Haraguchi T, Hiraoka Y, et al. Autophagosomes form at ER-mitochondria contact sites. Nature. 2013;495:389–93.
118. Ge L, Melville D, Zhang M, Schekman R. The ER-Golgi intermediate compartment is a key membrane source for the LC3 lipidation step of autophagosome biogenesis. Elife. 2013;2, e00947.
119. Ge L, Schekman R. The ER-Golgi intermediate compartment feeds the phagophore membrane. Autophagy. 2014;10:170–2.
120. Ge L, Zhang M, Schekman R. Phosphatidylinositol 3-kinase and COPII generate LC3 lipidation vesicles from the ER-Golgi intermediate compartment. Elife. 2014;3, e04135.
121. Ravikumar B, Moreau K, Jahreiss L, Puri C, Rubinsztein DC. Plasma membrane contributes to the formation of pre-autophagosomal structures. Nat Cell Biol. 2010;12:747–57.
122. Mari M, Griffith J, Rieter E, Krishnappa L, Klionsky DJ, Reggiori F. An Atg9-containing compartment that functions in the early steps of autophagosome biogenesis. J Cell Biol. 2010;190:1005–22.
123. Moreau K, Ravikumar B, Renna M, Puri C, Rubinsztein DC. Autophagosome precursor maturation requires homotypic fusion. Cell. 2011;146:303–17.
124. Orsi A, Razi M, Dooley HC, Robinson D, Weston AE, Collinson LM, Tooze SA. Dynamic and transient interactions of Atg9 with autophagosomes, but not membrane integration, are required for autophagy. Mol Biol Cell. 2012;23:1860–73.
125. Yamamoto H, Kakuta S, Watanabe TM, Kitamura A, Sekito T, Kondo-Kakuta C, Ichikawa R, Kinjo M, Ohsumi Y. Atg9 vesicles are an important membrane source during early steps of autophagosome formation. J Cell Biol. 2012;198:219–33.

126. Young AR, Chan EY, Hu XW, Kochl R, Crawshaw SG, High S, Hailey DW, Lippincott-Schwartz J, Tooze SA. Starvation and ULK1-dependent cycling of mammalian Atg9 between the TGN and endosomes. J Cell Sci. 2006;119:3888–900.
127. Hara T, Nakamura K, Matsui M, Yamamoto A, Nakahara Y, Suzuki-Migishima R, Yokoyama M, Mishima K, Saito I, Okano H, et al. Suppression of basal autophagy in neural cells causes neurodegenerative disease in mice. Nature. 2006;441:885–9.
128. Komatsu M, Waguri S, Chiba T, Murata S, Iwata J, Tanida I, Ueno T, Koike M, Uchiyama Y, Kominami E, et al. Loss of autophagy in the central nervous system causes neurodegeneration in mice. Nature. 2006;441:880–4.
129. Mizushima N, Levine B. Autophagy in mammalian development and differentiation. Nat Cell Biol. 2010;12:823–30.
130. Singh R, Kaushik S, Wang Y, Xiang Y, Novak I, Komatsu M, Tanaka K, Cuervo AM, Czaja MJ. Autophagy regulates lipid metabolism. Nature. 2009;458:1131–5.
131. Galluzzi L, Kepp O, Trojel-Hansen C, Kroemer G. Mitochondrial control of cellular life, stress, and death. Circ Res. 2012;111:1198–207.
132. Tsukamoto S, Kuma A, Murakami M, Kishi C, Yamamoto A, Mizushima N. Autophagy is essential for preimplantation development of mouse embryos. Science. 2008;321:117–20.
133. Kamada Y, Funakoshi T, Shintani T, Nagano K, Ohsumi M, Ohsumi Y. Tor-mediated induction of autophagy via an Apg1 protein kinase complex. J Cell Biol. 2000;150:1507–13.
134. Noda T, Ohsumi Y. Tor, a phosphatidylinositol kinase homologue, controls autophagy in yeast. J Biol Chem. 1998;273:3963–6.
135. Russell RC, Tian Y, Yuan H, Park HW, Chang YY, Kim J, Kim H, Neufeld TP, Dillin A, Guan KL. ULK1 induces autophagy by phosphorylating Beclin-1 and activating VPS34 lipid kinase. Nat Cell Biol. 2013;15:741–50.
136. Kim J, Kim YC, Fang C, Russell RC, Kim JH, Fan W, Liu R, Zhong Q, Guan KL. Differential regulation of distinct Vps34 complexes by AMPK in nutrient stress and autophagy. Cell. 2013;152:290–303.
137. Svenning S, Johansen T. Selective autophagy. Essays Biochem. 2013;55:79–92.
138. Birgisdottir AB, Lamark T, Johansen T. The LIR motif—crucial for selective autophagy. J Cell Sci. 2013;126:3237–47.
139. Kirkin V, Lamark T, Johansen T, Dikic I. NBR1 cooperates with p62 in selective autophagy of ubiquitinated targets. Autophagy. 2009;5:732–3.
140. Lamark T, Kirkin V, Dikic I, Johansen T. NBR1 and p62 as cargo receptors for selective autophagy of ubiquitinated targets. Cell Cycle. 2009;8:1986–90.
141. Wild P, Farhan H, McEwan DG, Wagner S, Rogov VV, Brady NR, Richter B, Korac J, Waidmann O, Choudhary C, et al. Phosphorylation of the autophagy receptor optineurin restricts Salmonella growth. Science. 2011;333:228–33.
142. Pandey UB, Nie Z, Batlevi Y, McCray BA, Ritson GP, Nedelsky NB, Schwartz SL, DiProspero NA, Knight MA, Schuldiner O, et al. HDAC6 rescues neurodegeneration and provides an essential link between autophagy and the UPS. Nature. 2007;447:859–63.
143. Lee JY, Nagano Y, Taylor JP, Lim KL, Yao TP. Disease-causing mutations in parkin impair mitochondrial ubiquitination, aggregation, and HDAC6-dependent mitophagy. J Cell Biol. 2010;189:671–9.
144. Jeong H, Then F, Melia Jr TJ, Mazzulli JR, Cui L, Savas JN, Voisine C, Paganetti P, Tanese N, Hart AC, et al. Acetylation targets mutant huntingtin to autophagosomes for degradation. Cell. 2009;137:60–72.
145. Kim I, Rodriguez-Enriquez S, Lemasters JJ. Selective degradation of mitochondria by mitophagy. Arch Biochem Biophys. 2007;462:245–53.
146. Gao F, Chen D, Si J, Hu Q, Qin Z, Fang M, Wang G. The mitochondrial protein BNIP3L is the substrate of PARK2 and mediates mitophagy in PINK1/PARK2 pathway. Hum Mol Genet. 2015;24:2528–38.
147. Zhang J, Loyd MR, Randall MS, Waddell MB, Kriwacki RW, Ney PA. A short linear motif in BNIP3L (NIX) mediates mitochondrial clearance in reticulocytes. Autophagy. 2012;8:1325–32.

148. Pallanck L, Greenamyre JT. Neurodegenerative disease: pink, parkin and the brain. Nature. 2006;441:1058.
149. Narendra D, Tanaka A, Suen DF, Youle RJ. Parkin is recruited selectively to impaired mitochondria and promotes their autophagy. J Cell Biol. 2008;183:795–803.
150. Okatsu K, Oka T, Iguchi M, Imamura K, Kosako H, Tani N, Kimura M, Go E, Koyano F, Funayama M, et al. PINK1 autophosphorylation upon membrane potential dissipation is essential for Parkin recruitment to damaged mitochondria. Nat Commun. 2012;3:1016.
151. Youle RJ, Narendra DP. Mechanisms of mitophagy. Nat Rev Mol Cell Biol. 2011;12:9–14.
152. Ponpuak M, Davis AS, Roberts EA, Delgado MA, Dinkins C, Zhao Z, Virgin HWT, Kyei GB, Johansen T, Vergne I, et al. Delivery of cytosolic components by autophagic adaptor protein p62 endows autophagosomes with unique antimicrobial properties. Immunity. 2010;32:329–41.
153. Szatmari Z, Kis V, Lippai M, Hegedus K, Farago T, Lorincz P, Tanaka T, Juhasz G, Sass M. Rab11 facilitates cross-talk between autophagy and endosomal pathway through regulation of Hook localization. Mol Biol Cell. 2014;25:522–31.
154. Gu F, Aniento F, Parton RG, Gruenberg J. Functional dissection of COP-I subunits in the biogenesis of multivesicular endosomes. J Cell Biol. 1997;139:1183–95.
155. Razi M, Chan EY, Tooze SA. Early endosomes and endosomal coatomer are required for autophagy. J Cell Biol. 2009;185:305–21.
156. Jager S, Bucci C, Tanida I, Ueno T, Kominami E, Saftig P, Eskelinen EL. Role for Rab7 in maturation of late autophagic vacuoles. J Cell Sci. 2004;117:4837–48.
157. Gutierrez MG, Munafo DB, Beron W, Colombo MI. Rab7 is required for the normal progression of the autophagic pathway in mammalian cells. J Cell Sci. 2004;117:2687–97.
158. Chen Y, Zhou F, Zou S, Yu S, Li S, Li D, Song J, Li H, He Z, Hu B, et al. A Vps21 endocytic module regulates autophagy. Mol Biol Cell. 2014;25:3166–77.
159. Liang C, Lee JS, Inn KS, Gack MU, Li Q, Roberts EA, Vergne I, Deretic V, Feng P, Akazawa C, et al. Beclin1-binding UVRAG targets the class C Vps complex to coordinate autophagosome maturation and endocytic trafficking. Nat Cell Biol. 2008;10:776–87.
160. Matsunaga K, Saitoh T, Tabata K, Omori H, Satoh T, Kurotori N, Maejima I, Shirahama-Noda K, Ichimura T, Isobe T, et al. Two Beclin 1-binding proteins, Atg14L and Rubicon, reciprocally regulate autophagy at different stages. Nat Cell Biol. 2009;11:385–96.
161. Zhong Y, Wang QJ, Li X, Yan Y, Backer JM, Chait BT, Heintz N, Yue Z. Distinct regulation of autophagic activity by Atg14L and Rubicon associated with Beclin 1-phosphatidylinositol-3-kinase complex. Nat Cell Biol. 2009;11:468–76.
162. Kim YM, Jung CH, Seo M, Kim EK, Park JM, Bae SS, Kim DH. mTORC1 phosphorylates UVRAG to negatively regulate autophagosome and endosome maturation. Mol Cell. 2015;57:207–18.
163. Filimonenko M, Stuffers S, Raiborg C, Yamamoto A, Malerod L, Fisher EM, Isaacs A, Brech A, Stenmark H, Simonsen A. Functional multivesicular bodies are required for autophagic clearance of protein aggregates associated with neurodegenerative disease. J Cell Biol. 2007;179:485–500.
164. Lee JA, Beigneux A, Ahmad ST, Young SG, Gao FB. ESCRT-III dysfunction causes autophagosome accumulation and neurodegeneration. Curr Biol. 2007;17:1561–7.
165. Rusten TE, Stenmark H. How do ESCRT proteins control autophagy? J Cell Sci. 2009;122:2179–83.
166. Urwin H, Authier A, Nielsen JE, Metcalf D, Powell C, Froud K, Malcolm DS, Holm I, Johannsen P, Brown J, et al. Disruption of endocytic trafficking in frontotemporal dementia with CHMP2B mutations. Hum Mol Genet. 2010;19:2228–38.
167. Fader CM, Sanchez D, Furlan M, Colombo MI. Induction of autophagy promotes fusion of multivesicular bodies with autophagic vacuoles in k562 cells. Traffic. 2008;9:230–50.
168. Murrow L, Malhotra R, Debnath J. ATG12-ATG3 interacts with Alix to promote basal autophagic flux and late endosome function. Nat Cell Biol. 2015;17:300–10.
169. Ward WF. Protein degradation in the aging organism. Prog Mol Subcell Biol. 2002;29:35–42.

170. Lopez-Otin C, Blasco MA, Partridge L, Serrano M, Kroemer G. The hallmarks of aging. Cell. 2013;153:1194–217.
171. Rubinsztein DC, Marino G, Kroemer G. Autophagy and aging. Cell. 2011;146:682–95.
172. Levine B, Klionsky DJ. Development by self-digestion: molecular mechanisms and biological functions of autophagy. Dev Cell. 2004;6:463–77.
173. Terman A, Gustafsson B, Brunk UT. Autophagy, organelles and ageing. J Pathol. 2007;211:134–43.
174. Nixon RA, Mathews PM, Cataldo AM. The neuronal endosomal-lysosomal system in Alzheimer's disease. J Alzheimers Dis. 2001;3:97–107.
175. Cuervo AM, Dice JF. Age-related decline in chaperone-mediated autophagy. J Biol Chem. 2000;275:31505–13.
176. Zhang C, Cuervo AM. Restoration of chaperone-mediated autophagy in aging liver improves cellular maintenance and hepatic function. Nat Med. 2008;14:959–65.
177. Smith JA, Leonardi T, Huang B, Iraci N, Vega B, Pluchino S. Extracellular vesicles and their synthetic analogues in aging and age-associated brain diseases. Biogerontology. 2015;16:147–85.
178. Wang AL, Lukas TJ, Yuan M, Du N, Tso MO, Neufeld AH. Autophagy, exosomes and drusen formation in age-related macular degeneration. Autophagy. 2009;5:563–4.
179. Lehmann BD, Paine MS, Brooks AM, McCubrey JA, Renegar RH, Wang R, Terrian DM. Senescence-associated exosome release from human prostate cancer cells. Cancer Res. 2008;68:7864–71.
180. Yu X, Riley T, Levine AJ. The regulation of the endosomal compartment by p53 the tumor suppressor gene. FEBS J. 2009;276:2201–12.
181. Mitsuhashi M, Taub DD, Kapogiannis D, Eitan E, Zukley L, Mattson MP, Ferrucci L, Schwartz JB, Goetzl EJ. Aging enhances release of exosomal cytokine mRNAs by Abeta1-42-stimulated macrophages. FASEB J. 2013;27:5141–50.
182. Dhahbi JM, Spindler SR, Atamna H, Yamakawa A, Boffelli D, Mote P, Martin DI. 5' tRNA halves are present as abundant complexes in serum, concentrated in blood cells, and modulated by aging and calorie restriction. BMC Genomics. 2013;14:298.
183. van Balkom BW, de Jong OG, Smits M, Brummelman J, den Ouden K, de Bree PM, van Eijndhoven MA, Pegtel DM, Stoorvogel W, Wurdinger T, et al. Endothelial cells require miR-214 to secrete exosomes that suppress senescence and induce angiogenesis in human and mouse endothelial cells. Blood. 2013;121(3997–4006):S3991–3915.
184. Pusic AD, Kraig RP. Youth and environmental enrichment generate serum exosomes containing miR-219 that promote CNS myelination. Glia. 2014;62:284–99.
185. Yuyama K, Sun H, Sakai S, Mitsutake S, Okada M, Tahara H, Furukawa J, Fujitani N, Shinohara Y, Igarashi Y. Decreased amyloid-beta pathologies by intracerebral loading of glycosphingolipid-enriched exosomes in Alzheimer model mice. J Biol Chem. 2014;289:24488–98.
186. Perrett RM, Alexopoulou Z, Tofaris GK. The endosomal pathway in Parkinson's disease. Mol Cell Neurosci. 2015;69:21–8.
187. Tofaris GK. Lysosome-dependent pathways as a unifying theme in Parkinson's disease. Movement Disord. 2012;27:1364–9.
188. Ginsberg SD, Mufson EJ, Counts SE, Wuu J, Alldred MJ, Nixon RA, Che S. Regional selectivity of rab5 and rab7 protein upregulation in mild cognitive impairment and Alzheimer's disease. J Alzheimers Dis. 2010;22:631–9.
189. Schulze H, Sandhoff K. Lysosomal lipid storage diseases. Cold Spring Harb Perspect Biol. 2011;3.
190. Eden ER, White IJ, Tsapara A, Futter CE. Membrane contacts between endosomes and ER provide sites for PTP1B-epidermal growth factor receptor interaction. Nat Cell Biol. 2010;12:267–72.
191. Toulmay A, Prinz WA. Lipid transfer and signaling at organelle contact sites: the tip of the iceberg. Curr. Opin Cell Biol. 2011;23:458–63.
192. Chen FW, Li C, Ioannou YA. Cyclodextrin induces calcium-dependent lysosomal exocytosis. PLoS One. 2010;5, e15054.

193. Strauss K, Goebel C, Runz H, Mobius W, Weiss S, Feussner I, Simons M, Schneider A. Exosome secretion ameliorates lysosomal storage of cholesterol in Niemann-Pick type C disease. J Biol Chem. 2010;285:26279–88.
194. Klein D, Bussow H, Fewou SN, Gieselmann V. Exocytosis of storage material in a lysosomal disorder. Biochem Biophys Res Commun. 2005;327:663–7.
195. Usenovic M, Tresse E, Mazzulli JR, Taylor JP, Krainc D. Deficiency of ATP13A2 leads to lysosomal dysfunction, alpha-synuclein accumulation, and neurotoxicity. J Neurosci. 2012;32:4240–6.
196. Kong SM, Chan BK, Park JS, Hill KJ, Aitken JB, Cottle L, Farghaian H, Cole AR, Lay PA, Sue CM, et al. Parkinson's disease-linked human PARK9/ATP13A2 maintains zinc homeostasis and promotes alpha-Synuclein externalization via exosomes. Hum Mol Genet. 2014;23:2816–33.
197. Tsunemi T, Hamada K, Krainc D. ATP13A2/PARK9 regulates secretion of exosomes and alpha-synuclein. J Neurosci. 2014;34:15281–7.
198. Dehay B, Martinez-Vicente M, Caldwell GA, Caldwell KA, Yue Z, Cookson MR, Klein C, Vila M, Bezard E. Lysosomal impairment in Parkinson's disease. Movement Disord. 2013;28:725–32.
199. Kalani A, Tyagi A, Tyagi N. Exosomes: mediators of neurodegeneration, neuroprotection and therapeutics. Mol Neurobiol. 2014;49:590–600.
200. Rajendran L, Honsho M, Zahn TR, Keller P, Geiger KD, Verkade P, Simons K. Alzheimer's disease beta-amyloid peptides are released in association with exosomes. Proc Natl Acad Sci U S A. 2006;103:11172–7.
201. Grey M, Dunning CJ, Gaspar R, Grey C, Brundin P, Sparr E, Linse S. Acceleration of alpha-synuclein aggregation by exosomes. J Biol Chem. 2015;290:2969–82.
202. Dinkins MB, Dasgupta S, Wang G, Zhu G, Bieberich E. Exosome reduction in vivo is associated with lower amyloid plaque load in the 5XFAD mouse model of Alzheimer's disease. Neurobiol Aging. 2014;35:1792–800.
203. Basso M, Pozzi S, Tortarolo M, Fiordaliso F, Bisighini C, Pasetto L, Spaltro G, Lidonnici D, Gensano F, Battaglia E, et al. Mutant copper-zinc superoxide dismutase (SOD1) induces protein secretion pathway alterations and exosome release in astrocytes: implications for disease spreading and motor neuron pathology in amyotrophic lateral sclerosis. J Biol Chem. 2013;288:15699–711.
204. Gomes C, Keller S, Altevogt P, Costa J. Evidence for secretion of Cu, Zn superoxide dismutase via exosomes from a cell model of amyotrophic lateral sclerosis. Neurosci Lett. 2007;428:43–6.
205. Grad LI, Yerbury JJ, Turner BJ, Guest WC, Pokrishevsky E, O'Neill MA, Yanai A, Silverman JM, Zeineddine R, Corcoran L, et al. Intercellular propagated misfolding of wild-type Cu/Zn superoxide dismutase occurs via exosome-dependent and -independent mechanisms. Proc Natl Acad Sci U S A. 2014;111:3620–5.
206. Levine B, Kroemer G. Autophagy in the pathogenesis of disease. Cell. 2008;132:27–42.
207. Komatsu M, Waguri S, Koike M, Sou YS, Ueno T, Hara T, Mizushima N, Iwata J, Ezaki J, Murata S, et al. Homeostatic levels of p62 control cytoplasmic inclusion body formation in autophagy-deficient mice. Cell. 2007;131:1149–63.
208. Guo JL, Lee VM. Cell-to-cell transmission of pathogenic proteins in neurodegenerative diseases. Nat Med. 2014;20:130–8.
209. Danzer KM, Kranich LR, Ruf WP, Cagsal-Getkin O, Winslow AR, Zhu L, Vanderburg CR, McLean PJ. Exosomal cell-to-cell transmission of alpha synuclein oligomers. Mol Neurodegener. 2012;7:42.
210. Rubinsztein DC, Gestwicki JE, Murphy LO, Klionsky DJ. Potential therapeutic applications of autophagy. Nat Rev Drug Discov. 2007;6:304–12.

Microglia: Features of Polarization and Aging

G. Jean Harry and Christopher A. McPherson

1 Introduction

Resident macrophages can be found in virtually all tissues, representing approximately 10–15 % of the total cells. These cells display different transcriptional profiles possibly representing distinct classes of macrophages [1]. In the central nervous system (CNS), microglia were described by Virchow as early as the mid-nineteenth century. These cells were morphologically classified as distinct from neurons and astrocytes [2] and further classified as distinct from oligodendrocytes [3]. In addition, rather than sharing a similar neuroepithelial origin as other CNS-specific cells, microglia originate from a primitive monocyte population derived from the yolk sack [4–8]. By the 1980s, their distinct macrophage-like features were established including their role in immune defense within the CNS parenchyma [9]. These morphologically heterogeneous cells [10–12] comprise approximately 20 % of the total cells in the brain [13] performing crucial physiological functions for tissue development, architecture refinement, and remodeling/repairing tissue.

Under normal conditions, microglia assume a neural specific phenotype due to the CNS environment [14] and retain a relative quiescent surveillance phenotype with fine processes extending into the surrounding microenvironment. This allows the cells to maintain constant surveillance of the brain parenchyma for tissue changes [15, 16]. When microglia sense a change in their surrounding microenvironment (injury, disease, pathology) they can rapidly change their morphology by

G.J. Harry, Ph.D. (✉) • C.A. McPherson
National Toxicology Program Laboratory, National Institute of Environmental Health Sciences, National Institutes of Health, Neurotoxicology Group,
111 T.W. Alexander Drive, MD E1-07, Research Triangle Park, NC 27709, USA
e-mail: harry@niehs.nih.gov

S.C. Bondy, A. Campbell (eds.), *Inflammation, Aging, and Oxidative Stress*,
Oxidative Stress in Applied Basic Research and Clinical Practice,
DOI 10.1007/978-3-319-33486-8_3

increasing soma size and retracting their normal elongated fine processes to shorter, coarser cytoplasmic processes displaying a bushy appearance. This can continue to progress to a fully amoeboid morphology depending on the nature and severity of the signal [17, 18]. They serve to mount an immune response to pathogens, contribute to cell-cell interactions that serve to trigger or facilitate programmed cell death during the development of various brain regions [19, 20], and maintain tissue homeostasis by clearing dying cells, debris, or aberrant proteins [21–28]. Upon removal of the stimulating factor, microglia can then downregulate and return to a more ramified phenotype. The question remains however, as to whether this downregulated cell is representative of the "normal" cells or whether the cell has been altered in some manner that may manifest in subsequent activation scenarios.

Maintenance and expansion of the resident microglia population is considered to result from local self-renewal rather than from the recruitment of circulating monocytes [29–31]. The limited replication and turnover of microglia make them a relatively stable population that is maintained throughout life. With injury to the blood-brain-barrier, blood-borne monocytes can enter the brain parenchyma and assume a brain macrophage phenotype [32–35]. The blood-brain-barrier undergoes several changes over the course of aging [36–38] that may contribute to the reported changes in microglia-related factors occurring with aging or with age-related neurodegenerative diseases. It is suggested that infiltrating cells would be more involved in severe inflammatory injuries as compared to a predominant focus of resident microglia on maintaining tissue homeostasis [39]. When compared, non-activated microglia and peritoneal macrophages show similar but also distinct profiles. In steady state conditions, microglia and other CNS-associated myeloid cells express surface markers CD11b, F4/80, Fc-gamma receptor 1 (CD64), and CD115 (Csf-1R), ionized calcium-binding adapter molecule 1 (Iba-1) and proto-oncogene tyrosine-protein kinase MER (MerTK) [1]. In a recent study conducted by Hickman et al., [40] several highly enriched genes classified as "sensome" genes were identified in microglia, which allowed the cells to sense and interact with their local environment. These sensome genes included those for putative purinergic receptors, *P2ry12* and *P2ry13*, transmembrane protein 119 (*Tmem119*), G-protein coupled receptor 34 (*Gpr34*), the 1-type lectin receptor, sialic acid-binding immunogloubulin-type lectin H (*Siglec-h*), triggering receptor expressed on myeloid cells 2 (*Trem2*), and the fractalkine receptor, *Cx3cr1*. Chiu et al. [41] and Butovsky et al. [42] reported many of these same genes as well as Sal-like 1 (*Sall1*) and Fc receptor-like S (*Fcrls*) specifically in quiescent microglia. In comparison, peripheral macrophages have been shown to express a significantly fewer number of these "sensome" genes suggesting a less complex response network outside of the CNS [40].

2 Cell Polarization

Macrophages can respond to endogenous stimuli generated following infection or injury. This represents a product of a cell-mediated immune response that is essential for a host defense but also one that requires tight regulation. In the absence of microorganisms, a similar but sterile inflammatory response occurs often as a result of trauma, ischemia-reperfusion injury, or chemical exposure [43–45]. In general the macrophage response is characterized based on the nature of the activating stimulus and the resulting production of factors [46]. A conceptual framework for such activation identifies two polar extremes of signals computed by macrophages, classical (M1) or alternative (M2) [47]. This characterization has been beneficial for describing immune responses occurring during acute infections, allergies, asthma, and even to obesity [48]. This conceptual framework has more recently been applied to characterize the microglial response [49, 50]. Lipopolysaccharide (LPS) and interferon gamma (IFNγ) are classic inducers of M1 activation [51]. LPS is recognized by toll-like receptor 4 (TLR4) upon which activation induces myeloid differentiation primary response gene 88 (MyD88) and MaL/Tirap (Toll-interleukin 1 receptor domain containing adaptor protein)-dependent pathways. These pathways lead to production of pro-inflammatory cytokines (e.g. IFNγ, interleukin (IL)-12, tumor necrosis factor (TNF)α, IL-6, and IL-1β), chemokines (e.g. chemokine [C-C motif] ligand 2 CCL2, chemokine [C-X-C motif] ligand 10 [CXCL10], and CXCL11), and antigen presentation molecules, such as major histocompatibility complex (MHC). Regulators of these profiles include nuclear factor of kappa light polypeptide gene enhancer (NF-kB), activator protein 1 (AP-1), interferon regulatory factors, signal transducers and activators of transcription 1 (STAT1), and early growth response (EGR) family members [52]. IFNγ signals through the IFNγ receptors (IFNγR1 and IFNγR2) for gene expression of cytokine receptors (IL-15R, IL-2RA, and IL-6R) and cell activation factors (CD36, CD38, CD69, and CD97). IFNγ also regulates inducible SH2-containing protein (CISH), N-myc-interactor (NMI), protein tyrosine phosphatase, receptor type, C (PTPRC), protein tyrosine phosphatase, receptor type, O (PTPRO), and suppressor of cytokine signaling 1 (SOCS1) [53, 54]. Mediators of IFNγ-induced signaling include STAT1, JAK2, and IRF1. There are similarities between gene expression after IFNγ and LPS stimulation of macrophages; however, these M1 responses are not considered homolgous [55]. Gene expression profiles following stimulation with a combination of the cytokines are different from LPS or profiles alone, as are those from stimuli other than LPS or IFNγ [52].

While the initial response of macrophages to injury has been known for some time, positive influences of the response to facilitate resolution of inflammation through anti-inflammatory factors to deactivate pro-inflammatory cell phenotypes and reestablish homeostasis [56–59] for tissue remodeling have been recognized more recently [60–63]. In the early 1990s, the concept of macrophage alternative activation was developed largely based on work showing a role for IL-4 in the induction of an alternative (M2) activation state [64]. Under this state expression of the anti-inflammatory cytokines (IL-4, IL-10, IL-13, and transforming growth factor beta (Tgfβ)) as well as, arginase-1 (Arg-1), CD206, and Chitinase-3-like-3 (*Ym-1* in rodents) was induced [50, 65]. Upon further study, subclasses of M2 activation have been identified and as with the M1 phenotype, they are dependent upon the inducting stimuli. The M2a activation state is induced by parasitic products or associated signals (IL-4 and IL-13) providing a longer-term function for resolution and repair [56, 66–69]. In this case, signaling occurs through IL-4 receptor alpha

(IL-4Rα) leading to inhibition of NF-kB signaling induced by M1 activation. M2b polarization is observed with triggering of Fc gamma receptors, TLRs, and immune complexes [51]. M2c polarization occurs in response to specific anti-inflammatory factors such as, IL-10, TGFβ, and glucocorticoids [56, 67, 70–72]. In addition, cells do not appear to be committed to one phenotype but rather can shift from M2b phenotype to a mixture of M1 and M2a/b [73]. M2 polarization of microglia is similar to peripheral macrophages [74–77] displaying different mRNA profiles for IL-4 and IL-10 stimulation including arginase 1 (*Arg1*), mannose receptor 1 (*Mrc1*), chitinase-like 3 (*Ym1*), found in inflammatory zone 1 (*Fizz1*), and peroxisome proliferator-activated receptor (*Ppar*) [78]. In addition, the specificity of such markers remains in question. For example the primary "marker" for M2, Arg1, is also induced in M1 macrophages and expressed in some resident and mycobacteria-infected macrophages [79]. While these associations have been demonstrated *in vitro*, a number of the M2 related products can be induced *in vivo* in sterile wounds in the absence IL-4 or IL-13 [80] suggesting an alternative stimulus.

2.1 *Macrophage Versus Microglia*

Recent work examining macrophage activation demonstrated that the environment within which the macrophage resides drives the selection and tissue specific functions [81]. A comparison between microglia and peripheral macrophages showed that peripheral macrophages displayed a more robust level of induction of phenotype-specific genes [82]. In addition, both cell types showed a greater induction of gene expression in response to M1 vs. M2 polarization [82]. Of interest was the observation that, in comparison to macrophages, M2 stimulated microglia increased expression of multiple genes that, with the exception of CD64, were also elevated in M1 microglia. Michelucci et al. [78] were the first to publish a wide-ranging transcription and functional profile using the M1/M2 differentiation spectrum in microglia. In this study they characterized the inflammatory and functional phenotype in the murine microglia cell line, MMGT12, and in murine primary microglia upon stimulation with LPS, IFN-γ, IL-4 or, IL-10. Transcription profile analysis showed that MMGT12 and primary microglial cells exposed to LPS or IFN-γ were characterized by a classical M1 phenotype (*Il1b*, *Il6*, *Tnfα*, *NosII*, *Cox2*, *Ccl2*, *Ccl20*, *Ccr2*) and those stimulated with IL-10 or IL-4 were characterized by M2-related mRNA expression (*Arg1*, *Cd206*, *Ym1*, *Fizz1*, *Ppar*). They also demonstrated a decreased phagocytic activity under M1 polarization and enhanced activity with IL-10 stimulation [78]. Upon further examination of an association between phenotype and Notch 1 and Hes1 signaling pathways, Michelucci et al. [78] demonstrated phagocytic activity suggesting that the heterogeneous microglia response is determined largely by the nature of the stimulus and intracellular signal transduction pathways activated. Additional work in monocyte populations supports the existence of multiple activation phenotypes [83] and that the roles may vary depending on the tissue environment [84]. For example, exposure of microglia to branched-chain amino acids results in an intermediate M1/M2 phenotype of an enhanced IL-10 expression and phagocytic activity but an increase in free radical production [85]. CNS biopsies obtained from multiple sclerosis patients synergistically displayed markers for both M1 and M2 macrophage populations including CD40, CD64, CD86, and CD32, mannose receptor, and CD163 [86]. Recently, work suggested that rather than a marker for M2, the induction of arginase 1 could be related to neurotoxicity based on arginine deprivation [87]. From work assessing transcriptional regulation during human macrophage activation, an extended version of the current M1 versus M2 polarization model has been proposed. This spectrum model contains nine distinct activation programs [88]. A distinct axis of response was demonstrated upon stimulation with IFNγ or IL-4, with a similar axis demonstrated between LPS and IL-13 polarization. With stimuli not directly linked to M1 or M2 polarization such as, free fatty acids, high-density lipoprotein, or combinations associated with chronic inflammation, a spectrum of signatures was displayed. In addition to immune related factors, the spectrum recruited clusters associated with cell death, biosynthetic processes of small molecules, and metabolic and catabolic process as critical aspects of macrophage activation.

2.2 *Polarization and Bioenergetics*

In their normal surveillance state, microglia are likely to rely on oxidative phosphorylation metabolism [89] similar to what is observed in M2 macrophages [90, 91]. In M2 macrophages, glucose consumption is significantly lower as compared to M1 [92] and the sedoheptulose kinase carbohydrate kinase-like protein is critical for regulating the pentose phosphate pathway [93]. Upon stimulation with TLR agonists, microglia switch from oxidative metabolism towards glycolytic metabolism to maintain mitochondrial function and ensure cell survival [94–96]. It has been proposed that a two-stage process occurs with M1 activation with cells in the initial stage capable of utilizing both oxidative and glycolytic metabolism then shifting to rely on glycolytic metabolism for survival. Under both stages activation of the pentose phosphate pathway occurs [96].

A key feature of M1 macrophages is their ROS production to facilitate killing of phagocytized bacteria [97]. Intracellular damage from ROS is limited due to the increased generation of nicotinamide adenine dinucleotide phosphate (NADPH) required for maintenance of reduced glutathione [98, 99] and also NO production [100, 101]; however, they can also cause extensive damage to surrounding tissue. While these inflammatory factors serve in the host defense process they can also lead to the creation of a toxic microenvironment can be created with the production and release of pro-inflammatory cytokines, chemokines, redox molecules (NOX, PHOX, iNOS), scavenger receptors (macrophage receptor with collagenous structure (MARCO), co-stimulatory proteins (CD40), and MHC-II [18, 46, 49, 51, 57, 65, 71, 102–104]. The outcome of a M1 polarizing event is dependent upon a number of features, not the least of which is the severity or length of the response and whether it includes production of iNOS, reactive oxygen species (ROS), or activation of NOD-like receptor family pyrin domain-containing 3 (NLRP3) inflammasome complex [105, 106]. NLRP3 facilitates caspase-1 activation for proteolytic processing and secretion of pro-inflammatory cytokines, IL-1β and IL-18 [107–110]. A role for inflammasome activation in CNS injury has recently being identified [106, 111–113]. However, the complexity of the induction state is also demonstrated as an alternative activation of NLRP3 inflammasome can come from the classical M2-inducer in microglia [114, 115].

3 Microglia During Aging

Macrophages responding to endogenous stimuli generated following infection or in a sterile inflammatory response following injury [43–45], represents a product of a cell-mediated immune response essential for a host defense but also one that requires tight regulation. Maintaining the normal functional capacity of microglia to rapidly return an injured tissue back to normal homeostasis is a critical component for a healthy tissue environment. This involves the ability of cells to rapidly respond in a manner that will allow for clearance of aberrant proteins, invading pathogens, synaptic remodeling, and providing appropriate signaling to extend these processes and provide trophic support. It has been hypothesized that changes in the aging microglia drive pathogenic progression of diseases or injury through a diminution of neuroprotective functions, increase in neurotoxicity, and dysregulation of responses to signals and perturbations [116–120]. It has been suggested that microglia may not only chronologically age but also as a result of cumulative activation in response to systemic infections, microbiome characterization, or mild CNS injuries over a lifetime [121–123]. One hypothesis for neurodegeneration is based upon the idea that interactions with Aβ peptide or other inflammatory stimuli in the disease brain drives the cells to an early and maintained alternative activation state that shifts the ability of the cells to respond in a normal manner [124]. It has also been considered that such age related changes could be exacerbated not only with genetic background such as apolipoprotein E [125] but also with predisposing factors such as cigarette smoking, high blood pressure, type 2 diabetes, and obesity [126–129]. Changes that occur with aging may also be associated with changes that occur in the periphery that could contribute to local CNS perturbations [121, 123, 130, 131]. These changes and shifts in susceptibility are considered relevant for with regards to neurodegenerative processes associated with diseases and injury [132, 133].

3.1 Senescence

Microglia exhibit telomere shortening and decreased telomerase activity with aging, lending support for the hypothesized occurrence of microglial replicative senescence in normal and pathological aging [118, 134]. Microglia from Alzheimer's disease brains are reported to have shorter telomeres, as compared to control brains

[118] contributing to the speculation that amyloid β promotes microglial deterioration and accelerates senescence [135]. If microglia were functionally impaired by senescence, they would likely demonstrate decreased production of neurotrophic factors, impaired phagocytosis, and impaired protein clearance cumulatively, resulting in increased neurotoxicity. While it has not been examined, it is also possible that the presence of senescent microglia could in and of itself contribute to brain aging. Baker et al. [136] demonstrated delayed onset of age-related pathologies with lifetime removal of $p16^{Ink4a}$-positive senescent cells in the adipose tissue, skeletal muscle, and eye. Furthermore, they demonstrated that late-life clearance of senescent cells attenuated the progression of already established age-related disorders. While this work was not conducted in the nervous system, one could envision that the increased presence of senescent or dysfunctional cells, regardless of type, would result in a potentially adverse "disease-related" environment.

Age-related neurodegeneration might not only be due to a loss of neuroprotective properties, but also the actual loss of microglia [137]. Microglia retain an ability to proliferate and self-renew in adulthood; however, this is primarily in response to injury signals and even then it is relatively infrequent [138, 139]. Whether or not this capability is retained with aging is not clear however, data suggests that the capacity is not lost but rather, may be increased under specific conditions [140, 141]. A number of studies have reported an age-related increase in number and density of immunoreactive microglia across various brain regions [142–145]. These cells also appear to become less evenly distributed within the tissue [144, 146]. Aging would shift the population toward dysfunctional and away from healthy protective immune cells and thus diminish the ability of the system to mount a defensive immune response when required. One hypothesis for a gradual increase in microglia with age is based on a compensatory mechanism to maintain a required or minimal level of cells to provide protection in the face of diminished functional microglial capability [147].

3.2 Morphology

It has been proposed that, with aging, microglia assume a more reactive/activated phenotype with cytoplasmic structures showing excessive beading and spheroid swellings may be reflective of dystrophy and senescence [134, 137, 141]. In rodents, microglia arborization is reduced in size and complexity [144–146] coupled with a higher variability in soma size [144] Surveillance features of microglia require that the cell's processes are highly motile [16] and this motility is decreased with age [144, 146]. Whether this is secondary to the decrease in process arborization or a direct change has not been determined. In the aged brain, microglia co-localization around degenerating neurons is similar to that seen at any age however, there is a higher incidence of clumping, irregular distribution, and accumulation of phagocytic debris such as, lysosomal lipopigments, cellular elements, vacuoles, and large vesicles [116, 144, 146, 148]. Age-related changes in the retina show smaller and less branched processes and slower process motility of surveillance microglia [145].

In addition, the dynamic microglia response to injury or to ATP was altered in that microglia demonstrated a slower acute response to injury as compared to microglia in the younger retina [145]. Of additional interest was the observation that these responses were maintained for longer in the aged retina, suggesting a deficit in the down-regulatory signaling for resolution of the injury response.

3.3 Migration

Amyloid β peptides can directly stimulate chemotaxic responses of microglia [149, 150]. *In vivo*, a rapid appearance of microglia occurs at sites of newly formed amyloid plaques [151] and in older mice, these localized cells show less movement of processes as compared to younger mice [152]. While morphological and migration differences are observed in the aged brain, how they relate to impaired functional changes in the cells and adverse outcome has not been clearly determined [153–155].

3.4 Phagocytic Activation

The process of recognizing, attaching, engulfing, and internalizing extracellular material for clearance and degradation is orchestrated in the CNS primarily by microglia. Early work of Vaughan and Peters [143] suggested that the accumulation of non-cellular inclusions and condensed debris within microglia was the result of an impaired ability to degrade the ingested material. The ability of microglia to recognize material for clearance is dependent upon receptor signaling. For example, receptors that recognize apoptotic cellular material such as, phosphatidylserine on the inner membrane leaflet, are important for phagocyte clearance processes and may stimulate an anti-inflammatory response [156]. Recent identification of such receptors included T-cell immunoglobulin-and mucin-domain-containing molecule-1 (TIM4) [157], the metabotropic P2Y6 receptor that recognizes the nucleotide UDP released from injured neurons [158], and TREM-2. TREM2 is a member of the immmunoreceptor tyrosine-based activation motif (ITAM) receptors and plays a role in inflammation and microglia phagocytic activity to facilitate debris clearance [159–161]. When microglial cells are isolated from aged brains a blunted phagocytic response is observed [124]. A dysfunction in microglia catabolic autophagy occurs during aging altering the ability of the cell to degrade unnecessary or dysfunctional proteins thru the lysosome [162]. An efficient removal of amyloid β by microglia is observed in early stages of Alzheimer's disease that declines in later disease stages [163, 164]. This occurred with no deficit in generalized phagocytic capability with equivalent clearance of bacteria observed in cells derived from both ages. In addition, aged microglia showed a diminished ability to degrade amyloid β [124, 165]. *In vivo*, microglia appear to engulf amyloid; however, subsequent degradation of the protein is not as clearly established [166]. This could lead to the overloading of the cells and an inability to continue to clear the protein. Additionally, microglia can use macropinocytosis [167] and various Aβ degrading enzymes [168–170] to clear soluble forms of Aβ. Thus, in the absence of clear data it is speculative to consider the hypothesis that, with aging, the inability of microglia to degrade and clear aberrantly formed proteins leads to an intracellular accumulation and diminished ability of the cells to clear further amounts of protein from the extracellular space. A critical role for microglia in Aβ clearance has been questioned by recent work showing that plaque deposition and amyloid-associated neuritic dystrophy in an amyloid precursor protein transgenic mouse model was not influenced by the ablation of microglia [171]. This work also called into question the role of microglia and microglia-mediated neuroinflammatory responses in Alzheimer's disease neurodegeneration.

3.5 Polarization

Within the framework of M1/M2 polarization, many studies report that both M1 and M2 markers are elevated in aged mice as compared to young [74, 172–174]. Aging of microglia is associated with an increase in a number of inflammatory markers such as MHC II antigens (HLA-DR) and CD68, CD11b/CR3, CD14, pattern recognition receptors, and Toll like receptors suggesting the potential for dysregulation [116, 117, 141, 148, 172, 175–179]. The elevated basal levels also translated to a shift in response to LPS challenge as compared to young animals [179]. These observations of an exaggerated response raised the possibility that microglia shift to a primed state in the aged brain [180] however, other studies have raised questions on the simplicity of changes that occur in in the microglia phenotype [181, 182]. Minogue et al. [182] found that microglia were not less active with age but rather the cells alter their phenotype and the changes occur in the deactivation state. With a peripheral LPS administration, mRNA levels of IL-4Rα as a marker of M2 were elevated in aged mice [74]. Microglia extracted from the brain by gradient and flow cytometry demonstrated a lower sensitivity to IL-4 than that observed in microglia extracted from young adult mice [74]. With traumatic brain injury, *I4a* mRNA levels are lower in the aged mouse [174]. With a cocktail of polarizing cytokines and under conditions of an intracerebral hemorrhage, elevations in mRNA levels for M2 factors were found blunted in aged animals [183, 184]. For cytokine secretion, the increase normally observed in aged microglia is dependent upon the nature of the stimulation [185–188]. In addition, with aging a decrease is observed in factors which downregulate microglia pro-inflammatory responses such as, fractalkine, fractalkine receptor CX3CR1 [189, 190] and CD200 [112, 191]. Effects of aging, on the microglial transcriptome were reported by Orre et al. [192]. Microglia isolated from 2.5 and 15–18 months old mice showed a decrease in the gene ontology (GO) category of "cytoskeletal protein binding" likely reflecting a decrease in motility of microglia with age. The distribution across categories was not as distinct for cytokine related factors in that genes in the GO category of inflammatory response were most enriched in young microglia while those associated with cytokine activity were enriched in the aged microglia. In 5-month-old and 24-month-old mice, sequencing of the microglia transcriptome found a prominent gene profile representative of pathways associated with neuroprotection and M2 alternative activation, as well as genes encoding receptors to sense bacterial and fungal ligands. Rather than an elevation in genes associated with pro-inflammatory responses, gene pathways associated with toxicity and sensing endogenous ligands from apoptotic cells were predominantly downregulated with aging [42]. These tdata suggest that microglia within the aging brain environment are actively shifting their phenotype to promote a protective environment. In comparison, in the retina, progressive changes in the expression profile of microglia were noted as a function of aging [193]. These genes were related to microglia supportive functions and immune activation with a notable increase in the expression of complement genes *C3* and *Cfb*. The authors speculated that the profile represented an age related decrease in neuronal support and dysregulation of complement signaling. In general, the detailed transcriptome data of microglia across aging and as compared to peripheral tissue macrophages identifies microglia as a unique population of cells and that these cells express functions beyond immune responses that may be more relevant in understanding the role of these cells in age related disorders [194].

4 Summary

Gaining a better understanding of microglia resident to the CNS and their functions will contribute to our understanding of the critical and necessary actions taken by the brain to maintain a healthy environment. If however, continuing to force the biological functions of microglia into discrete modes for experimental expediency will only serve to hinder and potentially blind us from such understanding.

Acknowledgment This work was supported by the Division National Toxicology Program, National Institute of Environmental Health Sciences, #1Z01ES101623 and ES021164.

References

1. Gautier EL, Shay T, Miller J, Greter M, Jakubzick C, Ivanov S, Helft J, Chow A, Elpek KG, Gordonov S, Mazloom AR, Ma'ayan A, Chua WJ, Hansen TH, Turley SJ, Merad M, Randolph GJ. Gene-expression profiles and transcriptional regulatory pathways that underlie the identity and diversity of mouse tissue macrophages. Nat Immunol. 2012;13:1118–28.
2. Cajal RY. Contribucion al conocimiento de la meuroglia del cerebro humano. Trab Lab Invest Biol. 1913;11:254.
3. del Rio Hortega P. Microglia. In: Penfield W, editor. Cytology and cellular pathology of the nervous system. New York: Hoeber; 1932. p. 481–558.

4. Alliot F, Godin I, Pessac B. Microglia derive from progenitors, originating from the yolk sac, and which proliferate in the brain. Brain Res Dev Br Res. 1999;117(2):145–52.
5. Ginhoux F, Greter M, Leboeuf M, Nandi S, See P, Gokhan S, Mehler MF, ConwaySJ NLG, Stanley ER, Samokhvalov IM, Merad M. Fate mapping analysis reveals that adult microglia derive from primitive macrophages. Science. 2010;330:841–5.
6. Schulz C, Gomez Perdiguero E, Chorro L, Szabo-Rogers H, Cagnard N, Kierdorf K, Prinz M, Wu B, Jacobsen SE, Pollard JW, Frampton J, Liu KJ, Geissmann F. A lineage of myeloid cells independent of Myb and hematopoietic stem cells. Science. 2012;336(6077):86–90.
7. Kierdorf K, Erny D, Goldmann T, Sander V, Schulz C, Perdiguero EG, Wieghofer P, Heinrich A, Riemke P, Hölscher C, Müller DN, Luckow B, Brocker T, Debowski K, Fritz G, Opdenakker G, Diefenbach A, Biber K, Heikenwalder M, Geissmann F, Rosenbauer F, Prinz M. Microglia emerge from erythromyeloid precursors via Pu.1- and Irf8-dependent pathways. Nat Neurosci. 2013;16(3):273–80.
8. Gomez Perdiguero E, Klapproth K, Schulz C, Busch K, Azzoni E, Crozet L, Garner H, Trouillet C, de Bruijn MF, Geissmann F, Rodewald HR. Tissue-resident macrophages originate from yolk-sac-derived erythro-myeloid progenitors. Nature. 2015;518(7540):547–51.
9. Kreutzberg GW. Microglia: a sensor for pathological events in the CNS. Trends Neurosci. 1996;19(8):312–8.
10. Lawson LJ, Perry VH, Dri P, Gordon S. Heterogeneity in the distribution and morphology of microglia in the normal adult mouse brain. Neuroscience. 1990;39:151–70.
11. Mittelbronn M, Dietz K, Schluesener JH, Meyermann R. Local distribution of microglia in the normal adult human central nervous system differs by up to one order of magnitude. Acta Neuropathol. 2001;101:249–55.
12. Harry GJ, Kraft AD. Microglia in the developing brain: a potential target with lifetime effects. Neurotoxicology. 2012;33(2):191–206.
13. Carson MJ, Doose JM, Melchior B, Schmid CD, Ploix C. CNS immune privilege: hiding in plain sight. Immunol Rev. 2006;213:48–65.
14. Schmid CD, Melchior B, Masek K, Puntambekar SS, Danielson PE, Lo DD, Sutcliffe JG, Carson MJ. Differential gene expression in LPS/IFNgamma activated microglia and macrophages: in vitro versus in vivo. J Neurochem. 2009;109 Suppl 1:117–25.
15. Davalos D, Grutzendler J, Yang G, Kim JV, Zuo Y, Jung S, Littman DR, Dustin ML, Gan WB. ATP mediates rapid microglial response to local brain injury in vivo. Nat Neurosci. 2005;8(6):752–8.
16. Nimmerjahn A, Kirchhoff F, Helmchen F. Resting microglial cells are highly dynamic surveillants of brain parenchyma in vivo. Science. 2005;308:1314–8.
17. Hanisch UK, Kettenmann H. Microglia: active sensor and versatile effector cells in the normal and pathologic brain. Nat Neurosci. 2007;10(11):1387–94.
18. Ransohoff RM, Perry VH. Microglial physiology: unique stimuli, specialized responses. Annu Rev Immunol. 2009;27:119–45.
19. Mallat M, Marin-Teva JL, Cheret C. Phagocytosis in the developing CNS: more than clearing the corpses. Curr Opin Neurobiol. 2005;15:101–7.
20. Marin-Teva JL, Cuadros MA, Martin-Oliva D, Navascues J. Microglia and neuronal cell death. Neuron Glia Biol. 2011;7(1):25–40.
21. Dalmau I, Vela JM, Gonzalez B, Finsen B, Castellano B. Dunamics of microglia in the developing rat brain. J Comp Neurol. 2003;458:144–57.
22. Peri F, Nusslein-Volhard C. Live images of neuronal degradation by microglia reveals a role for v0-ATPase al in phagocomal fusion in vivo. Cell. 2008;133:916–27.
23. Wake H, Moorhouse AJ, Jinno S, Kohsaka S, Nabekura J. Resting microglia directly monitor the functional state of synapses in vivo and determine the fate of ischemic terminals. J Neurosci. 2009;29:3974–80.
24. Nagamoto-Combs K, Morecraft RJ, Darling WG, Combs CK. Long-term gliosis and molecular changes in the cervical spinal cord of the rhesus monkey after traumatic brain injury. J Neurotrauma. 2010;27(3):565–85.

25. Sierra A, Encinas JM, Deudero JJ, Chancey JH, Enikolopov G, Overstreet-Wadiche LS, Tsirka SE, Maletic-Savatic M. Microglia shape adult hippocampal neurogenesis through apoptosis-coupled phagocytosis. Cell Stem Cell. 2010;7:483–95.
26. Paolicelli RC, Bolasco G, Pagani F, Maggi L, Scianni M, Panzanelli P, Giustetto M, Ferreira TA, Guiducci E, Bumas L, Ragozzino D, Gross CT. Suynaptic pruning by microglia is necessary for normal brain development. Science. 2011;333:1456–8.
27. Olah M, Biber K, Vinet J, Boddeke HW. Microglia phenotype diversity. CNS Neurol Disord Drug Targets. 2011;10(1):108–18.
28. Ueno M, Fujita Y, Tanaka T, Nakamura Y, Kikura J, Ishii M, Yamashita T. Layer V cortical neurons require microglial support for survival during postnatal development. Nat Neurosci. 2013;16:543–51.
29. Ajami B, Bennett JL, Krieger C, Tetzlaff W, Rossi FM. Local self-renewal can sustain CNS microglia maintenance and function throughout adult life. Nat Neurosci. 2007;10(12):1538–43.
30. Goldmann T, Wieghofer P, Müller PF, Wolf Y, Varol D, Yona S, Brendecke SM, Kierdorf K, Staszewski O, Datta M, Luedde T, Heikenwalder M, Jung S, Prinz M. A new type of microglia gene targeting shows TAK1 to be pivotal in CNS autoimmune inflammation. Nat Neurosci. 2013;16(11):1618–26.
31. Sieweke MH, Allen JE. Beyond stem cells: self-renewal of differentiated macrophages. Science. 2013;342(6161):1242974.
32. Vallieres L, Sawchenko PE. Bone marrow-derived cells that populate the adult mouse brain preserve their hematopoietic identity. J Neurosci. 2003;23(12):5197–207.
33. Mildner A, Schmidt H, Nitsche M, Merkler D, Hanisch UK, Mack M, Heikenwalder M, Brück W, Priller J, Prinz M. Microglia in the adult brain arise from Ly-6ChiCCR2+ monocytes only under defined host conditions. Nat Neurosci. 2007;10(12):1544–53.
34. Mildner A, Mack M, Schmidt H, Brück W, Djukic M, Zabel MD, Hille A, Priller J, Prinz M. CCR2+ Ly-6Chi monocytes are crucial for the effector phase of autoimmunity in the central nervous system. Brain. 2009;132:2487–500.
35. King IL, Dickendesher TL, Segal BM. Circulating Ly-6C+ myeloid precursors migrate to the CNS and play a pathogenic role during autoimmune demyelinating disease. Blood. 2009;113(14):3190–7.
36. Blau CW, Cowley TR, O'Sullivan J, Grehan B, Browne TC, Kelly L, Birch A, Murphy N, Kelly AM, Kerskens CM, Lynch MA. The age-related deficit in LTP is associated with changes in perfusion and blood-brain barrier permeability. Neurobiol Aging. 2012;33:1005. doi:10.1016/jneurobiolaging.2011.09.035.
37. Enciu AM, Gherghiceanu M, Popescu BO. Triggers and effectors of oxidative stress at blood-brain barrier level: relevance for brain ageing and neurodegeneration. Oxidative Med Cell Longevity 2013:297512. doi:10.1155/2013/297512.
38. Marques F, Sousa JC, Sousa N, Palha JA. Blood-brain barrier in aging and in Alzheimer's disease. Mol Neurodeg. 2013;8:38. doi:10.1186/1750-1326-8-38.
39. Ginhoux F, Jung S. Monocytes and macrophages: developmental pathways and tissue homeostasis. Nat Rev Immunol. 2014;14(6):392–404.
40. Hickman SE, Kingery ND, Ohsumi TK, Borowsky ML, Wang LC, Means TK, El Khoury J. The microglial sensome revealed by direct RNA sequencing. Nat Neurosci. 2013;16(12):1896–905.
41. Chiu IM, Morimoto ET, Goodarzi H, Liao JT, O'Keeffe S, Phatnani HP, Muratet M, Carroll MC, Levy S, Tavazoie S, Myers RM, Maniatis T. A neurodegeneration-specific gene-expression signature of acutely isolated microglia from an amyotrophic lateral sclerosis mouse model. Cell Rep. 2013;4:385–401.
42. Butovsky O, Jedrychowski MP, Moore CS, Cialic R, Lanser AJ, Gabriely G, Koeglsperger T, Dake B, Wu PM, Doykan CE, Fanek Z, Liu L, Chen Z, Rothstein JD, Ransohoff RM, Gygi SP, Antel JP, Weiner HL. Identification of a unique TGF-beta- dependent molecular and functional signature in microglia. Nat Neurosci. 2014;17(1):131–43.

43. Chen GY, Nunez G. Sterile inflammation: sensing and reacting to damage. Nat Rev Immunol. 2010;10(12):826–37.
44. Shechter R, Schwartz M. CNS sterile injury: just another wound healing? Trends Mol Med. 2013;19(3):135–43.
45. McPherson CA, Merrick BA, Harry GJ. In vivo molecular markers for pro- inflammatory cytokine M1 stage and resident microglia in trimethyltin-induced hippocampal injury. Neurotoxicity Res. 2014;25(1):45–56.
46. Mosser DM, Edwards JP. Exploring the spectrum of macrophage activation. Nat Rev Immunol. 2008;8(12):958–69.
47. Biswas SK, Mantovani A. Macrophage plasticity and interaction with lymphocyte subsets: cancer as a paradigm. Nat Immunol. 2010;11(10):889–96.
48. Chinetti-Gbaguidi G, Staels B. Macrophage polarization in metabolic disorders: functions and regulation. Curr Opin Lipidol. 2011;22(5):365–72.
49. Boche D, Perry VH, Nicoll JA. Review: activation patterns of microglia and their identification in the human brain. Neuropath Appl Neurobiol. 2013;39(1):3–18.
50. Colton CA. Heterogeneity of microglial activation in the innate immune response in the brain. J Neuroimm Pharmacol. 2009;4(4):399–418.
51. Martinez FO, Gordon S. (2014). The M1 and M2 paradigm of macrophage activation: time for reassessment. F1000Prime Rep. 6:13. doi:10.12703/P6-13.
52. Martinez FO, Gordon S, Locati M, Mantovani A. Transcriptional profiling of the human monocyte-to-macrophage differentiation and polarization: new molecules and patterns of gene expression. J Immunol. 2006;177(10):7303–11.
53. Hu X, Ivashkiv LB. Cross-regulation of signaling pathways by interferon-gamma: implications for immune responses and autoimmune diseases. Immunity. 2009;31:539–50.
54. Waddell SJ, Popper SJ, Rubins KH, Griffiths MJ, Brown PO, Levin M, Relman DA. Dissecting interferon-induced transcriptional progrrams in human peripheral blood cells. PLoS One. 2010; doi: http://dx.doi.org/10.1371/journal.pone.0009753.
55. Murray PJ, Allen JE, Biswas SK, Fisher EA, Gilroy DW, Goerdt S, Gordon S, Hamilton JA, Ivashkiv LB, Lawrence T, Locati M, Mantovani A, Martinez FO, Mege JL, Mosser DM, Natoli G, Saeij JP, Schultze JL, Shirey KA, Sica A, Suttles J, Udalova I, van Ginderachter JA, Vogel SN, Wynn TA. Macrophage activation and polarization: nomenclature and experimental guidelines. Immunity. 2014;41(1):14–20.
56. Gordon S. Alternative activation of macrophages. Nat Rev Immunol. 2003;3(1):23–35.
57. Gordon S, Martinez FO. Alternative activation of macrophages: mechanism and functions. Immunity. 2010;32(5):593–604.
58. Ortega-Gomez A, Perretti M, Soehnlein O. Resolution of inflammation: an integrated review. EMBO Mol Med. 2013;5(5):661–74.
59. Cherry JD, Olschowka JA, O'Banion MK. Neuroinflammation and M2 microglia: the good, the bad, and the inflamed. J Neuroinflammation. 2014;11:98–9. doi:10.1186/1742-2094-11-98.
60. Longbrake EE, Lai W, Ankeny DP, Popovich PG. Characterization and modeling of monocyte-derived macrophages after spinal cord injury. J Neurochem. 2007;102(4):1083–94.
61. Novak ML, Koh TJ. Macrophage phenotypes during tissue repair. J Leukocyte Biol. 2013;93(6):875–81.
62. Ruffell D, Mourkioti F, Gambardella A, Kirstetter P, Lopez RG, Rosenthal N, Nerlov C. A CREB-C/EBPbeta cascade induces M2 macrophage-specific gene expression and promotes muscle injury repair. Proc Natl Acad Sci U S A. 2009;106(41):17475–80.
63. Shechter R, Miller O, Yovel G, Rosenzweig N, London A, Ruckh J, Kim KW, Klein E, Kalchenko V, Bendel P, Lira SA, Jung S, Schwartz M. Recruitment of beneficial M2 macrophages to injured spinal cord is orchestrated by remote brain choroid plexus. Immunity. 2013;38(3):555–69.
64. Stein M, Keshav S, Harris N, Gordon S. Interleukin 4 potently enhances murine macrophage mannose receptor activity: a marker of alternative immunologic macrophage activation. J Exper Med. 1992;176(1):287–92.

65. Henkel JS, Beers DR, Zhao W, Appel SH. Microglia in ALS: the good, the bad, and the resting. J Neuroimmune Pharmacol. 2009;4(4):389–98.
66. Lawrence T, Natoli G. Transcriptional regulation of macrophage polarization: enabling diversity with identity. Nat Rev Immunol. 2011;11(11):750–61.
67. Mills CD. M1 and M2 macrophages: oracles of health and disease. Critical Rev Immunol. 2012;32(6):463–88.
68. Rutschman R, Lang R, Hesse M, Ihle JN, Wynn TA, Murray PJ. Cutting edge: Stat6- dependent substrate depletion regulates nitric oxide production. J Immunol. 2001;166(4):2173–7.
69. Wynn TA, Chawla A, Pollard JW. Macrophage biology in development, homeostasis and disease. Nature. 2013;496(7446):445–55.
70. Vodovotz Y, Bogdan C, Paik J, Xie QW, Nathan C. Mechanisms of suppression of macrophage nitric oxide release by transforming growth factor beta. J Exper Med. 1993;178(2):605–13.
71. Martinez FO, Sica A, Mantovani A, Locati M. Macrophage activation and polarization. Front Biosci. 2008;13:453–61.
72. Morris T, Stables M, Hobbs A, de Souza P, Colville-Nash P, Warner T, Newson J, Bellingan G, Gilroy DW. Effects of low-dose aspirin on acute inflammatory responses in humans. J Immunol. 2009;183(3):2089–96.
73. Lisi L, Stigliano E, Lauriola L, Navarra P, Dello RC. Proinflammatory-activated glioma cells induce a switch in microglial polarization and activation status, from a predominant M2b phenotype to a mixture of M1 and M2a/B polarized cells. ASN Neuro. 2014;6(3):171–83.
74. Fenn AM, Henry CJ, Huang Y, Dugan A, Godbout JP. Lipopolysaccharide- induced interleukin (IL)-4 receptor-alpha expression and corresponding sensitivity to the M2 promoting effects of IL-4 are impaired in microglia of aged mice. Brain Behav Immun. 2012;26(5):766–77.
75. Liu HC, Zheng MH, Du YL, Wang L, Kuang F, Qin HY, Zhang BF, Han H. N9 microglial cells polarized by LPS and IL4 show differential responses to secondary environmental stimuli. Cellular Immunol. 2012;278(1–2):84–90.
76. Chhor V, Le Charpentier T, Lebon S, Oré MV, Celador IL, Josserand J, Degos V, Jacotot E, Hagberg H, Sävman K, Mallard C, Gressens P, Fleiss B. Characterization of phenotype markers and neuronotoxic potential of polarised primary microglia in vitro. Brain Beh Immun. 2013;32:70–85.
77. Freilich RW, Woodbury ME, Ikezu T. Integrated expression profiles of mRNA and miRNA in polarized primary murine microglia. PLoS One. 2013;8(11), e79416.
78. Michelucci A, Heurtaux T, Grandbarbe L, Morga E, Heuschling P. Characterization of t he microglial phenotype under specific pro-inflammatory and anti-inflammatory conditions: Effects of oligomeric and fibrillar amyloid-beta. J Immunol. 2009;210(1–2):3–12.
79. El Kasmi KC, Qualls JE, Pesce JT, Smith AM, Thompson RW, Henao-Tamayo M, Basaraba RJ, König T, Schleicher U, Koo MS, Kaplan G, Fitzgerald KA, Tuomanen EI, Orme IM, Kanneganti TD, Bogdan C, Wynn TA, Murray PJ. Toll-like receptor-induced arginase 1 in macrophages thwarts effective immunity against intracellular pathogens. Nat Immunol. 2008;9(12):1399–406.
80. Crane MJ, Daley JM, van Houtte O, Brancato SK, Henry Jr WL, Albina JE. The monocyte to macrophage transition in the murine sterile wound. PLoS One. 2014;9(1), e86660. doi:10.1371/journal.pone.0086660.
81. Gosselin D, Link VM, Romanoski CE, Fonseca GJ, Eichenfield DZ, Spann NJ, Stender JD, Chun HB, Garner H, Geissmann F, Glass CK. Environment drives selection and function of enhancers controlling tissue-specific macrophage identities. Cell. 2014;159(6):1327–40.
82. Durafourt BA, Moore CS, Zammit DA, Johnson TA, Zaguia F, Guiot MC, Bar-Or A, Antel JP. Comparison of polarization properties of human adult microglia and blood- derived macrophages. Glia. 2012;60(5):717–27.
83. Chávez-Galán L, Olleros ML, Vesin D, Garcia I. Much More than M1 and M2 Macrophages, There are also CD169(+) and TCR(+) Macrophages. Front Immunol. 2015;6:263. doi:10.3389/fimmu.2015.00263.

84. Dey A, Allen J, Hankey-Giblin PA. Ontogeny and polarization of macrophages in inflammation: blood monocytes versus tissue macrophages. Front Immunol. 2015;5:683. doi:10.3389/fimmu.2014.00683.
85. De Simone R, Vissicchio F, Mingarelli C, De Nuccio C, Visentin S, Ajmone-Cat MA, Minghetti L. Branched-chain amino acids influence the immune properties of microglial cells and their responsiveness to pro-inflammatory signals. Biochim Biophys Acta. 2013;1832(5):650–9.
86. Vogel DY, Vereyken EJ, Glim JE, Heijnen PD, Moeton M, van der Valk P, Amor S, Teunissen CE, van Horssen J, Dijkstra CD. Macrophages in inflammatory multiple sclerosis lesions have an intermediate activation status. J Neuroinflammation. 2013;10:35.
87. Kan MJ, Lee JE, Wilson JG, Everhart AL, Brown CM, Hoofnagle AN, Jansen M, Vitek MP, Gunn MD, Colton CA. Arginine deprivation and immune suppression in a mouse model of Alzheimer's disease. J Neurosci. 2015;35(15):5969–82.
88. Xue J, Schmidt SV, Sander J, Draffehn A, Krebs W, Quester I, De Nardo D, Gohel TD, Emde M, Schmidleithner L, Ganesan H, Nino-Castro A, Mallmann MR, Labzin L, Theis H, Kraut M, Beyer M, Latz E, Freeman TC, Ulas T, Schultze JL. Transcriptome-based network analysis reveals a spectrum model of human macrophage activation. Immunity. 2014;40(2):274–88.
89. Cherry JD, Olschowka JA, O'Banion MK. Are "resting" microglia more "m2"? Front Immunol. 2014;5:594. doi:10.3389/fimmu.2014.00594.
90. Murray PJ, Wynn TA. Protective and pathogenic functions of macrophage subsets. Nat Rev Immunol. 2011;11:723–37.
91. Biswas SK, Mantovani A. Orchestration of metabolism by macrophages. Cell Metab. 2012;15:432–7.
92. Rodríguez-Prados JC, Través PG, Cuenca J, Rico D, Aragonés J, Martín-Sanz P, Cascante M, Boscá L. Substrate fate in activated macrophages: a comparison between innate, classic, and alternative activation. J Immunol. 2010;185(1):605–14.
93. Galvan-Peña S, O'Neill LA. Metabolic reprograming in macrophage polarization. Front Immunol. 2014;5:420.
94. Voloboueva LA, Emery JF, Sun X, Giffard RG. Inflammatory response of microglial BV-2 cells includes a glycolytic shift and is modulated by mitochondrial glucose- regulated protein 75/mortalin. FEBS Lett. 2013;587(6):756–62.
95. Gimeno-Bayon J, Lopez-Lopez A, Rodriguez MJ, Mahy N. Glucose pathways adaptation supports acquisition of activated microglia phenotype. J Neurosci. 2014;92(6):723–31.
96. Orihuela R, McPherson CA, Harry GJ. Microglial M1/M2 polarization and metabolic states. Br J Pharmacol. 2016;173(4):649–65. doi:10.1111/bph.13139.
97. West AP, Brodsky IE, Rahner C, Woo DK, Erdjument-Bromage H, Tempst P, Walsh MC, Choi Y, Shadel GS, Ghosh S. TLR signalling augments macrophage bactericidal activity through mitochondrial ROS. Nature. 2011;472(7344):476–80.
98. Kletzien RF, Harris PK, Foellmi LA. Glucose-6-phosphate dehydrogenase: a "housekeeping" enzyme subject to tissue-specific regulation by hormones, nutrients, and oxidant stress. FASEB J. 1994;8(2):174–81.
99. Salvemini F, Franze A, Iervolino A, Filosa S, Salzano S, Ursini MV. Enhanced glutathione levels and oxido resistance mediated by increased glucose-6-phosphate dehydrogenase expression. J Biol Chem. 1999;274(5):2750–7.
100. Bredt DS, Snyder SH. Isolation of nitric oxide synthetase, a calmodulin-requiring enzyme. Proc Natl Acad Sci U S A. 1990;87(2):682–5.
101. Knowles RG, Moncada S. Nitric oxide synthases in mammals. Biochem J. 1994;298(Pt 2):249–58.
102. Hanisch UK, Kettenmann H. Microglia: active sensor and versatile effector cells in the normal and pathologic brain. Nature Neurosci. 2007;10(11):1387–94.
103. Varnum MM, Ikezu T. The classification of microglial activation phenotypes on neurodegeneration and regeneration in Alzheimer's disease brain. Arch Immunol Therap Exper. 2012;60(4):251–66.

104. Colton C, Wilcock DM. Assessing activation states in microglia. CNS Neurol Dis Drug Targets. 2010;9(2):174–91.
105. Bordt EA, Polster BM. NADPH oxidase- and mitochondria-derived reactive oxygen species in proinflammatory microglial activation: a bipartisan affair? Free Radical Biol Med. 2014;76:34–46.
106. de Rivero Vaccari JC, Brand 3rd FJ, Berti AF, Alonso OF, Bullock MR, de Rivero Vaccari JP. Mincle signaling in the innate immune response after traumatic brain injury. J Neurotrauma. 2015;32(4):228–36.
107. Labbe K, Saleh M. Cell death in the host response to infection. Cell Death Differ. 2008;15(9):1339–49.
108. Salminen A, Ojala J, Suuronen T, Kaarniranta K, Kauppinen A. Amyloid-beta oligomers set fire to inflammasomes and induce Alzheimer's pathology. J Cell Mol Med. 2008;12(6a):2255–62.
109. Martinon F, Mayor A, Tschopp J. The inflammasomes: guardians of the body. Ann Rev Immunol. 2009;27:229–65.
110. Netea MG, van de Veerdonk FL, van der Meer JW, Dinarello CA, Joosten LA. Inflammasome-independent regulation of IL-1-family cytokines. Annu Rev Immunol. 2015;33:49–77.
111. Shi F, Yang L, Kouadir M, Yang Y, Wang J, Zhou X, Yin X, Zhao D. The NALP3 inflammasome is involved in neurotoxic prion peptide-induced microglial activation. J Neuroinflammation. 2012;9:73. doi:10.1186/1742-2094-9-73.
112. Frank M, Barrientos R, Biedenkapp J, Rudy J, Watkins L, Maier S. mRNA up-regulation of MHC II and pivotal pro-inflammatory genes in normal brain aging. Neurobiol Aging. 2006;27(5):717–22.
113. Walsh JG, Muruve DA, Power C. Inflammasomes in the CNS. Nat Rev Neurosci. 2014;15(2):84–97.
114. Cao L, Fei L, Chang TT, DeLeo JA. Induction of interleukin-1beta by interleukin- 4 in lipopolysaccharide-treated mixed glial cultures: microglial-dependent effects. J Neurochem. 2007;102(2):408–19.
115. Ferger AI, Campanelli L, Reimer V, Muth KN, Merdian I, Ludolph AC, Witting A. Effects of mitochondrial dysfunction on the immunological properties of microglia. J Neuroinflammation. 2010;7:45. doi:10.1186/1742-2094-7-45.
116. Perry VH, Matyszak MK, Fearn S. Altered antigen expression of microglia in theaged rodent CNS. Glia. 1993;7:60–7.
117. Sheffield L, Berman N. Microglial expression of MHC class II increases in normal aging of nonhuman primates. Neurobiol Aging. 1998;19(1):47–55.
118. Flanary BE, Sammons NW, Nguyen C, Walker D, Streit WJ. Evidence that aging and amyloid promote microglial cell senescence. Rejuvenat Res. 2007;10(1):61–74.
119. Lopes KO, Sparks DL, Streit WJ. Microglial dystrophy in the aged and Alzheimer's disease brain is associated with ferritin immunoreactivity. Glia. 2008;56(10):1048–60.
120. Luo XG, Ding JQ, Chen SD. Microglia in the aging brain: relevance to neurodegeneration. Mol Neurodeg. 2010;5:12. doi:10.1186/1750-1326-5-12.
121. Perry VH, Cunningham C, Holmes C. Systemic infections and inflammation affec chronic neurodegeneration. Nat Rev Immunol. 2007;7(2):161–7.
122. Krstic D, Knuesel I. Deciphering the mechanism underlying late-onset alzheimer disease. Nat Rev Neurol. 2013;9(1):25–34.
123. Erny D, de Angelis AL H, Jaitin D, Wieghofer P, Staszewski O, David E, Keren- Shaul H, Mahlakoiv T, Jakobshagen K, Buch T, Schwierzeck V, Utermöhlen O, Chun E, Garrett WS, McCoy KD, Diefenbach A, Staeheli P, Stecher B, Amit I, Prinz M. Host microbiota constantly control maturation and function in the CNS. Nat Neurosci. 2015;18:965–77.
124. Floden AM, Combs CK. Microglia demonstrate age-dependent interaction with amyloid-β fibrils. J Alzheimers Dis. 2011;25(2):279.
125. Harry GJ, Lefebvre d'Hellencourt C, Bruccoleri A, Schmechel D. Age-dependent cytokine responses: trimethyltin hippocampal injury in wild-type, APOE knockout, and APOE4 mice. Brain Behav Immun. 2000;14(4):288–304.

126. Qiu C, Kivipelto M, von Strauss E. Epidemiology of Alzheimer's disease: occurrence, determinants, and strategies toward intervention. Dialogues Clin Neurosci. 2009;11(2):111–28.
127. Lee EB. Obesity, leptin, and Alzheimer's disease. Ann N Y Acad Sci. 2011;1243(1):15–29.
128. Puig KL, Floden AM, Adhikari R, Golovko MY, Combs CK. Amyloid precursor protein and proinflammatory changes are regulated in brain and adipose tissue in a murine model of high fat diet-induced obesity. PLoS One. 2012;7(1), e30378.
129. Buckman LB, Hasty AH, Flaherty DK, Buckman CT, Thompson MM, Matlock BK, Weller K, Ellacot KL. Obesity induced by a high-fat diet is associated with increased immune cell entry into the central nervous system. Brain Behav Immun. 2013;35:33–42.
130. Franceschi C, Capri M, Monti D, Giunta S, Olivieri F, Sevini F, Panourgia MP, Invidia L, Celani L, Scurti M, Cevenini E, Castellani GC, Salvioli S. Inflammaging and anti- inflammaging: a systemic perspective on aging and longevity emerged from studies in humans. Mech Ageing Dev. 2007;128(1):92–105.
131. Perry VH. Contribution of systemic inflammation to chronic neurodegeneration. Acta Neuropathol. 2010;120(3):277–86.
132. Nakanishi H, Wu Z. Microglia-aging: roles of microglial lysosome- and mitochondria-derived reactive oxygen species in brain aging. Behav Brain Res. 2009;201(1):1–7.
133. Von Bernhardi R, Tichauer JE, Eugenı' NJ. Aging-dependent changes of microglial cells and their relevance for neurodegenerative disorders. J Neurochem. 2010;112(5):1099–114.
134. Flanary BE, Streit WJ. Effects of axotomy on telomere length, telomerase activity, and protein in activated microglia. J Neurosci Res. 2005;82(2):160–71.
135. Streit WJ. Microglia and Alzheimer's disease pathogenesis. J Neurosci Res. 2004;77(1):1–8.
136. Baker DJ, Wijshake T, Tchkonia T, LeBrasseur NK, Childs BG, van de Sluis B, Kirkland JL, van Deursen JM. Clearance of p16Ink4a-positive senescent cells delays ageing- associated disorders. Nature. 2011;479(7372):232–6.
137. Streit WJ, Miller KR, Lopes KO, Njie E. Microglial degeneration in the aging brain—bad news for neurons? Front Biosci. 2008;13:3423–38.
138. Lawson LJ, Perry VH, Gordon S. Turnover of resident microglia in the normal adult mouse brain. Neurosci. 1992;48(2):405–15.
139. Xu H, Chen M, Mayer EJ, Forrester JV, Dick AD. Turnover of resident retinal microglia in the normal adult mouse. Glia. 2007;55(11):1189–98.
140. Hua K, Schindler MK, Mcquail JA, Forbes ME, Riddle DR. Regionally distinct responses of microglia and glial progenitor cells to whole brain irradiation in adult and aging rats. PLoS One. 2012;7(12), e52728.
141. Conde JR, Streit WJ. Microglia in the aging brain. J Neuropath Exp Neurol. 2006;65:199–203.
142. Long JM, Kalehua AN, Muth NJ, Calhoun ME, Jucker M, Hengemihle JM, Ingram DK, Mouton PR. Stereological analysis of astrocyte and microglia in aging mouse hippocampus. Neurobiol Aging. 1998;19(5):497–503.
143. Vaughan DW, Peters A. Neuroglial cells in the cerebral cortex of rats from young adulthood to old age: an electron microscope study. J Neurocytol. 1974;3(4):405–29.
144. Tremblay ME, Zettel ML, Ison JR, Allen PD, Majewska AK. Effects of aging and sensory loss on glial cells in mouse visual and auditory cortices. Glia. 2012;60(4):541–58.
145. Damani MR, Zhao L, Fontainhas AM, Amaral J, Fariss RN, Wong WT. Age-related alterations in the dynamic behavior of microglia. Aging Cell. 2011;10(2):263–76.
146. Hefendehl JK, Neher JJ, Suhs RB, Kohsaka S, Skodras A, Jucker M. Homeostatic and injury-induced microglia behavior in the aging brain. Aging Cell. 2014;13(1):60–9.
147. Streit WJ, Xue Q-S. The Brain's aging immune system. Aging Dis. 2010;1(3):254–61.
148. Hart AD, Wyttenbach A, Perry VH, Teeling JL. Age related changes in microglial phenotype vary between CNS regions: Grey versus white matter differences. Brain Behav Immun. 2012;26(5):754–65.
149. Maeda K, Nakai M, Maeda S, Kawamata T, Yamaguchi T, Tanaka C. Possible different mechanism between amyloid-beta (25-35)-and substance P-induced chemotaxis of murine microglia. Gerontology. 1997;43(Suppl1):11–5.

150. Ard MD, Cole GM, Wei J, Mehrle AP, Fratkin JD. Scavenging of Alzheimer's amyloid beta-protein by microglia in culture. J Neurosci Res. 1996;43(2):190–202.
151. Meyer-Luehmann M, Spires-Jones TL, Prada C, Garcia-Alloza M, de Calignon A, Rozkalne A, Koenigsknecht-Talboo J, Holtzman DM, Bacskai BJ, Hyman BT. Rapid appearance and local toxicity of amyloid-(plaques in a mouse model of Alzheimer's disease. Nature. 2008;451(7179):720–4.
152. Koenigsknecht-Talboo J, Meyer-Luehmann M, Parsadanian M, Garcia-Alloza M, Finn MB, Hyman BT, Bacskai BJ, Holtzman DM. Rapid microglial response around amyloid pathology after systemic antibody administration in PDAPP mice. J Neurosci. 2008;28(52):14156–64.
153. Aguzzi A, Barres BA, Bennett ML. Microglia: scapegoat, saboteur, or something else? Science. 2013;339(6116):156–61.
154. Kettenmann H, Kirchhoff F, Verkhratsky A. Microglia: new roles for the synaptic stripper. Neuron. 2013;77(1):10–8.
155. Siskova Z, Tremblay ME. Microglia and synapse: interactions in health and neurodegeneration. Neural Plast. 2013:425845. doi: 10.1155/2013/425845.
156. Ravichandran KS. "Recruitment signals" from apoptotic cells: invitation to a quiet meal. Cell. 2003;113(7):817–20.
157. Miyanishi M, Tada K, Koike M, Uchiyama Y, Kitamura T, Nagata S. Identification of Tim4 as a phosphatidylserine receptor. Nature. 2007;450(7168):435–9.
158. Koizumi S, Shigemoto-Mogami Y, Nasu-Tada K, Shinozaki Y, Ohsawa K, Tsuda M, Joshi BV, Jacobson KA, Kohsaka S, Inoue K. UDP acting at P2Y6 receptors is a mediator of microglial phagocytosis. Nature. 2007;446(7139):1091–5.
159. Takahashi K, Rochford CD, Neumann H. Clearance of apoptotic neurons without inflammation by microglial triggering receptor expressed on myeloid cells-2. J Exp Med. 2005;201(4):647–57.
160. Takahashi K, Prinz M, Stagi M, Chechneva O, Neumann H. TREM2-transduced myeloid precursors mediate nervous tissue debris clearance and facilitate recovery in an animal model of multiple sclerosis. PLoS Med. 2007;4(4), e124.
161. Neumann H, Takahashi K. Essential role of the microglial triggering receptor expressed on myeloid cells-2 (TREM2) for central nervous tissue immune homeostasis. J Neuroimmunol. 2007;184(1-2):92–9.
162. Wong PM, Puente C, Ganley IG, Jiang X. The ULK1 complex: sensing nutrient signals for autophagy activation. Autophagy. 2013;9(2):124–37.
163. Solito E, Sastre M. Microglia function in Alzheimer's disease. Front Pharmacol. 2012;3:1–10.
164. Krabbe G, Halle A, Matyash V, Rinnenthal JL, Eom GD, Bernhardt U, Miller KR, Prokop S, Kettenmann H, Heppner FL. Functional impairment of microglia coincides with beta-amyloid deposition in mice with Alzheimer-like pathology. PLoS One. 2013;8(4), e60921.
165. Majumdar A, Cruz D, Asamoah N, Buxbaum A, Sohar I, Lobel P, Maxfield FR. Activation of microglia acidifies lysosomes and leads to degradation of Alzheimer amyloid fibrils. Mol Biol Cell. 2007;18(4):1490–6.
166. Prokop S, Miller KR, Heppner FL. Microglia actions in Alzheimer's disease. Acta Neuropathol. 2013;126(4):461–77.
167. Mandrekar S, Jiang Q, Lee CYD, Koenigsknecht-Talboo J, Holtzman DM, Landreth GE. Microglia mediate the clearance of soluble a through fluid phase macropinocytosis. J Neurosci. 2009;29(13):4252–62.
168. Miners JS, Barua N, Kehoe PG, Gill S, Love S. Ab-degrading enzymes: potential for treatment of Alzheimer disease. J Neuropathol Exp Neurol. 2011;70(11):944–59.
169. Shimizu E, Kawahara K, Kajizono M, Sawada M, Nakayama H. IL-4-induced selective clearance of oligomeric beta-amyloid peptide(1-42) by rat primary type 2 microglia. J Immunol. 2008;181(9):6503–13.
170. Qiu WQ. Insulin-degrading enzyme regulates extracellular levels of amyloid beta protein by degradation. J Biol Chem. 1998;273(49):32730–8.
171. Grathwohl SA, Kälin RE, Bolmont T, Prokop S, Winkelmann G, Kaeser SA, Odenthal J, Radde R, Eldh T, Gandy S, Aguzzi A, Staufenbiel M, Mathews PM, Wolburg H, Heppner FL,

Jucker M. Formation and maintenance of Alzheimer's disease beta-amyloid plaques in the absence of microglia. Nat Neurosci. 2009;12(11):1361–3.
172. Sierra A, Gottfried-Blackmore AC, McEwen BS, Bulloch K. Microglia derived from aging mice exhibit an altered inflammatory profile. Glia. 2007;55(4):412–24.
173. Henry CJ, Huang Y, Wynne AM, Godbout JP. Peripheral lipopolysaccharide (LPS) challenge promotes microglial hyperactivity in aged mice that is associated with exaggerated induction of both pro-inflammatory IL-1β and anti-inflammatory IL-10 cytokines. Brain Behav Immun. 2009;23(3):309–17.
174. Kumar A, Stoica BA, Sabirzhanov B, Burns MP, Faden AI, Loane DJ. Traumatic brain injury in aged animals increases lesion size and chronically alters microglial/macrophage classical and alternative activation states. Neurobiol Aging. 2013;34(5):1397–411.
175. Sheng JG, Mrak RE, Griffin WS. Enlarged and phagocytic, but not primed, interleukin-1 alpha-immunoreactive microglia increase with age in normal human brain. Acta Neuropathol. 1998;95(3):229–34.
176. Strohmeyer R, Ramirez M, Cole GJ, Mueller K, Rogers J. Association of factor H of the alternative pathway of complement with agrin and complement receptor 3 in the Alzheimer's disease brain. J Neuroimmunol. 2002;131(1–2):135–46.
177. Schindler M, Forbes M, Robbins M, Riddle D. Aging-dependent changes in the radiation response of the adult rat brain. Int J Radiat Oncol Biol Phys. 2008;70(3):826–34.
178. Harry GJ. Microglia during development and aging. Pharmacol Therap. 2013;139(3):313–26.
179. Wong AM, Patel NV, Patel NK, Wei M, Morgan TE, de Beer MC, de Villiers WJ, Finch CE. Macrosialin increases during normal brain aging are attenuated by caloric restriction. Neurosci Lett. 2005;390(2):76–80.
180. Dilger RN, Johnson RW. Aging, microglial cell priming, and the discordant central inflammatory response to signals from the peripheral immune system. J Leukocyte Biol. 2008;84(4):932–9.
181. Sheng JG, Mrak RE, Griffin WS. Enlarged and phagocytic, but not primed, interleukin-1 alpha-immunoreactive microglia increase with age in normal human brain. Acta Neuropathol. 1998;95(3):229–34.
182. Minogue AM, Jones RS, Kelly RJ, McDonald CL, Connor TJ, Lynch MA. Age- associated dysregulation of microglial activation is coupled with enhanced blood-brain barrier permeability and pathology in APP/PS1 mice. Neurobiol Aging. 2014;35(6):1442–52.
183. Lively S, Schlichter LC. Age-related comparisons of evolution of the inflammatory response after intracerebral hemorrhage in rats. Trans Stroke Res. 2012;3(S1):132–46.
184. Lee DC, Ruiz CR, Lebson L, Selenica ML, Rizer J, Hunt Jr JB, Rojiani R, Reid P, Kammath S, Nash K, Dickey CA, Gordon M, Morgan D. Aging enhances classical activation but mitigates alternative activation in the central nervous system. Neurobiol Aging. 2013;34(6):1610–20.
185. Ye SM, Johnson RW. Increased interleukin-6 expression by microglia from brain of aged mice. J Immunol. 1999;93(1–2):139–48.
186. Yu WH, Go L, Guinn BA, Fraser PE, Westaway D, McLaurin J. Phenotypic and functional changes in glial cells as a function of age. Neurobiol Aging. 2002;23(1):105–15.
187. Njie EG, Boelen E, Stassen FR, Steinbusch HW, Borchelt DR, Streit WJ. Ex vivo cultures of microglia from young and aged rodent brain reveal age-related changes in microglial function. Neurobiol Aging. 2012;33(1):195. doi:10.1016/j.neurobiolaging.2010.05.008.
188. Lai AY, Dibal CD, Armitage GA, Winship IR, Todd KG. Distinct activation profiles in microglia of different ages: a systematic study in isolated embryonic to aged microglial cultures. Neurosci. 2013;254:185–95.
189. Wynne AM, Henry CJ, Huang Y, Cleland A, Godbout JP. Protracted down-regulation of CX3CR1 on microglia of aged mice after lipopolysaccharide challenge. Brain Behav Immun. 2010;24(7):1190–201.
190. Bachstetter AD, Morganti JM, Jernberg J, Schlunk A, Mitchell SH, Brewster KW, Hudson CE, Cole MJ, Harrison JK, Bickford PC, Gemma C. Fractalkine and CX3CR1 regulate hippocampal neurogenesis in adult and aged rats. NBA. 2011;32(11):2030–44.

191. Cox FF, Carney D, Miller A-M, Lynch MA. CD200 fusion protein decreases microglial activation in the hippocampus of aged rats. Brain Behav Immun. 2012;26(5):789–96.
192. Orre M, Kamphuis W, Osborn LM, Melief J, Kooijman L, Huitinga I, Klooster J, Bossers K, Hol EM. Acute isolation and transcriptome characterization of cortical astrocytes and microglia from young and aged mice. Neurobiol Aging. 2014;35:1–14.
193. Ma W, Cojocaru R, Gotoh N, Gieser L, Villasmil R, Cogliati T, Swaroop A, Wong WT. Gene expression changes in aging retinalmicroglia:relationship to microglial support functions and regulation of activation. Neurobiol Aging. 2013;34:2310–21.
194. Wes PD, Holtman IR, Boddeke EWGM, Moller T, Eggen BJL. Next generation transcriptomics and genomics elucidate biological complexity of microglia in health and disease. Glia. 2015;64(2):197–213. doi:10.1002/glia.22866.

Modifiable Factors Influencing Telomere Length and Aging

Frej Y. Fyhrquist and Outi J. Saijonmaa

1 Introduction

Telomere length is considered a potential biomarker of biological aging, although telomere length does not fulfill strict criteria [1]. Thus, telomere length is positively related to longevity [2, 3] and healthy lifestyle [2–4], and inversely associated with harmful lifestyle [2–4] and age-related diseases, including cardiovascular disease, diabetes, and dementia [2–6].

Telomere shortening is one of several factors driving cellular senescence which, in turn, promotes tissue oxidative stress and inflammation [7, 8]. While a causal relation between telomere attrition and aging and age-related disease has been suggested [9], evidence of causality is still incomplete. Regardless of causality, both telomere attrition and aging can be modified, both positively and negatively, by principally similar factors, mainly connected with lifestyle and, to a lesser extent, with certain drugs. This chapter will discuss modification of telomere length in connection with aging.

1.1 Telomere Structure and Function

Telomeres are protective structures at the ends of chromosomal DNA comprising thousands of tandem repeats of the TTAGGG sequence (9–18 kb in humans) and associated nucleoproteins. Telomeres participate in the maintenance of genomic and cellular stability and replication. They protect the genome from degradation, unwanted recombination, and chromosomal fusion [10]. Telomeres shorten by

F.Y. Fyhrquist (✉) • O.J. Saijonmaa
Minerva Institute for Medical Research, Biomedicum U2, Helsinki 00290, Finland
e-mail: frej.fyhrquist@helsinki.fi

S.C. Bondy, A. Campbell (eds.), *Inflammation, Aging, and Oxidative Stress*,
Oxidative Stress in Applied Basic Research and Clinical Practice,
DOI 10.1007/978-3-319-33486-8_4

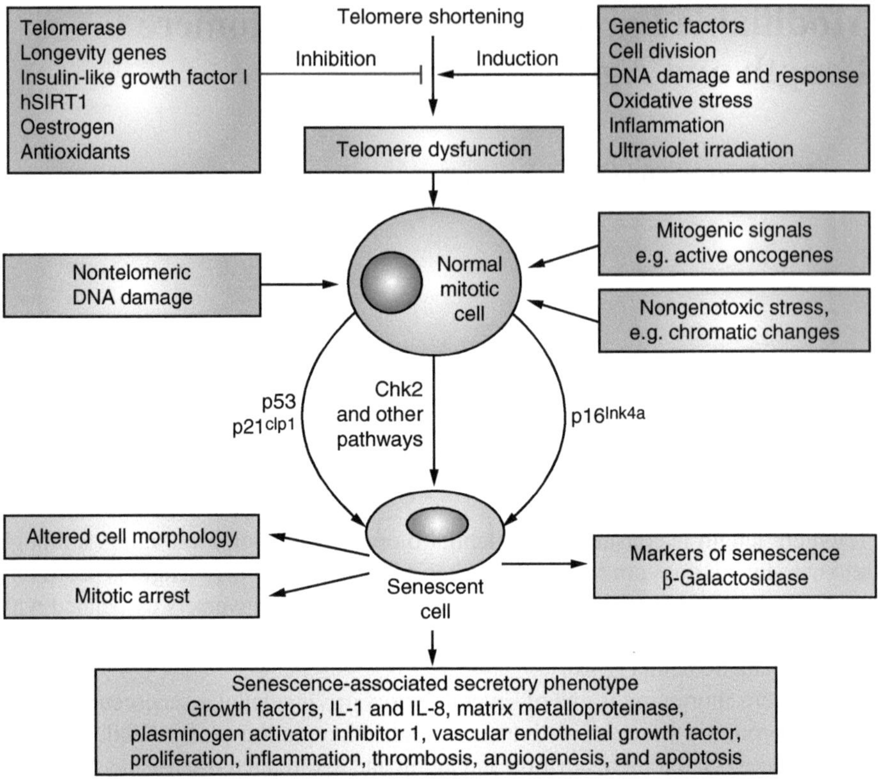

Fig. 1 The pathway from a normal mitotic cell to the senescent, aged cell. Factors affecting the rate of telomere shortening and telomere dysfunction are depicted. Other factors that induce progression of mitotic cells to senescence include strong mitogenic signals from oncogenes, nontelomeric DNA damage, and structural chromatin alterations. All these influences trigger the tumor suppressor pathways p16, $p21^{Cip1}$, and p53. Characteristics of senescent cells include mitotic arrest, altered morphology, and—in the senescence-associated secretory phenotype—many autocrine and endocrine activities potentially involved in tissue repair, proliferation, inflammation, and apoptosis. Chk2 is activated in response to DNA damage and is involved in cell-cycle arrest. *Chk2* checkpoint signaling kinase 2, *hSIRT1* NAD-dependent protein deacetylase sirtuin-1. With permission from [4]

30–150 base pairs with each cell division [11, 12] because of insufficient maintenance of the 3' overhang (single-stranded DNA). When a critical telomere length is reached, the protective nucleotide T-loop cannot function adequately, which leads to activation of the DNA damage repair system and cell-cycle inhibitors, including p16Ink4a, p21Cip1, and p53 (Fig. 1). The cell then enters replicative senescence, which is followed by apoptosis. Telomeres are considered to be a mitotic clock. Notably, not only shortening of the telomere, but also disruption of components and interactions within the shelterin complex can initiate telomere dysfunction and genomic instability [12, 13]. Consequently, telomere shortening is not necessarily a prerequisite for telomere dysfunction.

Telomerase is associated with the telomere complex, and promotes DNA synthesis to maintain telomere length. Germ cells, stem cells, and most cancer cells have a high level of telomerase activity to avoid senescence, whereas somatic cells have a low or undetectable level of telomerase activity. Telomerase consists of an RNA (TERC) component forming a template, and a catalytic subunit, telomerase reverse transcriptase (TERT), together generating new telomeric DNA repeats. Like telomere length, telomerase is under both genetic [14] and environmental control [15]. Epigenetic modifications have also been implicated in the transcriptional regulation of TERT [16].

1.2 Telomere Function and Cellular Senescense

Cellular senescence implies proliferative arrest. Two types of cellular senescence are recognized: stress-induced premature senescence and replicative senescence. Stress-induced premature senescence is promoted by oxidative stress, oncogenes, and irradiation [17, 18] and is not usually associated with telomere shortening. Replicative senescence is mainly a consequence of cell division leading to shortening and dysfunction of the telomeres. Thus, replicative senescence is driven by factors that accelerate telomere shortening, including genes, cell division, DNA damage, and oxidative stress. These factors elicit DNA damage response signals that arise from detectable nuclear foci or 'DNA-SCARS' (DNA segments with chromatin alterations reinforcing senescence, 7,19), and trigger the tumor-suppressor pathways p16 and p53 [15]. Senescent cells display irreversible proliferative arrest and altered morphology, including cellular enlargement, flattening, and vacuolization. Cells of the senescence-associated secretory phenotype (SASP) (Fig. 1) display autocrine and paracrine activities [7, 19]. They secrete IL-6 and IL-8, intercellular adhesion molecule 1, metalloproteases, monocyte attractants, plasminogen activator inhibitor 1, and vascular endothelial growth factor [7, 20]. Via these secretory activities, cells with the senescence-associated secretory phenotype contribute to both degenerative and proliferative age-related alterations by causing a chronic state of inflammation and oxidative stress, remodeling, and tissue repair [7, 21].

1.3 Measurement of Telomere Length

In human studies, telomere length is usually measured in DNA from leukocytes (leukocyte telomere length, LTL) from blood because they are easily obtainable. Measurement by Southern blot is considered the gold standard assay [22]. Another widely used method to assess telomere length is the quantitative polymerase chain reaction [23]. This method is well suited for the study of large populations.

When comparing LTL and telomere length in somatic cells within individuals a substantial synchrony is usually reported [24] which implies that LTL may serve as a proxy for telomere length in various tissues.

2 Regulation of Telomere Length

Telomere length is regulated by a large number of factors, including genetic and DNA damage, cell division, aging, oxidative stress, inflammation, and ultraviolet irradiation (Table 1). These factors are essentially non-modifiable, while modifiable factors are of vital interest for man in the pursuit for longevity and healthy aging. Modifiable factors affecting both telomere length, longevity, and health are mainly related to lifestyle and possibly to some drugs, hormones, and vitamins (Table 2). Modifiable factors affecting are mostly well recognized risk factors for age related disease, notably cardiovascular disease [1–6]. Certain drugs extending lifespan in animals such as worms and mice are under intense investigation [25] and offer promise in the hunt for means of achieving healthy aging.

2.1 Non-modifiable (Endogenous) Factors Regulating Telomere Length

Telomere length and function is strongly influenced by *genetic factors* [26] and is partly hereditable [26]. Genome wide association studies (GWAS) have identified loci associated with LTL located near *TERC* and a component of the telomere-maintenance complex [27–29]. However, GWAS have explained only about 1.6 % of inter individual variation in LTL.

Table 1 Endogenous (non-modifiable) factors affecting LTL[a] and aging

A. Shortening/aging		
Factor	LTL	Aging
– Genetic	Yes [26–30, 33]	Yes [25, 32]
– Male gender	Yes [1–5]	Yes
– Cell division	Yes [10, 11]	Yes [2]
– Age	Yes [1–5]	
– Oxidative stress	Yes [7, 19, 35, 36]	Yes [25]
– Inflammation	Yes [7, 8, 35, 36]	Yes [20]
B. Inhibition of shortening/ageing		
– Telomerase	Yes [2, 14, 17, 77]	Yes [50, 77]
– Estrogen	Yes [69]	

[a]*LTL* leukocyte telomere length

Table 2 Environmental (modifiable) factors affecting LTL[a] and aging

A. Shortening/aging		
Factor	LTL	Aging
– Smoking	Yes [37, 38]	Yes [44]
– Alcohol abuse	Yes [48]	Yes [48]
– Obesity	Yes [37–39]	Yes [43]
– Sedentary lifestyle	Yes [40, 54]	Yes [25, 57]
– UV irradiation	Yes [50]	Yes [50]
– Mental stress	Yes [51, 52]	
B. Inhibition of shortening/ageing		
– Healthy lifestyle	Yes [40, 54–59, 62]	Yes [25, 57, 58, 61]
– Vitamins C, D, and E	Yes [66–68]	Yes [61]
– Ω-3 fatty acids	Yes [65]	Yes [65]
– Healthy diet	Yes [71–73]	Yes [71]
– Caloric restriction (animals)	Yes [78–80]	Yes [78–80]
– Statin treatment	Yes [70]	Yes

[a]*LTL* leukocyte telomere length

Mutations in the telomerase components TERT and TERC are associated with short telomeres and premature aging in dyskeratosis congenita [30]. Mutations in the WRN gene, related to telomere maintenance, are found in Werner syndrome, which is characterized by premature aging and short telomeres [30]. Major clinical symptoms of Werner syndrome include common age- associated diseases, such as insulin-resistant diabetes mellitus, and atherosclerosis. Recent studies have reported the generation of induced pluripotent stem cells (iPSCs) from cells of patients with Werner syndrome, and they have concluded that reprogramming represses premature senescence phenotypes in these cells, [31]. These observations offer potential promises of treatment of telomeric dysfunction in cells from Werner syndrome patients by genetic reprogramming.

Sirtuins. The NAD-dependent protein deacetylase sirtuin (SIRT) 1–7 family of proteins promotes survival, stress resistance, and longevity. [32]. The presence of SIRT6 in human endothelial cells was reported to confer protection from telomere and genomic DNA damage thus preventing a decrease in replicative capacity and premature senescence and delay vascular aging [33]. Interestingly, the *SIRT1* locus was positively associated with both LTL and longevity in individuals from the Louisiana Healthy Aging Study [34], which suggests a link between telomere length and longevity.

Cell division.

Division of somatic cells with no or little active telomerase leads to telomere shortening by 30–150 base pairs per cell division because of the inability to maintain the length of the 3' overhang (single-stranded DNA). Thus, *aging* is the main cause of telomere shortening in somatic cells. Germ and stem cells, and most cancer cells have active telomerase and preserved telomere length through cell division [10, 11].

Oxidative stress, and *inflammation* are important causes of telomere shortening,and also implicated in cellular senescence and aging [35, 36]. Studies in mice with low-grade inflammation induced by knockout of the nfkb1 subunit of NF-kappa-ß indicated that systemic chronic inflammation can accelerate ageing via ROS-mediated exacerbation of telomere dysfunction and cell senescence [36]. Reactive oxygen species (ROS) are probably involved in both the induction and stabilisation of cellular senescence. Senescent cells of the secretory phenotype induce tissue inflammation (SASP) by secreting inflammatory molecules and inducing increases ROS generation, which in turn can accelerate telomere shortening [35, 36]. Accordingly, endogenous antioxidants, such as superoxide dismutase, inhibit ageing processes and telomere shortening [35].

2.2 *Modifiable (Environmental) Factors Regulating Telomere Length and Aging*

2.2.1 Associated with Telomere Shortening

Obesity is associated with short LTL [37–39]. Accordingly, short LTL is also related to *sedentary lifestyle* [40]. Further, short LTL was related to the *metabolic syndrome*, and associated with a higher metabolic risk profile and with less favorable trajectories of metabolic biomarkers over an observational period of 6 years [41]. Telomere attrition in obesity and related conditions (metabolic syndrome, sedentary lifestyle) is likely to be triggered by inflammation and oxidative stress [42], which also are major factors driving cellular and organismal aging [43].

Smoking is associated with short telomeres [37, 38], cardiovascular and pulmonary disease and cancer. A meta-analysis of 17 studies on the impact of smoking on all-cause mortality in people ≥60 years found that mortality was 83 % and 34 % higher in current and former smokers, respectively, than in never-smokers [44]. Telomere attrition in smokers is thought to be related to increased oxidative stress. Thus activation of NDPH and resultant reactive oxygen species production by cigarette smoke have been observed in isolated blood vessels and cultured vascular cells, including endothelial and smooth muscle cells [45]. However, in one study on cultured endothelial cells isolated from the internal mammary artery of smoking or nonsmoking patients with coronary artery disease (CAD), cellular senescence was independent of telomere length, but was related to oxidative damage and markers of inflammation [46]. In smokers, therefore, endothelial senescence might be both stress-induced and replicative. Importantly, in a recent large Danish population study on 55,568 individuals, employing a Mendelian randomization study and observational, genetic and mediation analysis, high tobacco consumption was *causally* associated with mortality but not with short telomeres [47]. The Authors concluded that "the pathway from high tobacco consumption to increased mortality is not likely to be mediated through telomere shortening".

Alcohol abuse has been related to short telomeres in a dose-dependent manner [48], possibly triggered by oxidative stress [49]. Obesity, sedentary life style, and alcohol abuse are well known to be associated with shortened lifespan and increased risk of age-related diseases. However, association does not prove causality.

UV irradiation by sunlight is known to promote cutaneous inflammation and oxidative stress, and skin aging and can also lead to telomere damage [50] Thus, avoiding excess sunlight can be expected to reduce both the rate of telomere shortening, aging, and cancer in human skin. However, while suggestive, this does not prove a causal relation.

Mental stress was reported to be associated with short LTL and reduction in leukocyte telomerase activity [51]. The hypothetical mechanisms whereby mental stress may explain telomere attrition have been reviewed elsewhere [52]. However, no association between LTL and life stress was found in a fairly large (n = 677) 30 year birth cohort, monitored for stressful events until age 25 and LTL measured at the age of 28–30 years [53]. Thus, the relation between life stress and telomere length justifies further studies, e.g. investigating telomere length both before and after stressful life experiences.

2.2.2 Associated with Maintenance of Telomere Length

Healthy lifestyle is associated with long telomeres [54–56], and also with reduced risk of age-related diseases and longevity [25, 57]. Healthy lifestyle implies positive modification of both aging and telomere length in a nutshell. *Optimal body weight* maintained through diet and physical activity remains a cornerstone in the prevention of chronic diseases and promoting healthy aging [58]. The factors driving telomere attrition in obesity are largely considered similar to factors promoting aging, e.g. oxidative stress and inflammation [25, 57, 58].

Rather few studies have reported on the impact of *lifestyle intervention* on telomere length. In a 5 years follow-up small-sized pilot study in men with prostatic cancer, with 10 men in the intervention group who followed a program of comprehensive lifestyle changes (diet, activity, stress management, and social support) an increase in LTL was noticed [59]. In the larger longitudinal Finnish Diabetes Prevention Study, comprising >300 subjects LTL increased in about two thirds of subjects, but LTL was not associated with development of diabetes and lifestyle intervention did not affect LTL [60]. The beneficial effect of life style intervention in terms of reduced mortality and age-related disease is well acknowledged [61]. This includes weight loss, physical activity, smoking, excessive alcohol consumption and notably their combination.

Physical activity is related to maintenance of telomere length. Thus, in a large study involving 2401 white twins (249 men and 2152 women), higher-intensity physical activity was associated with longer LTL [40]. Physical training was related to increased telomerase activity in the aorta and in circulating leukocytes of mice [62] and in leukocytes from track and field athletes compared with control subjects. Physical training is generally related to improved health and to longevity [25, 40, 54, 55, 58]. However, extreme physical training may not reduce mortality compared to modest physical activity [63]. Interestingly, in the Helsinki Business Men study [64] we observed that *modest* physical activity was associated with the longest LTL compared with low or high physical activity.

Dietary additives and vitamins. An increased level of *marine ω-3 fatty acid* was associated with a decreased rate of LTL shortening over 5 years in patient with CAD [65]. Furthermore, increased blood concentrations of *vitamin C, 25-hydroxyvitamin D, and vitamin E* were related to long LTL [66–68].

Estrogen activates telomerase and thereby inhibits telomere shortening [69], which is believed to explain that females have longer telomeres than age-matched males.

Statin treatment was associated with long LTL in a small observational study on patients with acute myocardial infarction [70]. However, this observation has not been confirmed. Intervention studies in humans on effects of drugs on telomere length are lacking, somewhat surprisingly.

Diet may influence telomere length [71]. Reports of *dietary* modification of telomere length are scanty. Adherence to Mediterranean diet in 217 elderly subjects was associated with longer LTL and higher telomerase activity in circulating leukocytes [72]. In a large study, comprising a subset of 32,825 women in the Nurses' Health Study [73], LTL was also positively associated with adherence to Mediterranean diet.

2.3 Modification of Telomere Length and Aging in Animal Studies

Genetically modified animal models have suggested causal links between telomere attrition, cellular senescence and organismal aging. Thus, mice with genetically shortened or lengthened telomeres exhibit decreased or increased lifespan, respectively [74, 75]. Recent evidence also indicates that aging can be partially reverted by telomerase activation. Thus, the premature aging of telomerase- deficient mice can be reverted when telomerase is genetically reactivated in these aged mice [76]. Moreover, normal physiological aging can be delayed without increasing the incidence of cancer in adult wild-type mice by pharmacological activation or systemic viral transduction of telomerase [77].

Caloric restriction (CR) without malnutrition can prolong life in e.g. worms, mice, and rats [25], and even in primates [78]. CR reduces the release of growth factors like growth hormone, insulin, and insulin-like growth factor 1 (IGF1), which have been shown to accelerate aging and increase mortality in many organisms [25]. Studies in mice show that CR attenuates age-associated telomere erosion in leukocytes and various tissues including lung, kidney-cortex and muscle fiber, and this effect synergizes with the increase in health span and longevity [79]. However, CR did not affect telomere dynamics in muscle biopsies from rhesus monkeys [80] suggesting that the anti-aging effect of CR may not be dependent on telomere attrition in primates. *Drugs extending lifespan* in mice, including resveratrol, rapamycin, spermidine, and metformin reviewed elsewhere [25], have not reportedly been tested for modification of telomere length.

2.4 Telomere Length and Age-Related Diseases

Clinical and epidemiological studies have shown that short LTL is associated with age-related diseases, including atherosclerosis, coronary artery disease, heart failure, and Alzheimer's disease [2–5]. Short LTL is also related to diabetes, both type 1 [81] and type 2 [82]. In a follow-up study on subjects with type 1 diabetes, progression of diabetic nephropathy was predicted by short LTL [83]. Further, short LTL is related to increased mortality in most studies [84, 85]. Telomere attrition is also related to cardiovascular risk factors (Table 2) and to aging itself. Again, whether telomere shortening is causally related to or caused by age-related diseases and risk factors is presently unclear.

3 Conclusions

Both telomere attrition and aging can be negatively modified by an unhealthy life style and positively affected by healthy life style and possibly by certain 'healthy' diets and drugs. The connection between aging and telomere biology is complex. While telomere attrition and dysfunction is considered one of several 'hallmarks of aging' [86], age by itself is the main cause of telomere shortening by each cell division. On the other hand, telomere attrition and dysfunction is affected by a number of factors independently of age (Fig. 1, Tables 1 and 2). Of these factors, some are modifiable, e.g. oxidative stress or inflammation, via improved life style and possibly certain diets and drugs, while most factors are non-modifiable. Research on aging and telomere biology is currently intense and will hopefully elucidate mechanisms of aging and its connections to telomere functions. Eventually, improved knowledge may be translated into more effective targeting of life style and development of new drugs promoting successful aging.

References

1. Wang JC, Bennett M. Aging and atherosclerosis: mechanisms, functional consequences, and potential therapeutics for cellular senescence. Circ Res. 2012;111:245–59.
2. Tzanetakou IP, Nzietchueng R, Perrea DN, Benetos A. Telomeres and their Role in Aging and Longevity. Curr Vasc Pharmacol. 2014;12:726–34.
3. Kovacic JC, Moreno P, Nabel EG, Hachinski V, Fuster V. Cellular senescence, vascular disease, and aging: part 2 of a 2-part review. Circulation. 2011;123:1990–2010.
4. Fyhrquist F, Saijonmaa O, Strandberg T. The roles of senescence and telomere shortening in cardiovascular disease. Nat Rev Cardiol. 2013;10:274–83.
5. Nilsson PM, Tufvesson H, Leosdottir M, Melander O. Telomeres and cardiovascular disease risk: an update 2013. Transl Res. 2013;162:371–80.
6. Haycock PC, Heydon EE, Kaptoge S, Butterworth AS, Thompson A, Willeit P. Leucocyte telomere length and risk of cardiovascular disease: systematic review and meta-analysis. BMJ. 2014;8:349. doi:10.1136/bmj.g4227.
7. Campisi J. Cellular senescence: putting the paradoxes in perspective. Curr Opin Genet Dev. 2011;21:107–12.
8. Correia-Melo C, Hewitt G, Passos JF. Telomeres, oxidative stress and inflammatory factors: partners in cellular senescence? Longevity & Healthspan. 2014;3(1):1. doi:10.1186/2046-2395-3-1.
9. Samani NJ, van der Harst P. Biological aging and cardiovascular disease. Heart. 2008;94:537–9.
10. de Lange T. Shelterin: the protein complex that shapes and safeguards human telomeres. Genes Dev. 2005;19:2001–10.
11. Blackburn EH, Greider CW, Szostak JW. Telomeres and telomerase: the path from maize, *Tetrahymena* and yeast to human cancer and aging. Nat Med. 2006;12:1133–8.
12. Cesare AJ, Reddel RR. Alternative lengthening of telomeres: models, mechanisms and implications. Nat Rev Genet. 2010;11:319–30.
13. Cesare AJ, Karlseder JA. Three-state model of telomere control over human proliferative boundaries. Curr Opin Cell Biol. 2012;24:731–8.
14. Savage SA, Stewart BJ, Eckert A, Kiley M, Liao JS, Channock SJ. Genetic variation, nucleotide diversity, and linkage disequilibrium in seven telomere stability genes suggest that these genes may be under constraint. Hum Mutat. 2005;26:343–50.
15. Jacobs TL, Epel ES, Lin J, Blackburn EH, Wolkowitz OM, Bridwell DA, Zanesco AP, Aichele SR, Sahdra BK, MacLean KA, King BG, Shaver PR, Rosenberg EL, Ferrer E, Wallace BA, Saron CD. Intensive mediation training, immune cell telomerase activity, and psychological mediators. Psychoneuroendocrinology. 2011;36:664–81.
16. Blasco MA. The epigenetic regulation of mammalian telomeres. Nat Rev Genet. 2007;8:299–309.
17. Blasco MA. Telomeres and human disease: ageing, cancer and beyond. Nat Rev Genet. 2005;6:611–22.
18. Calado RT, Young NS. Telomere diseases. N Engl J Med. 2009;361:2353–65.
19. von Zglinicki T, Saretzki G, Ladhoff J, d'Adda di Fagagna F, Jackson SP. Human cell senescence as a DNA damage response. Mech Ageing Dev. 2005;126:111–7.
20. Erusalimsky JD, Kurz DJ. Cellular senescence *in vivo*: its relevance in ageing and cardiovascular disease. Exp Gerontol. 2005;40:634–42.
21. Franceschi C, Campisi J. Chronic inflammation (inflammaging) and its potential contribution to age-associated diseases. J Gerontol A Biol Sci Med Sci. 2014;69(S1):54–9.
22. Aviv A, Hunt SC, Lin J, Cao X, Kimura M, Blackburn E. Impartial comparative analysis of measurement of leukocyte telomere length/DNA content by Southern blots and qPCR. Nucleic Acids Res. 2011;39, e134. doi:10.1093/nar/gkr634.
23. Cawthon RM. Telomere measurement by quantitative PCR. Nucleic Acids Res. 2002;30(10), e47.
24. Daniali L, Benetos A, Susser E, Kark JD, Labat C, Kimura M, Desai K, Granick M, Aviv A. Telomeres shorten at equivalent rates in somatic tissues of adults. Nat Commun. 2013;4:1597. doi:10.1038/ncomms2602.

25. de Cabo R, Carmona-Gutierrez D, Bernier M, Hall MN, Madeo F. The search for antiaging interventions: from elixirs to fasting regimens. Cell. 2014;157(7):1515–26. doi:10.1016/j.cell.2014.05.031.
26. Aviv A. Genetics of leukocyte telomere length and its role in atherosclerosis. Mutat Res. 2012;730:68–74.
27. Levy D, Neuhausen SL, Hunt SC, Kimura M, Hwang SJ, Chen W, Bis JC, Fitzpatrick AL, Smith E, Johnson AD, Gardner JP, Srinivasan SR, Schork N, Rotter JI, Herbig U, Psaty BM, Sastrasinh M, Murray SS, Vasan RS, Province MA, Glazer NL, Lu X, Cao X, Kronmal R, Mangino M, Soranzo N, Spector TD, Berenson GS, Aviv A. Genome-wide association identifies *OBFC1* as a locus involved in human leukocyte telomere biology. Proc Natl Acad Sci U S A. 2010;107:9293–8.
28. Mangino M, Hwang SJ, Spector TD, Hunt SC, Kimura M, Fitzpatrick AL, Christiansen L, Petersen I, Elbers CC, Harris T, Chen W, Srinivasan SR, Kark JD, Benetos A, El Shamieh S, Visvikis-Siest S, Christensen K, Berenson GS, Valdes AM, Viñuela A, Garcia M, Arnett DK, Broeckel U, Province MA, Pankow JS, Kammerer C, Liu Y, Nalls M, Tishkoff S, Thomas F, Ziv E, Psaty BM, Bis JC, Rotter JI, Taylor KD, Smith E, Schork NJ, Levy D, Aviv A. Genome-wide meta-analysis points to CTC1 and ZNF676 as genes regulating telomere homeostasis in humans. Hum Mol Genet. 2012;21:5385–94.
29. Codd V, Nelson CP, Albrecht E, Mangino M, Deelen J, Buxton JL, et al. Identification of seven loci affecting mean telomere length and their association with disease. Nat Genet. 2013;45:422–7.
30. Armanios M, Blackburn EH. The telomere syndromes. Nat Rev Genet. 2012;13:693–704.
31. Shimamoto A, Koutaro Y, Tahara H. Werner-Syndrome-specific induced pluripotent stem cells: recovery of telomere function by reprogramming. Front Genet. 2015. doi:10.3389/fgene.2015.00010.
32. Guarante L. The Franklin H. Epstein lecture: sirtuins, aging, and medicine. N Engl J Med. 2011;364:2235–44.
33. Cardus A, Uryga AK, Walters G, Erusalimsky JD. SIRT6 protects human endothelial cells from DNA damage, telomere dysfunction, and senescence. Cardiovasc Res. 2013;97:571–9.
34. Kim S, Bi X, Czarny-Ratajczak M, Dai J, Welsh DA, Myers L, Welsch MA, Cherry KE, Arnold J, Poon LW, Jazwinski SM. Telomere maintenance genes *SIRT1* and *XRCC6* impact age-related decline in telomere length but only *SIRT1* is associated with human longevity. Biogerontology. 2012;13:119–31.
35. von Zglinicki T. Oxidative stress shortens telomeres. Trends Biochem Sci. 2002;27:339–44.
36. Jurk D, Wilson C, Passos JF, Oakley F, Correia-Melo C, Greaves L, Saretzki G, Fox C, Lawless C, Anderson R, Hewitt G, Pender SL, Fullard N, Nelson G, Mann J, van de Sluis B, Mann DA, von Zglinicki T. Chronic inflammation induces telomere dysfunction and accelerates ageing in mice. Nat Commun. 2014;2:4172. doi:10.1038/ncomms5172.
37. Valdes AM, Andrew T, Gardner JP, Kimura M, Oelsner E, Cherkas LF, Aviv A, Spector TD. Obesity, cigarette smoking, and telomere length in women. Lancet. 2005;366:662–720.
38. Strandberg TE, Strandberg AY, Saijonmaa O, Tilvis RS, Pitkälä KH, Fyhrquist F. Association of telomere length in older men with mortality and midlife body mass index and smoking. J Gerontol A Biol Sci Med Sci. 2011;66:815–20.
39. Müezzinler A, Zaineddin AK, Brenner H. Body mass index and leukocyte telomere length in adults: a systematic review and meta-analysis. Obes Rev. 2014;15:192–201.
40. Cherkas LF, Hunkin JL, Kato BS, Richards JB, Gardner JP, Surdulescu GL, Kimura M, Lu X, Spector TD, Aviv A. The association between physical activity in leisure time and leukocyte telomere length. Arch Intern Med. 2008;168:154–8.
41. Révész D, Milaneschi Y, Verhoeven JE, Penninx BW. Telomere length as a marker of cellular ageing is associated with prevalence and progression of metabolic syndrome. J Clin Endocrinol Metab. 2014;99(12):4607–15. doi:10.1210/jc.2014-1851.
42. Furukawa S, Fujita T, Shimabukuro M, Iwaki M, Yamada Y, Nakajima Y, Nakayama O, Makishima M, Matsuda M, Shimomura I. Increased oxidative stress in obesity and its impact on metabolic syndrome. J Clin Invest. 2004;114:1752–61.

43. Donato AJ, Morgan RG, Walker AE, Lesniewski LA. Cellular and molecular biology of aging endothelial cells. J Mol Cell Cardiol. 2015;89:122–35. doi:10.1016/j.yjmcc.2015.01.021.
44. Gellert C, Schöttker B, Brenner H. Smoking and all-cause mortality in older people: systematic review and meta-analysis. Arch Intern Med. 2012;172:837–44.
45. Kim M, Han C, Lee M-Y. NADPH oxidase and the cardiovascular toxicity associated with smoking. Toxicol Res. 2014;30:149–57.
46. Farhat N, Thorin-Trescases N, Voghel G, Villeneuve L, Mamarbachi M, Perrault LP, Carrier M, Thorin E. Stress-induced senescence predominates in endothelial cells isolated from atherosclerotic chronic smokers. Can J Physiol Pharmacol. 2008;86:761–9.
47. Rode L, Bojesen SE, Weischer M, Nordestgaard BG. High tobacco consumption is causally associated with increased all-cause mortality in a general population sample of 55 568 individuals, but not with short telomeres: a Mendelian randomization study. Int J Epidemiol. 2014;43:1473–82.
48. Strandberg TE, Strandberg AY, Saijonmaa O, Tilvis RS, Pitkälä KH, Fyhrquist F. Association between alcohol consumption in healthy midlife and telomere length in older men. The Helsinki Businessmen Study. Eur J Epidemiol. 2012;27:815–22.
49. Comporti M, Signorini C, Leoncini S, Gardi C, Ciccoli L, Giardini A, Vecchio D, Arezzini B. Ethanol-induced oxidative stress: basic knowledge. Genes Nutr. 2010;5:101–9.
50. Kosmadaki MG, Gilchrest BA. The role of telomeres in skin aging/photoaging. Micron. 2004;35:155–9.
51. Epel ES, Blackburn EH, Lin J, Dhabhar FS, Adler NE, Morrow JD, Cawthon RM. Accelerated telomere shortening in response to life stress. Proc Natl Acad Sci U S A. 2004;101:17312–5.
52. Shalev I, Entringer S, Wadhwa PD, Wolkowitz OM, Puterman E, Lin J, Epel ES. Stress and telomere biology: a lifespan perspective. Psychoneuroendocrinology. 2013;38:1835–42.
53. Jodczyk S, Fergusson DM, Horwood LJ, Pearson JF, Kennedy MA. No association between mean telomere length and life stress observed in a 30 year birth cohort. PLoS One. 2014;9, e97102.
54. Njajou OT, Hsueh WC, Blackburn EH, Newman AB, Wu SH, Li R, Simonsick EM, Harris TM, Cummings SR, Cawthon RM. Association between telomere length, specific causes of death, and years of healthy life in health, aging, and body composition, a population-based cohort study. J Gerontol A Biol Sci Med Sci. 2009;64(8):860–4.
55. Krauss J, Farzaneh-Far R, Puterman E, Na B, Lin J, Epel E, Blackburn E, Whooley MA. Physical fitness and telomere length in patients with coronary heart disease: findings from the Heart and Soul Study. PLoS One. 2011;6, e26983.
56. Ornish D, Lin J, Daubenmier J, Weidner G, Epel E, Kemp C, Magbanua MJ, Marlin R, Yglecias L, Carroll PR, Blackburn EH. Increased telomerase activity and comprehensive lifestyle changes: a pilot study. Lancet Oncol. 2008;9:1048–57.
57. Kennedy BK, et al. Geroscience; linking aging to chronic disease. Cell. 2014;159:709–13.
58. Fontana L, Hu FB. Optimal body weight for health and longevity: bridging basic, clinical, and population research. Aging Cell. 2014;14:391–400.
59. Ornish D, Lin J, Chan JM, Epel E, Kemp C, Weidner G, Marlin R, Frenda SJ, Magbanua MJ, Daubenmier J, Estay I, Hills NK, Chainani-Wu N, Carroll PR, Blackburn EH. Effect of comprehensive lifestyle changes on telomerase activity and telomere length in men with biopsy-proven low-risk prostate cancer: 5-year follow-up of a descriptive pilot study. Lancet Oncol. 2013;14:1112–20.
60. Hovatta I, de Mello VD, Kananen L, Lindström J, Eriksson JG, Ilanne-Parikka P, Keinänen-Kiukaanniemi S, Peltonen M, Tuomilehto J, Uusitupa M. Leukocyte telomere length in the Finnish Diabetes Prevention Study. PLoS One. 2012;7(4):e34948.
61. Rizzuto D, Fratiglioni L. Life style factors related to mortality and survival: a mini-review. Gerontology. 2014;60:327–35.
62. Werner C, Fürster T, Widmann T, Pöss J, Roggia C, Hanhoun M, Scharhag J, Büchner N, Meyer T, Kindermann W, Haendeler J, Böhm M, Laufs U. Physical exercise prevents cellular senescence in circulating leukocytes and in the vessel wall. Circulation. 2009;120:2438–47.

63. Schnohr P, O'Keefe JH, Marott JL, Lange P, Jensen GB. Dose of jogging and long-term mortality: the Copenhagen City Heart Study. J Am Coll Cardiol. 2015;65:411–9. doi:10.1016/j.
64. Savela S, Saijonmaa O, Strandberg TE, Koistinen P, Strandberg AY, Tilvis RS, Pitkälä KH, Miettinen TA, Fyhrquist F. Physical activity and telomere length measured in old age. Exp Gerontol. 2013;48:81–4.
65. Farzaneh-Far R, Lin J, Epel ES, Harris WS, Blackburn EH, Whooley MA. Association of marine omega-3 fatty acid levels with telomeric aging in patients with coronary heart disease. JAMA. 2010;303:250–7.
66. Richards BJ, Valdes AM, Gardner JP, Paximadas D, Kimura M, Nessa A, Lu X, Surdulescu GL, Swaminathan R, Spector TD, Aviv A. Higher vitamin D concentrations are associated with longer leukocyte telomere length in women. Am J Clin Nutr. 2007;86:1420–5.
67. Xu Q, Parks CG, DeRoo LA, Cawthon RM, Sandler DP, Chen H. Multivitamin use and telomere length in women. Am J Clin Nutr. 2009;89:1857–63.
68. Sen A, Marsche G, Freudenberger P, Schallert M, Toeglhofer AM, Nagl C, Schmidt R, Launer LJ, Schmidt H. Association between higher plasma lutein, zeaxanthin, and vitamin C concentrations and longer telomere length: results of the Austrian Stroke Prevention Study. J Am Geriatr Soc. 2014;62(2):222–9. doi:10.1111/jgs.12644.
69. Kyo S, Takakura M, Kanaya T, Zhuo W, Fujimoto K, Nishio Y, Orimo A, Inoue M. Estrogen activates telomerase. Cancer Res. 1999;59:5917–21.
70. Saliques S, Teyssier J-R, Vergely C, Lorgis L, Lorin J. Circulating leukocyte telomere length and oxidative stress: a new target for statin therapy. Atherosclerosis. 2011;219:753–60.
71. Paul L. Diet, nutrition and telomere length. J Nutr Biochem. 2011;20:895–901.
72. Boccardi V, Esposito A, Rizzo MR, Marfella R, Barbieri M, Polisso G. Mediterranean diet, telomere maintenance and health status among elderly. PLoS One. 2013;8:1–6.
73. Crous-Bou M, Fung TF, Prescott J, Julin B, Du M, Sun Q, Rexrode KM, Hu FB, De Vivo I. Mediterranean diet and telomere length in Nurses' health study: population based cohort study. BMJ. 2014;349:6674. doi:10.1136/bmj.g6674.
74. Armanios M, Alder JK, Parry EM, Karim B, Strong MA, Greider CW. Short telomeres are sufficient to cause the degenerative defects associated with aging. Am J Hum Genet. 2009;85(6):823–32.
75. Tomás-Loba A, Flores I, Fernández-Marcos PJ, Cayuela ML, Maraver A, Tejera A, Borrás C, Matheu A, Klatt P, Flores JM, Viña J, Serrano M, Blasco MA. Telomerase reverse transcriptase delays aging in cancer-resistant mice. Cell. 2008;135(4):609–22.
76. Jaskelioff M, Muller FL, Paik JH, Thomas E, Jiang S, Adams AC, Sahin E, Kost-Alimova M, Protopopov A, Cadiñanos J, Horner JW, Maratos-Flier E, Depinho RA. Telomerase reactivation reverses tissue degeneration in aged telomerase-deficient mice. Nature. 2011;469(7328):102–6.
77. de Jesus BB, Schneeberger K, Vera E, Tejera A, Harley CB, Blasco MA. The telomerase activator TA-65 elongates short telomeres and increases health span of adult/old mice without increasing cancer incidence. Aging Cell. 2011;10:604–21.
78. Colman RJ, Anderson RM, Johnson SC, Kastman EK, Kosmatka KJ, Beasley TM, Allison DB, Cruzen C, Simmons HA, Kemnitz JW, Weindruch R. Caloric restriction delays disease onset and mortality in rhesus monkeys. Science. 2009;325:201–4.
79. Vera E, de Jesus B, Foronda M, Flores JM, Blasco MA. Telomerase reverse transcriptase synergizes with calorie restriction to increase health span and extend mouse longevity. PLoS One. 2013;8, e53760.
80. Smith DL, Mattison JA, Desmond RA, Gardner JP, et al. Telomere dynamics in Rhesus monkeys: no apparent effect of caloric restriction. J Gerontol Biol Sci Med Sci. 2011;66A:1163–8.
81. Jeanclos E, Krolewski A, Skurnick J, Kimura M, Aviv H, Warram JH, Aviv A. Shortened telomere length in white blood cells of patients with IDDM. Diabetes. 1998;47(3):482–6.
82. Willeit P, Raschenberger J, Heydon EE, Tsimikas S, Haun M, Mayr A, Weger S, Witztum JL, Butterworth AS, Willeit J, Kronenberg F, Kiechl S. Leucocyte telomere length and risk of type 2 diabetes mellitus: new prospective cohort study and literature-based meta-analysis. PLoS One. 2014;9(11):112483.

83. Fyhrquist F, Tiitu A, Saijonmaa O, Forsblom C, Groop PH, FinnDiane Study Group. Telomere length and progression of diabetic nephropathy in patients with type 1 diabetes. J Intern Med. 2010;267:278–86.
84. Cawthon RM, Smith KR, O'Brien SA, Kerber RA. Association between telomere length in blood and mortality in people aged 60 years or older. Lancet. 2003;361:393–5.
85. Rode L, Nordestgaard BG, Bojesen SE. Peripheral blood leukocyte telomere length and mortality among 64 637 individuals from the general population. J Natl Cancer Inst. 2015;107(6):107. doi:10.1093/jnci/djv074.
86. López-Otín C, Blasco MA, Partridge L, Serrano M, Kroemer G. The hallmarks of aging. Cell. 2013;153(6):1194–217.

Tyrosine Kinase Inhibitors and Neurodegenerative Disorders

Syed Jafar Mehdi, Steven W. Barger, Merle G. Paule, Syed F. Ali, and Syed Z. Imam

1 Introduction

Cellular signaling is important for many biological processes including growth, differentiation, adhesion, motility and apoptosis. Signal transduction pathways play a major role in coordinating the complex functions of human body. Protein phosphorylation is one of the important mechanisms in signal transduction pathways, which is carried out by protein kinases. These kinases regulate the fundamental processes of proliferation, differentiation, migration, metabolism and anti-apoptotic signaling of the cell. Protein kinases fall into two major classes: the serine/threonine kinases and tyrosine kinases; respectively, they catalyze the phosphorylation of serine/threonine or tyrosine amino acid residues in intact proteins.

S.J. Mehdi
Department of Geriatrics, University of Arkansas for Medical Sciences,
Little Rock, AR 72205, USA

S.W. Barger
Geriatric Research Education and Clinical Center, Central Arkansas Veterans Healthcare System, Little Rock, AR 72205, USA

Department of Geriatrics, University of Arkansas for Medical Sciences,
Little Rock, AR 72205, USA

M.G. Paule • S.F. Ali • S.Z. Imam (✉)
Division of Neurotoxicology, National Center for Toxicological Research/US FDA,
Jefferson, AR 72079, USA
e-mail: Syed.Imam@fda.hhs.gov

S.C. Bondy, A. Campbell (eds.), *Inflammation, Aging, and Oxidative Stress*,
Oxidative Stress in Applied Basic Research and Clinical Practice,
DOI 10.1007/978-3-319-33486-8_5

Tyrosine kinases can be further divided into receptor tyrosine kinases and cellular tyrosine kinases. Receptor tyrosine kinases contain an extracellular ligand-binding domain, a transmembrane domain and an intracellular catalytic domain. Dimerization of two receptor tyrosine kinases upon ligand binding results in autophosphorylation of the tyrosine residues of either the intracellular domains of the receptor or of an accessory protein [1–3]. This leads to activation of the signal transduction cascade within the cell. On the other hand, non-receptor tyrosine kinases relay intracellular signals; they reside in the cytoplasm or in the nucleus rather than as transmembrane proteins and are typically activated by interaction with a protein that is not an extracellular ligand [4]. Tyrosine kinases catalyze the transfer of the γ phosphate group from adenosine triphosphate to the hydroxylated phenolic carbon of a tyrosine in target proteins.

Dysregulation of kinase activity has been reported in various diseases such as vascular disorders, neurological disorders, inflammatory diseases and cancer [5, 6]. This has generated an intense interest of researchers in the pursuit of protein kinases as drug targets. Because all these disorders are associated with receptor tyrosine kinase activation, they are considered to be the key targets for inhibitors. Therefore, researches have been focused on the design or identification of inhibitors that can affect a wide range of targeted kinases. Yaish et al. [7] originally coined the term "tyrosine phosphorylation inhibitor" in their initial description of compounds inhibiting the catalytic activity of the epidermal growth factor receptor (EGFR), but "tyrosine kinase inhibitor" (TKI) finds more common usage now. Due to the prevalent role that tyrosine kinases play in mitogenic signaling, TKIs are often useful as anti-cancer drugs. Several TKIs targeting various tyrosine kinases have been generated and proven to be effective as anti-tumor agents and anti-leukemic agents [8, 9]. Imatinib was developed against chronic myeloid leukemia (CML), and later gefitinib and erlotinib was generated aiming at the EGF receptor [10, 11]. The identification of sunitinib as an inhibitor of the receptors for FGF, PDGF and VEGF is also based on early studies on TKIs targeted at VEGF receptors [12]. About 30 % of the current pharmaceutical industry's effort is engaged in the development of protein kinase inhibitors, especially TKIs.

Most of the kinase-targeted drugs and their potential kinase targets have been investigated mainly for non-central nervous system (CNS) disorders. Kinase-targeted drugs have seen less utility for CNS disease indications in comparison to other diseases such as cancer. About half of the oncology pipelines target protein kinases; but few, if any, of CNS drugs in current usage are protein kinase inhibitors. In recent years, researchers have shown a greater interest in the development of kinase-targeted therapeutics for CNS indications [13–21]. Also, the established successes in other disease indications have encouraged the researchers to develop small-molecule kinase modulators for CNS disorders such as Alzheimer's and Parkinson's diseases. A growing number of studies have aimed to develop inhibitors of protein kinases as therapeutic strategies to regulate the pathology of neurological disorders. In this chapter, we will mostly focus on the role of tyrosine-kinases inhibitors in neurodegenerative diseases as a potential therapeutic agent.

2 Tyrosine Kinase Inhibitors and Parkinson's Disease

Parkinson's disease (PD) is the second most common movement disorder. It is characterized by the progressive loss of dopaminergic neurons due to neuronal apoptosis in the substantia nigra pars compacta and depletion of dopamine (DA) in the striatum, which lead to pathological and clinical abnormalities [22]. The neuropathology of PD includes the appearance of Lewy bodies, neuronal inclusions of protein aggregates comprising primarily α-synuclein. Indeed, mutations in the α-synuclein gene causes a small number of hereditary PD cases. Demographically, mutations in other genes make more significant contributions to familial PD; notable are parkin, an ubiquitin E3 ligase, and PINK1, a still enigmatic protein that probably participates in mitophagy and perhaps other aspects of autophagy. Nevertheless, the vast majorities of PD cases are sporadic and show risk assessments more likely to reflect environmental exposures and/or the risks associated with advanced age.

The tyrosine kinase c-Abl is involved in the regulation of several cellular processes that may be relevant to PD. c-Abl also plays a major role in the development of the CNS [23] by controling neurogenesis, neurite outgrowth and neuronal plasticity. It is a 120-kDa protein that belongs to the cytoplasmic tyrosine-kinase family. c-Abl possesses sequential SH3 and SH2 domains followed by a core catalytic domain with tyrosine-kinase activity [24, 25]. Moreover, c-Abl has been detected in the nucleus and has a unique myristoylated N-terminal region that negatively regulates its kinase activity [24]. Recently, several studies using various experimental model systems have shown that c-Abl is activated in neurodegenerative diseases such as PD [26, 27]. Researchers have found that (a) the c-Abl protein level is upregulated in the postmortem striatum of PD patients [28], and c-Abl phosphorylation at Y412 is also increased in the substantia nigra [26, 28] and striatum [26] of PD patients; (b) c-Abl phosphorylates parkin and impairs its E3 ligase activity, leading to loss of dopaminergic neurons in the substantia nigra [26, 27]; (c) the inhibition of c-Abl activity by imatinib/Gleevec [29], nilotinib/Tasigna [30] or bafetinib/INNO-406 [31] restores the level of DA and protects against the loss of dopaminergic neurons in the substantia nigra of WT mice [26, 28]. More recently, Hebron et al. [28] showed that c-Abl also regulates the clearance of α-synuclein (α-syn) (Fig. 1).

Issues related to PD are often investigated in animal models that utilize either *i)* manipulation of genes mutated in familial PD, *ii)* toxins specific for PD-affected brain regions, or *iii)* a combination of genetic and toxin-triggered stresses. The most extensively used toxin in such models is 1-methyl-4-phenyl-1,2,3,6-tetrahydropyridine (MPTP), which is metabolized by astrocytes to 1-methyl-4-phenylpyridinium (MPP^+); the latter has preferential toxicity for DA neurons and therefore compromises the function and survival of the substantia nigra. MPP^+ is also applied directly to cultured neurons to study neurotoxicity *in vitro*. Imatinib, a specific c-Abl inhibitor has been applied to these toxin models with promising results [26, 27]. MPTP and MPP^+ lead to impairment of parkin's E3 ligase activity and accumulation of parkin substrates, and these effects were mitigated by imatinib.

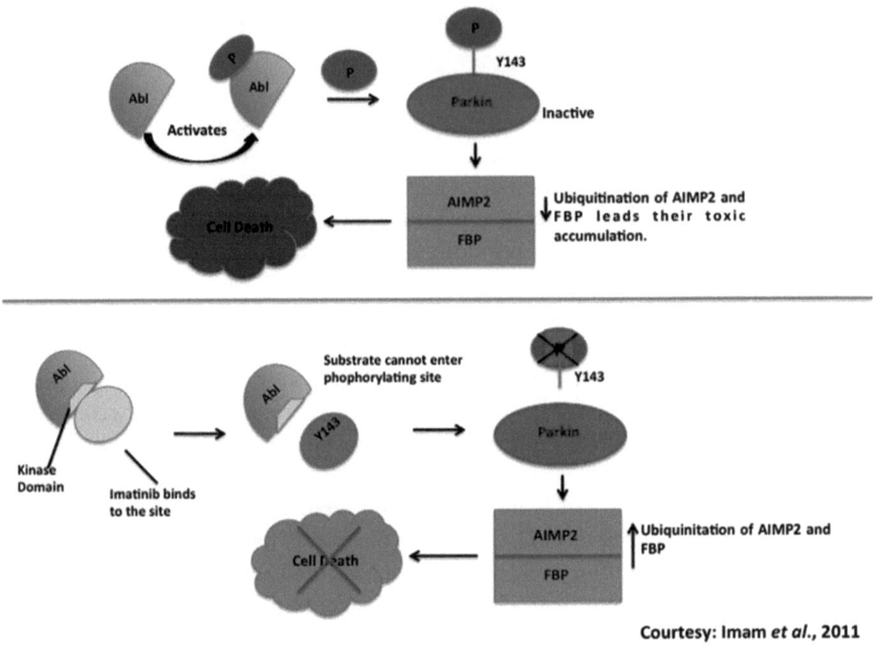

Fig. 1 Schematic representation of imatinib mode of action. During sporadic PD, activated c-Abl tyrosine phosphorylates parkin, resulting in loss of ubiquitin-ligase activity and leading to accumulation of toxic parkin substrates and neuronal death. Imatinib prevents parkin tyrosine phosphorylation by binding to c-Abl, restoring parkin's E3 ligase activity and cytoprotective function, thus protecting from cell death and PD

MPTP-induced DA neuronal loss was also reduced in c-Abl conditional-knockout mice. Consistent with the implicated role of parkin regulation, the protective effects of imatinib were lost in parkin-knockout mice [26].

PD-related progress has also been achieved with a pair of second-generation c-Abl tyrosine kinase inhibitors, compounds that are more selective and potent—with moderate brain penetration—in comparison to the other c-Abl inhibitors [32]. One of these newer drugs, nilotinib, inhibited α-synuclein accumulation in genetic animal models involving either viral induction of wild-type α-synuclein or transgenesis of α-synuclein bearing a familial PD mutation [28]. Nilotinib was also reported to protect against MPTP-induced loss of DA neurons, prevent the MPTP-induced reduction in DA levels and its metabolites, inhibit the MPTP-induced reduction in striatal DA terminal density and restore the behavioural deficits induced by MPTP [33]. INNO-406, an irreversible second-generation Abl-kinase inhibitor, has been reported capable of preventing the progression of DA neuronal damage in an MPTP-induced mouse model of PD [31]. That study showed that INNO-406 penetrates into the brain of mice PD and prevents MPTP-induced activation of c-Abl and tyrosine phosphorylation of parkin in a significant manner. In addition, INNO-406 prevents dopaminergic neuronal and terminal damage and preserves dopamine content in the MPTP-mouse model of PD very efficiently.

Another tyrosine kinase associated with the pathology of PD is Src. Src protein tyrosine kinases comprise a group of non-receptor membrane-associated tyrosine kinases which contribute to a vast range of physiological functions, including maintenance of cell homeostasis, intercellular contacts, regulation of cell shape, cell migration and manipulation of acute inflammatory responses-related signal transduction [34, 35]. The role of Src protein tyrosine kinases in inflammatory responses has been demonstrated convincingly [36]. This may present a therapeutic opportunity in PD and other neurodegenerative disease, most of which include inflammation-related events such as the production of superoxide by nicotinamide adenine dinucleotide phosphate (NADPH) oxidase in microglia [37]. FLZ is a novel synthetic derivative of squamosamide from a Chinese herb. Studies have shown that FLZ has strong protective effects in an MPTP-induced PD mouse model [38] and remarkably protects against apoptosis of dopaminergic neurons poisoned with 6-hydroxydopamine (6-OHDA) [39] or MPP^+ [40], indicating potent neuroprotective effects of FLZ in both *in vivo* and *in vitro* PD models. Tai et al. [41] investigated the neuroprotective and anti-inflammatory properties of FLZ in *in vivo* and *in vitro* PD models, focusing on the specific anti-inflammatory mechanism by which FLZ provided relief in an inflammation-induced animal PD model. In that study, FLZ significantly improved the movement disorders of rats, reduced dopaminergic neuronal loss and inhibited microglial over-activation in PD model rats. Mechanistic analysis indicated that those neuroprotective effects were at least partially due to inhibition of Src activity.

3 Tyrosine Kinase Inhibitors and Alzheimer's Disease

Alzheimer's disease (AD) is a neurodegenerative condition that is characterized clinically by mnemestic dementia and pathologically by β-amyloid deposits, neurofibrillary tangles (NFT) of the microtubule protein Tau, and cerebrocortical atrophy [42, 43]. The amyloid cascade hypothesis for AD states that brain accumulation of Aβ triggers a cascade of pathogenic events leading to alterations in the microtubule-associated protein Tau and neuronal death and dysfunction [42]. Recent autopsy reports from AD patients who underwent experimental Aβ immunization have shown a significant decrease in amyloid deposition in certain brain areas without meaningful amelioration of Tau pathology or loss of synapses and neurons [44, 45]. These observations may indicate that therapies aimed exclusively at Aβ will not work, at least in symptomatic stages of the disease process, and highlight the need for therapeutic strategies that tackle both amyloid and Tau alterations to successfully halt and/or reverse the pathology and cognitive decline in AD.

Recent studies have indicated a role for c-Abl in AD pathogenesis. Alvarez et al. [46] reported that upon treatment of primary neurons with amyloid-*β* peptide (A*β*) fibrils, c-Abl is activated and forms a protein complex with p73 prior to cell death. The study also revealed that pharmacologic inhibition of c-Abl prevents the neuronal cell death induced by A*β* [46]. Recent studies have also suggested a role for c-Abl in NFT formation by direct phosphorylation of Tau [47]. Moreover, c-Abl may induce Tau phosphorylation indirectly through activation of the kinase's downstream targets, cdk5 and GSK-3, which have been shown to phosphorylate Tau [48].

A role for c-Abl in AD pathogenesis is also implied by studies in which c-Abl activity or levels were diminished. Inhibition of c-Abl with imatinib reduced neuronal loss, cognitive impairments, and A*β* deposition and prevented tau phosphorylation in a transgenic mouse model of AD [49]. That study also showed that the Aβ-induced pathology *in vivo* is associated with changes in the levels of c-Abl, as well as levels and phosphorylation state of p73, a c-Abl substrate that is also a structural and functional homologue of the p53 tumour-suppressor protein. Cancino et al. [49] showed that imatinib protects against the toxicity of A*β* fibrils both at the behavioural and morphological levels. In addition, that study also demonstrated that an increase in a particular form of p73 induced by Aβ depends on c-Abl activity and is associated with the induction of apoptosis in neurons. These researchers provided evidence that the c-Abl/p73 signal is activated by Aβ deposits *in vivo* and suggested that this signaling pathway has a pathogenic role in AD. In another study, they demonstrated that imatinib reduces *tau* phosphorylation and phospho-Cdk5 levels in a transgenic mouse model of AD [50]. This finding supports a role for c-Abl in Cdk5 activation and *tau* phosphorylation induced by Aβ in AD mouse models. It may also participate in perturbation of epigenetic mechanisms contributing to AD. Evidence shows that histone deacetylase 2 (HDAC2) is increased in AD and that its activation worsens neuronal and synaptic function. A study conducted by Gonzalez-Zuňiga [51] demonstrated that c-Abl tyrosine phosphorylation of HDAC2 stabilizes the latter and enhances its activity. The study showed that c-Abl knockdown cells show a decrease in HDAC2 levels, while c-Abl overexpression increases them. Moreover, c-Abl inhibition with imatinib prevented the Aβ-induced increase in HDAC2 levels in a transgenic mouse model of AD (Fig. 2).

Recent studies have also indicated a role in AD for Fyn, a member of the Src family of non-receptor tyrosine kinases (SFKs). Earlier, Shirazi and Wood [52] reported that a subset of neurons from AD brain exhibited strong Fyn immunoreactivity compared with control brains and that these neurons were also positive for abnormally phosphorylated Tau protein. Though controversial, some evidence indicates that oligomeric aggregates of Aβ bind cellular prion protein (PrP_c), which then mediates AD-related cellular pathology. PrP_c is enriched in post-synaptic densities, and Aβ-PrP_c interaction leads to Fyn kinase activation; this was even demonstrated with soluble Aβ assemblies derived from human AD brain. In addition to its role in Aβ signaling, Fyn is involved in Tau phosphorylation. Fyn physically associates with Tau and can phosphorylate tyrosine residues, including Tyr18, near the amino terminus [53]. These findings implicate Fyn in the pathogenesis of AD, and its interaction with both Aβ and Tau renders Fyn a unique therapeutic target that addresses both of the major pathologic hallmarks of AD. Saracatinib (AZD0530), a small molecule tyrosine kinase inhibitor with high potency for Src and Fyn, has been identified as a potential therapeutic compound and is currently under clinical trials for the treatment of AD [54].

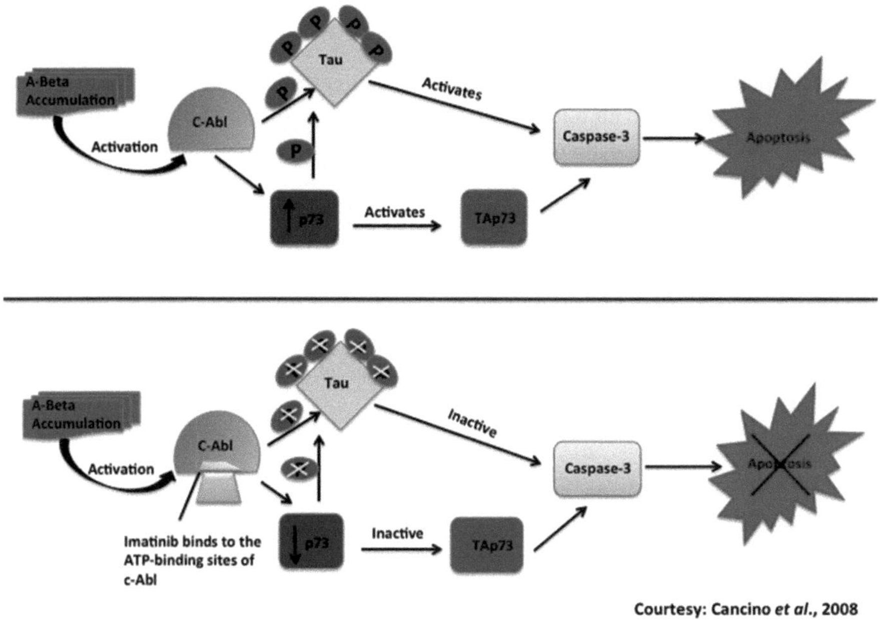

Fig. 2 Schematic representation of imatinib mode of action. Aβ accumulation induces p73 phosphorylation, increases levels of the latter's full-length isoform (TAp73) and thus enhances its pro-apoptotic function. Imatinib binds to the ATP- binding sites of c-Abl, reduces tau phosphorylation and produces a marked decline in phosphorylated p73 and prevents cell death and AD pathology

4 Tyrosine Kinase Inhibitors and Down Syndrome

Down syndrome (DS), the most common genetic cause of intellectual disability, is a manifestation of trisomy of Chromosome 21. MNB/DYRK1A (Minibrain/ dual specificity tyrosine phosphorylation-regulated kinase 1A) has possibly been the most extensively studied Chr21 gene during the last decade due to the remarkable correlation of its functions in the brain with important DS neuropathologies, such as neuronal deficits, dendrite atrophy, spine dysgenesis, precocious Alzheimer-like neurodegeneration, and cognitive deficits [55]. The flavonol epigallocatechin gallate (EGCG) is a potent and selective inhibitor of DYRK1A activity [56]. Studies have shown that prenatal EGCG treatment could partially rescue brain alterations in neonatal transgenic mice overexpressing Dyrk1A. It also normalizes the levels of some synaptic plasticity-related proteins in the hippocampus of adult Dyrk1A transgenic mice, suggesting possible cognitive effects [57]. It was recently reported that EGCG significantly restores cognitive function both in DS mouse models and in humans [58]. It also significantly reversed cognitive deficits in a pilot study in DS individuals with effects on memory recognition, working memory and quality of life.

5 Concluding Remarks

Since the prevalence, morbidity, and mortality of age-related neurological disorders increase dramatically with age, the ongoing expansion of the population of elderly signals continuing major increases in its impact. Current therapies control symptoms for only a limited time and fail completely to impact on progression of the disease. Neuronal cell death pathway mediated by activated tyrosine kinases leading to neuronal cell death can therapeutically be targeted by various tyrosine kinase inhibitors. Such inhibitor molecules, some of them already in clinic for other indications, can be repositioned for ameliorating cellular injury in age-related neurological disorders. Successful use of these inhibitors may open the way for a new therapeutic approach that would be broadly applicable to age-related neurological disorders.

References

1. Pawson T. Regulation and targets of receptor tyrosine kinases. Eur J Cancer. 2002;38(5):S3–10.
2. Heldin CH. Dimerization of cell surface receptors in signal transduction. Cell. 1995;80:213–23.
3. Arteaga CL. The epidermal growth factor receptor: from mutant oncogene in nonhuman cancers to therapeutic target in human neoplasia. J Clin Oncol. 2001;19:32S–40.
4. Schlessinger J. Cell signaling by receptor tyrosine kinases. Cell. 2000;103:211–25.
5. Catapano LA, Manji HK. Kinases as drug targets in the treatment of bipolar disorder. Drug Disc Today. 2008;13:295–302.
6. Zhang J, Yang PL, Gray NS. Targeting cancer with small molecule kinase inhibitors. Nature Rev Cancer. 2009;9:28–39.
7. Yaish P, Gazit A, Gilon C, Levitzki A. Blocking of EGF-dependent cell proliferation by EGF receptor kinase inhibitors. Science. 1988;242(4880):933–5.
8. Anafi M, Gazit A, Zehavi A, Ben-Neriah Y, Levitzki A. Tyrphostin-induced inhibition of p210bcr-abl tyrosine kinase activity induces K562 to differentiate. Blood. 1993;82(12):3524–9.
9. Meydan N, Grunberger T, Dadi H, Shahar M, Arpaia E, Lapidot Z, Leeder JS, Freedman M, Cohen A, Gazit A, Levitzki A, Roifman CM. Inhibition of acute lymphoblastic leukaemia by a Jak-2 inhibitor. Nature. 1996;379(6566):645–8.
10. Druker BJ, Tamura S, Buchdunger E, Ohno S, Segal GM, Fanning S, Zimmermann J, Lydon NB. Effects of a selective inhibitor of the Abl tyrosine kinase on the growth of Bcr-Abl positive cells. Nat Med. 1996;2(5):561–6.
11. Sordella R, Bell DW, Haber DA, Settleman J. Gefitinib-sensitizing EGFR mutations in lung cancer activate anti-apoptotic pathways. Science. 2004;305(5687):1163–7.
12. Strawn LM, McMahon G, App H, Schreck R, Kuchler WR, Longhi MP, Hui TH, Tang C, Levitzki A, Gazit A, Chen I, Keri G, Orfi L, Risau W, Flamme I, Ullrich A, Hirth KP, Shawver LK. Flk-1 as a target for tumor growth inhibition. Cancer Res. 1996;56(15):3540–5.
13. Borders AS, de Almeida L, Van Eldik LJ, Watterson DM. The p38α mitogen-activated protein kinase as a central nervous system drug discovery target. BMC Neurosci. 2008;9:S12.
14. Martinez A, Perez DI. GSK-3 inhibitors: a ray of hope for the treatment of Alzheimer's disease? J Alzheimer Dis. 2008;15:181–91.

15. Jope RS, Yuskaitis CJ, Beurel E. Glycogen synthase kinase-3 (GSK3): inflammation, diseases, and therapeutics. Neurochem Res. 2007;32:577–95.
16. Forde JE, Dale TC. Glycogen synthase kinase 3: a key regulator of cellular fate. Cell Mol Life Sci. 2007;64:1930–44.
17. Kockeritz L, Doble B, Patel S, Woodgett JR. Glycogen synthase kinase-3—an overview of an over-achieving protein kinase. Curr Drug Targets. 2006;7:1377–88.
18. Kubo T, Yamaguchi A, Iwata N, Yamashita T. The therapeutic effects of Rho-ROCK inhibitors on CNS disorders. Ther. Clin. Risk Manag. 2008;4:605–15.
19. Martinez A. Preclinical efficacy on GSK-3 inhibitors: towards a future generation of powerful drugs. Med Res Rev. 2008;28:773–96.
20. Mueller BK, Mack H, Teusch N. Rho kinase, a promising drug target for neurological disorders. Nat Rev Drug Discov. 2005;4:387–98.
21. Schumacher AM, Velentza AV, Watterson DM. Death associated protein kinase as a potential therapeutic target. Exp Opin Ther Targets. 2002;6:497–506.
22. Cookson MR. The biochemistry of Parkinson's disease. Annu Rev Biochem. 2005;74:29–52.
23. Moresco EM, Koleske AJ. Regulation of neuronal morphogenesis and synaptic function by Abl family kinases. Curr Opin Neurobiol. 2003;13:535–44.
24. Hantschel O, Superti-Furga G. Regulation of the c-Abl and Bcr-Abl tyrosine kinases. Nat Rev Mol Cell Biol. 2004;5:33–44.
25. Sirvent A, Benistant C, Roche S. Oncogenic signaling by tyrosine kinases of the SRC family in advanced colorectal cancer. Am J Cancer Res. 2012;2:357–71.
26. Ko HS, Lee Y, Shin JH, Karuppagounder SS, Gadad BS, Koleske AJ, Pletnikova O, Troncoso JC, Dawson VL, Dawson TM. Phosphorylation by the c-Abl protein tyrosine kinase inhibits Parkin's ubiquitination and protective function. Proc Natl Acad Sci U S A. 2010;107:16691–6.
27. Imam SZ, Zhou Q, Yamamoto A, Valente AJ, Ali SF, Bains M, Roberts JL, Kahle PJ, Clark RA, Li S. Novel regulation of parkin function through c-Abl-mediated tyrosine phosphorylation: implications for Parkinson's disease. J Neurosci. 2011;31:157–63.
28. Hebron ML, Lonskaya I, Moussa CE. Nilotinib reverses loss of dopamine neurons and improves motor behavior via autophagic degradation of alpha-synuclein in Parkinson's disease models. Hum Mol Genet. 2013;22:3315–28.
29. Lin YL, Meng Y, Jiang W, Roux B. Explaining why Gleevec is a specific and potent inhibitor of Abl kinase. Proc Natl Acad Sci U S A. 2013;110:1664–9.
30. Deremer DL, Ustun C, Natarajan K. Nilotinib: a second-generation tyrosine kinase inhibitor for the treatment of chronic myelogenous leukemia. Clin Therap. 2008;30:1956–75.
31. Imam SZ, Trickler W, Kimura S, Binienda ZK, Paule MG, Slikker Jr W, Li S, Clark RA, Ali SF. Neuroprotective efficacy of a new brain-penetrating C-Abl inhibitor in a murine Parkinson's disease model. PLoS One. 2013;8, e65129.
32. Soverini S, Martinelli G, Rosti G, Iacobucci I, Baccarani M. Advances in treatment of chronic myeloid leukemia with tyrosine kinase inhibitors: the evolving role of Bcr-Abl mutations and mutational analysis. Pharmacogenomics. 2012;13:1271–84.
33. Karuppagounder SS, Brahmachari S, Lee Y, Dawson VL, Dawson TM, Ko HS. The c-Abl inhibitor, nilotinib, protects dopaminergic neurons in a preclinical animal model of Parkinson's disease. Sci Rep. 2014;4:4874.
34. Lowell CA. Src-family kinases: rheostats of immune cell signaling. Mol Immunol. 2004;41:631–43.
35. Thomas SM, Brugge JS. Cellular functions regulated by Src family kinases. Annu Rev Cell Dev Biol. 1997;13:513–609.
36. Okutani D, Lodyga M, Han B, Liu M. Src protein tyrosine kinase family and acute inflammatory responses. Am J Physiol Lung Cell Mol Physiol. 2006;291:L129–41.
37. Hunot S, Boissiere F, Faucheux B, Brugg B, Mouatt-Prigent A, Agid Y, Hirsch EC. Nitric oxide synthase and neuronal vulnerability in Parkinson's disease. Neuroscience. 1996;72:355–63.

38. Feng W, Wei H, Liu GT. Pharmacological study of the novel compound FLZ against experimental Parkinson's models and its active mechanism Mol. Neurobiol. 2005;31:295–300.
39. Zhang D, Zhang JJ, Liu GT. The novel squamosamide derivative FLZ protects against 6-hydroxydopamine-induced apoptosis through inhibition of related signal transduction in SH-SY5Y cells. Eur J Pharmacol. 2007;561:1–6.
40. Kong XC, Zhang D, Qian C, Liu GT, Bao XQ. FLZ, a novel HSP27 and HSP70 inducer, protects SH-SY5Y cells from apoptosis caused by MPP(+). Brain Res. 2011;1383:99–107.
41. Tai W, Ye X, Bao X, Zhao B, Wang X, Zhang D. Inhibition of Src tyrosine kinase activity by squamosamide derivative FLZ attenuates neuroinflammation in both *in vivo* and *in vitro* Parkinson's disease models. Neuropharmacology. 2013;75:201–12.
42. Hardy JA, Higgins GA. Alzheimer's disease: the amyloid cascade hypothesis. Science. 1992;256:184–5.
43. Iqbal K, Alonso AC, Chen S, Chohan MO, El-Akkad E, Gong CX, Khatoon S, Li B, Liu F, Rahman A, Tanimukai H, Grundke-Iqbal I. Tau pathology in Alzheimer's disease and other tauopathies. Biochim Biophys Acta. 2005;1739:198–210.
44. Nicoll JA, Wilkinson D, Holmes C, Steart P, Markham H, Weller RO. Neuropathology of human Alzheimer disease after immunization with amyloid- beta peptide: a case report. Nat Med. 2003;9:448–52.
45. Ferrer I, Boada-Rovira M, Sanchez-Guerra ML, Rey MJ, Costa-Jussa F. Neuropathology and pathogenesis of encephalitis following amyloid-beta immunization in Alzheimer's disease. Brain Pathol. 2004;14:11–20.
46. Alvarez AR, Sandoval PC, Leal NR, Castro PU, Kosik KS. Activation of the neuronal c-Abl tyrosine kinase by amyloid-beta-peptide and reactive oxygen species. Neurobiol Dis. 2004;17:326–36.
47. Derkinderen P, Scales TM, Hanger DP, Leung KY, Byers HL, Ward MA, Lenz C, Price C, Bird IN, Perera T, Kellie S, Williamson R, Noble W, Van Etten RA, Leroy K, Brion JP, Reynolds CH, Anderton BH. Tyrosine 394 is phosphorylated in Alzheimer's paired helical filament tau and in fetal tau with c-Abl as the candidate tyrosine kinase. J Neurosci. 2005;25:6584–93.
48. Klafki HW, Staufenbiel M, Kornhuber J, Wiltfang J. Therapeutic approaches to Alzheimer's disease. Brain. 2006;129:2840–55.
49. Cancino GI, Toledo EM, Leal NR, Hernandez DE, Yévenes LF, Inestrosa NC, Alvarez AR. STI571 prevents apoptosis, tau phosphorylation and behavioural impairments induced by Alzheimer's beta-amyloid deposits. Brain. 2008;131:2425–42.
50. Cancino GI, Arce KP, Castro PU, Toledo EM, von Bernhardi R, Alvarez AR. c-Abl tyrosine kinase modulates tau pathology and Cdk5 phosphorylation in AD transgenic mice. Neurobiol. Aging. 2011;32:1249–61.
51. Gonzalez-Zuñiga M, Contreras PS, Estrada LD, Chamorro D, Villagra A, Zanlungo S, Seto E, Alvarez AR. c-Abl Stabilizes HDAC2 Levels by Tyrosine Phosphorylation Repressing Neuronal Gene Expression in Alzheimer's Disease. Mol Cell. 2014;56(1):163–73.
52. Shirazi SK, Wood JG. The protein tyrosine kinase, fyn, in Alzheimer's disease pathology. Neuroreport. 1993;4:435–7.
53. Lee G, Thangavel R, Sharma VM, Litersky JM, Bhaskar K, Fang SM, Do LH, Andreadis A, Van Hoesen G, Ksiezak-Reding H. Phosphorylation of tau by fyn: implications for Alzheimer's disease. J Neurosci. 2004;24:2304–12.
54. Nygaard HB, van Dyck CH, Strittmatter SM. Fyn kinase inhibition as a novel therapy for Alzheimer's disease. Alzheimers Res Ther. 2014;6(1):8. doi:10.1186/alzrt238.
55. Park J, Song WJ, Chung KC. Function and regulation of Dyrk1A: towards understanding Down syndrome. Cell Mol Life Sci. 2009;66(20):3235–40.
56. Bain J, McLauchlan H, Elliott M, Cohen P. The specificities of protein kinase inhibitors: an update. Biochem J. 2003;371:199–204.
57. Guedj F, Sebrie C, Rivals I, Ledru A, Paly E, Bizot JC, Smith D, Rubin E, Gillet B, Arbones M, Delabar JM. Green tea polyphenols rescue of brain defects induced by overexpression of DYRK1A. PLoS One. 2009;4, e4606.

58. De la Torre R, De Sola S, Pons M, Duchon A, de Lagran MM, Farré M, Fitó M, Benejam B, Langohr K, Rodriguez J, Pujadas M, Bizot JC, Cuenca A, Janel N, Catuara S, Covas MI, Blehaut H, Herault Y, Delabar JM, Dierssen M. Epigallocatechin-3-gallate, a DYRK1A inhibitor, rescues cognitive deficits in Down syndrome mouse models and in humans. Mol Nutr Food Res. 2014;58(2):278–88.

Part II
Role of Inflammatory and Oxidative Processes in Age Related Diseases

Oxidative and Inflammatory Pathways in Age-Related Chronic Disease Processes

Arezoo Campbell and Parrisa Solaimani

1 Introduction

Aging during the lifespan of mammalian organisms is known as senescence. Due to the advent of antibiotics, therapeutic interventions, and other healthcare improvements, the lifespan of humans has substantially improved (Fig. 1). While advances in science and technology have increased longevity, these improvements have also had negative consequences. An increased prevalence of age-related disorders poses a significant financial burden on society. According to the American Heart Association, the 2011 estimated annual cost of cardiovascular disease and stroke amounted to approximately $320.1 billion [2], while the 2015 cost of Alzheimer's disease and other dementias are estimated at $226 billion [3].

One major contributor to the underlying cause of age-related chronic disorders is a complex set of pathophysiological abnormalities referred to as the metabolic syndrome. It is characterized by abdominal obesity, elevated triglycerides, high blood pressure, elevated fasting glucose, and low levels of the good cholesterol, high density lipoprotein [4]. In today's society in particular, a sedentary life-style and high calorie diet has led to a higher prevalence of this syndrome. Two factors which are associated with the metabolic syndrome are a proinflammatory state [5] and an increase in oxidative stress [6]; both of which have been associated with the aging process and are the subject of this chapter.

A. Campbell, Ph.D. (✉) • P. Solaimani, Ph.D.
Department of Pharmaceutical Sciences, College of Pharmacy, Western University of Health Sciences, 309 E. Second Street, Pomona, CA 91766-1854, USA
e-mail: acampbell@westernu.edu

S.C. Bondy, A. Campbell (eds.), *Inflammation, Aging, and Oxidative Stress*, Oxidative Stress in Applied Basic Research and Clinical Practice,
DOI 10.1007/978-3-319-33486-8_6

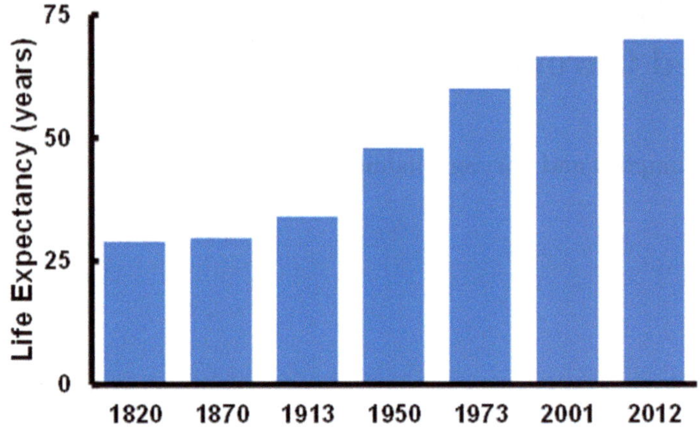

Fig. 1 Global average life expectancy between the years of 1820–2012. Data from Max Roser [1]

2 Oxidative Stress and Aging

Senescence is accompanied by an increase in oxidative stress which results in modifications to macromolecules such as lipids (lipid peroxidation), proteins (protein oxidation), and DNA [7]. In 1956, Dr. Denham Harman proposed the 'free radical theory of aging' in which he indicated that the accumulation of damage by oxidative mediators throughout the lifespan of an organism brings about the aging phenomenon [8]. Free radical refers to a molecule with an unpaired valence electron in its atomic orbital which makes it unstable and chemically reactive. Examples of reactive oxygen species (ROS) include the hydroxyl radical ($OH^{\bullet}$) and superoxide (O_2^-) [9]. Although not a free radical, hydrogen peroxide (H_2O_2) is also considered a reactive oxygen species because it readily forms hydroxyl free radicals by the Fenton reaction [10]. Free radicals are naturally detoxified by endogenous antioxidants such as glutathione (GSH), catalase, and superoxide dismutase (SOD).

The free radical theory of aging was tested as part of a ZENITH study sponsored by the European Commission on 'Quality of Life and Management of Living Resources'. Oxidative stress markers were evaluated in late-middle-aged (55–70 years) compared to aged (70–85 years) free-living individuals. A decrease in plasma antioxidant levels was observed in the aged population, which may be a result of heightened free radical formation sequestering the antioxidants. However, lipid peroxidation and protein oxidation were also decreased in the older population indicating less oxidative damage had occurred in this group [11]. The fact that the older group had survived and were 'free-living' (did not need institutionalization), may have selected for a healthier subpopulation of aged individuals. Furthermore, the proximity of the age range between the two groups may have masked overt senescence-associated changes in oxidative markers.

In a cross-sectional study a wider age-gap was compared. When healthy young (25–29 years) individuals were compared to healthy old (>70 years) subjects, a marker for oxidative damage to lipids was shown to be increased in the healthy old population. These results contrasted what was observed in the ZENITH study. However, similar to the ZENITH study, the total antioxidant status was diminished in the older group [12]. The question from these studies arose as to whether the increase in ROS is contributing to aging or is merely a consequence of the aging process [13]. The answer may not be so straightforward. Not only do free radicals and their byproducts contribute to tissue damage, but impaired tissue can also generate oxidative intermediates. Thus senescence-associated tissue damage may lead to a positive feedback loop that augments free radical formation [14–17].

2.1 *Contribution of Mitochondrion to Oxidative Stress*

Mitochondria provide chemical energy for the cell via oxidative phosphorylation (OXPHOS). Within the inner membrane of the mitochondria there is an electron transport chain which consists of four complexes; three of the complexes (I, III, & IV) pump protons into the intermembrane space to form a gradient. This proton gradient allows conversion of ADP to ATP, the cells chemical energy unit, by the enzyme ATP synthase [18]. An endogenous by-product of the electron transport chain is the production of H_2O_2 (hydrogen peroxide), which may mediate the activation of a number of kinases involved in other mitochondrial processes [19]. These processes include cell proliferation, differentiation, apoptosis, and adaption to stress [20–22]. Thus, even though oxidative damage occurs as a result of excessive free radical production, the formation of reactive oxygen species is a consequence of normal mitochondrion function and aerobic respiration.

Although hydrogen peroxide can generate the highly reactive hydroxyl radical, the molecule regulates receptor tyrosine kinases and as such has an important function in cell signaling cascades [19]. As an example, H_2O_2 is an insulinomimetic agent and this activity was associated with enhanced phosphorylation of protein tyrosine [23]. Hydrogen peroxide also plays an important role in the regulation of the stress-related mitogen activated protein (MAP) kinase pathway [19]. The activation of MAP kinases leads to the translocation of the transcription factor nuclear factor kB (NF-kB). This transcription factor not only mediates immune responses, but also functions in cellular survival [24].

ROS also plays an important physiological role in the innate immune response. The respiratory burst (also referred to as the oxidative burst) is an important mechanism by which immune competent cells, such as neutrophils and macrophages, destroy pathogens [25]. The respiratory burst relies on the enzyme NADPH oxidase (NOX) which reduces the oxygen molecule to yield superoxide (O_2^-):

$$\mathrm{NADPH} + 2\mathrm{O}_2 \rightarrow \mathrm{NADP}^+ + 2\mathrm{O}_2^{\,-} + \mathrm{H}^+$$

The superoxide can then be converted to hydrogen peroxide by the function of superoxide dismutase:

$$2H^{+} + 2O_2 \rightarrow H_2O_2 + O_2$$

Although superoxide and hydrogen peroxide may not have direct antimicrobial action, they do give rise to other more potent oxygen metabolites that do have antimicrobial function which play an important role in the respiratory burst [26].

Normally, excess levels of potentially damaging ROS are detoxified by endogenous sources of antioxidants. Exogenous sources of antioxidants, such as vitamin E, vitamin C and beta-carotene, can also detoxify oxidative mediators. However, as noted above, reactive oxygen species have normal biological roles and as such their modulation should be carefully determined. For example, in clinical studies supplementation with exogenous antioxidants has not always shown to be protective against heart disease or cancer [27–31]. In some cases, antioxidant supplementation has actually been connected to adverse outcomes. For instance, vitamin E supplementation was associated with an increased risk for heart failure [30]. Furthermore, fatal coronary heart disease was enhanced in men who received beta carotene or a combination of beta carotene and vitamin E [32]. In another study, it was observed that supplementation with beta carotene may actually increase lung cancer in cigarette smokers [33]. Thus, nonspecific antioxidant therapy may not necessarily be protective against chronic diseases which have been linked to heightened oxidative stress. More research is needed to understand how to achieve the necessary balance between minimizing the damaging effects of ROS without compromising the biological roles of these molecules.

3 Inflammation in Aging

The immune system consists of an interconnected system of cellular and molecular factors that provide a means of protecting an organism against both exogenous and endogenous stressors. The first line of defense is the innate or natural immunity which involves the inflammatory response. The function of inflammation is to recruit immune competent cells to an area that has been exposed to potentially damaging substances or pathogens. Recognition of foreign molecules (antigens) initiates the innate immune response. This allows the recruitment of immune-competent cells which can then eliminate the foreign antigen. The key participants in the innate inflammatory response are a plethora of cytokines and chemokines which work together to orchestrate a response appropriate for the type of danger stimuli encountered [25]. The main proinflammatory cytokines, Interleukin-6 (IL-6), tumor necrosis factor alpha (TNF-α), and interleukin-1 (IL-1) have been shown to be elevated with aging [34–38].

The innate inflammatory response is necessary to protect against potentially pathogenic conditions. It is of short duration and once the stress factor is eliminated, or the more specific adaptive immune response has been triggered, the process is stopped. Transient inflammation is natural and protects an organism against a variety of exogenous and endogenous stressors [25]. However, it appears that as an organism ages, the capability to heal and fix damaged tissue declines. This may be due to the decrease in tissue regeneration capability of the organism which has been associated with a decline in the adult stem cell niche [39]. The lack of the ability to replace damaged tissue may also be partially due to the inability of an aging immune system to properly coordinate a protective response. This process is referred to as Immunosenescence.

Immunosenescence is characterized by a number of changes. The adaptive immune response is characterized by the functions of B and T cells [25]. T cell development occurs in the thymus. As an individual ages, the thymus gland shrinks (thymic involution) which then leads to production of fewer T cells [40]. Amongst the affected cell population are regulatory T cells which play an important role in controlling immune responses. It has been shown that the suppressive function of T cells decline as an organism ages [41]. Changes in T cell population also underlie the inability to properly resist infections and decreased response to immunizations [42, 43].

The decline in the adaptive immune response in the elderly may shift the response more towards innate immunity [44]. The result is an enhanced baseline inflammation which has been coined "inflammaging" [45]. This increase in proinflammatory cytokines in the elderly is thought to be involved in chronic age-associated diseases [36]. The overall tendency for the lack of appropriate immune response may predispose the elderly to heightened damage.

3.1 Association Between Oxidative Stress and Inflammatory Pathways

An increase in oxidative stress causes damage to macromolecules which, if not fixed, can cause cellular damage. Thus, it is not surprising that heightened oxidative stress, if sufficient to cause damage, may elicit an inflammatory response. Not only does free radical mediated damage have the capability to initiate an inflammatory response, but as described above, phagocyte associated respiratory burst is an important oxidative mechanism used by the innate immune response to clear pathogenic stimuli. Thus oxidative stress and inflammatory responses are intricately interconnected and both processes have the ability to activate and propagate each other.

4 Oxidative Stress and Inflammation in Age-Related Chronic Diseases

As described above, mitochondria are an important source of ROS. Mitochondria are unique organelles within the cell because they contain their own specific DNA which is maternally inherited [46]. The genetics of cellular bioenergetics is complex because it utilizes not only the maternally inherited mitochondrial DNA (mtDNA), but also the biparentally inherited nuclear DNA (nDNA) [47]. Compared to nDNA, mtDNA has a high mutation rate that accumulates over time and disrupts the ability of mitochondria to function efficiently, which may lead to an enhanced ROS production [48, 49]. It has been proposed that the maternal ovarian system has a method of obliterating the gametes with the most deleterious mtDNA mutations. The mutations that are not dangerous to the organism are allowed to pass to the next generation. In this manner, non-lethal, potentially beneficial mutations are introduced into the general population. It has been hypothesized that these mutations allow for physiological changes appropriate for the survival of an organism in a constantly changing environment [50].

Although these mutations may allow an organism to adapt to its ever changing environment, there may also be negative consequences. For example, predisposition to age-related disease may be a consequence of inappropriate changes in the mtDNA. This may occur more often in today's society where access to rapid transportation has allowed humans to migrate to new environments that do not favorably select for the same mtDNA and epigenomic factors which were beneficial in the previous surrounding [51]. As the organism ages, there is an increase in the number of mtDNA mutations which tend to accumulate over time and disrupt the ability of cellular mitochondria to function efficiently, leading to enhanced ROS formation and oxidative stress [50]. The tissue with the greatest demand for energy production is most vulnerable. These include the heart and the brain, predisposing an organism to cardiovascular and neurodegenerative diseases respectively.

4.1 Cardiovascular Diseases

According to the American Heart Association, cardiovascular disease is a complex set of abnormalities that results mainly from an accumulation of "plaque" on the arterial walls. This process is referred to as atherosclerosis. An increase in circulating levels of low density lipoprotein (LDL) cholesterol initiates the response. Recruitment of many cell types to the arterial wall, including monocytes, macrophages, lymphocytes, smooth muscle cells, and endothelial cells, leads to an inflammatory response that contributes to the formation of the atherosclerotic plaque [52]. The plaque reduces the diameter of the vessels and impedes blood flow. A fibrous cap stabilizes the plaque, but if it ruptures, it can form a thrombosis (blood clot). If the clot occludes heart vessels, this leads to a heart attack. If the arteries in the brain are blocked, the result is an ischemic stroke. The lack of oxygen to the area will result in cell death and tissue damage.

Other than age and gender, underlying conditions such as high blood pressure, dyslipidemia, and diabetes mellitus predispose individuals to cardiovascular disease. Lifestyle risk factors include a history of cigarette smoking, obesity, sedentary life style, and a high fat diet. These risk factors influence inflammatory/oxidative stress pathways in a manner that contributes to disease pathogenesis [53]. Obese female mice show accelerated aging and a higher mortality rate that correlates with an increase in markers of oxidative stress and vascular abnormalities [54]. In aged mice, cellular senescence has been associated with vascular dysfunction. Antioxidants were shown to be protective against vascular dysfunction, which suggests that oxidative stress mediates the endothelial impairment that characterizes cardiovascular disease [55]. Evidence for the correlation between oxidative stress and cardiovascular abnormalities is further demonstrated by the finding that antioxidant status is significantly diminished in hypertensive patients, especially if they are also diabetic [56].

There is a strong proinflammatory and pro-thrombotic state in patients with cardiovascular disorders [5]. This is not surprising considering the pathophysiology of atherosclerosis. LDL is retained in vascular walls by binding to artery wall proteoglycans [57]. Accumulation and modification of LDL to oxidized LDL (ox-LDL) leads to endothelial activation and expression of vascular cell adhesion molecules which allow migration of immune cells into the vascular wall endothelium. Monocytes differentiate into macrophages that engulf ox-LDL and transform into foam cells. T cells are also recruited and differentiate into proinflammatory cytokine producing T helper 1 (TH1) cells. The self-propagating inflammatory response and factors released by immune cells eventually weaken the plaque, leading to rupture and thrombosis [58]. As previously discussed, the immune response is crucial to the survival of organisms. Although large scale clinical trials are in place to determine the protective effect of anti-inflammatory therapy in cardiovascular disease, the potential risk versus benefits need to be carefully determined [59].

4.2 Neurodegenerative Diseases

The brain is an organ which is highly protected by the presence of the blood brain barrier (BBB). As such, historically it was thought that this organ was immunologically privileged. However, many studies provide clear evidence that the brain is capable of mounting an immune response and defending itself against foreign antigens and cellular debris resulting from normal age-related deterioration [60]. Microglial cells are the main immune-competent cells in the brain but both astrocytes and oligodendrocytes have also been shown to be important participants in protecting the brain by mounting a proinflammatory response. Although the immune response in the brain is protective, chronic inflammation can be particularly hazardous to the brain as a consequence of BBB dysfunction.

In neurodegenerative diseases, such as Alzhiemer's disease (AD), chronic neuroinflammation induced by cytokines has been recognized as one of the major mechanisms of disease pathology. This neuroinflammation, which is primarily driven by the brain's microglia, escalates with disease progression [61, 62]. The most potent inflammatory cytokines are Interleukin-1-alpha (IL-1α), interleukin-1-beta (IL-1β), and interleukin-6 (IL-6). These cytokines are the products of activated microglia and astrocytes and are up-regulated in AD brains [63]. Inflammatory cytokines such as these are involved in loosening tight junctions which leads to a loss of BBB integrity [64].

Mitochondrial dysfunction and oxidative stress are commonly observed in patients with neurodegenerative diseases, specifically with defects in mitochondrial oxidative phosphorylation (OXPHOS) in complex IV leading to reduced membrane potential, reduced ATP production, and increased ROS production [65]. In addition, patients with AD have been shown to have an accumulation of specific mutations in the mitochondrial DNA which leads to a decline in mitochondrial copy number and function. This decline in mitochondrial function is also correlated with an accumulation of the disease marker amyloid beta Aβ peptide. Elevated Aβ levels facilitate the entry of the peptide into the mitochondrion where it inhibits specific enzymes during OXPHOS [66–70]. This results in an increase in ROS production which can then lead to extensive oxidative damage that leaves neurons vulnerable to cell death [71]. Enhancement in both oxidative and inflammatory pathways appears to contribute to the pathogenesis of neurodegenerative diseases.

5 Summary and Conclusions

The formation of free radicals is a natural process which forms redox active molecules that normally function in different cellular pathways, such as the response to stress. However, over time, the accrual of environmental insults and genetic susceptibility factors may not only exhaust antioxidant capacity, but also lead to abnormal mitochondrial bioenergetics. The subsequent enhancement in free radical formation may lead to age-related accumulation of cell damage. This in turn may trigger an innate immune response that initially promotes cell survival and protection. However, if free radical-mediated damage cannot be adequately repaired, the initial innate immune response may not be resolved and this may lead to a chronic inflammatory state. Chronic inflammation has been associated with age-related chronic disorders.

Mitochondrial dysfunction has been implicated in a number of age related disorders including neurodegenerative and cardiovascular diseases [18]. While many factors appear to contribute to pathogenesis of these age-related diseases, markers of oxidative stress and inflammation are consistently shown to be upregulated. Whether these two events contribute to, or are a consequence, of the aging phenomenon is not well understood. In this review we have discussed the interrelationship of ROS formation and inflammation. Although both events are heightened in aging, inflammation may be a protective mechanism and excess ROS formation just a natural consequence of the aging process that leads to changes in cellular bioenergetics. There is a fine balance between the potential of inflammatory and oxidative events to be either protective or deleterious. A combination of genetic predisposition, high calorie diet, sedentary lifestyle, and environmental insults can tilt the scale to deleterious pathways while exercise and a diet rich in antioxidants can shift the balance to more protective avenues.

References

1. Max Roser (2015). 'Life Expectancy'. OurWorldInData.org. http://ourworldindata.org/data/population-growth-vital-statistics/life-expectancy/.
2. Mozaffarian D, Benjamin EJ, Go AS, Arnett DK, Blaha MJ, Cushman M, de Ferranti S, Després JP, Fullerton HJ, Howard VJ, Huffman MD, Judd SE, Kissela BM, Lackland DT, Lichtman JH, Lisabeth LD, Liu S, Mackey RH, Matchar DB, McGuire DK, Mohler 3rd ER, Moy CS, Muntner P, Mussolino ME, Nasir K, Neumar RW, Nichol G, Palaniappan L, Pandey DK, Reeves MJ, Rodriguez CJ, Sorlie PD, Stein J, Towfighi A, Turan TN, Virani SS, Willey JZ, Woo D, Yeh RW, Turner MB, American Heart Association Statistics Committee and Stroke Statistics Subcommittee. Heart disease and stroke statistics—2015 update: a report from the American Heart Association. Circulation. 2015;131:229–322.
3. Alzheimer's Association. Alzheimer's disease facts and figures. Alzheimer's Dementia. 2015;11(3):332.
4. Grundy SM, Cleeman JI, Daniels SR, Donato KA, Eckel RH, Franklin BA, Gordon DJ, Krauss RM, Savage PJ, Smith Jr SC, Spertus JA, Costa F. Diagnosis and management of the metabolic syndrome: an American Heart Association/National Heart, Lung, and Blood Institute scientific statement: Executive Summary. Crit Pathw Cardiol. 2005;4:198–203.
5. Grundy SM, Brewer BH, Cleeman JI, Smith SC, Lenfant C. Definition of Metabolic Syndrome: Report of the National Heart, Lung, and Blood Institute/American Heart Association Conference on Scientific Issues Related to Definition. Circulation. 2004;109:433–8.
6. Bonomini F, Rodella LF, Rezzani R. Metabolic syndrome, aging and involvement of oxidative stress. Aging Dis. 2015;6:109–20.
7. Bellomo G. Cell damage by oxygen free radicals. Cytotechnology. 1991;5:71–3.
8. Harman D. Aging: a theory based on free radical and radiation chemistry. J Gerontol. 1956;11:298–300.
9. Lobo V, Patil A, Phatak A, Chandra N. Free radicals, antioxidants, and functional foods: Impact on human health. Pharmacogn Rev. 2010;4:118–26.
10. Floyd RA, Lewis CA. Hydroxyl free radical formation from hydrogen peroxide by ferrous iron-nucleotide complexes. Biochemistry. 1983;22:2645–9.
11. Andriollo-Sanchez M, Hininger-Favier I, Meunier N, Venneria E, O'Connor JM, Maiani G, Coudray C, Roussel AM. Age-related oxidative stress and antioxidant parameters in middle-aged and older European subjects: the ZENITH study. Eur J Clin Nutr. 2005;59 Suppl 2:S58–62.
12. Mendoza-Núñez VM, Ruiz-Ramos M, Sánchez-Rodríguez MA, Retana-Ugalde R, Muñoz-Sánchez JL. Aging-related oxidative stress in healthy humans. Tohoku J Exp Med. 2007;213:261–8.
13. Fukagawa NK. Aging: is oxidative stress a marker or is it causal? Proc Soc Exp Biol Med. 1999;222:293–8.
14. Awasthi YC, Sharma R, Sharma A, Yadav S, Singhal SS, Chaudhary P, Awathi S. Self-regulatory role of 4-hydroxynonenal in signaling for stress-induced programmed cell death. Free Radic Biol Med. 2008;45:111–8.
15. Balaban RS, Nemoto S, Finkel T. Mitochondria, oxidants, and aging. Cell. 2005;120:483–95.
16. Cadenas E, Davies KJ. Mitochondrial free radical generation, oxidative stress, and aging. Free Radic Biol Med. 2000;29:222–30.
17. Lee J, Giordano S, Zhang J. Autophagy, mitochondria, and oxidative stress: cross-talk and redox signaling. Biochem J. 2012;441:523–40.
18. Nunnari J, Suomalainen A. Mitochondria: in sickness and in health. Cell. 2012;148:1145–59.
19. Truong TH, Carroll KS. Redox regulation of protein kinases. Crit Rev Biochem Mol Biol. 2013;48:332–56.
20. Green DR, Reed JC. Mitochondria and apoptosis. Science. 1998;281:1309–12.
21. Hamanaka RB, Chandel NS. Mitochondrial reactive oxygen species regulate cellular signaling and dictate biological outcomes. Trends Biochem Sci. 2010;35:505–13.

22. Mitra K. Mitochondrial fission-fusion as an emerging key regulator of cell proliferation and differentiation. Bioessays. 2013;35:955–64.
23. Heffetz D, Bushkin I, Dror R, Zick Y. The insulinomimetic agents H2O2 and vanadate stimulate protein tyrosine phosphorylation in intact cells. J Biol Chem. 1990;265:2896–902.
24. Hatada EN, Krappman D, Scheidereit C. NF-kB and the innate immune response. Curr Opin Immunol. 2000;12:52–8.
25. Parham P. The immune system. 4th ed. New York: Garland Sciences; 2014.
26. Dahlgren C, Karlsson A. Respiratory burst in human neutrophils. J Immunological Meth. 1999;232:3–14.
27. Cook NR, Albert CM, Gaziano JM, Zaharris E, MacFadyen J, Danielson E, Buring JE, Manson JE. A randomized factorial trial of vitamins C and E and beta carotene in the secondary prevention of cardiovascular events in women: results from the Women's Antioxidant Cardiovascular Study. Arch Intern Med. 2007;167:1610–8.
28. Hennekens CH, Buring JE, Manson JE, Stampfer M, Rosner B, Cook NR, Belanger C, LaMotte F, Gaziano JM, Ridker PM, Willett W, Peto R. Lack of effect of long-term supplementation with beta carotene on the incidence of malignant neoplasms and cardiovascular disease. N Engl J Med. 1996;334:1145–9.
29. Lee IM, Cook NR, Gaziano JM, Gordon D, Ridker PM, Manson JE, Hennekens CH, Buring JE. Vitamin E in the primary prevention of cardiovascular disease and cancer: the Women's Health Study: a randomized controlled trial. JAMA. 2005;294:56–65.
30. Lonn E, Bosch J, Yusuf S, Sheridan P, Pogue J, Arnold JM, Ross C, Arnold A, Sleight P, Probstfield J, Dagenais GR, HOPE and HOPE-TOO Trial Investigators. Effects of long-term vitamin E supplementation on cardiovascular events and cancer: a randomized controlled trial. JAMA. 2005;16:1338–47.
31. Virtamo J, Edwards BK, Virtanen M, Taylor PR, Malila N, Albanes D, Huttunen JK, Hartman AM, Hietanen P, Mäenpää H, Koss L, Nordling S, Heinonen OP. Effects of supplemental alpha-tocopherol and beta-carotene on urinary tract cancer: incidence and mortality in a controlled trial (Finland). Cancer Causes Control. 2000;11:933–9.
32. Rapola JM, Virtamo J, Ripatti S, Huttunen JK, Albanes D, Taylor PR, Heinonen OP. Randomised trial of alpha-tocopherol and beta-carotene supplements on incidence of major coronary events in men with previous myocardial infarction. Lancet. 1997;349:171–1720.
33. Albanes D, Heinonen OP, Taylor PR, Virtamo J, Edwards BK, Rautalahti M, Hartman AM, Palmgren J, Freedman LS, Haapakoski J, Barrett MJ, Pietinen P, Malila N, Tala E, Liippo K, Salomaa ER, Tangrea JA, Teppo L, Askin FB, Taskinen E, Erozan Y, Greenwald P, Huttunen JK. Alpha-Tocopherol and beta-carotene supplements and lung cancer incidence in the alpha-tocopherol, beta-carotene cancer prevention study: effects of base-line characteristics and study compliance. J Natl Cancer Inst. 1996;88:1560–70.
34. Cevenini E, Caruso C, Candore G, Capri M, Nuzzo D, Duro G, Rizzo C, Colonna-Romano G, Lio D, Di Carlo D, Palmas MG, Scurti M, Pini E, Franceschi C, Vasto S. Age-related inflammation: the contribution of different organs, tissues and systems. How to face it for therapeutic approaches. Curr Pharm Des. 2010;16:609–18.
35. Franceschi C, Capri M, Monti D, Giunta S, Olivieri F, Sevini F, Panourgia MP, Invidia L, Celani L, Scurti M, Cevenini E, Castellani GC, Salvioli S. Inflammaging and anti-inflammaging: a systemic perspective on aging and longevity emerged from studies in humans. Mech Ageing Dev. 2007;128:92–105.
36. Michaud M, Balardy L, Moulis G, Gaudin C, Peyrot C, Vellas B, Cesari M, Nourhashemi F. Proinflammatory cytokines, aging, and age-related diseases. J Am Med Dir Assoc. 2013;14:877–82.
37. Roubenoff R, Harris TB, Abad LW, Wilson PW, Dallal GE, Dinarello CA. Monocyte cytokine production in an elderly population: effect of age and inflammation. J Gerontol A Biol Sci Med Sci. 1998;53:20–6.
38. Vasto S, Candore G, Balistreri CR, Caruso M, Colonna-Romano G, Grimaldi MP, Listi F, Nuzzo D, Lio D, Caruso C. Inflammatory networks in ageing, age-related diseases and longevity. Mech Ageing Dev. 2007;128:83–91.

39. Rando TA. Stem cells, ageing and the quest for immortality. Nature. 2006;441:1080–6.
40. Ki S, Park D, Selden HJ, Seita J, Chung H, Kim J, Iyer VR, Ehrlich LI. Global transcriptional profiling reveals distinct functions of thymic stromal subsets and age-related changes during thymic involution. Cell Rep. 2014;9:402–41.
41. Tsaknaridis L, Spencer L, Culbertson N, Hicks K, LaTocha D, Chou YK, Whitham RH, Bakke A, Jones RE, Offner H, Bourdette DN, Vandenbark AA. Functional assay for human CD4+CD25+ Treg cells reveals an age-dependent loss of suppressive activity. J Neurosci Res. 2003;76:129–40.
42. Pawelec G, Goldeck D, Derhovanessian E. Inflammation, ageing and chronic disease. Curr Opin Immunol. 2014;29:23–8.
43. Poland GA, Ovsyannikova IG, Kennedy RB, Lambert ND, Kirkland JL. A systems biology approach to the effect of aging, immunosenescence and vaccine response. Curr Opin Immunol. 2014;29:62–8.
44. Müller L, Pawelec G. As we age: does slippage of quality control in the immune system lead to collateral damage? Ageing Res Rev. 2015;23:116–23.
45. Franceschi C, Bonafè M, Valensin S, Olivieri F, De Luca M, Ottaviani E, De Benedictis G. Inflamm-aging. An evolutionary perspective on immunosenescence. Ann N Y Acad Sci. 2000;908:244–54.
46. Giles RE, Blanc H, Cann HM, Wallace DC. Maternal inheritance of human mitochondrial DNA. Proc Natl Acad Sci USA. 1980;77:6715–9.
47. Wallace DC. Mitochondrial genetics: a paradigm for aging and degenerative diseases? Science. 1992;256:628–32.
48. Fan W, Lin CS, Potluri P, Procaccio V, Wallace DC. mtDNA lineage analysis of mouse L-cell lines reveals the accumulation of multiple mtDNA mutants and intermolecular recombination. Genes Dev. 2012;26:384–94.
49. Nachman MW, Brown WM, Stoneking M, Aquadro CF. Nonneutral mitochondrial DNA variation in humans and chimpanzees. Genetics. 1996;142:953–63.
50. Wallace DC. Mitochondrial DNA mutations in disease and aging. Env Mol Mutagenesis. 2010a;51:440–50.
51. Wallace DC. Bioenergetics and the epigenome: Interface between the environment and genes in common disease. Dev Disab Res Rev. 2010b;16:114–9.
52. Chinetti-Gbaguidi G, Colin S, Staels B. Macrophage subsets in atherosclerosis. Nat Rev Cardiol. 2015;12:10–7.
53. Mozaffarian D, Wilson PWF, Kannel WB. Beyond established and novel risk factors lifestyle risk factors for cardiovascular disease. Circulation. 2008;117:3031–8.
54. Zhang Y, Fischer KE, Soto V, Liu Y, Sosnowska D, Richardson A, Salmon AB. Obesity-induced oxidative stress, accelerated functional decline with age and increased mortality in mice. Arch Biochem Biophys. 2015;576:39–48.
55. Bhayadia R, Schmidt BM, Melk A, Hömme M. Senescence-induced oxidative stress causes endothelial dysfunction. J Gerontol A Biol Sci Med Sci. 2015;71(2):161–9.
56. Okoduwa SI, Umar IA, Ibrahim S, Bello F, Habila N. Age-dependent alteration of antioxidant defense system in hypertensive and type-2 diabetes patients. J Diabetes Metab Disord. 2015;14:32.
57. Skålén K, Gustafsson M, Rydberg EK, Hultén LM, Wiklund O, Innerarity TL, Borén J. Subendothelial retention of atherogenic lipoproteins in early atherosclerosis. Nature. 2002;417:70–754.
58. Hansson GK, Libby P. The immune response in atherosclerosis: a double-edged sword. Nat Rev Immunol. 2006;6:508–19.
59. Ridker PM, Lüscher TF. Anti-inflammatory therapies for cardiovascular disease. Eur Heart J. 2014;35:1782–91.
60. Campbell A. Inflammation, neurodegenerative diseases, and environmental exposures. Ann NY Acad Sci. 2004;1035:117–32.
61. Heppner FL, Ransohoff RM, Becher B. Immune attack: the role of inflammation in Alzheimer disease. Nat Rev Neurosci. 2015;16(6):358–72.

62. Zhang F, Jiang L. Neuroinflammation in Alzheimer's disease. Neuropsychiatr Dis Treat. 2015;11:243–56.
63. McGeer EG, McGeer PL. Innate immunity in Alzheimer's disease: a model for local inflammatory reactions. Mol Interv. 2001;1(1):22–9.
64. Steinman L. Inflammatory cytokines at the summits of pathological signal cascades in brain diseases. Sci Signal. 2013;6(258):3.
65. Coskun PE, Wyrembak J, Derbereva O, Melkonian G, Doran E, Lott IT, Head E, Cotman CW, Wallace DC. Systemic mitochondrial dysfunction and the etiology of Alzheimer's disease and down syndrome dementia. J Alzheimers Dis. 2010;20 Suppl 2:293–310.
66. Caspersen C, Wang N, Yao J, Sosunov A, Chen X, Lustbader JW, Xu HW, Stern D, McKhann G, Yan SD. Mitochondrial Abeta: a potential focal point for neuronal metabolic dysfunction in Alzheimer's disease. FASEB J. 2005;19:2040–1.
67. Gibson GE, Karuppagounder SS, Shi Q. Oxidant-induced changes in mitochondria and calcium dynamics in the pathophysiology of Alzheimer's disease. Ann N Y Acad Sci. 2008;1147:221–32.
68. Lustbader JW, Cirilli M, Lin C, Xu HW, Takuma K, Wang N, Caspersen C, Chen X, Pollak S, Chaney M, Trinchese F, Liu S, Gunn-Moore F, Lue LF, Walker DG, Kuppusamy P, Zewier ZL, Arancio O, Stern D, Yan SS, Wu H. ABAD directly links Abeta to mitochondrial toxicity in Alzheimer's disease. Science. 2004;304:448–52.
69. Takuma K, Yao J, Huang J, Xu H, Chen X, Luddy J, Trillat AC, Stern DM, Arancio O, Yan SS. ABAD enhances Abeta-induced cell stress via mitochondrial dysfunction. FASEB J. 2005;19:597–8.
70. Yan SD, Stern DM. Mitochondrial dysfunction and Alzheimer's disease: role of amyloid-beta peptide alcohol dehydrogenase (ABAD). Int J Exp Pathol. 2005;86:161–71.
71. Lakatos A, Derbeneva O, Younes D, Keator D, Bakken T, Lvova M, Brandon M, Guffanti G, Reglodi D, Saykin A, Weiner M, Macciardi F, Schork N, Wallace DC, Potkin SG, Initiative A's D N. Association between mitochondrial DNA variations and Alzheimer's disease in the ADNI cohort. Neurobiol Aging. 2010;31(8):1355–63.

Aging and Microglial Activation in Neurodegenerative Diseases

Esteban A. Oyarzabal, Yen-Yu Ian Shih, and Jau-Shyong Hong

Biological aging describes the gradual degradation of physiological function which is a consequence of accumulating spontaneous and environmentally derived mutations and adducts. Aging lowers cellular fitness and drives senescence in the later years of life. Some hallmarks associated with aging include genomic instability; increased protein aggregates with reduced proteolytic capabilities; hypometabolism; and primed inflammatory responses—together increasing the vulnerability of aged cells to insults, disease and death. According to the World Health Organization, better living and

E.A. Oyarzabal (✉)
Neuropharmacology Section, Neurobiology Laboratory, National Institute of Environmental Health Sciences, National Institutes of Health, P.O. Box 12233; MD: F1-01, 111 T.W. Alexander Dr., Research Triangle Park, NC 27709, USA

Department of Neurology, University of North Carolina at Chapel Hill, 130 Mason Farm Rd, CB# 7513, Chapel Hill, NC 27599, USA

Department of Biomedical Research Imaging Center, University of North Carolina at Chapel Hill, 130 Mason Farm Rd, CB# 7513, Chapel Hill, NC 27599, USA
e-mail: oyarzaba@ad.unc.edu

Y.-Y. I. Shih
Department of Neurology, University of North Carolina at Chapel Hill, 130 Mason Farm Rd, CB# 7513, Chapel Hill, NC 27599, USA

Department of Biomedical Research Imaging Center, University of North Carolina at Chapel Hill, 130 Mason Farm Rd, CB# 7513, Chapel Hill, NC 27599, USA

Department of Biomedical Engineering, University of North Carolina at Chapel Hill, 130 Mason Farm Rd, CB# 7513, Chapel Hill, NC 27599, USA

J.-S. Hong
Neuropharmacology Section, Neurobiology Laboratory, National Institute of Environmental Health Sciences, National Institutes of Health, P.O. Box 12233; MD: F1-01, 111 T.W. Alexander Dr., Research Triangle Park, NC 27709, USA

S.C. Bondy, A. Campbell (eds.), *Inflammation, Aging, and Oxidative Stress*, Oxidative Stress in Applied Basic Research and Clinical Practice,
DOI 10.1007/978-3-319-33486-8_7

working conditions, improved access to healthcare, and trends to reduce harmful habits (e.g. improper hygiene, poor nutrition, smoking and sedentary lifestyles) have contributed to shifting the world average life expectancy from 31 to 71.5 years of age within the century. Consequently, with greater longevity the prevalence of late-onset neurodegenerative disorders is on the rise—particularly among developed countries. As population trends project the world average life expectancy to reach 87 years by 2030 with an elderly population (>65 years of age) growing from 6 to 12 % of the total population, understanding how aging increases the risk of developing a late-onset neurodegenerative disorder is paramount. This chapter will describe the current state of research regarding the risk of aging-related oxidative stress and neuroinflammation on the development of neurodegenerative diseases.

1 Biological Aging in the Brain: An Introduction

Cellular senescence in the brain is primarily attributed to age-related shifts towards **hypometabolic states** (rate-of-living theory [1]) and **redox status** (oxidative stress theory [2]). These conditions are interrelated since aging results in mitochondrial dysfunction that decreases cellular energy production and leads to increasing levels of intracellular reactive oxidative species (ROS). Overtime this gradual shift results in the development of oxidative stress, senescence and inevitably death. Variability in life expectancies across cultures, races and individuals are attributed to a graded combinations of environmental exposures—that accelerate the generation of ROS promoting states of redox/hypometabolism (e.g., alcohol consumption, cigarette smoking, diet, radiation, toxicant and microbial exposures, stress and physical trauma)—and the fidelity and function of the endogenous damage repair mechanisms (DNA repair complexes, anti-oxidant scavenger enzymes, anti-inflammatory factors and regulators).

In the aging brain, the gradual shift towards hypometabolism occurs sporadically and diffusely, resulting in microclusters of neurons with insufficient energy to maintain their physiological status and as a consequence of this undergo stress and release Danger Associated Molecular Patterns (DAMPs) to activate local microglia and subsequently astrocytes to provide trophic support. When neurons are transiently stressed, neuroinflammation is acute and protective; yet if stress is sustained, it can become detrimental by promoting chronic neuroinflammation that accelerates the shift towards states of hypometabolism and redox, thus accelerating aging (see **inflamm-aging** review [3]). Furthermore, this shift can be further accelerated with additional environmental hits (see **multiple-hit hypothesis** [4]) that can exacerbate the inflammatory response—resulting in even faster aging (Fig. 1). To survive, stressed neurons activate survival signaling that direct their energy use to vital processes while limiting energy expenditure on non-essential processes such as synaptic maintenance and remodeling. These adaptive changes, along with minor age-related neurodegeneration, are thought to result in the distinct behavioral phenotypes observed during aging.

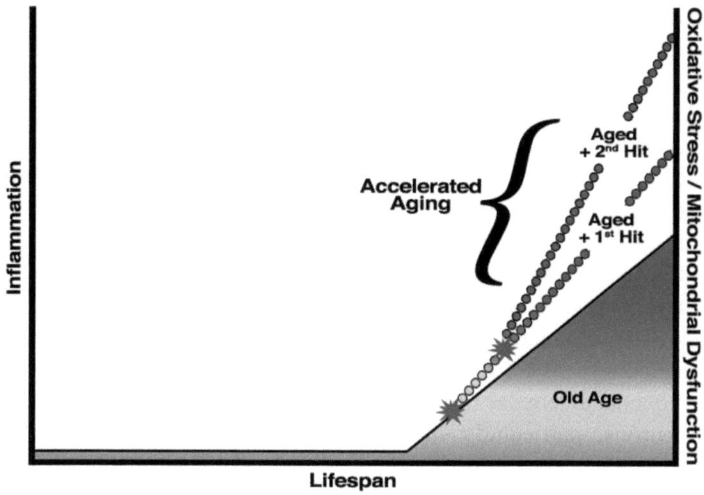

Fig. 1 A depiction of the relationship between inflammation, oxidative stress and mitochondrial dysfunction with respect to aging. Exposures, that increase the levels of inflammation, oxidative stress and mitochondrial dysfunction, are thought to accelerate the rate at which cells undergo senescence—thus accelerate aging

2 Mitochondrial Dysfunction and Hypometabolism in Aging and Neurodegenerative Diseases

The brain requires tremendous expenditures of energy to maintain and restore the ion gradients and to synthesize, transport and reuptake neurotransmitters [5, 6]. Glucose is the primary source of fuel for the brain, entering through glucose transporter on endothelial cells and astrocyte endfeet, where the bulk of it is converted within astrocytes to pyruvate through glycolysis. This pyruvate is catalyzed by lactate dehydrogenase to lactate and shuttled through monocarboxylic acid transporters from astrocytes to neurons where it is reconverted to pyruvate and broken down through the TCA and oxidative phosphorylation (OXPHOS) cycles to generate the energy required for neuronal survival. The rate of this neurometabolism—as measured by glucose and oxygen utilization—is tightly coupled with changes in local cerebral blood flow and blood volume [7–9]. During aging the brain undergoes a gradual decline in glucose and oxygen utilization [10, 11] becoming more prone to metabolic syndrome [12]—which together pathologically impact neuronal function.

The proton-motive forces of OXPHOS, that generate the bulk of the energy in the brain, leak superoxide anion radicals ($O_2^{-\cdot}$) as a by-product of mitochondrial respiration from complexes I and III into the mitochondrial matrix and the intermembrane space, respectively. Since the brain accounts for 20 % of body's total metabolism, neuronal mitochondria are thought to produce tenfold more oxidative by-products than other mitochondria found throughout the body [13]. Though mitochondrial superoxide can directly attack mitochondrial DNA (mtDNA) or form the potent oxidant peroxynitrite when bound to nitric oxide [14], it is largely reduced into hydrogen peroxide through spontaneous dismutation or through scavenger enzymes known as superoxide dismutase (Mn-SOD within the mitochondrial matrix and Cu, Zn-SOD within the intermembrane space) and reduced to water by glutathione (GSH) and thioredoxin (Trx). Hydrogen peroxide is innocuous at the low to moderate levels produced during neural activity in healthy adults and is thought to dose-dependently trigger gene transcription of redox protective mechanism in neurons when concentrations become deregulated. Aging mitochondria are known to produce high concentrations of hydrogen peroxide, resulting in severe oxidative damage by interacting with enzymatic co-factors such as Fe^{2+} and Cu^{+} to generate hydroxyl radical ($OH^{\cdot}$) through Fenton chemistry—directly attacking mtDNA and peroxidizing membrane lipids to produce 4-hydroxynonenal aldehydes or arachidonic acid to produce isoprostane.

Aged neurons, particularly their mitochondria, are vulnerable to oxidative damage due to their markedly reduced levels of antioxidant defenses resulting in DNA damage (i.e., adduct formation, double-strand breaks) and oxidative damage to the mitochondria resulting in the loss of their membrane potential—further driving the cells into oxidative stress. Oxidative attacks on mtDNA occur ten-times more frequently than nuclear DNA [15] due its proximity to the source of mitochondrial ROS and lack introns, histones and efficient DNA repair systems to protect and repair DNA [16]. As mtDNA damage accumulates with aging, afflicting the coding regions for the electron transfer complexes I, III, and IV [17–20]—leading to functional reductions in ATP generation by decreasing mitochondrial inner membrane potential and impairing calcium buffering required for turnover efficiency in the citric acid cycle [21] to produce an "electron bottleneck" leading to 4 to 5-fold greater levels of ROS [22]. ROS accumulation in mitochondria spills into the cytosol through voltage-dependent anion channels [23], attacking biomolecules such as lipids, nuclear DNA (nDNA) and post-translated proteins and enzymes. Beyond just accumulating with aging, damage to nDNA of post-mitotic neurons is also neuronal activity-driven [24]. Though age-related peroxidation of nDNA consistently afflicts only 1–2 % of neuronal genes, the affected genes are required for neuronal functions such as calcium regulation, neurotransmitter and neurotropic factor synthesis and synapse integrity [25, 26]. To support this, aged gerbils administered the free-radical spin-trap reagent N-tert-butyl-a-phenylnitrone (PBN) showed reduced biomolecule oxidation and aging-related behavioral defects [27].

Mitochondrial dysfunction occurs more frequently in neurodegenerative diseases, resulting in hypometabolic shifts being among the earliest clinically detectable hallmark of neurodegeneration [28]. Glucose metabolism is impaired in the hippocampus and frontal cortices in Alzheimer's disease patients [29] and in the striatum of Parkinson's disease patients [30] compared to healthy individuals of the same age. These patients have significantly greater losses in respiratory chain complex activities and higher levels of mtDNA mutations [31]. For instance, compared to healthy brains, brains from patients with Parkinson's disease or Alzheimer's disease have 17 times more mtDNA deletions in striatal neurons [32] and 12 times more mtDNA deletions in frontal cortex neurons [33], respectively. The importance of mitochondrial dysfunction in neurodegenerative models is best highlighted in that toxicant-induced inhibition of the electron transport chain complex I by MPTP or rotenone are widely used to induce Parkinsonianism in primates and rodents. While α-synuclein and β-amyloid deposition is accelerated with mitochondrial failure, recent findings have shown that they can directly impair mitochondrial function and produce ROS [34–36]. Though a direct link between bioenergetic failure and neurodegeneration is implicated, it is difficult to tease out the additional contributions of oxidative stress and inflammation—since all three are interrelated.

3 Mitochondrial Dysfunction and Oxidative Stress in Aging and Neurodegenerative Diseases

Oxidative stress is thought to occur gradually with aging when the intracellular production/accumulation of ROS supersedes the antioxidant capacity of the cell.

Despite neurons being extremely vulnerable to oxidative stress, the defense mechanisms to neutralize ROS are heterogeneous among neuronal populations through enzymatic (e.g., glutathione peroxidase, glutathione reductase, superoxide dismutases and catalase) and non-enzymatic pathways (e.g., glutathione, ubiquinol, uric acid) [37]. This, as well as tonic firing [38], is thought to explain why dopaminergic, noradrenergic and serotonergic neuronal populations are selectively degenerated with aging [39, 40]. During oxidative stress, mitochondrial function is regulated by an energy-redox loop, whereby complex I becomes glutathionylated resulting in lower energy-transducing efficiency, increased generation of ROS and greater mitochondrial fragmentation [41]. These changes gradually prevent the autophagy of dysfunctional mitochondria [42, 43] leading to their continued utilization and inevitably to oxidative stress-driven neurodegeneration known as the **mitochondrial-lysosomal axis theory of aging** [44].

Stressed neurons prioritize their energy to upregulate a gamut of ROS sequestering genes and release DAMPs such a α-synuclein [45], β-amyloid [46], μ-calpain [47], neuromelanin [48], HSP-70 [49], HMGB-1 [50] to activate and recruit local microglia and astrocytes to assist them with anti-oxidative buffering and if needed phagocytosis of dying neurons. Though neuroinflammation can rescue neurons undergoing oxidative stress, it becomes less probable with aging. Aged microglia, as will be described in greater detail below, display lower activation thresholds and constitutively generate basal levels of inflammatory mediators and oxidative products that can themselves result in collateral damage of injured neurons leading to even greater levels of intracellular ROS. Higher basal levels of ROS were detected in the substantia nigra pars compacta of 22 months old mice than that of 2-month old mice. Unlike typical aging, when 2-month old mice were treated with a systemic injection of the inflammagen lipopolysaccharide (LPS) showed marked acceleration in the

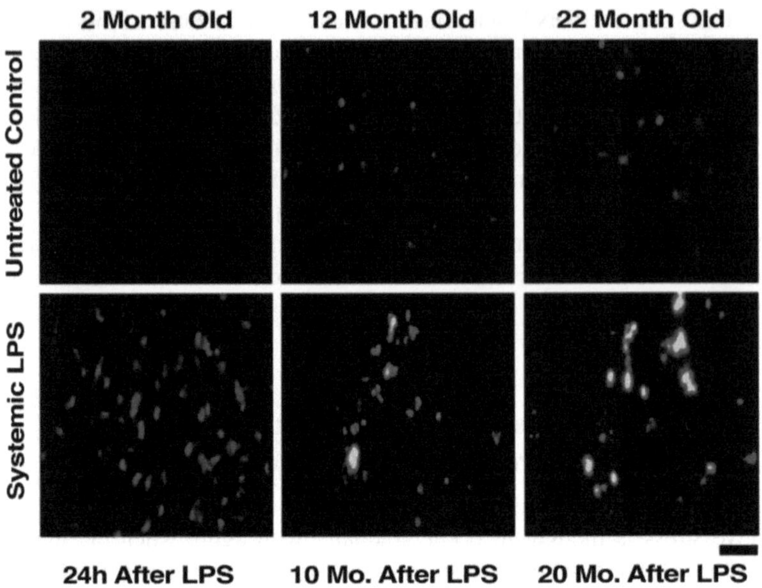

Fig. 2 As mice age from 2 months of age to 22 months of age, they have elevated basal levels of intracellular superoxide within the substantia nigra pars compacta, as assessed with dihydroethidium in the untreated control group. Yet, when 2 month old mice were administered a systemic injection of lipopolysaccharide, the basal levels of intracellular superoxide within the substantia nigra pars compacta were significantly augmented and the accumulation rate was also accelerated. The data from this figure was modified with permission from [51]

accumulation of intracellular ROS—reaching greater levels than that observed in 22 month old aged mice by 24 h after the immunological challenge (Fig. 2; [51]). To determine the levels of intracellular ROS, mice were injected with dihydroethidium—a compound that becomes oxidized *in vivo* into fluorescent ethidium in a superoxide concentration-dependent manner. Together these findings further support the multiple-hit hypothesis of accelerated aging. *In situ* evaluation of the major DNA peroxidation adduct 8-oxo-deoxyguanosine, a quantitative marker of oxidative stress, showed aged microglia and neurons had greater levels of DNA peroxidation [52]—and that peroxidation occurred more frequently in mtDNA rather than nDNA.

The oxidative stress theory of aging suggests that aging is a result of (1) increased production of oxidative species (e.g., through increased rates of or dysfunction in mitochondrial respiration, heavy metal accumulation, presence of inflammation) while (2) reducing endogenous anti-oxidant defenses (e.g., through persistent oxidative stress, reduced transcription, epigenetic modifications, reduced presence of required co-factor to donate electrons) resulting in redox-induced cellular senescence and death. This theory has been supported by epidemiological data and experimental evidence in laboratory animals by achieving significant improvements in longevity with:

(i) antioxidant rich diets [53], overexpression of the antioxidant enzymes SOD1, SOD2 and catalase [54] and pharmacological suppression of enzymes that generate oxidative species known to participate in cellular senescence [55] that are inversely correlated with the rates of mitochondrial superoxide and hydrogen peroxide production;
(ii) caloric restriction [56] and endocrine regulation of metabolism through insulin-like signaling [57] or gonadotropin-releasing hormone [58] that decreases metabolic rate and production of mitochondrial superoxide and hydrogen peroxide;
(iii) anti-inflammatory minocycline, IL1 receptor antagonists, exogenous CX3CL1, CD200 fusion protein FGL and IKK/NF-kB inhibitors [59];
(iv) proper functioning nucleotide repair proteins [60]; and
(v) exercise [61].

The shift towards oxidative stress is exacerbated in a variety of neurodegenerative disease, such as Alzheimer's disease since many of the free radical scavenging enzymes that provide antioxidant defense—namely SOD1, catalase, glutathione peroxidase and glutathione reductase—have age-related reductions in activity in hippocampus and cortex [16, 62]. Heme oxygenase-1 (HO-1), a microsomal enzyme that degrades heme into the antioxidant bilirubin, is commonly used as a marker of oxidative stress since it is rapidly expressed in response to oxidative and nitrosative stresses. In Alzheimer's disease, the expression of HO-1 is upregulated in response to pathological alterations in amyloid precursor protein (APP) and the generation of neurofibrillary tangles [63, 64]. Likewise, the expression of 8-oxo-deoxyguanosine was significantly elevated in post-mortem brains from patients with Alzheimer's disease and Parkinson's disease as compared to age-matched controls [65, 66]—suggesting that beyond accelerating the rate of mitochondrial dysfunction, neurodegenerative diseases also have greater oxidative stress in nuclei with degenerating neurons.

Neurons undergoing oxidative stress alter APP, tau and α-synuclein processing—resulting in the aberrant amyloidopathies, tauopathies and synucleopathies hallmarks associated with both Alzheimer's disease and Parkinson's disease. This partly occurs through c-Jun and p38-MAPK mediated increase in β-secretase that cleaves APP into amyloidgenic peptide β-amyloid 40–42 [67] and GSK-3β mediated phosphorylation of tau [67]—three signaling pathways that are active in neurons during oxidative stress. Oxidative stress also leads to greater lipid peroxidation end products such as isoprostanes and 4-hydroxy-trans-nonenal that directly oxidize proteins into difficult to digest aggregates. Since aged mice have impaired proteolytic degradation attributed to a loss in lysosome acidification [68] and a decline in proteostasis due to the accumulation of undigestable proteins [69], protein aggregates may interact with mitochondrial function generating ROS that can stimulate the redox-sensitive transcription factor NF-kB generating a vicious cycle of oxidative stress within aged neurons. Interestingly, the threshold at which the vicious cycle is induced could only be achieved in aged neurons, since a study found that an intracranial injection of fibrillar β-amyloid triggered sustained inflammation, neurofibrillary tangle formation and neuronal loss in aged rhesus monkeys but not in young adults [70]. This supports that oxidative stress and aging-related lysosomal dysfunctions are required for neurodegenerative progression and pathogenesis.

4 Uncontrolled Neuroinflammation in Aging and Neurodegenerative Diseases

Neuroinflammation generally serves to protect and restore homeostasis to the brain following an insult. This is achieved by endogenous innate immune cells of the brain known as microglia, that detect distress and injury signals through pattern recognition receptors (PRRs) resulting in their activation and recruitment to the insult. Upon activation, microglia serve as housekeepers to clear the insult (e.g., oxidative burst for microbial infections, phagocytosis for cellular debris, environmental exposures and 'prion-like' protein aggregates, and glial scar formation for stroke and head trauma) and subsequently self-regulate their own inactivation by releasing pro-resolution factors to quench the inflammation once the distress signals are no longer detected. Factors such as the severity and distribution of the insult, genetic and environmental predisposition to alter the inflammatory responses and age dictate the extent and duration of the staged neuroinflammatory response. Deregulated immune resolution can occur when a severe enough insult is coupled with preexisting susceptibilities that limit an appropriate response to resolve the insult, resulting in pathological chronic neuroinflammation that contributes to collateral neurodegeneration near 'dysfunctional microglia' persistently undergoing oxidative bursts [71].

Aging microglia share many similar features to pathological microglia found in brain injuries and diseases—displaying amoeboid morphologies with enlarged somas and thick, shortened processes that lose their arbor-like complexity [72, 73] with severe alterations in their abilities to host an adequate response to insults. Yet, detailed evaluations of aged microglia have characterized them as senescent or dystrophic rather than pathologically activated due to their greater density of cytoplasmic vacuoles and inclusions, deramified dendritic arbors with fragmented processes, enlarged peri-nuclear cytoplasms with beading suggesting cytorrhexis, membrane blebbing, and excessive accumulation of ferratin and neuromelanin [74, 75]. As microglia age, one notable feature is that they accumulate undegradable lipofuscin, polymeric chains of cross-linked proteins that form intracellularly during oxidative stress [76] or accumulated from phagocytized dysfunctional neurons within their lysosomes. Together, the morphological changes observed in dystrophic microglia impair the dynamic nature of microglial processes utilized in immune surveillance [77] and synaptic remodeling [78–80] and thus likely influence their function.

Functionally, aged microglia display delayed response times, slower rates of motility, low phagocytic activity, exaggerated inflammatory responses and greater proliferative capacity to insults compared to younger microglia [81–91]. This was verified *in vivo* by examining how aged microglia respond to exogenous ATP to simulate cell lysis or laser injury using two-photon imaging. Prior to injury, aged microglia display slower process dynamics, which were maintained even after insult. Aged microglia responded to the injuries with delayed ramification and motility and lingered at the site of insult far longer than microglial responding to the same insult in younger animals [82]. Age also impaired the rate and degree of microglial phagocytosis of β-amyloid in a model of Alzeihmer's disease [92, 93] and of myelin in an experimental autoimmune encephalomyelitis model of multiple sclerosis [94]. These findings suggest that the functional changes in aging microglia afflict both the ability to detect and respond to the initial immunoactivating stimuli, but also may alter their ability to resolve insults.

Gene expression levels of the PRRs TLR1, TLR2, TLR4, TLR5, TLR7, and CD14—used to detect insults—are up-regulated with age and support the exaggerated inflammatory responses observed in aged microglia [95, 96]. Paradoxically, the expression levels of microglial P2 purinergic receptors [82, 97], integrins CD11b and CD11c [88], scavenger receptors CD68 [84, 98] and RAGE [88], TREM2 [99] and FcγRs [100] involved in ATP-mediated chemotaxis, adhesion and phagocytosis of opsonised debris and pathological protein aggregates are up-regulated, rather than down-regulated, with aging. Age-related hypometabolism could partly explain this contradiction between gene expression and altered function in microglia, since even though the microglial machinery for motility and phagocytosis are present, they are highly energy-intensive processes [101]. Another explanation is that repeated insults generate large amounts of debris, many of which are undigestable with age, that overwhelm microglia and render them dysfunctional over time [102]. However, the expression of degrading enzymes such as IDE, neprilysin, and matrix metalloprotease 9 (MMP9) [92] and the rate of authophagy in aging microglia [101] are consistently decreased with the functional impairment of phagocytosis by aged microglia.

Aged microglia express increased levels of effector molecules associated with their activated states, including elevated basal levels of the pro-inflammatory cytokines IL-1β, IL-6 and TNF-α [84, 103–109] coupled with a minor decrease in the expression of anti-inflammatory cytokines IL-10 [104, 110, 111]. Likewise, healthy elderly individuals show increased basal levels of inflammation as detected on positron emission tomography (PET) using [^{11}C]-PK11195 [112]. This imbalance between inflammatory and anti-inflammatory factors potentiate age-related neuroinflammatory responses [113–116], more importantly this study suggests that many aged microglia displaying graded yet chronic states of 'semi-activation' or 'para-inflammation' within the brain due to the aging process [117]. In support of this theory, many studies found aged brains to be leakier to environmental factors circulating in the blood and more abundant in persistent inflammagens within the parenchyma, thus capable of being chronically stimulated into a maladaptive activated phenotype [118–120].

Aged microglia in 'semi-activated' states have reduced activation thresholds, a phenomenon known as priming, resulting in exaggerated inflammatory responses to additional insults [105]. *In situ* hybridization for MHC II, a marker of primed microglia, showed ~2 % of adult microglia in mice were MHC II positive, whereas ~25 % of aged microglia were MHC II positive in the absent of an insult [121]. Priming was confirmed in aged microglia to generate larger, more sustained immune responses to insults resulting in more collateral degeneration and dysfunction compared to similar lesions on younger animals. Insults ranging from infectious agents and their cellular components [105, 107, 121–127]; hemorrhagic stroke [128, 129], physiological stress [130, 131], and mechanical- or toxicant-induced neurological injury [132–134] all significant potentiated the release of IL-1β, IL-6, and TNF-α in aged rodent brains. Suppressing inflammation with minocycline in aged mice prior to LPS stimulation attenuated the priming-induced amplification of pro-inflammatory factors released by microglia [135].

Though the age-related shift of microglia into primed states can explain the altered sensitization and reactivity to immune challenges, the **theory of replicative senescence** suggests that since quiescent microglia are rarely thought to replicate compared to activated microglia [136, 137], it is only microglia in chronic states of activation that may reach their lifetime replication limits (as defined by telomere attrition with each replication; [138]). To accommodate for the age-related loss in microglial turnover, the resident brain macrophage population can be steadily replaced with infiltrating monocytes that adapt to the brain environment yet express slightly altered phenotypes that could be perceived as priming [139]. Though monocyte-derived microglia-like macrophages can be identified as $CD11b^{+}$ $CD45^{high}$ cells through flow cytometry, it is nearly impossible to differentiate these two distinct populations through conventional histological methods due to the similarity in expression of markers, thus more work needs to be done to verify this theory.

The pathway by which IL-1β is released in its active form from microglia has recently been determined to require two environmental cues [140]—a priming signal usually initiated that activate the NF-kB pathway to transcribe immature Pro-IL-1β and a second signal to activate NLRP3 inflammasome (e.g., ATP through P2X7/Pannexin, Cathepsin B released from phagolysosome rupture from indigestible lipid crystals and misfolded proteins and direct stimulation by cytosolic viral vectors) to activate caspase-1 to catalyze Pro-IL-1β into its active form [141]. Though inflammasome activation within microglia has been associated with cognitive impairment and dementia [142, 143], recent findings support that aging is associated with increases in the phosphorylation of NF-kB signaling and the increased expression of inflammasome assembly genes and activation of caspase 1 [144]. Evolutionarily, the two-step activation process required to generate active IL-1β in the brain likely served as an additional checkpoint to regulate neuroinflammation from becoming pathogenic. Yet, since aging shifts microglia into primed, dystrophic state associated with basal levels of NF-kB signaling (produced by distress signals or ROS from dysfunctional mitochondria), only the second signal is required for NLRP3 inflammasome formation and signaling in aging brains [145]. Since a large population of aging microglia release autocrine pro-inflammatory mediators such as ATP and TNF-α and store α-synuclein aggregates and β-amyloid fibrils that are implicated in lysosomal damage that release cathepsins, the secondary signals to form the inflammasome may also be present in aged microglia [146, 147]. Interestingly, 95 % of the isolated of primed microglia (MHC II positive) showed co-expression of IL-1β once stimulated with LPS, compared to 31 % of the non-primed microglia [121]. Since the release of IL-1β requires inflammasome formation, this finding suggesting that the population of primed aged microglia have sufficient secondary signals to induce inflammasome assembly.

Several soluble and membrane bound ligands interacting with microglial receptors may modulate the activation state of microglia [148]. For instance, astrocytes and neurons are thought to possess immunosuppressive functions within the brain to regulate microglia and infiltrating leukocytes [149]—limiting the collateral damage they may cause during inflammation on the brain's network of non-regenerating, post-mitotic neurons that are highly vulnerable to lipid peroxidation. Age-related modifications (e.g., loss of immunosuppressive factor expression on astrocytes and neurons, loss of expression of their respective receptors on microglia) are also thought partake in the exaggerated inflammatory responses observed in primed, aged microglia. Though astrocytes secrete S100β, TGFβ and neurotrophins that suppress microglial activation [150], their expression in aged astrocyte seems to be either unaltered or overexpressed compared to younger astrocytes—and thus are negligible with regards to aging-related shifts in modulating activation. Neurons, on the other hand, gradually lose many of the secreted and cell-to-cell contact factors that suppress microglial activation with aging (see Table 1; [110, 151–173]). Though activated adult microglia can undergo immunosuppression through M1/M2 polarization in the presence of IL-4 or IL-10, one nuance of activated aged microglia is that they remain in the M1 activation state upon treatment with either of these anti-inflammatory factors—suggesting that aging impairs the microglial response to IL-4 and IL-10 regardless of the expression of levels of either the ligand or receptor.

Aged microglia have enhanced leukocyte transmigration into the brain, most likely a result of age-related increased IL-1β- and TNF-α-mediated upregulation of endothelial cell adhesion molecules [174] and secretion of the chemokines monocyte chemotactic protein (MCP)-1 and macrophage inflammatory protein-1 (MIP)-1α by reactive astrocytes [175, 176]. In support of this, aged microglia express more MHC class II complexes that bind to T cell receptor (TCR) on naive $CD4^+$ T cells and co-stimulatory molecule CD80 and CD86 that bind to T-cell CD28 [177]. Though T cells are rarely found in the brain parenchyma of young adults [178], observation in mice indicate that lymphocyte extravasation in brain increases after 12 months of age and accumulate near activated microglia and astrocytes that secrete IL-1β and TNF-α [179]. Likewise, dendritic cells, which are rarely found within the brain parenchyma in young adults, accumulate within the brain in an age-dependent manner [179]. The adaptive immune cells are active participants in amplification of the brain immune response that has been observed with advanced age.

Neuroinflammation serves as a primary pathological hallmark of neurodegenerative diseases such as Alzheimer's disease, Parkinson's disease and multiple sclerosis. Despite their different etiologies and specific nuclei of degeneration, many of the immunologically-active endogenous factors released by aged, degenerating neurons that maintain chronic neuoinflammation are shared among many diseases. Our group has previously reported how chronically activated microglia may participate in exacerbating neurodegeneration through collateral damage. For instance, anti-inflammatory interventions that inhibit the formation of superoxide by NOX2 inhibition [169, 180–191] and through antioxidant natural compounds [192–195] have been shown to be effective at preventing dopaminergic degenerations by reducing inflammation and oxidative stress in *in vitro* models of Parkinson's disease.

Table 1 A list of immunosupressent neuronal ligands and their respective receptors on microglia that have altered expression with aging

	Neuronal ligand	Microglial receptor	Function	Age-related changes	Sources
Cell adhesion and contact inhibition	CD22	CD45	Suppresses microglial activation and proliferation	Gradual loss of receptor	[151–153]
	CD47	SIRPα (CD172a), ECM glycoprotein thrombospondin	Suppresses microglial activation and phagocytosis	Gradual loss of ligand	[154, 155]
	CD200 (OX-2)	CD200R	Suppresses microglial activation	Gradual loss of ligand	[110]
	NCAM; Polysialylated NCAM	NCAM, Singlec-11	Suppresses microglial activation	Gradual loss of ligands and Singlec-11	[156–159]
	FasL (CD95L)	Fas (CD95)	Suppresses microglial activation while promoting apoptosis	Gradual loss of ligand	[160, 161]
Cytokines and chemokines	IL10	IL10R	Promote M1/M2 Polorization	Gradual loss of ligand	[111]
	CX3CL1	CX3CR1	Suppresses microglial activation and chemotaxis	Gradual loss of ligand and receptor	[162, 163]
Transmitters and peptides	NE	α1A, α2A, β1, β2	Suppresses microglial activation and chemotaxis	Gradual loss of ligand	[164, 165]
	VIP, PACAP	VPAC1, VPAC2	Suppresses microglial activation	Gradual loss of ligand and VPAC2	[166–169]
Neurotrophins	NGF	p75, NTR, TrkA	Suppresses microglial activation	Cortical decrease of ligand	[170, 171]
	BDNF	p75, NTR, TrkA	Suppresses microglial activation	Gradual loss of ligand	[170, 171]
	Neurotrophin-3	11p75, NTR, TrkB, TrkC	Suppresses microglial activation	Cortical decrease of ligand, systematic decline in pathological models	[170, 171]

5 Behavioral Deficits in Aging and Neurodegeneration

Aging is associated with cognitive decline, anxiety, depression, reduced sociability and locomotor retardation. Though changes to neurons including their loss in neurocircuit, Long-term potentiation/long-term depression-mediated synaptic plasticity and alterations in neurotransmission and firing rates can modulate behavior, aging-related neuronal dysfunction and degeneration is thought to be almost entirely attributed to hypometabolism, neuroinflammation and oxidative stress. Microbial infections [196, 197] and inflammatory cytokines [198, 199] can induce transient and protective sickness behavior with cognitive deficits, lethargy, locomotor retardation, reduced appetite, loss of pleasure from rewarding activities, sleep disturbances and increased pain sensitivity that together prioritize the body's energy consumption to fend infectious pathogens. At the cellular and molecular level, sickness behavior occurs in response to central or systemic inflammation resulting in central production of IL-1β and subsequent hyperactivation of the hypothalamic-pituitary-adrenal (HPA) axis involved in modulatory changes of glucocorticoid levels in the blood to suppress systemic inflammation [200].

In the elderly, systemic inflammation can be linked prolonged deficits in cognitive function with faster rates of decline [201, 202], functional disability [203], frailty [204], and mortality [205]. These findings are strikingly similar to the prolonged behaviors observed in adult mice with primed microglia that received repeated exposures to LPS [206] or had a chronic bacterial infection [207]— showing sustained altered cognition, disrupted in food intake and reward, retarded exploration and social interactions, increased pain sensitivity, and disrupted sleep patterns. Repeated/chronic release of IL-1β by aged microglia with activated inflammasomes can sustain a prolonged state of sickness behavior. Repeated hyperactivation of the HPA axis produces activity-dependent concentration of extracellular ATP [208, 209], further attracting and polarizing nearby microglia to release IL-1β through an inflammasome-dependent manner to generate a feed-forward loop.

Beyond the extended sickness behavior, the prolonged secretion of the pro-inflammatory cytokines IL-1β, IL-6 and TNF-α by aged microglia is also associated with depressive behaviors and impaired cognition [103, 122, 210]. Aging-related depressive behaviors are a result of TNF-α-mediated loss of serotonergic neurons [211] observed with aging and persistent neuroinflammation that result in a drastic decline in blood tryptophan required for serotonin production in the brain [212]. Neuroinflammation-mediates increased enzymatic activity by indoleamine 2,3-dioxygenase (IDO), that reduces the bioavailability of serotonin in brain [213]. Aged mice injected with LPS show amplified IDO expression with prolonged depressive phenotypes compared to similarly treated adult as measured by resignation behavior in trail suspension test and forced swim test. In support of this, inhibition of IDO reversed insult-mediated depressive behavior in aged mice.

Beyond upregulation of genes associated with inflammation and oxidative stress, aged brains have a decreased expression in genes associated with synaptic function and neurotropic support [84, 103, 122, 210]. Though all aging-related behavioral deficits are partly attributed to inflammation-driven demyelination [214] and synaptic [215] and dendritic [216] stripping of stressed neurons. These features are salient in cognitive impairments association with persistent IL-1β and IL-6 release by aged microglia, which alter synaptic integrity and function in the hippocampus impairing spatial learning and memory [215] and can be reversed through suppressing microglial activation or antagonizing the receptors for IL-1β and IL-6 [217–222]. To further support this, the degree of cognitive impairment was associated with the extent of microglial activation, as determined by immunoreactivity to MHCII and iNOS, in white matter [223]. Aged microglia also impair synaptoplasticity by inhibiting LTP [224] through AMPA receptor endocytosis [225], eventually resulting in dendritic atrophy in the hippocampus [216]. Furthermore, aging and chronic neuroinflammation are associated with loss of trophic factors such as BDNF and NGF [170, 171, 226] that alter the development of synaptic LTP [225], adult neurogenesis [227, 228] and contribute to altered learning and memory.

Recently, inflammation within the hypothalamus—thought to modulate many of the features of metabolic syndrome—has been implicated in aging throughout the body. Dystrophic hypothalamic microglia are thought to drive the cognitive decline and muscle weakness with age-related increases in basal NF-kB signaling resulting in the generation of pro-inflammatory factors that can stimulate neurons to activate upstream IKK-β and subsequently NF-kB signaling. In support of this, IKK-β knockout mice showed less dystrophy in aged hypothalamic microglia and had not only improved spatial memory, muscle strength, bone mass, skin thickness but also had increased their life expectancy by 10% compared to age-matched wild-type mice [59]. Inflammation-driven increases in neuronal IKK-β signaling within the hypothalamus can accelerate aging by suppressing the release of gonadotropin-releasing hormone (GnRH), a factor that regulates sex hormones, reproduction and stimulates adult neurogenesis that was shown to gradually lose expression within the hypothalamus with age. Recovery studies show that GnRH replacement therapy in aged mice improved physical and cognitive deficits observed with typical aging in age-match control mice [59].

As expected, neurodegenerative diseases not only display accelerated behavioral changes associated with typical aging, but they also display additional phenotypes associated with their respective degenerating neurocircuits. In the case of Alzheimer's disease, though environment, lifestyle and genetics may vary from case to case, progression between the earliest recognizable symptoms of memory loss typically occur 3–10 years before dementia and the health complication that lead to death. The symptoms of Alzheimer's disease rapidly progress with advanced aging and as of yet studies have shown improved cognitive function in patients and animal models of Alzheimer's disease with mild and moderate stages using organic anti-inflammatory compounds such as reservatrol, propolis, ratanasampil and curcumin [229–231], high dietary levels of anti-oxidants such as Vitamin E and exercise [232]. Furthermore, these compounds reduce oxidative damage, amyloidopathy, and the expression of proinflammatory cytokines.

Parkinson's disease patients experience resting tremors, spine rigidity, loss of fine motor control and irregular changes to gate. Although genetic (e.g., A53T, Parkin) and direct toxicant models (e.g., MPTP, 6-OHDA, Roteone) of Parkinson's disease in animal models recapitulate the motor deficits as evaluated by methamphetamine-induced circling behavior, rotorod, open-field studies and gait analysis, low-grade chronic neuroinflammation resulting from a single systemic injection of LPS results in the development of many of the early non-motor symptoms associated with Parkinson's diseases. Preliminary data from a battery of behavioral tests conducted on LPS treated mice show that these mice develop many of these pre-motor symptoms including constipation, changes in behavior towards a more stoic nature with greater anhedonia and impulsivity toward riskier behavior, reduced spatial memory and a greater propensity to startle. Interestingly, suppressing the generation of superoxide by NOX2 with DPI [233, 234] was shown to improve many of these behaviors. Additionally, exercise [235] and caloric restrictions [236] have also been shown to improve motor deficits in MPTP-induced models of Parkinson's disease as well.

6 Conclusion

Aging increases the probability of mitochondrial dysfunction, oxidative stress and neuroinflammation resulting in neuronal and microglial senescence and inevitably in altered behavior. The role of microglia in aging is paramount, since immunologically activated aged microglia are the central source of ROS in the brain, leading to collateral damage of neighboring neurons. The repeated on slot of cytotoxic factors released by these microglia increase neuronal intracellular ROS, impacting the functions of mitochondria and authophagy and impairing their plasticity and function. Therapeutically targeting either the release of proinflammatory factors or oxidative products improve longevity by reducing mtDNA damage, increased anti-oxidant buffering and reducing inflammation. Determining how aging contributes to the development of age-associated neurodegenerative disorders will be important in the upcoming years, since better designed therapies that treat these diseases as a series of interrelated dysfunctions will be required to halt neurodegeneration.

References

1. Pearl R. The rate of living, being an account of some experimental studies on the biology of life duration. New York: A.A. Knopf; 1928. 8 p.l., 185, 1 p.
2. Harman D. Aging: a theory based on free radical and radiation chemistry. J Gerontol. 1956;11(3):298–300.
3. Franceschi C, et al. Inflamm-aging. An evolutionary perspective on immunosenescence. Ann N Y Acad Sci. 2000;908:244–54.
4. Lourbopoulos A, Erturk A, Hellal F. Microglia in action: how aging and injury can change the brain's guardians. Front Cell Neurosci. 2015;9:54.
5. Alle H, Roth A, Geiger JR. Energy-efficient action potentials in hippocampal mossy fibers. Science. 2009;325(5946):1405–8.

6. Attwell D, Laughlin SB. An energy budget for signaling in the grey matter of the brain. J Cereb Blood Flow Metab. 2001;21(10):1133–45.
7. Figley CR, Stroman PW. The role(s) of astrocytes and astrocyte activity in neurometabolism, neurovascular coupling, and the production of functional neuroimaging signals. Eur J Neurosci. 2011;33(4):577–88.
8. Magistretti PJ, Pellerin L. Cellular bases of brain energy metabolism and their relevance to functional brain imaging: evidence for a prominent role of astrocytes. Cereb Cortex. 1996;6(1):50–61.
9. Raichle ME, Mintun MA. Brain work and brain imaging. Annu Rev Neurosci. 2006;29:449–76.
10. Marchal G, et al. Regional cerebral oxygen consumption, blood flow, and blood volume in healthy human aging. Arch Neurol. 1992;49(10):1013–20.
11. Petit-Taboue MC, et al. Effects of healthy aging on the regional cerebral metabolic rate of glucose assessed with statistical parametric mapping. Neuroimage. 1998;7(3):176–84.
12. Morley JE. The metabolic syndrome and aging. J Gerontol A Biol Sci Med Sci. 2004;59(2):139–42.
13. Yudkoff M, et al. The ketogenic diet and brain metabolism of amino acids: relationship to the anticonvulsant effect. Annu Rev Nutr. 2007;27:415–30.
14. Bao L, et al. Mitochondria are the source of hydrogen peroxide for dynamic brain-cell signaling. J Neurosci. 2009;29(28):9002–10.
15. Bohr VA, Dianov GL. Oxidative DNA damage processing in nuclear and mitochondrial DNA. Biochimie. 1999;81(1–2):155–60.
16. Zemlan FP, Thienhaus OJ, Bosmann HB. Superoxide dismutase activity in Alzheimer's disease: possible mechanism for paired helical filament formation. Brain Res. 1989;476(1): 160–2.
17. Kwong LK, Sohal RS. Age-related changes in activities of mitochondrial electron transport complexes in various tissues of the mouse. Arch Biochem Biophys. 2000;373(1):16–22.
18. Navarro A, Boveris A. Rat brain and liver mitochondria develop oxidative stress and lose enzymatic activities on aging. Am J Physiol Regul Integr Comp Physiol. 2004;287(5): R1244–9.
19. Navarro A, Boveris A. Brain mitochondrial dysfunction in aging: conditions that improve survival, neurological performance and mitochondrial function. Front Biosci. 2007;12: 1154–63.
20. Navarro A, et al. Hippocampal mitochondrial dysfunction in rat aging. Am J Physiol Regul Integr Comp Physiol. 2008;294(2):R501–9.
21. LaFrance R, et al. Age-related changes in regional brain mitochondria from Fischer 344 rats. Aging Cell. 2005;4(3):139–45.
22. Yin F, Boveris A, Cadenas E. Mitochondrial energy metabolism and redox signaling in brain aging and neurodegeneration. Antioxid Redox Signal. 2014;20(2):353–71.
23. Han D, et al. Voltage-dependent anion channels control the release of the superoxide anion from mitochondria to cytosol. J Biol Chem. 2003;278(8):5557–63.
24. Jurk D, et al. Postmitotic neurons develop a p21-dependent senescence-like phenotype driven by a DNA damage response. Aging Cell. 2012;11(6):996–1004.
25. Jiang CH, et al. The effects of aging on gene expression in the hypothalamus and cortex of mice. Proc Natl Acad Sci U S A. 2001;98(4):1930–4.
26. Lee CK, Weindruch R, Prolla TA. Gene-expression profile of the ageing brain in mice. Nat Genet. 2000;25(3):294–7.
27. Carney JM, Floyd RA. Protection against oxidative damage to CNS by alpha-phenyl-tert-butyl nitrone (PBN) and other spin-trapping agents—a novel series of nonlipid free-radical scavengers. J Mol Neurosci. 1991;3(1):47–57.
28. Gibson GE, et al. Cause and consequence: mitochondrial dysfunction initiates and propagates neuronal dysfunction, neuronal death and behavioral abnormalities in age-associated neurodegenerative diseases. Biochim Biophys Acta. 2010;1802(1):122–34.
29. Mosconi L, Pupi A, De Leon MJ. Brain glucose hypometabolism and oxidative stress in preclinical Alzheimer's disease. Ann N Y Acad Sci. 2008;1147:180–95.

30. Eidelberg D, et al. Striatal hypometabolism distinguishes striatonigral degeneration from Parkinson's disease. Ann Neurol. 1993;33(5):518–27.
31. Calabrese V, et al. Mitochondrial involvement in brain function and dysfunction: relevance to aging, neurodegenerative disorders and longevity. Neurochem Res. 2001;26(6):739–64.
32. Ikebe S, et al. Increase of deleted mitochondrial DNA in the striatum in Parkinson's disease and senescence. Biochem Biophys Res Commun. 1990;170(3):1044–8.
33. Corral-Debrinski M, et al. Marked changes in mitochondrial DNA deletion levels in Alzheimer brains. Genomics. 1994;23(2):471–6.
34. Anandatheerthavarada HK, et al. Mitochondrial targeting and a novel transmembrane arrest of Alzheimer's amyloid precursor protein impairs mitochondrial function in neuronal cells. J Cell Biol. 2003;161(1):41–54.
35. Lustbader JW, et al. ABAD directly links Abeta to mitochondrial toxicity in Alzheimer's disease. Science. 2004;304(5669):448–52.
36. Martin LJ, et al. Parkinson's disease alpha-synuclein transgenic mice develop neuronal mitochondrial degeneration and cell death. J Neurosci. 2006;26(1):41–50.
37. Floyd RA. Antioxidants, oxidative stress, and degenerative neurological disorders. Proc Soc Exp Biol Med. 1999;222(3):236–45.
38. Roselli F, Caroni P. From intrinsic firing properties to selective neuronal vulnerability in neurodegenerative diseases. Neuron. 2015;85(5):901–10.
39. Mukherjee J, et al. Brain imaging of 18F-fallypride in normal volunteers: blood analysis, distribution, test-retest studies, and preliminary assessment of sensitivity to aging effects on dopamine D-2/D-3 receptors. Synapse. 2002;46(3):170–88.
40. Wong DF, et al. Effects of age on dopamine and serotonin receptors measured by positron tomography in the living human brain. Science. 1984;226(4681):1393–6.
41. Taylor ER, et al. Reversible glutathionylation of complex I increases mitochondrial superoxide formation. J Biol Chem. 2003;278(22):19603–10.
42. Cardenas C, et al. Essential regulation of cell bioenergetics by constitutive InsP3 receptor Ca^{2+} transfer to mitochondria. Cell. 2010;142(2):270–83.
43. Pletjushkina OY, et al. Effect of oxidative stress on dynamics of mitochondrial reticulum. Biochim Biophys Acta. 2006;1757(5-6):518–24.
44. Liot G, et al. Complex II inhibition by 3-NP causes mitochondrial fragmentation and neuronal cell death via an NMDA- and ROS-dependent pathway. Cell Death Differ. 2009;16(6): 899–909.
45. Gao HM, et al. Neuroinflammation and alpha-synuclein dysfunction potentiate each other, driving chronic progression of neurodegeneration in a mouse model of Parkinson's disease. Environ Health Perspect. 2011;119(6):807–14.
46. Qin L, et al. Microglia enhance beta-amyloid peptide-induced toxicity in cortical and mesencephalic neurons by producing reactive oxygen species. J Neurochem. 2002;83(4):973–83.
47. Levesque S, et al. Reactive microgliosis: extracellular micro-calpain and microglia-mediated dopaminergic neurotoxicity. Brain. 2010;133(Pt 3):808–21.
48. Zhang W, et al. Neuromelanin activates microglia and induces degeneration of dopaminergic neurons: implications for progression of Parkinson's disease. Neurotox Res. 2011;19(1): 63–72.
49. El Mezayen R, et al. Endogenous signals released from necrotic cells augment inflammatory responses to bacterial endotoxin. Immunol Lett. 2007;111(1):36–44.
50. Gao HM, et al. HMGB1 acts on microglia Mac1 to mediate chronic neuroinflammation that drives progressive neurodegeneration. J Neurosci. 2011;31(3):1081–92.
51. Qin L, et al. NADPH oxidase and aging drive microglial activation, oxidative stress, and dopaminergic neurodegeneration following systemic LPS administration. Glia. 2013;61(6): 855–68.
52. Hayashi Y, et al. Reverse of age-dependent memory impairment and mitochondrial DNA damage in microglia by an overexpression of human mitochondrial transcription factor a in mice. J Neurosci. 2008;28(34):8624–34.
53. Pandey KB, Rizvi SI. Plant polyphenols as dietary antioxidants in human health and disease. Oxid Med Cell Longev. 2009;2(5):270–8.

54. Sun J, Tower J. FLP recombinase-mediated induction of Cu/Zn-superoxide dismutase transgene expression can extend the life span of adult Drosophila melanogaster flies. Mol Cell Biol. 1999;19(1):216–28.
55. Calabrese V, Bates TE, Stella AM. NO synthase and NO-dependent signal pathways in brain aging and neurodegenerative disorders: the role of oxidant/antioxidant balance. Neurochem Res. 2000;25(9–10):1315–41.
56. Sohal RS, Weindruch R. Oxidative stress, caloric restriction, and aging. Science. 1996;273(5271):59–63.
57. Wolkow CA, et al. Regulation of C. elegans life-span by insulinlike signaling in the nervous system. Science. 2000;290(5489):147–50.
58. Wang L, et al. Gonadotropin-releasing hormone receptor system: modulatory role in aging and neurodegeneration. CNS Neurol Disord Drug Targets. 2010;9(5):651–60.
59. Zhang G, et al. Hypothalamic programming of systemic ageing involving IKK-beta, NF-kappaB and GnRH. Nature. 2013;497(7448):211–6.
60. Borgesius NZ, et al. Accelerated age-related cognitive decline and neurodegeneration, caused by deficient DNA repair. J Neurosci. 2011;31(35):12543–53.
61. Garcia-Mesa Y, et al. Oxidative stress is a central target for physical exercise neuroprotection against pathological brain aging. J Gerontol A Biol Sci Med Sci. 2016;71(1):40–9.
62. Pappolla MA, et al. Immunohistochemical evidence of oxidative [corrected] stress in Alzheimer's disease. Am J Pathol. 1992;140(3):621–8.
63. Smith MA, et al. Heme oxygenase-1 is associated with the neurofibrillary pathology of Alzheimer's disease. Am J Pathol. 1994;145(1):42–7.
64. Xing S, et al. Early induction of oxidative stress in a mouse model of Alzheimer's disease with heme oxygenase activity. Mol Med Rep. 2014;10(2):599–604.
65. Gackowski D, et al. Oxidative stress and oxidative DNA damage is characteristic for mixed Alzheimer disease/vascular dementia. J Neurol Sci. 2008;266(1–2):57–62.
66. Nakabeppu Y, et al. Oxidative damage in nucleic acids and Parkinson's disease. J Neurosci Res. 2007;85(5):919–34.
67. Piccini A, et al. Beta-amyloid 1-42 induces physiological transcriptional regulation of BACE1. J Neurochem. 2012;122(5):1023–31.
68. Majumdar A, et al. Activation of microglia acidifies lysosomes and leads to degradation of Alzheimer amyloid fibrils. Mol Biol Cell. 2007;18(4):1490–6.
69. Taylor RC, Dillin A. Aging as an event of proteostasis collapse. Cold Spring Harb Perspect Biol. 2011;3(5):1–17.
70. Geula C, et al. Aging renders the brain vulnerable to amyloid beta-protein neurotoxicity. Nat Med. 1998;4(7):827–31.
71. Polazzi E, Monti B. Microglia and neuroprotection: from in vitro studies to therapeutic applications. Prog Neurobiol. 2010;92(3):293–315.
72. Choi JH, et al. Age-related changes in ionized calcium-binding adapter molecule 1 immunoreactivity and protein level in the gerbil hippocampal CA1 region. J Vet Med Sci. 2007;69(11):1131–6.
73. Hwang IK, et al. Comparison of ionized calcium-binding adapter molecule 1 immunoreactivity of the hippocampal dentate gyrus and CA1 region in adult and aged dogs. Neurochem Res. 2008;33(7):1309–15.
74. Beach TG, et al. Marked microglial reaction in normal aging human substantia nigra: correlation with extraneuronal neuromelanin pigment deposits. Acta Neuropathol. 2007;114(4):419–24.
75. Streit WJ. Microglia and Alzheimer's disease pathogenesis. J Neurosci Res. 2004;77(1):1–8.
76. Brunk UT, Terman A. Lipofuscin: mechanisms of age-related accumulation and influence on cell function. Free Radic Biol Med. 2002;33(5):611–9.
77. Nimmerjahn A, Kirchhoff F, Helmchen F. Resting microglial cells are highly dynamic surveillants of brain parenchyma in vivo. Science. 2005;308(5726):1314–8.
78. Schafer DP, et al. Microglia sculpt postnatal neural circuits in an activity and complement-dependent manner. Neuron. 2012;74(4):691–705.

79. Tremblay ME, Lowery RL, Majewska AK. Microglial interactions with synapses are modulated by visual experience. PLoS Biol. 2010;8(11):e1000527.
80. Wake H, et al. Resting microglia directly monitor the functional state of synapses in vivo and determine the fate of ischemic terminals. J Neurosci. 2009;29(13):3974–80.
81. Conde JR, Streit WJ. Microglia in the aging brain. J Neuropathol Exp Neurol. 2006;65(3):199–203.
82. Damani MR, et al. Age-related alterations in the dynamic behavior of microglia. Aging Cell. 2011;10(2):263–76.
83. DiPatre PL, Gelman BB. Microglial cell activation in aging and Alzheimer disease: partial linkage with neurofibrillary tangle burden in the hippocampus. J Neuropathol Exp Neurol. 1997;56(2):143–9.
84. Godbout JP, et al. Exaggerated neuroinflammation and sickness behavior in aged mice following activation of the peripheral innate immune system. FASEB J. 2005;19(10): 1329–31.
85. Huang Y, et al. Exaggerated sickness behavior and brain proinflammatory cytokine expression in aged mice in response to intracerebroventricular lipopolysaccharide. Neurobiol Aging. 2008;29(11):1744–53.
86. Miller KR, Streit WJ. The effects of aging, injury and disease on microglial function: a case for cellular senescence. Neuron Glia Biol. 2007;3(3):245–53.
87. Ogura K, Ogawa M, Yoshida M. Effects of ageing on microglia in the normal rat brain: immunohistochemical observations. Neuroreport. 1994;5(10):1224–6.
88. Perry VH, Matyszak MK, Fearn S. Altered antigen expression of microglia in the aged rodent CNS. Glia. 1993;7(1):60–7.
89. Rogers J, et al. Expression of immune system-associated antigens by cells of the human central nervous system: relationship to the pathology of Alzheimer's disease. Neurobiol Aging. 1988;9(4):339–49.
90. Streit WJ. The role of microglia in brain injury. Neurotoxicology. 1996;17(3–4):671–8.
91. Yan BC, et al. Differences in neuronal damage and gliosis in the hippocampus between young and adult gerbils induced by long duration of transient cerebral ischemia. J Neurol Sci. 2014;337(1–2):129–36.
92. Hickman SE, Allison EK, El Khoury J. Microglial dysfunction and defective beta-amyloid clearance pathways in aging Alzheimer's disease mice. J Neurosci. 2008;28(33):8354–60.
93. Lee S, et al. CX3CR1 deficiency alters microglial activation and reduces beta-amyloid deposition in two Alzheimer's disease mouse models. Am J Pathol. 2010;177(5):2549–62.
94. Zhao C, Li WW, Franklin RJ. Differences in the early inflammatory responses to toxin-induced demyelination are associated with the age-related decline in CNS remyelination. Neurobiol Aging. 2006;27(9):1298–307.
95. Letiembre M, et al. Innate immune receptor expression in normal brain aging. Neuroscience. 2007;146(1):248–54.
96. VanGuilder HD, et al. Concurrent hippocampal induction of MHC II pathway components and glial activation with advanced aging is not correlated with cognitive impairment. J Neuroinflammation. 2011;8:138.
97. Crain JM, Nikodemova M, Watters JJ. Expression of P2 nucleotide receptors varies with age and sex in murine brain microglia. J Neuroinflammation. 2009;6:24.
98. Wong AM, et al. Macrosialin increases during normal brain aging are attenuated by caloric restriction. Neurosci Lett. 2005;390(2):76–80.
99. Frank S, et al. TREM2 is upregulated in amyloid plaque-associated microglia in aged APP23 transgenic mice. Glia. 2008;56(13):1438–47.
100. Nimmerjahn F, Ravetch JV. Fcgamma receptors as regulators of immune responses. Nat Rev Immunol. 2008;8(1):34–47.
101. Nakanishi H, Wu Z. Microglia-aging: roles of microglial lysosome- and mitochondria-derived reactive oxygen species in brain aging. Behav Brain Res. 2009;201(1):1–7.
102. Li W. Phagocyte dysfunction, tissue aging and degeneration. Ageing Res Rev. 2013;12(4): 1005–12.

103. Chen J, et al. Neuroinflammation and disruption in working memory in aged mice after acute stimulation of the peripheral innate immune system. Brain Behav Immun. 2008;22(3): 301–11.
104. Maher FO, Nolan Y, Lynch MA. Downregulation of IL-4-induced signalling in hippocampus contributes to deficits in LTP in the aged rat. Neurobiol Aging. 2005;26(5):717–28.
105. Njie EG, et al. Ex vivo cultures of microglia from young and aged rodent brain reveal age-related changes in microglial function. Neurobiol Aging. 2012;33(1):195.e1–12.
106. Sheng JG, Mrak RE, Griffin WS. Enlarged and phagocytic, but not primed, interleukin-1 alpha-immunoreactive microglia increase with age in normal human brain. Acta Neuropathol. 1998;95(3):229–34.
107. Sierra A, et al. Microglia derived from aging mice exhibit an altered inflammatory profile. Glia. 2007;55(4):412–24.
108. Xie Z, et al. Aging and glial responses to lipopolysaccharide in vitro: greater induction of IL-1 and IL-6, but smaller induction of neurotoxicity. Exp Neurol. 2003;182(1):135–41.
109. Ye SM, Johnson RW. Increased interleukin-6 expression by microglia from brain of aged mice. J Neuroimmunol. 1999;93(1–2):139–48.
110. Frank MG, et al. mRNA up-regulation of MHC II and pivotal pro-inflammatory genes in normal brain aging. Neurobiol Aging. 2006;27(5):717–22.
111. Ye SM, Johnson RW. An age-related decline in interleukin-10 may contribute to the increased expression of interleukin-6 in brain of aged mice. Neuroimmunomodulation. 2001;9(4): 183–92.
112. Schuitemaker A, et al. Microglial activation in healthy aging. Neurobiol Aging. 2012;33(6): 1067–72.
113. Heyen JR, et al. Interleukin (IL)-10 inhibits IL-6 production in microglia by preventing activation of NF-kappaB. Brain Res Mol Brain Res. 2000;77(1):138–47.
114. Loane DJ, et al. Interleukin-4 mediates the neuroprotective effects of rosiglitazone in the aged brain. Neurobiol Aging. 2009;30(6):920–31.
115. Sawada M, et al. Interleukin-10 inhibits both production of cytokines and expression of cytokine receptors in microglia. J Neurochem. 1999;72(4):1466–71.
116. Strle K, et al. Interleukin-10 in the brain. Crit Rev Immunol. 2001;21(5):427–49.
117. Medzhitov R. Origin and physiological roles of inflammation. Nature. 2008;454(7203): 428–35.
118. Gao HM, Hong JS. Why neurodegenerative diseases are progressive: uncontrolled inflammation drives disease progression. Trends Immunol. 2008;29(8):357–65.
119. Mrak RE, Griffin WS. Glia and their cytokines in progression of neurodegeneration. Neurobiol Aging. 2005;26(3):349–54.
120. Puntener U, et al. Long-term impact of systemic bacterial infection on the cerebral vasculature and microglia. J Neuroinflammation. 2012;9:146.
121. Henry CJ, et al. Peripheral lipopolysaccharide (LPS) challenge promotes microglial hyperactivity in aged mice that is associated with exaggerated induction of both pro-inflammatory IL-1beta and anti-inflammatory IL-10 cytokines. Brain Behav Immun. 2009;23(3):309–17.
122. Abraham J, et al. Aging sensitizes mice to behavioral deficits induced by central HIV-1 gp120. Neurobiol Aging. 2008;29(4):614–21.
123. Barrientos RM, et al. Time course of hippocampal IL-1 beta and memory consolidation impairments in aging rats following peripheral infection. Brain Behav Immun. 2009;23(1):46–54.
124. Barrientos RM, et al. Characterization of the sickness response in young and aging rats following E. coli infection. Brain Behav Immun. 2009;23(4):450–4.
125. Norden DM, Godbout JP. Review: microglia of the aged brain: primed to be activated and resistant to regulation. Neuropathol Appl Neurobiol. 2013;39(1):19–34.
126. Sparkman NL, et al. Effects of intraperitoneal lipopolysaccharide on Morris maze performance in year-old and 2-month-old female C57BL/6J mice. Behav Brain Res. 2005;159(1): 145–51.
127. Frank MG, et al. Aging sensitizes rapidly isolated hippocampal microglia to LPS ex vivo. J Neuroimmunol. 2010;226(1–2):181–4.

128. Wasserman JK, Schlichter LC. White matter injury in young and aged rats after intracerebral hemorrhage. Exp Neurol. 2008;214(2):266–75.
129. Wasserman JK, Yang H, Schlichter LC. Glial responses, neuron death and lesion resolution after intracerebral hemorrhage in young vs. aged rats. Eur J Neurosci. 2008;28(7):1316–28.
130. Rosczyk HA, Sparkman NL, Johnson RW. Neuroinflammation and cognitive function in aged mice following minor surgery. Exp Gerontol. 2008;43(9):840–6.
131. Buchanan JB, et al. Cognitive and neuroinflammatory consequences of mild repeated stress are exacerbated in aged mice. Psychoneuroendocrinology. 2008;33(6):755–65.
132. Conde JR, Streit WJ. Effect of aging on the microglial response to peripheral nerve injury. Neurobiol Aging. 2006;27(10):1451–61.
133. Sandhir R, Onyszchuk G, Berman NE. Exacerbated glial response in the aged mouse hippocampus following controlled cortical impact injury. Exp Neurol. 2008;213(2):372–80.
134. Sugama S, et al. Age-related microglial activation in 1-methyl-4-phenyl-1,2,3,6-tetrahydropyridine (MPTP)-induced dopaminergic neurodegeneration in C57BL/6 mice. Brain Res. 2003;964(2): 288–94.
135. Kohman RA, et al. Effects of minocycline on spatial learning, hippocampal neurogenesis and microglia in aged and adult mice. Behav Brain Res. 2013;242:17–24.
136. Ajami B, et al. Local self-renewal can sustain CNS microglia maintenance and function throughout adult life. Nat Neurosci. 2007;10(12):1538–43.
137. Lawson LJ, Perry VH, Gordon S. Turnover of resident microglia in the normal adult mouse brain. Neuroscience. 1992;48(2):405–15.
138. Flanary BE, Streit WJ. Progressive telomere shortening occurs in cultured rat microglia, but not astrocytes. Glia. 2004;45(1):75–88.
139. Kofler J, Wiley CA. Microglia: key innate immune cells of the brain. Toxicol Pathol. 2011;39(1):103–14.
140. Simi A, et al. Mechanisms of regulation for interleukin-1beta in neurodegenerative disease. Neuropharmacology. 2007;52(8):1563–9.
141. van de Veerdonk FL, et al. Inflammasome activation and IL-1beta and IL-18 processing during infection. Trends Immunol. 2011;32(3):110–6.
142. Chakraborty S, et al. Inflammasome signaling at the heart of central nervous system pathology. J Neurosci Res. 2010;88(8):1615–31.
143. Mawhinney LJ, et al. Heightened inflammasome activation is linked to age-related cognitive impairment in Fischer 344 rats. BMC Neurosci. 2011;12:123.
144. Cribbs DH, et al. Extensive innate immune gene activation accompanies brain aging, increasing vulnerability to cognitive decline and neurodegeneration: a microarray study. J Neuroinflammation. 2012;9:179.
145. Tschopp J, Schroder K. NLRP3 inflammasome activation: the convergence of multiple signalling pathways on ROS production? Nat Rev Immunol. 2010;10(3):210–5.
146. Codolo G, et al. Triggering of inflammasome by aggregated alpha-synuclein, an inflammatory response in synucleinopathies. PLoS One. 2013;8(1):e55375.
147. Salminen A, et al. Inflammation in Alzheimer's disease: amyloid-beta oligomers trigger innate immunity defence via pattern recognition receptors. Prog Neurobiol. 2009;87(3): 181–94.
148. Lucin KM, Wyss-Coray T. Immune activation in brain aging and neurodegeneration: too much or too little? Neuron. 2009;64(1):110–22.
149. Lynch MA. The multifaceted profile of activated microglia. Mol Neurobiol. 2009;40(2): 139–56.
150. Hailer NP, et al. Astrocytic factors deactivate antigen presenting cells that invade the central nervous system. Brain Pathol. 1998;8(3):459–74.
151. Mott RT, et al. Neuronal expression of CD22: novel mechanism for inhibiting microglial proinflammatory cytokine production. Glia. 2004;46(4):369–79.
152. Tan J, et al. CD45 opposes beta-amyloid peptide-induced microglial activation via inhibition of p44/42 mitogen-activated protein kinase. J Neurosci. 2000;20(20):7587–94.
153. Veeranna, et al. Declining phosphatases underlie aging-related hyperphosphorylation of neurofilaments. Neurobiol Aging 2011;32(11):2016–29.

154. Floden AM, Combs CK. Microglia demonstrate age-dependent interaction with amyloid-beta fibrils. J Alzheimers Dis. 2011;25(2):279–93.
155. Junker A, et al. MicroRNA profiling of multiple sclerosis lesions identifies modulators of the regulatory protein CD47. Brain. 2009;132(Pt 12):3342–52.
156. Lyons A, et al. CD200 ligand receptor interaction modulates microglial activation in vivo and in vitro: a role for IL-4. J Neurosci. 2007;27(31):8309–13.
157. Meuth SG, et al. CNS inflammation and neuronal degeneration is aggravated by impaired CD200-CD200R-mediated macrophage silencing. J Neuroimmunol. 2008;194(1–2):62–9.
158. Chang RC, et al. Immune modulatory effects of neural cell adhesion molecules on lipopolysaccharide-induced nitric oxide production by cultured glia. Brain Res Mol Brain Res. 2000;81(1-2):197–201.
159. Hickman SE, et al. The microglial sensome revealed by direct RNA sequencing. Nat Neurosci. 2013;16(12):1896–905.
160. Linnemann D, et al. Expression of NCAM mRNA and polypeptides in aging rat brain. Int J Dev Neurosci. 1993;11(1):71–81.
161. Wang Y, Neumann H. Alleviation of neurotoxicity by microglial human Siglec-11. J Neurosci. 2010;30(9):3482–8.
162. Choi C, Benveniste EN. Fas ligand/Fas system in the brain: regulator of immune and apoptotic responses. Brain Res Brain Res Rev. 2004;44(1):65–81.
163. Zuliani C, et al. Control of neuronal branching by the death receptor CD95 (Fas/Apo-1). Cell Death Differ. 2006;13(1):31–40.
164. Bachstetter AD, et al. Fractalkine and CX 3 CR1 regulate hippocampal neurogenesis in adult and aged rats. Neurobiol Aging. 2011;32(11):2030–44.
165. Wynne AM, et al. Protracted downregulation of CX3CR1 on microglia of aged mice after lipopolysaccharide challenge. Brain Behav Immun. 2010;24(7):1190–201.
166. Delgado M, Ganea D. Neuroprotective effect of vasoactive intestinal peptide (VIP) in a mouse model of Parkinson's disease by blocking microglial activation. FASEB J. 2003;17(8):944–6.
167. Delgado M, Leceta J, Ganea D. Vasoactive intestinal peptide and pituitary adenylate cyclase-activating polypeptide inhibit the production of inflammatory mediators by activated microglia. J Leukoc Biol. 2003;73(1):155–64.
168. Kallo I, et al. Ageing and the diurnal expression of mRNAs for vasoactive intestinal peptide and for the VPAC2 and PAC1 receptors in the suprachiasmatic nucleus of male rats. J Neuroendocrinol. 2004;16(9):758–66.
169. Yang S, et al. Pituitary adenylate cyclase-activating polypeptide (PACAP) 38 and PACAP4-6 are neuroprotective through inhibition of NADPH oxidase: potent regulators of microglia-mediated oxidative stress. J Pharmacol Exp Ther. 2006;319(2):595–603.
170. Katoh-Semba R, et al. Age-related changes in levels of brain-derived neurotrophic factor in selected brain regions of rats, normal mice and senescence-accelerated mice: a comparison to those of nerve growth factor and neurotrophin-3. Neurosci Res. 1998;31(3):227–34.
171. Neumann H, et al. Neurotrophins inhibit major histocompatibility class II inducibility of microglia: involvement of the p75 neurotrophin receptor. Proc Natl Acad Sci U S A. 1998;95(10):5779–84.
172. Farber K, Pannasch U, Kettenmann H. Dopamine and noradrenaline control distinct functions in rodent microglial cells. Mol Cell Neurosci. 2005;29(1):128–38.
173. Gyoneva S, Traynelis SF. Norepinephrine modulates the motility of resting and activated microglia via different adrenergic receptors. J Biol Chem. 2013;288(21):15291–302.
174. Weng NP. Aging of the immune system: how much can the adaptive immune system adapt? Immunity. 2006;24(5):495–9.
175. Blau CW, et al. The age-related deficit in LTP is associated with changes in perfusion and blood-brain barrier permeability. Neurobiol Aging. 2012;33(5):1005.e23–35.
176. Kumagai N, et al. Involvement of pro-inflammatory cytokines and microglia in an age-associated neurodegeneration model, the SAMP10 mouse. Brain Res. 2007;1185:75–85.
177. Griffin R, et al. The age-related attenuation in long-term potentiation is associated with microglial activation. J Neurochem. 2006;99(4):1263–72.

178. Hickey WF. Basic principles of immunological surveillance of the normal central nervous system. Glia. 2001;36(2):118–24.
179. Stichel CC, Luebbert H. Inflammatory processes in the aging mouse brain: participation of dendritic cells and T-cells. Neurobiol Aging. 2007;28(10):1507–21.
180. Gao HM, et al. Critical role of microglial NADPH oxidase-derived free radicals in the in vitro MPTP model of Parkinson's disease. FASEB J. 2003;17(13):1954–6.
181. Qian L, et al. Microglia-mediated neurotoxicity is inhibited by morphine through an opioid receptor-independent reduction of NADPH oxidase activity. J Immunol. 2007;179(2): 1198–209.
182. Zhang W, et al. 3-Hydroxymorphinan is neurotrophic to dopaminergic neurons and is also neuroprotective against LPS-induced neurotoxicity. Faseb J. 2004;18(15):395–7.
183. Qian L, et al. Sinomenine, a natural dextrorotatory morphinan analog, is anti-inflammatory and neuroprotective through inhibition of microglial NADPH oxidase. J Neuroinflammation. 2007;4:23.
184. Liu B, Du LN, Hong JS. Naloxone protects rat dopaminergic neurons against inflammatory damage through inhibition of microglia activation and superoxide generation. J Pharmacol Exp Ther. 2000;293(2):607–17.
185. Li GR, et al. Femtomolar concentrations of dextromethorphan protect mesencephalic dopaminergic neurons from inflammatory damage. Faseb J. 2005;19(6):489–96.
186. Block ML, et al. Potent regulation of microglia-derived oxidative stress and dopaminergic neuron survival: substance P vs. dynorphin. Faseb J. 2006;20(2):251–8.
187. Qin L, et al. Microglial NADPH oxidase mediates leucine enkephalin dopaminergic neuroprotection. Ann N Y Acad Sci. 2005;1053:107–20.
188. Hu XM, et al. Clozapine protects dopaminergic neurons from inflammation-induced damage by inhibiting microglial overactivation. J Neuroimmune Pharmacol. 2012;7(1):187–201.
189. Wu HM, et al. Novel neuroprotective mechanisms of memantine: increase in neurotrophic factor release from astroglia and anti-inflammation by preventing microglial activation. Neuropsychopharmacology. 2009;34(10):2344–57.
190. Ossola B, et al. Amantadine protects dopamine neurons by a dual action: reducing activation of microglia and inducing expression of GNDF in astroglia. Neuropharmacology. 2011;61(4): 574–82.
191. Zhang F, et al. Fluoxetine protects neurons against microglial activation-mediated neurotoxicity. Parkinsonism Relat Disord. 2012;18 Suppl 1:S213–7.
192. Li FQ, et al. Inhibition of microglial activation by the herbal flavonoid baicalein attenuates inflammation-mediated degeneration of dopaminergic neurons. J Neural Transm. 2005;112(3):331–47.
193. Yang SF, et al. Curcumin protects dopaminergic neuron against LPS induced neurotoxicity in primary rat neuron/glia culture. Neurochem Res. 2008;33(10):2044–53.
194. Zhang F, et al. Resveratrol protects dopamine neurons against lipopolysaccharide-induced neurotoxicity through its anti-inflammatory actions. Mol Pharmacol. 2010;78(3):466–77.
195. Zhang D, et al. Squamosamide derivative FLZ protects dopaminergic neurons against inflammation-mediated neurodegeneration through the inhibition of NADPH oxidase activity. J Neuroinflammation. 2008;5:21.
196. Reichenberg A, et al. Cytokine-associated emotional and cognitive disturbances in humans. Arch Gen Psychiatry. 2001;58(5):445–52.
197. Brydon L, et al. Peripheral inflammation is associated with altered substantia nigra activity and psychomotor slowing in humans. Biol Psychiatry. 2008;63(11):1022–9.
198. Kammula US, White DE, Rosenberg SA. Trends in the safety of high dose bolus interleukin-2 administration in patients with metastatic cancer. Cancer. 1998;83(4):797–805.
199. Capuron L, et al. Neurobehavioral effects of interferon-alpha in cancer patients: phenomenology and paroxetine responsiveness of symptom dimensions. Neuropsychopharmacology. 2002;26(5):643–52.
200. Reyes EP, et al. LPS-induced c-Fos activation in NTS neurons and plasmatic cortisol increases in septic rats are suppressed by bilateral carotid chemodenervation. Adv Exp Med Biol. 2012;758:185–90.

201. Mooijaart SP, et al. Circulating interleukin-6 concentration and cognitive decline in old age: the PROSPER study. J Intern Med. 2013;274(1):77–85.
202. Reuben DB, et al. Peripheral blood markers of inflammation predict mortality and functional decline in high-functioning community-dwelling older persons. J Am Geriatr Soc. 2002;50(4):638–44.
203. Cohen HJ, et al. The association of plasma IL-6 levels with functional disability in community-dwelling elderly. J Gerontol A Biol Sci Med Sci. 1997;52(4):M201–8.
204. Ershler WB, Keller ET. Age-associated increased interleukin-6 gene expression, late-life diseases, and frailty. Annu Rev Med. 2000;51:245–70.
205. Bettcher BM, Kramer JH. Inflammation and clinical presentation in neurodegenerative disease: a volatile relationship. Neurocase. 2013;19(2):182–200.
206. Kubera M, et al. A new animal model of (chronic) depression induced by repeated and intermittent lipopolysaccharide administration for 4 months. Brain Behav Immun. 2013;31: 96–104.
207. Moreau M, et al. Inoculation of Bacillus Calmette-Guerin to mice induces an acute episode of sickness behavior followed by chronic depressive-like behavior. Brain Behav Immun. 2008;22(7):1087–95.
208. Fontainhas AM, et al. Microglial morphology and dynamic behavior is regulated by ionotropic glutamatergic and GABAergic neurotransmission. PLoS One. 2011;6(1):e15973.
209. Li Y, et al. Reciprocal regulation between resting microglial dynamics and neuronal activity in vivo. Dev Cell. 2012;23(6):1189–202.
210. Godbout JP, et al. Aging exacerbates depressive-like behavior in mice in response to activation of the peripheral innate immune system. Neuropsychopharmacology. 2008;33(10): 2341–51.
211. Hochstrasser T, et al. Inflammatory stimuli reduce survival of serotonergic neurons and induce neuronal expression of indoleamine 2,3-dioxygenase in rat dorsal raphe nucleus organotypic brain slices. Neuroscience. 2011;184:128–38.
212. Capuron L, et al. Association between decreased serum tryptophan concentrations and depressive symptoms in cancer patients undergoing cytokine therapy. Mol Psychiatry. 2002;7(5):468–73.
213. Andre C, et al. Spatio-temporal differences in the profile of murine brain expression of proinflammatory cytokines and indoleamine 2,3-dioxygenase in response to peripheral lipopolysaccharide administration. J Neuroimmunol. 2008;200(1-2):90–9.
214. Pang Y, Cai Z, Rhodes PG. Effects of lipopolysaccharide on oligodendrocyte progenitor cells are mediated by astrocytes and microglia. J Neurosci Res. 2000;62(4):510–20.
215. Deng XH, et al. Lipopolysaccharide induces paired immunoglobulin-like receptor B (PirB) expression, synaptic alteration, and learning-memory deficit in rats. Neuroscience. 2012;209: 161–70.
216. Richwine AF, et al. Architectural changes to CA1 pyramidal neurons in adult and aged mice after peripheral immune stimulation. Psychoneuroendocrinology. 2008;33(10):1369–77.
217. Lawson MA, et al. Intracerebroventricular administration of lipopolysaccharide induces indoleamine-2,3-dioxygenase-dependent depression-like behaviors. J Neuroinflammation. 2013;10(1):87.
218. Choi DY, et al. Obovatol attenuates LPS-induced memory impairments in mice via inhibition of NF-kappaB signaling pathway. Neurochem Int. 2012;60(1):68–77.
219. Rosi S, et al. Memantine protects against LPS-induced neuroinflammation, restores behaviorally-induced gene expression and spatial learning in the rat. Neuroscience. 2006;142(4):1303–15.
220. Zarifkar A, et al. Agmatine prevents LPS-induced spatial memory impairment and hippocampal apoptosis. Eur J Pharmacol. 2010;634(1–3):84–8.
221. Chapman TR, et al. Synaptic correlates of increased cognitive vulnerability with aging: peripheral immune challenge and aging interact to disrupt theta-burst late-phase long-term potentiation in hippocampal area CA1. J Neurosci. 2010;30(22):7598–603.
222. Frank MG, et al. IL-1RA blocks E. coli-induced suppression of Arc and long-term memory in aged F344xBN F1 rats. Brain Behav Immun. 2010;24(2):254–62.

223. Sloane JA, et al. Increased microglial activation and protein nitration in white matter of the aging monkey. Neurobiol Aging. 1999;20(4):395–405.
224. Hennigan A, Trotter C, Kelly AM. Lipopolysaccharide impairs long-term potentiation and recognition memory and increases p75NTR expression in the rat dentate gyrus. Brain Res. 2007;1130(1):158–66.
225. Zhang J, et al. Microglial CR3 activation triggers long-term synaptic depression in the hippocampus via NADPH oxidase. Neuron. 2014;82(1):195–207.
226. Guan Z, Fang J. Peripheral immune activation by lipopolysaccharide decreases neurotrophins in the cortex and hippocampus in rats. Brain Behav Immun. 2006;20(1):64–71.
227. Belarbi K, et al. Chronic neuroinflammation impacts the recruitment of adult-born neurons into behaviorally relevant hippocampal networks. Brain Behav Immun. 2012;26(1):18–23.
228. Jakubs K, et al. Inflammation regulates functional integration of neurons born in adult brain. J Neurosci. 2008;28(47):12477–88.
229. Ji HF, Zhang HY. Multipotent natural agents to combat Alzheimer's disease. Functional spectrum and structural features. Acta Pharmacol Sin. 2008;29(2):143–51.
230. Mishra S, Palanivelu K. The effect of curcumin (turmeric) on Alzheimer's disease: an overview. Ann Indian Acad Neurol. 2008;11(1):13–9.
231. Rezai-Zadeh K, et al. Green tea epigallocatechin-3-gallate (EGCG) modulates amyloid precursor protein cleavage and reduces cerebral amyloidosis in Alzheimer transgenic mice. J Neurosci. 2005;25(38):8807–14.
232. Smith JC, et al. Semantic memory functional MRI and cognitive function after exercise intervention in mild cognitive impairment. J Alzheimers Dis. 2013;37(1):197–215.
233. Qian L, et al. NADPH oxidase inhibitor DPI is neuroprotective at femtomolar concentrations through inhibition of microglia over-activation. Parkinsonism Relat Disord. 2007;13: S316–20.
234. Wang Q, et al. Post-treatment with an ultra-low dose of NADPH oxidase inhibitor diphenyleneiodonium attenuates disease progression in multiple Parkinson's disease models. Brain. 2015;138(Pt 5):1247–62.
235. Tillerson JL, et al. Exercise induces behavioral recovery and attenuates neurochemical deficits in rodent models of Parkinson's disease. Neuroscience. 2003;119(3):899–911.
236. Maswood N, et al. Caloric restriction increases neurotrophic factor levels and attenuates neurochemical and behavioral deficits in a primate model of Parkinson's disease. Proc Natl Acad Sci U S A. 2004;101(52):18171–6.

Ambient Particles and Cerebrovascular Disease

Mark R. Miller and Anoop S.V. Shah

Abbreviations

BBB	Blood brain barrier
CDNP	Combustion-derived nanoparticles
CI	Confidence interval
CNS	Central nervous system
DEP	Diesel exhaust particulate
eNOS	Endothelial nitric oxide synthase
ICAM-1	Intracellular adhesion molecule-1
NAD(P)H	Nicotinamide adenine dinucleotide phosphate (reduced)
NMDA	*N*-Methyl-D-aspartate
NO	Nitric oxide
NO_2	Nitrogen dioxide
O_3	Ozone
PM	Particulate matter
$PM_{0.1}$	Particulate matter with a diameter of less than 0.1 micrometers (nanoparticles)
PM_{10}	Particulate matter with a diameter of less than 10 micrometers
$PM_{2.5}$	Particulate matter with a diameter of less than 2.5 micrometers
RR	Relative risk
SO_2	Sulphur dioxide
SPSHR	Stroke-prone spontaneously hypertensive rat
TNF_{α}	Tumour necrosis factor alpha

M.R. Miller (✉) • A.S.V. Shah
University/BHF Centre of Cardiovascular Science, University of Edinburgh,
Queens Medical Research Institute, 47 Little France Crescent, Edinburgh EH16 4TJ, UK
e-mail: mark.miller@ed.ac.uk

S.C. Bondy, A. Campbell (eds.), *Inflammation, Aging, and Oxidative Stress*,
Oxidative Stress in Applied Basic Research and Clinical Practice,
DOI 10.1007/978-3-319-33486-8_8

t-PA	Tissue plasminogen activator
TRPV1	Transient receptor potential cation channel subfamily V member 1
UFP	Ultrafine particle
VCAM-1	Vascular adhesion molecule-1

1 Introduction

The health effects of air pollution are striking, with estimates suggesting that it may be responsible for up to seven million deaths worldwide per year [1–3]. Over the last decade it is becoming clear that the cardiovascular effects of air pollution are especially important to both mortality and long-term morbidity. Perhaps this is unsurprising given the high prevalence of cardiovascular disease in many different socioeconomic settings, and the broad physicochemical similarities between many sources air pollution and smoking, which is an undeniably important risk factor for cardiovascular disease. Our understanding of the biological mechanisms of air pollution are rapidly growing, however, the emphasis of much of this research has surrounded the cardiac effects and systemic vasculature with particular relation to the coronary circulation. Nevertheless, there is now a wealth of epidemiological evidence linking air pollution and stoke. However, the biological mechanisms for how air pollution affects the cerebral vasculature have received much less attention.

Air pollution is an expansive term to cover a wide range of different sources and types of pollutants. Urban air pollution consists of solid matter ('particles'), volatile liquids and gases. Levels of gaseous pollutants, such as nitrogen dioxide and ozone, in urban environments do represent a risk to human health, however, it is the particles in air pollution that are most consistently associated with cardiovascular disease. This review will look at the evidence for the link between ambient particles in the air and cerebrovascular disease. It will provide an overview of the epidemiological evidence for airborne particulate matter (PM) and the incidence of stroke, before moving onto possible biological mechanisms for these associations.

2 Background

2.1 Cerebrovascular Disease: What Is Stroke?

Stroke, (also referred to as cerebrovascular accident or cerebrovascular insult) is the neurological response to a region of damage of the brain due to an interruption in blood flow. The majority of strokes are ischemic in nature associated with a sudden arrest of the blood supply to the brain primarily due to a thromboembolic event. A smaller proportion (30 %) of strokes occur due to an intracranial bleed leading to a haemorrhagic stroke.

Stroke accounts for five million deaths per annum [4]. In 2013, stroke was the second most frequent cause of death after coronary artery disease, accounting for 6.4 million deaths (12 % of the total) [5]. Overall two thirds of strokes occurred in those over 65 years old [5, 6]. The epidemiology of strokes is changing, with an increasing incidence in the low- and middle-income nations accounting for two-thirds of all strokes [6]. The global burden of stroke-related mortality and disability is therefore high and continues to rise especially in nations where health systems are not well established. This has been primarily attributed to a change in risk factors associated with stroke, including an ageing population compounded by the accumulation of risk factors such as increased smoking, hypertension and obesity, especially in low- and middle-income countries [7, 8]. While there is increasing recognition of the association of air pollution with stroke, pollution is not considered as a classical 'risk factor' in terms of patient care.

Importantly, stroke is the most common cause of major disability [4]. Symptoms and severity of the stroke vary considerable depending on the region of the brain affected by ischaemia/haemorrhage, the length of the ischaemic period, extent of haemorrhage, degree of reperfusion should it occur, health status of the individual, etc. Thus symptoms can range from small degrees of sensory deprivation/abnormalities, e.g. dizziness, impaired vision, and severe headache, to more debilitating consequences such as loss of sensation/movement in one side of the body/face, loss of control of bodily functions, impaired ability to reason or speak, as well as coma or death. Prognosis is similarly variable, however, a broad assessment of the condition as a whole suggests that approximately 50 % of those who suffer stroke live less than one year afterwards [9]. Rapidity of diagnosis and treatment is important, however, therapies are limited. For ischaemic stroke aspirin is used to prevent further blood clotting, and in some causes infusion of recombinant tissue plasminogen activator (t-PA) can be used to attempt to remove existing clots. Surgical intervention can also be used in some scenarios. Otherwise, reducing risk factors (lifestyle and pharmacologically, e.g. anticoagulents, aspirin, clopidogrel, statins, blood pressure lowering drugs) with rehabilitation (both within a stroke unit and long-term as an out-patient) to overcome the symptoms of the stroke remain the foremost therapeutic option.

2.2 Particulate Matter in the Air

Particulate matter in air pollution is categorised according to size: PM_{10} (or "coarse" particles) are particles with a diameter of 10 micrometers or less, $PM_{2.5}$ ("fine particles") cover diameters of less than 2.5 μm, and $PM_{0.1}$ ("ultrafine" particles or "nanoparticles") are restricted to particles with a diameter less than 100 nm [10]. In urban environments, coarse particles principally derive from airborne crustal elements, road dust/brake wear, construction dusts or mechanical abrasion, e.g. breakwear. Fine fractions in urban environments chiefly arise from the combustion of fossil fuels, e.g. from power stations and vehicle emissions. Combustion-derived nanoparticles (CDNPs) from vehicle exhausts dominate the source of ultrafine particles, and certain fuel types such as diesel exhaust are especially rich in nanoparticles.

The ability of the particle to cause harm depends on numerous factors: concentration, the dose and period of exposure, where in the body the particle has access to, the ability of the body to clear/defend against the particle, and the particle composition and reactivity. For a given particle, the a general rule is that the smaller the particle the more potential harm it may cause due to the greater reactive surface area for a given mass, and ability to penetrate deep into lung. Accordingly, associations with cardiovascular disease tend to be more consistent for $PM_{2.5}$ than PM_{10}. Ultrafine particles readily penetrate the alveolar spaces of the lung, and their small size may mean that they have the capacity to cross the alveolar-capillary wall and gain access to the bloodstream to reach systemic organs. Discussion of other factors influencing particle toxicity are beyond the scope of this review, however, it is important to emphasis the composition of the particle is key to its biological actions once it accesses the biological system of interest. CDNPs can be viewed as having a largely elemental carbon core, onto which other chemical species adhere to - in particular a vast cocktail of surface chemicals that are rich in organic carbon species and reactive transition metals. Both of these constituents, as well as the general electrochemical properties of the carbon surface, are believed to be major drivers of inflammation and oxidative stress that are hallmarks of the detrimental actions of CDNPs. This will be discussed in Sect. 4.5.

Finally, although this review concentrates on the particulate components on air pollution, we do not wish to overlook gaseous co-pollutants. Indeed epidemiological studies frequently link cerebrovascular events with gases such as nitrogen dioxide (NO_2) and ozone (O_3). The consistency of these associations and underlying biology is less understood, and as such will not be discussed in any depth here. However, it should be mentioned that that these two constituents of air pollution cannot be entirely separated, and the chemical interaction between particles and gases may synergistically increase the oxidative reactivity of the particle exposure [11–14].

3 Air Pollution and Cerebrovascular Disease

3.1 Epidemiological and Clinical Evidence

Clinical and epidemiological evidence gathered over the last six decades have established a long-standing and close temporal relationship with adverse health effects. These include the extreme episodes of the Meuse Valley fog in 1930 [15] and the London fog incident in 1952 [16], to recent studies implicating ambient air pollution as a major perpetrator of adverse health effects [17–19]. Globally cardiovascular disease including coronary disease and cerebrovascular disease remain the primary cause of mortality and morbidity, and ambient particulate pollution alone is responsible for approximately 3.2 million deaths worldwide [4, 20].

3.1.1 Long-Term Exposure to Ambient Air Pollution and Cardiovascular Events

Many elegant longitudinal studies have evaluated the association between long-term exposure to fine particulate pollution and cardiovascular disease. Miller et al. investigated post-menopausal women and showed a 35 % increased risk of cerebrovascular event (RR 1.35, 95 % CI 1.08–1.68) for every 10 μg/m^3 increment in $PM_{2.5}$ concentrations, compared to 21 % increased risk for coronary disease. Futhermore, there was a striking 83 % increased risk of death caused by a cerebrovascular event (RR 1.83, 95 % CI 1.11–3.00) per 10 μg/m^3 increment of $PM_{2.5}$. Several other large cohort studies in Europe have shown similar positive associations between long-term exposure to ambient air pollution and coronary and cerebrovascular events [21–23]. Staffogia et al. showed that even in high-income nations where fine particulate pollutant concentrations meet international standards, small increases in $PM_{2.5}$ were associated with a 19 % increase in the risk of cerebrovascular disease [23]. A few studies have compared both exposure to $PM_{2.5}$ and PM_{10} have shown stronger associations with exposure to smaller particulates [24].

3.1.2 Short-Term Exposure to Ambient Air Pollution and Cardiovascular Events

Patients with cardiovascular disease share many modifiable and non-modifiable risk factors including ethnicity, family history, hypertension and smoking [25–27]. One of the key differences between these risk factors and environmental air pollution is that exposure to pollutants is, for many, unavoidable. Therefore whist the individual risk estimates for exposure to ambient air pollution are relatively small compared to other cardiovascular risk factors, since the exposure can be assumed to be 100 %, the overall population attributable risk is significant [25].

In the early 1980s, Knox et al. proposed that air pollution was a means by which weather conditions could contribute to the relationship between weather conditions and stroke [28]. Later that decade, indoor coal fumes were highlighted as a risk factor for stroke, independent of age, blood pressure and cigarette smoking [29]. Schwartz et al. (1994) looked at the relationship daily PM recordings and cause of death in the US between 1973 and 1980. They noticed an weak, but sizeable (RR = 1.19) association between PM and stroke, discussing the findings in relation to those of the London smog episode in 1952 [30]. Subsequently, there has been a plethora of studies considering these associations with more recent data; these are reviewed in detail elsewhere [31]. Of note, though, Dominici et al. investigated the effect of short-term increases in particulate pollution and risk of admission from cardiopulmonary disease across one of the largest cohorts, involving 204 US counties [19]. The study showed that a reduction of 10 μg/m^3 of $PM_{2.5}$ concentrations would reduce approximately 1 in 10 stroke admissions. Diesel exhaust is likely to be a prominent contributor to the effects of urban PM on stroke in Europe [32].

Meta-analysis of ecological studies has already shown a consistent effect of both gaseous and particulate air pollution and adverse cardiac events [17, 23, 33, 34]. A recent meta-analysis has now shown that both incident stroke and mortality form stroke is associated with both gaseous and particulate pollutants [18] (Fig. 1). Across 103 studies evaluating short-term exposure to air pollution involving 6.2 million incident strokes or stroke deaths, there was a 1.1 % increment in the risk of stroke per 10 μg/m^3 increment in $PM_{2.5}$ levels. The associations were strongest on the day of the event and showed a close temporal relationship between exposure and outcome [18]. Interestingly, exposure to PM_{10} showed weaker associations with stroke hospitalisation and mortality compared to $PM_{2.5}$ consistent with observations from other meta-analyses for stroke [35] and long-term exposure studies [24]. The underlying reasons explaining these differences remain unclear, especially when compared to associations with other pathologies like heart failure where the adverse effects of the larger particles was more striking. These differences in particle size and adverse stroke events likely reflect differences in particle composition. However,

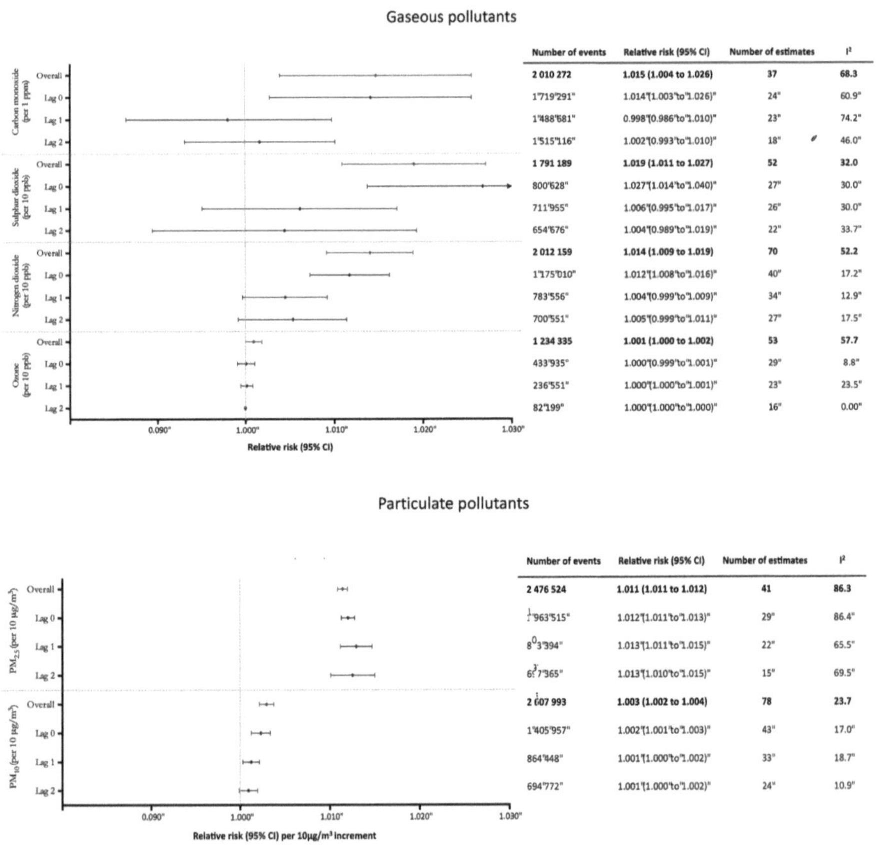

Fig. 1 Meta-analysis of the association of air pollutants and stroke worldwide, stratified by time lag (days). Adapted from Shah et al. 2015 [18]

mechanistic pathways will also be an influence, with larger particles exerting local pulmonary effects, whereas fine or ultra-fine particles cause additional systemic cardiovascular effects through alternative means [10]. The adverse effect of larger particles in patients with heart failure was more striking, possibly reflecting indirect biological pathways such as the adverse effect of sympathetic nervous activation on the failing heart [10, 33]. The effect estimates noticed for particulate matter were consistent with meta-estimates observed for coronary artery disease [17], although a recent study in Germany suggests that the effect size of PM could be greater for stroke than coronary artery disease [36]. Of concern, a recent study performed in Stockholm found associations for both PM_{10} and NO_x with stroke that, while the effect size was small, occurred at low levels of these pollutants [37], confirming previous studies demonstrating these associations at PM levels below that suggested by the US Environmental Protection Agency [38, 39].

3.1.3 Clinical Studies in Humans Exploring the Link Between Air Pollution and Stroke

Unlike in cardiac pathologies [40, 41], controlled exposure studies in patients with stroke remain sparse. However a few longitudinal studies do reveal potential mechanisms linking exposure to air pollution and stroke. Wellenius et al. showed that sub-acute exposure to $PM_{2.5}$ over a few weeks was associated with alteration in cerebral haemodynamics including increased cerebrovascular resistance and reduced cerebral blood flow [42]. In a further cross-sectional analysis Newman et al. showed a positive association between presence of carotid artery stenosis and exposure to $PM_{2.5}$, with a twofold increased risk of stenosis per 10 μg/m^3 of $PM_{2.5}$. This association was persistent despite adjustment for known risk factors for atherosclerosis including age, sex, ethnicity, body mass index, cholesterol levels and hypertension [43]. Biological mechanisms which could account for these observations are discussed in [44] and below.

3.2 Studies in Cells and Animals

There are limited options to perform insightful controlled exposure studies to examine the relationship between air pollution in stroke in man. Thus it falls to preclinical studies to dissect mechanisms more precisely (for reviews of various experimental models of stroke, see [45–47]). Yet there are surprisingly few cellular or animal studies that explore the mechanistic actions of PM in stroke, largely due to the difficulty of recreating the multiple pathophysiological processes of stroke with 'simple' experimental models. Cell culture studies are a key means to investigate the specific cellular processes of particle toxicology. Cultures of astrocytes, neurons, microglia can be used to investigate the cellular response to ischaemia, with additional levels of complexity provided through co-cultures, conditioned media or slices of brain tissue bathed in physiological, hypoxic or glucose-depleted buffers. *In vivo* models of stroke clearly have a greater physiological relevance, whereby a cerebroarterial insult is provided through surgical means, pharmacological tools or genetic modifications. Although the insult itself will not necessarily reflect the stimuli that induces stroke physiologically, these models have been employed to provide useful insight into the action of PM on the response to cerebrovascular injury.

3.2.1 Cell Studies

The ultimate consequence of stroke is the loss of blood flow to an area of the brain, causing ischaemic damage to the cells downstream. Ischaemia causes a series of biomolecular cascades that can culminate in neuropathological changes: alterations in glutamatergic activity, cellular calcium redistribution, changes in neuronal excitability, neuroinflammation, ionic imbalance, oedema, oxidative stress, changes to mitochondrial function, neuronal apoptosis, upregulation of matrix metalloproteinases, cell death, etc. CDNPs have been shown to promote many of these pathways of ischaemic damage. For example, cortical neurons were shown to be especially susceptible to oxygen and glucose deprivation in the presence of PM_{10} from a Chinese city [48]. Particle toxicity varied with season, suggesting differences in particle source and/or composition across the year. Davis et al. 2013 tested the ultrafine fraction of PM on hippocampal slices *in vitro*, demonstrating a number of alterations in synaptic function mediated via glutamatergic transmission and nitrosylation of *N*-methyl-D-aspartate (NMDA) channels [49]. Inflammation and oxidative stress are key drivers of particle-induced changes in neuronal preparations *in vitro*, which coincides with the inflammatory responses in the blood to PM_{10} exposure in man (see Sect. 4.5). The glutamatergic effects of ultrafine PM in cultured neuronal and glial cells were closely associated with inflammatory cytokines and the ability of the PM itself to generate free radicals [50]. Diesel exhaust particulate (DEP) altered the levels of dopamine and its metabolites in PC-12 cells via, and potentially contributing to, reactive oxygen species generation [51]. Both oxidative stress and inflammation are key pathways in the response to urban nanoparticles in (co-)cultured brain cells, although the balance of molecular pathways depends on the interaction between cells and the species origin of the cells [52]. Primary cultures of microvascular cells can also be used to model cerebral arteries. In particular the endothelial cells that line the inner surface of arteries that are crucial to vascular health, and an early target in the disease processes underlying cerebrovascular disease. Reactive oxygen species once again are a prominent mechanism of DEP action in this cell type [53, 54].

3.2.2 Animal Studies

Several animal models are used to investigate stroke. Focal or global ischaemia can be generated by permanent or temporary surgical ligation, or advancement of an intravascular filament/emboli into the cerebral artery. Injury can be produced through electrothrombosis or photochemical injury. Alternatively, stroke-prone spontaneously hypertensive rats (SPSHR) or infusion of potent vasoconstrictors can be also used. Infusion of collagenase provides a model of haemorrhagic stroke. The endpoint studied is usually infarct (or haemotoma) size or neurological defect (e.g. sensorimotor/behavioural). Histochemical staining of the brain or changes in blood brain barrier function leucocyte infiltration or platelet adhesion provide additional mechanistic insight.

In relation to air pollution, several groups have studied the effect of inhaled particles on the central actions (e.g. neurotoxicity and neurodegeneration) of exposure (see [55, 56] for review). However, these models reflect downstream actions of stroke as opposed the vascular effects that instigate them. Nevertheless, many investigate mechanisms that provide a closer link between the vasculature and downstream pathology induced by PM. Guo et al. repeatedly instilled rats with PM_{10} from a coal-burning China city [48]. Electron microscopy analysis of brain slices from the animals demonstrated morphological changes in synapses that correlated with dysregulation of the vascular endothelium and inflammation of the cortex. The biological actions were associated with carbon content and levels polyaromatic hydrocarbons in the brain. Parallel *in vitro* models of brain slices in deoxygenated buffers showed a similar susceptibility to PM. Morgan et al. [50] collected urban ultrafine particles (UFPs) and demonstrated they had the capacity to generate free radicals in the absence of cells. The UFPs were resuspended and given to rats by aerosolisation or directly to brain slices *in vitro. In vivo* exposure changed glutamatergic function in neuronal and glial cells, and increased levels of inflammatory cytokines. Similar pathways were activated in embryonic neuron cultures *in vitro*, leading the authors to conclude that urban nanoparticles have the capacity to detrimentally affect the response to cerebral ischaemia via both direct and indirect means.

Intratracheal instillation of animals with a suspension of PM is a useful way to explore specifically the particulate components of air pollutants in a manner that resembles inhalation exposure. Walleborn et al. made a comparison of the response to instillation of oil-combustion PM in healthy versus SPSHR [57]. While stroke itself was not explored, the SPSHR showed a greater increase in cardiac ferritin and expression of genes linked to oxidative stress. Interestingly, the levels of lung inflammation were similar in both strains. Further evidence of the role of oxidative stress is the upregulation of haemoxygenase expression (a key antioxidant pathway) in the brain after instillation of PM_{10} in mice [58]. The content of transition metals in inhaled PM from urban environments correlated with biological action, suggesting a role for metal-derived free radical generation [56, 59].

PM has also been shown to affect blood-brain barrier (BBB) function [60]. DEP upregulates the efflux transporter P-glycoprotein in isolated cerebral arteries [61]. The effect could be prevented through inhibition of NAD(P)H oxidase or TNF_{α} blockade, again indicating a causative role for both oxidative stress and inflammation. In a detailed study, Oppenheim et al. (2013) used Apolipoprotein-E knockout mice to show that mixed vehicle exhaust altered BBB function *in vivo* [62]. Exposure was associated with decreased levels of tight junction proteins (occludin and claudin-5), increased matrix metalloproteinase activity in cerebral microvessels, and various indications of inflammation were evident in the parenchyma. Interestingly serum from these animals had similar effects on an *in vitro* model. Taken as a whole, the study provides strong evidence for a mode of action; PM inhalation leads to a blood-borne mediator, that disrupts the integrity of the vasculature in the blood brain barrier, resulting in increased sensitivity to a proinflammatory/pro-oxidative insult within the brain.

4 Biological Mechanisms: How Could Particles Affect the Cerebral Circulation?

The epidemiological studies clearly show that airborne PM increases the risk of stroke, and preclinical studies have shed light on the possible means through which particles can exacerbate the response to injury. But there is a missing link: what is(/are) the mechanism(s) by which ambient PM *causes* stroke? Essentially, what needs to be established is how inhalation of particles promotes the blockage or rupture of cerebral arteries? The link from lung to artery is a reoccurring question in the field of particle toxicology and a block that we have only chipped slowly away at over the last few decades. While the cerebral circulation is a challenging vascular bed to address, many inferences can be drawn based on our knowledge of how inhaled particles affect other systemic arteries (see Fig. 2 for an overview).

4.1 From the Lung to the Artery

The means in which inhaled particles enter and deposit in the alveoli of the lungs have been well characterised. But how then do inhaled particles have cardiovascular actions: what is the 'signal' that links the lung to the heart and blood vessels? This question has hindered research in this field for many years, although three hypotheses have emerged. The traditional ('inflammation') hypothesis is that inhaled particles are ingested by alveolar macrophages, activating these cells to an extent that induces a significant pulmonary inflammation. Inflammatory mediators then 'spill-over' into the circulation and alter cardiovascular function 'indirectly' [63]. Alternatively, particles can stimulate sensory receptors on the alveolar surface providing a neural stimulus that results in changes in autonomic function, altering cardiovascular homeostasis, particularly that of the heart [64, 65]. More recently it has been hypothesised that the minute size of nanoparticles allows them to cross ("translocate") into the blood themselves and directly interact with the vasculature [66]. There is evidence for and against each hypothesis, and it is likely that all three pathways, and others, play a role to some extent; the degree of which will depend on which facet of the cardiovascular system is being investigated [67]. None of these pathways is specifically tailored to the subject of cerebrovascular disease, yet all three clearly have relevance. Specific examples of these linking mechanisms are discussed in the sections below.

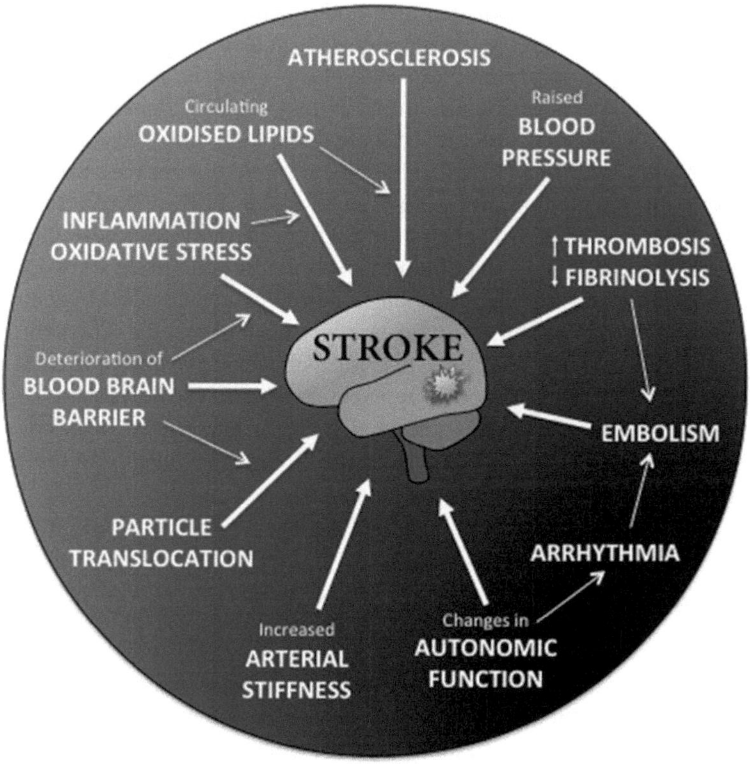

Fig. 2 Schematic of the multiple mechanisms through which inhaled particles may predispose or cause stroke. Note, the many interactions between these pathways, of which the diagram shows only a small selection

4.2 Before the Lung

The translocation hypothesis for the cardiovascular effects of nanoparticles is rapidly gaining ground. Undoubtedly particle translocation between organ systems can occur, what remains to be established is the exact mechanism for these processes and their fate of particles once they enter the systemic circulation. The bulk of attention has been the transit of particles across the alveolar wall, however, for many years the possibility of particle translocation to central regions has been postulated to occur prior to the lung, at levels of the nasal passage, particularly the olfactory bulb. The olfactory bulb is a multi-layered structure that allows a neural network to transmit the sense of smell from the nasal cavity to the brain. The structure has a blood capillary network and the presence of several channel structures is such that there is a means by which molecules can be delivered to the brain that would not otherwise cross the blood brain barrier; a feature which has been exploited for intranasal delivery of pharmaceuticals [68]. Due to the high dose of particles in the nasal cavities and the route of air-flow, a proportion of particles with a low nanometer size will deposit in this area and potentially could pass through this structure [69–71]. Furthermore, particles have been identified in the olfactory bulbs of a small number of autopsies from individuals living in areas of very high air pollution [72]. Indeed, evidence from animal studies demonstrates a clear inflammatory response in specific brain regions following inhalation/intranasal exposure to ultrafine PM [73–78]. Inflammation may disrupt the structure of the nasal respiratory-epithelium, allowing greater access of particle/cytokines across this region [60]. The degree of penetration after this barrier is less clear, in particular to what extent particles may enter blood vessels at this juncture to allow them to be carried to other central structures (although there is good evidence of penetration of some constituents of PM at the very least; [79]). Regardless, the direct detrimental effects of particles on neuronal cells, discussed above, suggests that changes in cell excitability could be transmitted to other brain structures. Several reviews have been written on the role of this potential route in age-related diseases such as Alzheimer's and Parkinson's disease [60, 80–82], but the relevance of this pathway for stroke has not been determined.

4.3 Arrhythmia

Activation of the autonomic nervous system by inhaled particles is probably the most clearly defined of the three hypotheses for the cardiovascular effects of inhaled particles. Particles are known to induce cardiac effects which can be blocked using pharmacological inhibitors of alveolar sensory receptors (such as TRPV1 receptors) or blockade of the autonomic nervous system [83, 84]. There is an extensive literature showing that inhaled urban particles reduces several parameters of heart rate variability [65, 85]; a parameter that is heavily regulated by the autonomic nervous system and a predictor of acute coronary events, and to some extent stroke itself [86]. Furthermore, several studies have demonstrated that particles change the sensitivity of baroreceptors that regulate reflex responses to vasodilation to maintain blood pressure homeostasis [87–89]. Some areas of the cerebral circulation have a form of autonomic innervation [90], but at present it is not known if this pathway contributes to the association of ambient particles with stroke.

Of immediate relevance to stroke, though, is the ability of ambient PM to induce arrhythmias—irregular heart beating/contraction. Atrial fibrillation, in particular, has been estimated to increase the risk of stroke by fivefold [91]. Several factors are likely to contribute to this association, however, thrombotic disturbances are especially important [92], e.g. increased blood coagulability and/or embolization of static thrombus, resulting in obstruction of downstream cerebral arteries. Urban PM and DEP have been shown to increase the frequency and duration of arrhythmia in various animal models of susceptibility (e.g. [83, 93–96]). However, there is limited evidence for a link between arrhythmia and acute exposure to air pollution in man, based on controlled exposures of a number of different particle types [97]. Whether such effects impact the prevalence of stroke requires further investigation.

4.4 Blood Pressure and Arterial Stiffness

Urban air pollution has been shown to be associated with an increase in blood pressure. Several mechanisms contribute to this, including loss of endogenous endothelium-derived vasodilator function, changes to baroreceptor autoregulation and increasing the synthesis and sensitivity to blood-borne vasoconstrictors such as endothelin and angiotensin II [98–101]. The acute increases in blood pressure are generally of a low magnitude (<5 mmHg) following controlled exposures in man, although there is clearly potential for much larger changes (up to 30 mmHg, or even greater) in animal studies that typical employ a higher dose of particulate (see [102, 103]). Nevertheless the prevalence of cardiovascular disease around the world, means that even a small increase in blood pressure will be associated a vast increase in the incidence of cardiovascular morbidity and mortality. Raised blood pressure is also a known risk factor for stroke, especially haemorrhage stroke, by increasing the propensity for susceptible vessels to rupture. Disentangling the direct and indirect mechanisms in which raised blood pressure increases the incidence and severity of stroke remains challenging [104], and beyond the scope of the current review. Nevertheless, despite controversy in regards to its effectiveness, lowering of blood pressure remains one of the main therapeutic strategies once a stroke has occurred [105–109].

Similarly, blood vessel fragility is also influenced by the distensibility (or lack of plasticity) of arteries that is evident with increasing age and cardiovascular disease. Arterial stiffness is a term for vascular 'aging' due to alterations in smooth muscle cell phenotype and wall structure. Arterial stiffness can be measured in peripheral arteries (e.g. brachial, ankle, carotid, femoral) through tonometry, measuring changes in the blood pulse wave propagation that provides a surrogate of the arterial stiffness in other vascular beds. In particular, peripheral measures of arterial stiffness have been shown to correlation with white matter hyperintensities, lacunar infarcts and cognitive decline [110]. Measures of arterial stiffness using ambulatory blood pressure have also been shown to be associated with stroke [111]. Although it is not possible to directly measure arterial stiffness in cerebral arteries, air pollution has been shown to increase indices of peripheral arterial stiffness. Mehta et al. (2014) demonstrated that short-term exposure to air pollution ($PM_{2.5}$ and sulphate within emissions) was associated with arterial stiffness in elderly men living in the city [112]. There is also an increase in arterial stiffness in children when stratified by the distance of their residence from a major road [113]. Controlled exposures to PM from vehicle exhaust [114] and wood smoke [115] have also been shown to increase arterial stiffness.

4.5 Blood-Borne Mediators

Inflammatory processes, as well as a range of other candidate blood-borne mediators, will have a wide-ranging effects on the cardiovascular system, many of which will reach and affect the cerebral circulation. Some are known to be directly linked to atherothrombotic disease that underlies the majority of ischaemic vessel disease, as well as being risk factors for aneurysm or exacerbation of the haemorrhagic stroke. Examples of these pathways are discussed below.

4.5.1 Inflammation

Inflammation plays a role in many diseases/conditions, and stroke is no different [116]. Inflammation of the cerebral arteries will lead to inflammatory cell infiltration into the vessel wall, impairing endothelial cell function and contribute to pathophysiology, e.g. risk of thrombosis, exacerbation of atherosclerosis and susceptibility to aneurysm. Similarly, exposure to urban PM has been shown to increase levels of circulating leucocytes and cytokines such as tumour necrosis factor alpha (TNF_{α}), interleukin-6, as well as acute-phase response proteins such as C-reactive protein (CRP), serum amyloid A and fibrinogen [117–120]. Other inflammatory molecules are also frequently upregulated, including adhesion molecules (ICAM-1, VCAM-1, P-selectin). Pulmonary exposure to PM in animals increases similar markers of inflammation in lung and the brain parenchyma [121]. Interestingly, intracerebral administration of pharmacological agents to block inflammatory pathways decreases the effects of inhaled $PM_{2.5}$ in a mouse model of diabetes [122]. Although causality cannot be implied, inflammatory mediators and oxidative stress provide plausible mechanisms to link particle exposure and cardiovascular disease [58]. Whether inflammation alone can account for the diverse cardiovascular actions of inhaled PM is uncertain [67], however, inflammation will undoubtedly be associated with a worsening of cardiovascular function and disease [123]. This will also be true of cerebrovascular disease where inflammation is considered to be a risk factor for both prevalence and severity of stroke.

4.5.2 Oxidative Stress

Inflammation and oxidative stress go hand-in-hand in most disease processes, often in a synergistic manner. This is especially true of ambient particles which have the capacity to generate oxygen-derived free radicals from the particle surface itself, and a propensity for inflammatory cell activation following engulfment of particles. The sources and biological actions of particle-derived oxidative stress have been reviewed previously [67, 102]. Oxidative stress itself plays many and varied roles in several stages of the pathophysiological actions of stroke, in response to both ischaemia and reperfusion, should it occur (see [124–126]). Blood vessel function is notably affected by oxidative stress, through the scavenging of nitric oxide (NO); a mediator produced by the vascular endothelium that controls blood vessel tone. NO is a key mediator of homeostasis in cerebral blood vessels, as well as having a functional role in other cells within the central nervous system (CNS) [127]. Cells within the CNS have antioxidant protection mechanisms like all other cells, however, the cellular reserves of these are considered to be lower than many other cell types [128]. Furthermore, the brain has high levels of oxygen consumption and contains high levels of various fatty acids that are highly susceptible to oxidative stress. Oxidised lipids and arachidonic acid metabolites (such as isoprostanes), among others, are carried by the blood and there is the increasing evidence that these molecules are active species (as opposed to just biomarkers of oxidative stress) [129]. Certainly there appears to be a close relationship between oxidative stress and lipid modification in the brain and systemic organs of PM_{10}-exposed mice [58]. Accordingly the depletion of antioxidants is a frequent observation in cellular models of cerebrovascular damage. Interestingly deficiency of antioxidants is also associated with atrial arrhythmia, another risk factor for stroke [130]. The role of oxidative stress is key to many of processes of stroke, however, at present antioxidant therapies have proved relatively unsuccessful, most likely due to the delay in starting therapy after critical oxidative damage has occurred [125, 126].

4.5.3 Lipids

The oxidation of blood-borne lipoproteins is a key step in the initiation of atherosclerosis (see below). DEP has been shown to oxidised low-density lipoprotein and potentiate atherosclerosis [54]. Indeed there is a synergistic effect of PM and low density lipoproteins in terms of regulating the gene expression of multiple pathways associated with cardiovascular disease. Furthermore, ambient particles are associated with an increase in plasma lipoprotein-associated phospholipase A_2 (an enzyme that generates various lipid products that can trigger an inflammatory cascade); an independent risk factor for stroke [131]. Interestingly, there is increasing recognition of the influence of blood lipid profile on age-related conditions such as Alzheimer's disease ([132]; see also previous chapters). Accordingly, statins have been shown to reduce the risk of stroke; the efficacy of which will undoubtedly reflect their pleiotropic effects (e.g. anti-inflammatory, anti-platelet, dampening oxidative damage, atherosclerotic plaque stabilisation) in addition to being lipid lowering [104, 133]. Finally there is the possibility that lipid modification may also affect the structural integrity of the blood brain barrier [134], granting better access to the brain for unwanted blood-borne mediators, and perhaps particles themselves.

4.6 Thrombosis

High levels of PM are also associated with a change in the balance of thrombosis/thrombolysis to augment clotting. This possibility was first raised from the observation that PM was associated with increased levels of fibrinogen [135]. Subsequently, a variety of factors in the coagulation cascade, platelet aggregability, or other aspects of hemostasis solidified this observation [136, 137]. Raised blood coagulability will increase the likelihood of occlusive thrombus and embolization that underlies ischaemic stroke. Similarly, platelet activation interacts with a number of other risk factors/causes of stroke, and antiplatelet drugs reduce the risk of ischaemic stroke (although counter-indicated in haemorrhagic stroke). Exposure to urban air pollution is associated with platelet hyperactivity [119, 138–140], and controlled exposure to vehicle exhaust increases the extent of platelet-rich thrombosis [141]. In an elegant study, Nemmar et al. demonstrated that pulmonary exposure to DEP exacerbated the extent of photochemical-induced thrombosis within cerebral microvessels *in vivo* [142]. Administration of an antioxidant cysteine pro-drug prevented the action of DEP.

One of the main therapeutic treatments for ischaemic stroke is administration of the recombinant t-PA. Controlled exposures to diesel exhaust have shown that there is not only a loss of vascular endothelium-derived NO (which inhibits platelet aggregability and platelet-leucocyte interactions, as well as controlling vascular tone) but also a loss of the release of endogenous t-PA from endothelial cells [143]. Since t-PA is far more effective at removing thrombosis if present at the initiation of the clot [144], as opposed to later on, exposure to air pollution in the early onset of ischaemic stroke may have significant effects on the severity of the stroke and effectiveness of other treatments.

4.7 From Car to Carotid

Atherosclerosis is the build up of lipid- and inflammatory cell-rich plaques at the bifurcations/curvature of large arteries. Atherosclerotic plaques are not uncommon in arteries leading to the CNS, e.g. the carotid and vertebral arteries. However, large vessel disease can also extend to the cerebral circulation itself, such as the middle cerebral artery or other major branches from the Circle of Willis [145]. Erosion of atherosclerotic plaques may result in the build up of thrombus on their surface, or immobilisation of the clot to block smaller arteries downstream in the cerebral circulation [146]. There is now a wealth of evidence linking long-term exposure to air pollutants to the prevalence of atherosclerosis. Animal studies have demonstrated that the particles in vehicle emissions increase both the size of atherosclerotic plaques and indicators of plaque vulnerability to rupture [147, 148]. This topic is reviewed in detail in Chap. "Jesus Araujo, on particles and atherosclerosis" of this book and [149].

Of interest, the early evidence for the proatherosclerotic effects of air pollution in man were demonstrated through measurement of plaques in the carotid artery in residents of high pollution areas or adjacent to major roads, [150–152]. The accessibility of the carotid artery to non-invasive imaging by ultrasound is ultimately the reason for this, although since the carotid is the major artery leading the cerebral circulation there is a clear relevance for stroke. While occlusion of the major carotid arteries is rarer than that of occlusion of the smaller downstream arteries [153], partial occlusion of the carotid arteries can be symptomatic, and its presence increases the likelihood of embolism from carotid plaque erosion.

For further discussion of the experimental models and mechanisms underlying the CNS effects of air pollution, see the excellent reviews by Genc et al. [79] and Calderon-Garciduenas and colleagues [82, 154].

5 Summary and Conclusions

Stroke remains the second most common cause of death and third most important cause of disability worldwide. Over the last 20 years many environmental studies have evaluated the association between outdoor air pollution and stroke with varying conclusions. Long-term and recent meta-analysis of short-term ecological studies have shown that both acute and chronic exposure to ambient pollutants are associated with cerebrovascular disease. Additionally, the diversity of location of studies provides insight into commonalities and highlights potentially susceptible populations in countries of varying socioeconomic status [18, 155]. Overall, air pollution is now a recognized risk factor for cardiorespiratory mortality—indeed several types of air pollutants are listed in the top-ten risk factors for all cause mortality worldwide [20]. On the evidence presented above, can we be even more implicit, and classify airborne PM as a risk factor for stroke?

Epidemiological studies have consistently highlighted small particulate matter in air pollution as a major risk factor for stroke. That is not to say that other constituents of air pollution are not detrimental, and human and animal studies have suggested an association for gases such as NO_2 [156], SO_2 [157] and O_3 [158], as well as less well-recognised larger particles in Asian dust storms [159, 160]. However, meta-analyses emphasis $PM_{2.5}$ as the major constituent of urban air pollution both in terms of consistency and magnitude of the relative risk of stroke. Ultrafine particles cannot currently be measured on a population level, however, evidence suggests that this proportion of PM could be especially likely to pose a risk of cerebrovascular events [161].

The exact mechanisms and biological plausibility on how exposure relates to outcome remains unclear. Identification of biological mechanisms would be greatly aided by more accurate models to assess the vascular events that precede stroke, and through the use of *in vivo* models of co-morbidity such as hypertension, atherosclerosis and lipid imbalance. However, we can gather much insight from what we know of how inhaled particles affect other arterial beds. Mechanistic pathways are likely to be many and act in concert [162]. They include particle translocation (into the blood stream and possibly across barriers to the CNS), changes in autonomic function, inflammation, oxidative stress, changes to blood coagulability, increased blood pressure and exacerbation of atherosclerosis to name but a few.

On a positive note, air quality is undoubtedly improving. While there is a long-way to go, there is an ever-increasing activity on the control of emissions, and also identifying which ones are harmful and why. With better techniques, biological models and predictive toxicological assays our understanding of particle toxicology is rapidly increasing. And accordingly, new interventions and initiatives to reduce these emissions will undoubtedly improve cardiovascular health, and hopefully reduce the incidence and severity of cerebrovascular disease.

Acknowledgements and Funding MRM is funded by a British Heart Foundation (BHF) Special Project Grant (SP/15/8/31575). ASVP is funded by a BHF Programme grant (RG/05/003) and BHF Clinical Research Scholarship (SS/CH/09/002).

Disclosures: All authors declare that they have no competing interests.

References

1. Brook RD, Rajagopalan S, Pope 3rd CA, Brook JR, Bhatnagar A, Diez-Roux AV, Holguin F, Hong Y, Luepker RV, Mittleman MA, Peters A, Siscovick D, Smith Jr SC, Whitsel L, Kaufman JD. Particulate matter air pollution and cardiovascular disease: an update to the scientific statement from the American Heart Association. Circulation. 2010;121(21): 2331–78.
2. World_Health_Organization. Air quality and health. In: WHO Factsheets. 2011. http://www.who.int/mediacentre/factsheets/fs313/en/.
3. World_Health_Organization. 7 Million premature deaths annually linked to air pollution. 2014. http://www.who.int/mediacentre/news/releases/2014/air-pollution/en/.
4. Lozano R, Naghavi M, Foreman K, Lim S, Shibuya K, Aboyans V, Abraham J, Adair T, Aggarwal R, Ahn SY, Alvarado M, Anderson HR, Anderson LM, Andrews KG, Atkinson C, Baddour LM, Barker-Collo S, Bartels DH, Bell ML, Benjamin EJ, Bennett D, Bhalla K, Bikbov B, Bin Abdulhak A, Birbeck G, Blyth F, Bolliger I, Boufous S, Bucello C, Burch M, Burney P, Carapetis J, Chen H, Chou D, Chugh SS, Coffeng LE, Colan SD, Colquhoun S, Colson KE, Condon J, Connor MD, Cooper LT, Corriere M, Cortinovis M, de Vaccaro KC, Couser W, Cowie BC, Criqui MH, Cross M, Dabhadkar KC, Dahodwala N, De Leo D, Degenhardt L, Delossantos A, Denenberg J, Des Jarlais DC, Dharmaratne SD, Dorsey ER, Driscoll T, Duber H, Ebel B, Erwin PJ, Espindola P, Ezzati M, Feigin V, Flaxman AD, Forouzanfar MH, Fowkes FG, Franklin R, Fransen M, Freeman MK, Gabriel SE, Gakidou E, Gaspari F, Gillum RF, Gonzalez-Medina D, Halasa YA, Haring D, Harrison JE, Havmoeller R, Hay RJ, Hoen B, Hotez PJ, Hoy D, Jacobsen KH, James SL, Jasrasaria R, Jayaraman S, Johns N, Karthikeyan G, Kassebaum N, Keren A, Khoo JP, Knowlton LM, Kobusingye O, Koranteng A, Krishnamurthi R, Lipnick M, Lipshultz SE, Ohno SL, Mabweijano J, MacIntyre MF, Mallinger L, March L, Marks GB, Marks R, Matsumori A, Matzopoulos R, Mayosi BM, McAnulty JH, McDermott MM, McGrath J, Mensah GA, Merriman TR, Michaud C, Miller M, Miller TR, Mock C, Mocumbi AO, Mokdad AA, Moran A, Mulholland K, Nair MN, Naldi L, Narayan KM, Nasseri K, Norman P, O'Donnell M, Omer SB, Ortblad K, Osborne R, Ozgediz D, Pahari B, Pandian JD, Rivero AP, Padilla RP, Perez-Ruiz F, Perico N, Phillips D, Pierce K, Pope CA, 3rd, Porrini E, Pourmalek F, Raju M, Ranganathan D, Rehm JT, Rein DB, Remuzzi G, Rivara FP, Roberts T, De Leon FR, Rosenfeld LC, Rushton L, Sacco RL, Salomon JA, Sampson U, Sanman E, Schwebel DC, Segui-Gomez M, Shepard DS, Singh D, Singleton J, Sliwa K, Smith E, Steer A, Taylor JA, Thomas B, Tleyjeh IM, Towbin JA, Truelsen T, Undurraga EA, Venketasubramanian N, Vijayakumar L, Vos T, Wagner GR, Wang M, Wang W, Watt K, Weinstock MA, Weintraub R, Wilkinson JD, Woolf AD, Wulf S, Yeh PH, Yip P, Zabetian A, Zheng ZJ, Lopez AD, Murray CJ, AlMazroa MA, Memish ZA. Global and regional mortality from 235 causes of death for 20 age groups in 1990 and 2010: a systematic analysis for the Global Burden of Disease Study 2010. Lancet. 2012;380(9859):2095–128.
5. Mortality GBD, Causes of Death Collaborators. Global, regional, and national age-sex specific all-cause and cause-specific mortality for 240 causes of death, 1990-2013: a systematic analysis for the Global Burden of Disease Study 2013. Lancet. 2015;385(9963):117–71.
6. Feigin VL, Forouzanfar MH, Krishnamurthi R, Mensah GA, Connor M, Bennett DA, Moran AE, Sacco RL, Anderson L, Truelsen T, O'Donnell M, Venketasubramanian N, Barker-Collo S, Lawes CM, Wang W, Shinohara Y, Witt E, Ezzati M, Naghavi M, Murray C, Global Burden

of Diseases, Injuries, and Risk Factors Study 2010 (GBD 2010) and the GBD Stroke Experts Group. Global and regional burden of stroke during 1990-2010: findings from the Global Burden of Disease Study 2010. Lancet. 2014;383(9913):245–54.
7. Mateen FJ, Brook RD. Air pollution as an emerging global risk factor for stroke. JAMA. 2011;305(12):1240–1.
8. O'Donnell MJ, Xavier D, Liu L, Zhang H, Chin SL, Rao-Melacini P, Rangarajan S, Islam S, Pais P, McQueen MJ, Mondo C, Damasceno A, Lopez-Jaramillo P, Hankey GJ, Dans AL, Yusoff K, Truelsen T, Diener HC, Sacco RL, Ryglewicz D, Czlonkowska A, Weimar C, Wang X, Yusuf S, INTERSTROKE investigators. Risk factors for ischaemic and intracerebral haemorrhagic stroke in 22 countries (the INTERSTROKE study): a case-control study. Lancet. 2010;376(9735):112–23.
9. Donnan GA, Fisher M, Macleod M, Davis SM. Stroke Lancet. 2008;371(9624):1612–23.
10. Donaldson K, Duffin R, Langrish JP, Miller MR, Mills NL, Poland CA, Raftis J, Shah A, Shaw CA, Newby DE. Nanoparticles and the cardiovascular system: a critical review. Nanomedicine. 2013;8(3):403–23.
11. Quan C, Sun Q, Lippmann M, Chen LC. Comparative effects of inhaled diesel exhaust and ambient fine particles on inflammation, atherosclerosis, and vascular dysfunction. Inhal Toxicol. 2010;22(9):738–53.
12. Kurhanewicz N, McIntosh-Kastrinsky R, Tong H, Walsh L, Farraj AK, Hazari MS. Ozone co-exposure modifies cardiac responses to fine and ultrafine ambient particulate matter in mice: concordance of electrocardiogram and mechanical responses. Part Fibre Toxicol. 2014;11(1):54.
13. Jariyasopit N, Zimmermann K, Schrlau J, Arey J, Atkinson R, Yu TW, Dashwood RH, Tao S, Simonich SL. Heterogeneous reactions of particulate matter-bound PAHs and NPAHs with $NO_3/N2O_5$, OH radicals, and O_3 under simulated long-range atmospheric transport conditions: reactivity and mutagenicity. Environ Sci Technol. 2014;48(17):10155–64.
14. Wang G, Jiang R, Zhao Z, Song W. Effects of ozone and fine particulate matter (PM(2.5)) on rat system inflammation and cardiac function. Toxicol Lett. 2013;217(1):23–33.
15. Nemery B, Hoet PH, Nemmar A. The Meuse Valley fog of 1930: an air pollution disaster. Lancet. 2001;357(9257):704–8.
16. Stone R. Air pollution. Counting the cost of London's killer smog. Science. 2002;298(5601): 2106–7.
17. Mustafic H, Jabre P, Caussin C, Murad MH, Escolano S, Tafflet M, Perier MC, Marijon E, Vernerey D, Empana JP, Jouven X. Main air pollutants and myocardial infarction: a systematic review and meta-analysis. JAMA. 2012;307(7):713–21.
18. Shah AS, Lee KK, McAllister DA, Hunter A, Nair H, Whiteley W, Langrish JP, Newby DE, Mills NL. Short term exposure to air pollution and stroke: systematic review and meta-analysis. BMJ. 2015;350:h1295.
19. Dominici F, Peng RD, Bell ML, Pham L, McDermott A, Zeger SL, Samet JM. Fine particulate air pollution and hospital admission for cardiovascular and respiratory diseases. JAMA. 2006;295(10):1127–34.
20. Lim SS, Vos T, Flaxman AD, Danaei G, Shibuya K, Adair-Rohani H, Amann M, Anderson HR, Andrews KG, Aryee M, Atkinson C, Bacchus LJ, Bahalim AN, Balakrishnan K, Balmes J, Barker-Collo S, Baxter A, Bell ML, Blore JD, Blyth F, Bonner C, Borges G, Bourne R, Boussinesq M, Brauer M, Brooks P, Bruce NG, Brunekreef B, Bryan-Hancock C, Bucello C, Buchbinder R, Bull F, Burnett RT, Byers TE, Calabria B, Carapetis J, Carnahan E, Chafe Z, Charlson F, Chen H, Chen JS, Cheng AT, Child JC, Cohen A, Colson KE, Cowie BC, Darby S, Darling S, Davis A, Degenhardt L, Dentener F, Des Jarlais DC, Devries K, Dherani M, Ding EL, Dorsey ER, Driscoll T, Edmond K, Ali SE, Engell RE, Erwin PJ, Fahimi S, Falder G, Farzadfar F, Ferrari A, Finucane MM, Flaxman S, Fowkes FG, Freedman G, Freeman MK, Gakidou E, Ghosh S, Giovannucci E, Gmel G, Graham K, Grainger R, Grant B, Gunnell D, Gutierrez HR, Hall W, Hoek HW, Hogan A, Hosgood HD, 3rd, Hoy D, Hu H, Hubbell BJ, Hutchings SJ, Ibeanusi SE, Jacklyn GL, Jasrasaria R, Jonas JB, Kan H, Kanis JA, Kassebaum N, Kawakami N, Khang YH, Khatibzadeh S, Khoo JP, Kok C, Laden F, Lalloo R, Lan Q,

Lathlean T, Leasher JL, Leigh J, Li Y, Lin JK, Lipshultz SE, London S, Lozano R, Lu Y, Mak J, Malekzadeh R, Mallinger L, Marcenes W, March L, Marks R, Martin R, McGale P, McGrath J, Mehta S, Mensah GA, Merriman TR, Micha R, Michaud C, Mishra V, Mohd Hanafiah K, Mokdad AA, Morawska L, Mozaffarian D, Murphy T, Naghavi M, Neal B, Nelson PK, Nolla JM, Norman R, Olives C, Omer SB, Orchard J, Osborne R, Ostro B, Page A, Pandey KD, Parry CD, Passmore E, Patra J, Pearce N, Pelizzari PM, Petzold M, Phillips MR, Pope D, Pope CA, 3rd, Powles J, Rao M, Razavi H, Rehfuess EA, Rehm JT, Ritz B, Rivara FP, Roberts T, Robinson C, Rodriguez-Portales JA, Romieu I, Room R, Rosenfeld LC, Roy A, Rushton L, Salomon JA, Sampson U, Sanchez-Riera L, Sanman E, Sapkota A, Seedat S, Shi P, Shield K, Shivakoti R, Singh GM, Sleet DA, Smith E, Smith KR, Stapelberg NJ, Steenland K, Stockl H, Stovner LJ, Straif K, Straney L, Thurston GD, Tran JH, Van Dingenen R, van Donkelaar A, Veerman JL, Vijayakumar L, Weintraub R, Weissman MM, White RA, Whiteford H, Wiersma ST, Wilkinson JD, Williams HC, Williams W, Wilson N, Woolf AD, Yip P, Zielinski JM, Lopez AD, Murray CJ, Ezzati M, AlMazroa MA, Memish ZA. A comparative risk assessment of burden of disease and injury attributable to 67 risk factors and risk factor clusters in 21 regions, 1990-2010: a systematic analysis for the Global Burden of Disease Study 2010. Lancet. 2012;380(9859):2224–60.

21. Beelen R, Raaschou-Nielsen O, Stafoggia M, Andersen ZJ, Weinmayr G, Hoffmann B, Wolf K, Samoli E, Fischer P, Nieuwenhuijsen M, Vineis P, Xun WW, Katsouyanni K, Dimakopoulou K, Oudin A, Forsberg B, Modig L, Havulinna AS, Lanki T, Turunen A, Oftedal B, Nystad W, Nafstad P, De Faire U, Pedersen NL, Ostenson CG, Fratiglioni L, Penell J, Korek M, Pershagen G, Eriksen KT, Overvad K, Ellermann T, Eeftens M, Peeters PH, Meliefste K, Wang M, Bueno-de-Mesquita B, Sugiri D, Kramer U, Heinrich J, de Hoogh K, Key T, Peters A, Hampel R, Concin H, Nagel G, Ineichen A, Schaffner E, Probst-Hensch N, Kunzli N, Schindler C, Schikowski T, Adam M, Phuleria H, Vilier A, Clavel-Chapelon F, Declercq C, Grioni S, Krogh V, Tsai MY, Ricceri F, Sacerdote C, Galassi C, Migliore E, Ranzi A, Cesaroni G, Badaloni C, Forastiere F, Tamayo I, Amiano P, Dorronsoro M, Katsoulis M, Trichopoulou A, Brunekreef B, Hoek G. Effects of long-term exposure to air pollution on natural-cause mortality: an analysis of 22 European cohorts within the multicentre ESCAPE project. Lancet. 2014;383(9919):785–95.
22. Cesaroni G, Forastiere F, Stafoggia M, Andersen ZJ, Badaloni C, Beelen R, Caracciolo B, de Faire U, Erbel R, Eriksen KT, Fratiglioni L, Galassi C, Hampel R, Heier M, Hennig F, Hilding A, Hoffmann B, Houthuijs D, Jockel KH, Korek M, Lanki T, Leander K, Magnusson PK, Migliore E, Ostenson CG, Overvad K, Pedersen NL, J JP, Penell J, Pershagen G, Pyko A, Raaschou-Nielsen O, Ranzi A, Ricceri F, Sacerdote C, Salomaa V, Swart W, Turunen AW, Vineis P, Weinmayr G, Wolf K, de Hoogh K, Hoek G, Brunekreef B, Peters A. Long term exposure to ambient air pollution and incidence of acute coronary events: prospective cohort study and meta-analysis in 11 European cohorts from the ESCAPE Project. BMJ. 2014;348:f7412.
23. Stafoggia M, Cesaroni G, Peters A, Andersen ZJ, Badaloni C, Beelen R, Caracciolo B, Cyrys J, de Faire U, de Hoogh K, Eriksen KT, Fratiglioni L, Galassi C, Gigante B, Havulinna AS, Hennig F, Hilding A, Hoek G, Hoffmann B, Houthuijs D, Korek M, Lanki T, Leander K, Magnusson PK, Meisinger C, Migliore E, Overvad K, Ostenson CG, Pedersen NL, Pekkanen J, Penell J, Pershagen G, Pundt N, Pyko A, Raaschou-Nielsen O, Ranzi A, Ricceri F, Sacerdote C, Swart WJ, Turunen AW, Vineis P, Weimar C, Weinmayr G, Wolf K, Brunekreef B, Forastiere F. Long-term exposure to ambient air pollution and incidence of cerebrovascular events: results from 11 European cohorts within the ESCAPE project. Environ Health Perspect. 2014;122(9):919-25.
24. Lipsett MJ, Ostro BD, Reynolds P, Goldberg D, Hertz A, Jerrett M, Smith DF, Garcia C, Chang ET, Bernstein L. Long-term exposure to air pollution and cardiorespiratory disease in the California teachers study cohort. Am J Respir Crit Care Med. 2011;184(7):828–35.
25. Nawrot TS, Perez L, Kunzli N, Munters E, Nemery B. Public health importance of triggers of myocardial infarction: a comparative risk assessment. Lancet. 2011;377(9767):732–40.

26. Shah AS, Bhopal R, Gadd S, Donohoe R. Out-of-hospital cardiac arrest in South Asian and white populations in London: database evaluation of characteristics and outcome. Heart. 2010;96(1):27–9.
27. Shah KS, Shah AS, Bhopal R. Systematic review and meta-analysis of out-of-hospital cardiac arrest and race or ethnicity: black US populations fare worse. Eur J Prev Cardiol. 2014;21(5):619–38.
28. Knox EG. Meteorological associations of cerebrovascular disease mortality in England and Wales. J Epidemiol Community Health. 1981;35(3):220–3.
29. Zhang ZF, Yu SZ, Zhou GD. Indoor air pollution of coal fumes as a risk factor of stroke. Shanghai Am J Public Health. 1988;78(8):975–7.
30. Schwartz J. What are people dying of on high air pollution days? Environ Res. 1994;64(1): 26–35.
31. Ljungman PL, Mittleman MA. Ambient air pollution and stroke. Stroke. 2014;45(12): 3734–41.
32. Le Tertre A, Medina S, Samoli E, Forsberg B, Michelozzi P, Boumghar A, Vonk JM, Bellini A, Atkinson R, Ayres JG, Sunyer J, Schwartz J, Katsouyanni K. Short-term effects of particulate air pollution on cardiovascular diseases in eight European cities. J Epidemiol Community Health. 2002;56(10):773–9.
33. Shah AS, Langrish JP, Nair H, McAllister DA, Hunter AL, Donaldson K, Newby DE, Mills NL. Global association of air pollution and heart failure: a systematic review and meta-analysis. Lancet. 2013;382(9897):1039–48.
34. Yang WS, Wang X, Deng Q, Fan WY, Wang WY. An evidence-based appraisal of global association between air pollution and risk of stroke. Int J Cardiol. 2014;175(2):307–13.
35. Wang Y, Eliot MN, Wellenius GA. Short-term changes in ambient particulate matter and risk of stroke: a systematic review and meta-analysis. J Am Heart Assoc. 2014;3(4):pii: e000983.
36. Hoffmann B, Weinmayr G, Hennig F, Fuks K, Moebus S, Weimar C, Dragano N, Hermann DM, Kalsch H, Mahabadi AA, Erbel R, Jockel KH. Air quality, stroke, and coronary events: results of the Heinz Nixdorf Recall Study from the Ruhr Region. Dtsch Arztebl Int. 2015;112(12):195–201.
37. Korek MJ, Bellander TD, Lind T, Bottai M, Eneroth KM, Caracciolo B, de Faire UH, Fratiglioni L, Hilding A, Leander K, Magnusson PK. Pedersen NL. Pershagen G, Penell JC. Traffic-related air pollution exposure and incidence of stroke in four cohorts from Stockholm. J Expo Sci Environ Epidemiol: Ostenson CG; 2015.
38. Kettunen J, Lanki T, Tiittanen P, Aalto PP, Koskentalo T, Kulmala M, Salomaa V, Pekkanen J. Associations of fine and ultrafine particulate air pollution with stroke mortality in an area of low air pollution levels. Stroke. 2007;38(3):918–22.
39. Lisabeth LD, Escobar JD, Dvonch JT, Sanchez BN, Majersik JJ, Brown DL, Smith MA, Morgenstern LB. Ambient air pollution and risk for ischemic stroke and transient ischemic attack. Ann Neurol. 2008;64(1):53–9.
40. Langrish JP, Li J, Wang S, Lee MM, Barnes GD, Lei GG, Miller MR, Cassee FR, Boon NA, Donaldson K, Li J, Mills NL, Jiang L, Newby DE. Reducing particulate air pollution exposure in patients with coronary heart disease improved cardiovascular health. Environ Health Perspect. 2012;120(3):367–72.
41. Mills NL, Tornqvist H, Gonzalez MC, Vink E, Robinson SD, Soderberg S, Boon NA, Donaldson K, Sandstrom T, Blomberg A, Newby DE. Ischemic and thrombotic effects of dilute diesel-exhaust inhalation in men with coronary heart disease. N Engl J Med. 2007;357(11):1075–82.
42. Wellenius GA, Boyle LD, Wilker EH, Sorond FA, Coull BA, Koutrakis P, Mittleman MA, Lipsitz LA. Ambient fine particulate matter alters cerebral hemodynamics in the elderly. Stroke. 2013;44(6):1532–6.
43. Newman JD, Thurston GD, Cromar K, Guo Y, Rockman CB, Fisher EA, Berger JS. Particulate air pollution and carotid artery stenosis. J Am Coll Cardiol. 2015;65(11):1150–1.
44. Joubert J, Cumming TB, McLean AJ. Diversity of risk factors for stroke: the putative roles and mechanisms of depression and air pollution. J Neurol Sci. 2007;262(1-2):71–6.

45. Macrae IM. Preclinical stroke research—advantages and disadvantages of the most common rodent models of focal ischaemia. Br J Pharmacol. 2011;164(4):1062–78.
46. Mergenthaler P, Meisel A. Do stroke models model stroke? Dis Model Mech. 2012;5(6):718–25.
47. Woodruff TM, Thundyil J, Tang SC, Sobey CG, Taylor SM, Arumugam TV. Pathophysiology, treatment, and animal and cellular models of human ischemic stroke. Mol Neurodegener. 2011;6(1):11.
48. Guo L, Li B, Miao JJ, Yun Y, Li GK, Sang N. Seasonal variation in air particulate matter (PM10) exposure-induced ischemia-like injuries in the rat brain. Chem Res Toxicol. 2015;28(3):431–9.
49. Davis DA, Akopian G, Walsh JP, Sioutas C, Morgan TE, Finch CE. Urban air pollutants reduce synaptic function of CA1 neurons via an NMDA/NO pathway in vitro. J Neurochem. 2013;127(4):509–19.
50. Morgan TE, Davis DA, Iwata N, Tanner JA, Snyder D, Ning Z, Kam W, Hsu YT, Winkler JW, Chen JC, Petasis NA, Baudry M, Sioutas C, Finch CE. Glutamatergic neurons in rodent models respond to nanoscale particulate urban air pollutants in vivo and in vitro. Environ Health Perspect. 2011;119(7):1003–9.
51. Kim YD, Lantz-McPeak SM, Ali SF, Kleinman MT, Choi YS, Kim H. Effects of ultrafine diesel exhaust particles on oxidative stress generation and dopamine metabolism in PC-12 cells. Environ Toxicol Pharmacol. 2014;37(3):954–9.
52. Campbell A, Daher N, Solaimani P, Mendoza K, Sioutas C. Human brain derived cells respond in a type-specific manner after exposure to urban particulate matter (PM). Toxicol In Vitro. 2014;28(7):1290–5.
53. Lawal AO, Zhang M, Dittmar M, Lulla A, Araujo JA. Heme oxygenase-1 protects endothelial cells from the toxicity of air pollutant chemicals. Toxicol Appl Pharmacol. 2015;284(3): 281–91.
54. Gong KW, Zhao W, Li N, Barajas B, Kleinman M, Sioutas C, Horvath S, Lusis AJ, Nel A, Araujo JA. Air-pollutant chemicals and oxidized lipids exhibit genome-wide synergistic effects on endothelial cells. Genome Biol. 2007;8(7):R149.
55. Costa LG, Cole TB, Coburn J, Chang YC, Dao K, Roque P. Neurotoxicants are in the air: convergence of human, animal, and in vitro studies on the effects of air pollution on the brain. Biomed Res Int. 2014;2014:736385.
56. Lucchini RG, Dorman DC, Elder A, Veronesi B. Neurological impacts from inhalation of pollutants and the nose-brain connection. Neurotoxicology. 2012;33(4):838–41.
57. Wallenborn JG, Schladweiler MC, Nyska A, Johnson JA, Thomas R, Jaskot RH, Richards JH, Ledbetter AD, Kodavanti UP. Cardiopulmonary responses of Wistar Kyoto, spontaneously hypertensive, and stroke-prone spontaneously hypertensive rats to particulate matter (PM) exposure. J Toxicol Environ Health A. 2007;70(22):1912–22.
58. Rizzo AM, Corsetto PA, Farina F, Montorfano G, Pani G, Battaglia C, Sancini G, Palestini P. Repeated intratracheal instillation of PM10 induces lipid reshaping in lung parenchyma and in extra-pulmonary tissues. PLoS One. 2014;9(9), e106855.
59. Ejaz S, Anwar K, Ashraf M. MRI and neuropathological validations of the involvement of air pollutants in cortical selective neuronal loss. Environ Sci Pollut Res Int. 2014;21(5):3351–62.
60. Peters A, Veronesi B, Calderon-Garciduenas L, Gehr P, Chen LC, Geiser M, Reed W, Rothen-Rutishauser B, Schurch S, Schulz H. Translocation and potential neurological effects of fine and ultrafine particles a critical update. Part Fibre Toxicol. 2006;3:13.
61. Hartz AM, Bauer B, Block ML, Hong JS, Miller DS. Diesel exhaust particles induce oxidative stress, proinflammatory signaling, and P-glycoprotein up-regulation at the blood-brain barrier. Faseb J. 2008;22(8):2723–33.
62. Oppenheim HA, Lucero J, Guyot AC, Herbert LM, McDonald JD, Mabondzo A, Lund AK. Exposure to vehicle emissions results in altered blood brain barrier permeability and expression of matrix metalloproteinases and tight junction proteins in mice. Part Fibre Toxicol. 2013;10:62.

63. Seaton A, MacNee W, Donaldson K, Godden D. Particulate air pollution and acute health effects. Lancet. 1995;345(8943):176–8.
64. Pope 3rd CA, Verrier RL, Lovett EG, Larson AC, Raizenne ME, Kanner RE, Schwartz J, Villegas GM, Gold DR, Dockery DW. Heart rate variability associated with particulate air pollution. Am Heart J. 1999;138(5 Pt 1):890–9.
65. Pieters N, Plusquin M, Cox B, Kicinski M, Vangronsveld J, Nawrot TS. An epidemiological appraisal of the association between heart rate variability and particulate air pollution: a meta-analysis. Heart. 2012;98(15):1127–35.
66. Oberdorster G, Sharp Z, Atudorei V, Elder A, Gelein R, Lunts A, Kreyling W, Cox C. Extrapulmonary translocation of ultrafine carbon particles following whole-body inhalation exposure of rats. J Toxicol Environ Health A. 2002;65(20):1531–43.
67. Miller MR. The role of oxidative stress in the cardiovascular actions of particulate air pollution. Biochem Soc Trans. 2014;42(4):1006–11.
68. Hanson LR, Frey WH, 2nd. Intranasal delivery bypasses the blood-brain barrier to target therapeutic agents to the central nervous system and treat neurodegenerative disease. BMC Neurosci. 2008;9 Suppl 3:S5.
69. Balasubramanian SK, Poh KW, Ong CN, Kreyling WG, Ong WY, Yu LE. The effect of primary particle size on biodistribution of inhaled gold nano-agglomerates. Biomaterials. 2013;34(22):5439–52.
70. Hopkins LE, Patchin ES, Chiu PL, Brandenberger C, Smiley-Jewell S, Pinkerton KE. Nose-to-brain transport of aerosolised quantum dots following acute exposure. Nanotoxicology. 2014;8(8):885–93.
71. Oberdorster G, Sharp Z, Atudorei V, Elder A, Gelein R, Kreyling W, Cox C. Translocation of inhaled ultrafine particles to the brain. Inhal Toxicol. 2004;16(6-7):437–45.
72. Calderon-Garciduenas L, Franco-Lira M, Henriquez-Roldan C, Osnaya N, Gonzalez-Maciel A, Reynoso-Robles R, Villarreal-Calderon R, Herritt L, Brooks D, Keefe S, Palacios-Moreno J, Villarreal-Calderon R, Torres-Jardon R, Medina-Cortina H, Delgado-Chavez R, Aiello-Mora M, Maronpot RR, Doty RL. Urban air pollution: influences on olfactory function and pathology in exposed children and young adults. Exp Toxicol Pathol. 2010;62(1):91–102.
73. Guerra R, Vera-Aguilar E, Uribe-Ramirez M, Gookin G, Camacho J, Osornio-Vargas AR, Mugica-Alvarez V, Angulo-Olais R, Campbell A, Froines J, Kleinman TM, De Vizcaya-Ruiz A. Exposure to inhaled particulate matter activates early markers of oxidative stress, inflammation and unfolded protein response in rat striatum. Toxicol Lett. 2013;222(2):146–54.
74. Cassee FR, Campbell A, Boere AJ, McLean SG, Duffin R, Krystek P, Gosens I, Miller MR. The biological effects of subacute inhalation of diesel exhaust following addition of cerium oxide nanoparticles in atherosclerosis-prone mice. Environ Res. 2012;115:1–10.
75. Gerlofs-Nijland ME, van Berlo D, Cassee FR, Schins RP, Wang K, Campbell A. Effect of prolonged exposure to diesel engine exhaust on proinflammatory markers in different regions of the rat brain. Part Fibre Toxicol. 2010;7:12.
76. Kleinman MT, Araujo JA, Nel A, Sioutas C, Campbell A, Cong PQ, Li H, Bondy SC. Inhaled ultrafine particulate matter affects CNS inflammatory processes and may act via MAP kinase signaling pathways. Toxicol Lett. 2008;178(2):127–30.
77. Campbell A, Oldham M, Becaria A, Bondy SC, Meacher D, Sioutas C, Misra C, Mendez LB, Kleinman M. Particulate matter in polluted air may increase biomarkers of inflammation in mouse brain. Neurotoxicology. 2005;26(1):133–40.
78. Bos I, De Boever P, Emmerechts J, Buekers J, Vanoirbeek J, Meeusen R, Van Poppel M, Nemery B, Nawrot T, Panis LI. Changed gene expression in brains of mice exposed to traffic in a highway tunnel. Inhal Toxicol. 2012;24(10):676–86.
79. Genc S, Zadeoglulari Z, Fuss SH, Genc K. The adverse effects of air pollution on the nervous system. J Toxicol. 2012;2012:782462.
80. Moulton PV, Yang W. Air pollution, oxidative stress, and Alzheimer's disease. J Environ Public Health. 2012;2012:472751.
81. Calderon-Garciduenas L, Franco-Lira M, Torres-Jardon R, Henriquez-Roldan C, Barragan-Mejia G, Valencia-Salazar G, Gonzalez-Maciel A, Reynoso-Robles R, Villarreal-Calderon R,

Reed W. Pediatric respiratory and systemic effects of chronic air pollution exposure: nose, lung, heart, and brain pathology. Toxicol Pathol. 2007;35(1):154–62.

82. Block ML, Calderon-Garciduenas L. Air pollution: mechanisms of neuroinflammation and CNS disease. Trends Neurosci. 2009;32(9):506–16.
83. Robertson S, Thomson AL, Carter R, Stott HR, Shaw CA, Hadoke PW, Newby DE, Miller MR, Gray GA. Pulmonary diesel particulate increases susceptibility to myocardial ischemia/reperfusion injury via activation of sensory TRPV1 and beta1 adrenoreceptors. Part Fibre Toxicol. 2014;11(1):12.
84. Gao Y, Lv J, Lin Y, Li X, Wang L, Yin Y, Liu Y. Effects of beta-adrenoceptor subtypes on cardiac function in myocardial infarction rats exposed to fine particulate matter (PM 2.5). Biomed Res Int. 2014;2014:308295.
85. Weichenthal S. Selected physiological effects of ultrafine particles in acute cardiovascular morbidity. Environ Res. 2012;115:26–36.
86. Cygankiewicz I, Zareba W. Heart rate variability. Handb Clin Neurol. 2013;117:379–93.
87. Robertson S, Gray GA, Duffin R, McLean SG, Shaw CA, Hadoke PW, Newby DE, Miller MR. Diesel exhaust particulate induces pulmonary and systemic inflammation in rats without impairing endothelial function ex vivo or in vivo. Part Fibre Toxicol. 2012;9:9.
88. Bartoli CR, Wellenius GA, Diaz EA, Lawrence J, Coull BA, Akiyama I, Lee LM, Okabe K, Verrier RL, Godleski JJ. Mechanisms of inhaled fine particulate air pollution-induced arterial blood pressure changes. Environ Health Perspect. 2009;117(3):361–6.
89. Perez CM, Hazari MS, Farraj AK. Role of autonomic reflex arcs in cardiovascular responses to air pollution exposure. Cardiovasc Toxicol. 2015;15(1):69–78.
90. Goadsby PJ. Autonomic nervous system control of the cerebral circulation. Handb Clin Neurol. 2013;117:193–201.
91. Lip GY, Lane DA. Stroke prevention in atrial fibrillation: a systematic review. JAMA. 2015;313(19):1950–62.
92. Watson T, Shantsila E, Lip GY. Mechanisms of thrombogenesis in atrial fibrillation: Virchow's triad revisited. Lancet. 2009;373(9658):155–66.
93. Carll AP, Lust RM, Hazari MS, Perez CM, Krantz QT, King CJ, Winsett DW, Cascio WE, Costa DL, Farraj AK. Diesel exhaust inhalation increases cardiac output, bradyarrhythmias, and parasympathetic tone in aged heart failure-prone rats. Toxicol Sci. 2013;131(2):583–95.
94. Kim JB, Kim C, Choi E, Park S, Park H, Pak HN, Lee MH, Shin DC, Hwang KC, Joung B. Particulate air pollution induces arrhythmia via oxidative stress and calcium calmodulin kinase II activation. Toxicol Appl Pharmacol. 2012;259(1):66–73.
95. Wang T, Lang GD, Moreno-Vinasco L, Huang Y, Goonewardena SN, Peng YJ, Svensson EC, Natarajan V, Lang RM, Linares JD, Breysse PN, Geyh AS, Samet JM, Lussier YA, Dudley S, Prabhakar NR, Garcia JG. Particulate matter induces cardiac arrhythmias via dysregulation of carotid body sensitivity and cardiac sodium channels. Am J Respir Cell Mol Biol. 2012;46(4):524–31.
96. Nadziejko C, Fang K, Narciso S, Zhong M, Su WC, Gordon T, Nadas A, Chen LC. Effect of particulate and gaseous pollutants on spontaneous arrhythmias in aged rats. Inhal Toxicol. 2004;16(6-7):373–80.
97. Langrish JP, Watts SJ, Hunter AJ, Shah AS, Bosson JA, Unosson J, Barath S, Lundback M, Cassee FR, Donaldson K, Sandstrom T, Blomberg A, Newby DE, Mills NL. Controlled exposures to air pollutants and risk of cardiac arrhythmia. Environ Health Perspect. 2014;122(7):747–53.
98. Lund AK, Lucero J, Lucas S, Madden MC, McDonald JD, Seagrave JC, Knuckles TL, Campen MJ. Vehicular emissions induce vascular MMP-9 expression and activity associated with endothelin-1-mediated pathways. Arterioscler Thromb Vasc Biol. 2009;29(4):511–7.
99. Sun Q, Yue P, Ying Z, Cardounel AJ, Brook RD, Devlin R, Hwang JS, Zweier JL, Chen LC, Rajagopalan S. Air pollution exposure potentiates hypertension through reactive oxygen species-mediated activation of Rho/ROCK. Arterioscl Thromb Vasc Biol. 2008;28(10):1760–6.
100. Li Z, Carter JD, Dailey LA, Huang YC. Pollutant particles produce vasoconstriction and enhance MAPK signaling via angiotensin type I receptor. Environ Health Perspect. 2005;113(8):1009–14.

101. Langrish JP, Lundback M, Mills NL, Johnston NR, Webb DJ, Sandstrom T, Blomberg A, Newby DE. Contribution of endothelin 1 to the vascular effects of diesel exhaust inhalation in humans. Hypertension. 2009;54(4):910–5.
102. Miller MR, Shaw CA, Langrish JP. From particles to patients: oxidative stress and the cardiovascular effects of air pollution. Future Cardiol. 2012;8(4):577–602.
103. Brook RD. Why physicians who treat hypertension should know more about air pollution. J Clin Hypertens. 2007;9(8):629–35.
104. Malfitano AM, Marasco G, Proto MC, Laezza C, Gazzerro P, Bifulco M. Statins in neurological disorders: an overview and update. Pharmacol Res. 2014;88:74–83.
105. Jordan JD, Morbitzer KA, Rhoney DH. Acute treatment of blood pressure after ischemic stroke and intracerebral hemorrhage. Neurol Clin. 2015;33(2):361–80.
106. Anderson CS, Qureshi AI. Implications of INTERACT2 and other clinical trials: blood pressure management in acute intracerebral hemorrhage. Stroke. 2015;46(1):291–5.
107. Bath PM, Krishnan K. Interventions for deliberately altering blood pressure in acute stroke. Cochrane Database Syst Rev. 2014;10, CD000039.
108. Arnan MK, Burke GL, Bushnell C. Secondary prevention of stroke in the elderly: focus on drug therapy. Drugs Aging. 2014;31(10):721–30.
109. Feldstein CA. Lowering blood pressure to prevent stroke recurrence: a systematic review of long-term randomized trials. J Am Soc Hypertens. 2014;8(7):503–13.
110. Singer J, Trollor JN, Baune BT, Sachdev PS, Smith E. Arterial stiffness, the brain and cognition: a systematic review. Ageing Res Rev. 2014;15:16–27.
111. Kollias A, Stergiou GS, Dolan E, O'Brien E. Ambulatory arterial stiffness index: a systematic review and meta-analysis. Atherosclerosis. 2012;224(2):291–301.
112. Mehta AJ, Zanobetti A, Koutrakis P, Mittleman MA, Sparrow D, Vokonas P, Schwartz J. Associations between short-term changes in air pollution and correlates of arterial stiffness: The Veterans Affairs Normative Aging Study, 2007-2011. Am J Epidemiol. 2014;179(2): 192–9.
113. Iannuzzi A, Verga MC, Renis M, Schiavo A, Salvatore V, Santoriello C, Pazzano D, Licenziati MR, Polverino M. Air pollution and carotid arterial stiffness in children. Cardiol Young. 2010;20(2):186–90.
114. Lundback M, Mills NL, Lucking A, Barath S, Donaldson K, Newby DE, Sandstrom T, Blomberg A. Experimental exposure to diesel exhaust increases arterial stiffness in man. Part Fibre Toxicol. 2009;6:7.
115. Unosson J, Blomberg A, Sandstrom T, Muala A, Boman C, Nystrom R, Westerholm R, Mills NL, Newby DE, Langrish JP, Bosson JA. Exposure to wood smoke increases arterial stiffness and decreases heart rate variability in humans. Part Fibre Toxicol. 2013;10:20.
116. del Zoppo G, Ginis I, Hallenbeck JM, Iadecola C, Wang X, Feuerstein GZ. Inflammation and stroke: putative role for cytokines, adhesion molecules and iNOS in brain response to ischemia. Brain Pathol. 2000;10(1):95–112.
117. Hoffmann B, Moebus S, Dragano N, Stang A, Mohlenkamp S, Schmermund A, Memmesheimer M, Brocker-Preuss M, Mann K, Erbel R, Jockel KH. Chronic residential exposure to particulate matter air pollution and systemic inflammatory markers. Environ Health Perspect. 2009;117(8):1302–8.
118. Delfino RJ, Staimer N, Tjoa T, Gillen DL, Polidori A, Arhami M, Kleinman MT, Vaziri ND, Longhurst J, Sioutas C. Air pollution exposures and circulating biomarkers of effect in a susceptible population: clues to potential causal component mixtures and mechanisms. Environ Health Perspect. 2009;117(8):1232–8.
119. Cascio WE, Cozzi E, Hazarika S, Devlin RB, Henriksen RA, Lust RM, Van Scott MR, Wingard CJ. Cardiac and vascular changes in mice after exposure to ultrafine particulate matter. Inhal Toxicol. 2007;19 Suppl 1:67–73.
120. Tsai DH, Amyai N, Marques-Vidal P, Wang JL, Riediker M, Mooser V, Paccaud F, Waeber G, Vollenweider P, Bochud M. Effects of particulate matter on inflammatory markers in the general adult population. Part Fibre Toxicol. 2012;9:24.

121. Farina F, Sancini G, Battaglia C, Tinaglia V, Mantecca P, Camatini M, Palestini P. Milano summer particulate matter (PM10) triggers lung inflammation and extra pulmonary adverse events in mice. PLoS One. 2013;8(2), e56636.
122. Liu C, Fonken LK, Wang A, Maiseyeu A, Bai Y, Wang TY, Maurya S, Ko YA, Periasamy M, Dvonch T, Morishita M, Brook RD, Harkema J, Ying Z, Mukherjee B, Sun Q, Nelson RJ, Rajagopalan S. Central IKKbeta inhibition prevents air pollution mediated peripheral inflammation and exaggeration of type II diabetes. Part Fibre Toxicol. 2014;11:53.
123. Donaldson K, Stone V, Seaton A, MacNee W. Ambient particle inhalation and the cardiovascular system: potential mechanisms. Environ Health Perspect. 2001;109 Suppl 4:523–7.
124. Radak D, Resanovic I, Isenovic ER. Link between oxidative stress and acute brain ischemia. Angiology. 2014;65(8):667–76.
125. Manzanero S, Santro T, Arumugam TV. Neuronal oxidative stress in acute ischemic stroke: sources and contribution to cell injury. Neurochem Int. 2013;62(5):712–8.
126. Adibhatla RM, Hatcher JF. Tissue plasminogen activator (tPA) and matrix metalloproteinases in the pathogenesis of stroke: therapeutic strategies. CNS Neurol Disord Drug Targets. 2008;7(3):243–53.
127. Garry PS, Ezra M, Rowland MJ, Westbrook J, Pattinson KT. The role of the nitric oxide pathway in brain injury and its treatment—from bench to bedside. Exp Neurol. 2015;263: 235–43.
128. Dringen R. Metabolism and functions of glutathione in brain. Prog Neurobiol. 2000;62(6): 649–71.
129. Basu S. Bioactive eicosanoids: role of prostaglandin F(2alpha) and F(2)-isoprostanes in inflammation and oxidative stress related pathology. Mol Cells. 2010;30(5):383–91.
130. Thompson J, Nitiahpapand R, Bhatti P, Kourliouros A. Vitamin D deficiency and atrial fibrillation. Int J Cardiol. 2015;184C:159–62.
131. Bruske I, Hampel R, Baumgartner Z, Ruckerl R, Greven S, Koenig W, Peters A, Schneider A. Ambient air pollution and lipoprotein-associated phospholipase A(2) in survivors of myocardial infarction. Environ Health Perspect. 2011;119(7):921–6.
132. Poirier J, Miron J, Picard C, Gormley P, Theroux L, Breitner J, Dea D. Apolipoprotein E and lipid homeostasis in the etiology and treatment of sporadic Alzheimer's disease. Neurobiol Aging. 2014;35 Suppl 2:S3–10.
133. O'Regan C, Wu P, Arora P, Perri D, Mills EJ. Statin therapy in stroke prevention: a meta-analysis involving 121,000 patients. Am J Med. 2008;121(1):24–33.
134. Lin YL, Chang HC, Chen TL, Chang JH, Chiu WT, Lin JW, Chen RM. Resveratrol protects against oxidized LDL-induced breakage of the blood-brain barrier by lessening disruption of tight junctions and apoptotic insults to mouse cerebrovascular endothelial cells. J Nutr. 2010;140(12):2187–92.
135. Seaton A, Soutar A, Crawford V, Elton R, McNerlan S, Cherrie J, Watt M, Agius R, Stout R. Particulate air pollution and the blood. Thorax. 1999;54(11):1027–32.
136. Cozzi E, Wingard CJ, Cascio WE, Devlin RB, Miles JJ, Bofferding AR, Lust RM, Van Scott MR, Henriksen RA. Effect of ambient particulate matter exposure on hemostasis. Transl Res. 2007;149(6):324–32.
137. Kilinc E, Van Oerle R, Borissoff JI, Oschatz C, Gerlofs-Nijland ME, Janssen NA, Cassee FR, Sandstrom T, Renne T, Ten Cate H, Spronk HM. Factor XII activation is essential to sustain the procoagulant effects of particulate matter. J Thromb Haemost. 2011;9(7):1359–67.
138. Radomski A, Jurasz P, Alonso-Escolano D, Drews M, Morandi M, Malinski T, Radomski MW. Nanoparticle-induced platelet aggregation and vascular thrombosis. Br J Pharmacol. 2005;146(6):882–93.
139. Tablin F, den Hartigh LJ, Aung HH, Lame MW, Kleeman MJ, Ham W, Norris JW, Pombo M, Wilson DW. Seasonal influences on CAPs exposures: differential responses in platelet activation, serum cytokines and xenobiotic gene expression. Inhal Toxicol. 2012;24(8):506–17.
140. Forestier M, Al-Tamimi M, Gardiner EE, Hermann C, Meyer SC, Beer JH. Diesel exhaust particles impair platelet response to collagen and are associated with GPIbalpha shedding. Toxicol In Vitro. 2012;26(6):930–8.

141. Lucking AJ, Lundback M, Mills NL, Faratian D, Barath SL, Pourazar J, Cassee FR, Donaldson K, Boon NA, Badimon JJ, Sandstrom T, Blomberg A, Newby DE. Diesel exhaust inhalation increases thrombus formation in man. Eur Heart J. 2008;29(24): 3043–51.
142. Nemmar A, Al-Salam S, Dhanasekaran S, Sudhadevi M, Ali BH. Pulmonary exposure to diesel exhaust particles promotes cerebral microvessel thrombosis: protective effect of a cysteine prodrug l-2-oxothiazolidine-4-carboxylic acid. Toxicology. 2009;263(2-3):84–92.
143. Mills NL, Tornqvist H, Robinson SD, Gonzalez M, Darnley K, MacNee W, Boon NA, Donaldson K, Blomberg A, Sandstrom T, Newby DE. Diesel exhaust inhalation causes vascular dysfunction and impaired endogenous fibrinolysis. Circulation. 2005;112(25):3930–6.
144. Wardlaw JM, Murray V, Berge E, del Zoppo G, Sandercock P, Lindley RL, Cohen G. Recombinant tissue plasminogen activator for acute ischaemic stroke: an updated systematic review and meta-analysis. Lancet. 2012;379(9834):2364–72.
145. Ritz K, Denswil NP, Stam OC, van Lieshout JJ, Daemen MJ. Cause and mechanisms of intracranial atherosclerosis. Circulation. 2014;130(16):1407–14.
146. Virmani R, Finn AV, Kolodgie FD. Carotid plaque stabilization and progression after stroke or TIA. Arterioscler Thromb Vasc Biol. 2009;29(1):3–6.
147. Miller MR, McLean SG, Duffin R, Lawal AO, Araujo JA, Shaw CA, Mills NL, Donaldson K, Newby DE, Hadoke PW. Diesel exhaust particulate increases the size and complexity of lesions in atherosclerotic mice. Part Fibre Toxicol. 2013;10:61.
148. Campen MJ, Lund AK, Knuckles TL, Conklin DJ, Bishop B, Young D, Seilkop S, Seagrave J, Reed MD, McDonald JD. Inhaled diesel emissions alter atherosclerotic plaque composition in ApoE(-/-) mice. Toxicol Appl Pharmacol. 2010;242(3):310–7.
149. Moller P, Mikkelsen L, Vesterdal LK, Folkmann JK, Forchhammer L, Roursgaard M, Danielsen PH, Loft S. Hazard identification of particulate matter on vasomotor dysfunction and progression of atherosclerosis. Crit Rev Toxicol. 2011;41(4):339–68.
150. Kunzli N, Jerrett M, Garcia-Esteban R, Basagana X, Beckermann B, Gilliland F, Medina M, Peters J, Hodis HN, Mack WJ. Ambient air pollution and the progression of atherosclerosis in adults. PLoS One. 2010;5(2), e9096.
151. Kunzli N, Jerrett M, Mack WJ, Beckerman B, LaBree L, Gilliland F, Thomas D, Peters J, Hodis HN. Ambient air pollution and atherosclerosis in Los Angeles. Environ Health Perspect. 2005;113(2):201–6.
152. Perez L, Wolf K, Hennig F, Penell J, Basagana X, Foraster M, Aguilera I, Agis D, Beelen R, Brunekreef B, Cyrys J, Fuks KB, Adam M, Baldassarre D, Cirach M, Elosua R, Dratva J, Hampel R, Koenig W, Marrugat J, De Faire U, Pershagen G, Probst-Hensch NM, de Nazelle A, Nieuwenhuijsen MJ, Rathmann W, Rivera M, Seissler J, Schindler C, Thierry J, Hoffmann B, Peters A, Kunzli N. Air pollution and atherosclerosis: a cross-sectional analysis of four European cohort studies in the ESCAPE study. Environ Health Perspect. 2015;123(6): 597–605. doi:10.1289/ehp.1307711.
153. Flaherty ML, Flemming KD, McClelland R, Jorgensen NW, Brown Jr RD. Population-based study of symptomatic internal carotid artery occlusion: incidence and long-term follow-up. Stroke. 2004;35(8):e349–52.
154. Calderon-Garciduenas L, Calderon-Garciduenas A, Torres-Jardon R, Avila-Ramirez J, Kulesza RJ, Angiulli AD. Air pollution and your brain: what do you need to know right now. Prim Health Care Res Dev. 2014;1–17.
155. Finkelstein MM, Jerrett M, Sears MR. Environmental inequality and circulatory disease mortality gradients. J Epidemiol Community Health. 2005;59(6):481–7.
156. Li H, Xin X. Nitrogen dioxide (NO(2)) pollution as a potential risk factor for developing vascular dementia and its synaptic mechanisms. Chemosphere. 2013;92(1):52–8.
157. Sang N, Yun Y, Li H, Hou L, Han M, Li G. SO2 inhalation contributes to the development and progression of ischemic stroke in the brain. Toxicol Sci. 2010;114(2):226–36.
158. Henrotin JB, Zeller M, Lorgis L, Cottin Y, Giroud M, Bejot Y. Evidence of the role of short-term exposure to ozone on ischaemic cerebral and cardiac events: the Dijon Vascular Project (DIVA). Heart. 2010;96(24):1990–6.

159. Kang JH, Liu TC, Keller J, Lin HC. Asian dust storm events are associated with an acute increase in stroke hospitalisation. J Epidemiol Community Health. 2013;67(2):125–31.
160. Yang CY, Chen YS, Chiu HF, Goggins WB. Effects of Asian dust storm events on daily stroke admissions in Taipei. Taiwan Environ Res. 2005;99(1):79–84.
161. Andersen ZJ, Olsen TS, Andersen KK, Loft S, Ketzel M, Raaschou-Nielsen O. Association between short-term exposure to ultrafine particles and hospital admissions for stroke in Copenhagen. Denmark Eur Heart J. 2010;31(16):2034–40.
162. Bernal-Pacheco O, Roman GC. Environmental vascular risk factors: new perspectives for stroke prevention. J Neurol Sci. 2007;262(1-2):60–70.

Particle-Induced Inflammation and Cardiovascular Diseases

Jesus A. Araujo and Gajalakshmi Ramanathan

1 Exposure to Air Pollutants Lead to Enhanced Cardiovascular Diseases

1.1 Particulate Matter and Cardiovascular Effects: General Concepts

Numerous epidemiological studies support the association of air pollution with adverse health effects leading to increased morbidity and mortality of worldwide significance [1–3]. Air pollution-related deaths are mostly due to cardiovascular and cerebrovascular diseases [4], which together accounted for 80 % of all deaths worldwide (40 % each) attributed to ambient air pollution in 2012 [5]. This is quite significant since heart disease remained as the top leading cause of death in the US [6] and in the world [7] in 2011. Exposure to air pollution is recognized as a relevant cardiovascular risk factor given its potential to affect large numbers of people around the globe, but it is frequently overlooked. The nature of the cardiovascular effects of air pollution and potential mechanisms involved have been the focus of

J.A. Araujo, M.D., Ph.D. (✉)
Division of Cardiology, Department of Medicine, David Geffen School of Medicine at UCLA, 10833 Le Conte Avenue, CHS 43-264, P.O. Box 951679, Los Angeles, CA 90095, USA

Department of Environmental Health Sciences, Fielding School of Public Health, UCLA, Los Angeles, CA, USA
e-mail: JAraujo@mednet.ucla.edu

G. Ramanathan
Division of Cardiology, Department of Medicine, David Geffen School of Medicine at UCLA, 10833 Le Conte Avenue, CHS 43-264, P.O. Box 951679, Los Angeles, CA 90095, USA

S.C. Bondy, A. Campbell (eds.), *Inflammation, Aging, and Oxidative Stress*, Oxidative Stress in Applied Basic Research and Clinical Practice,
DOI 10.1007/978-3-319-33486-8_9

two Concensus Statements from the American Heart Association in 2004 [1] and 2010 [3], and an Expert Consensus document from the European Society of Cardiology in 2015 [8] targeted to the healthcare professionals and researchers, which underline the importance of this topic for the general community.

Air pollutants are within a complex mixture of compounds with gaseous (ozone, carbon monoxide, sulfur and nitrogen oxides) and particle phases, although the cardiovascular effects have been mostly ascribed to the particulate matter (PM) components [2, 3, 9]. Indeed, ambient PM pollution ranked within the top ten leading risk factors associated with the global burden of disease worldwide, resulting in 3.0% of the global disability-adjusted life-years [10]. Ambient particles can be classified according to their aerodynamic diameter in size fractions such as PM_{10} ("thoracic" particles, <10 μm), $PM_{2.5\text{-}10}$ ("coarse" particles, 2.5–10 μm), $PM_{2.5}$ (fine particles, <2.5 μm) and UFP (ultrafine particles, <0.1 μm) that are derived from various sources and by a variety of processes characteristic of each size fraction [11]. Interestingly, although the associations appear to be stronger with PM components, two meta-analyses, one with 34 studies (with 17 time-series and 17 case-crossover designs) and the other with 35 studies, revealed that with the exception of ozone, all the main air pollutants that were studied (PM_{10}, $PM_{2.5}$, carbon monoxide, nitrogen dioxide and sulfur dioxide) were significantly associated with an increased risk for myocardial infarction (MI) [12] and heart failure hospitalizations [13].

Air pollution-mediated cardiovascular actions include a large range of effects on various processes such as atherosclerosis, platelet aggregability and thrombosis, vasoreactivity, arrhythmias and possibly cardiac systolic function that are likely responsible for the association with increased risk for myocardial infarction [12], cerebrovascular events [14] and heart failure hospitalizations [13]. Three main mechanisms have been proposed to explain how inhalation of ambient pollutants could result in systemic cardiovascular effects such as: (1) activation of pulmonary receptors resulting in autonomic nervous system imbalance and the development of dysrhythmias, (2) induction of pulmonary and systemic inflammation, (3) access of particles, gases or their chemical constituents to the systemic circulation. This chapter focuses on the inflammatory effects induced by particulate matter that lead to promotion of atherosclerosis and ischemic heart disease.

1.2 Exposure to Particulate Matter Associates with Atherosclerosis in Human Studies

Exposure to ambient PM positively associates with various measures of subclinical atherosclerosis in humans (Table 1). The first study was reported in 2005, where Kunzli et al. reported that the degree of carotid intima-medial thickness (CIMT) correlated with an increase of 5.9% per every 10 μg/m^3 rise in $PM_{2.5}$ levels [15] in 798 individuals studied with a cross-sectional design (Table 1). Data from the Multi-Ethnic Study of Atherosclerosis (MESA) further supports the association between PM and atherosclerosis [16]. Diez Roux et al. reported that PM_{10} exposures assessed over long-term (20-year means and 2001 mean) and 20-year $PM_{2.5}$ exposures correlated with a 1–3 % increase in CIMT per 21 μg/m^3 increase in PM_{10} or 12.5 μg/m^3 increase in $PM_{2.5}$, respectively [16]. In addition, Allen et al. reported that $PM_{2.5}$ exposures were correlated with an increased risk for aortic calcification in a related study [17] (Table 1). Data from the Heinz Nixdorf Recall (HNR) study is also supportive, a population-based cohort of 4814 participants where exposure to $PM_{2.5}$, PM_{10} and distance to a major road and high traffic were associated with an increase in coronary artery calcium scores (CACS) and/or CIMT [18, 19]. Thus, subjects living within 101 to 200 m, 51 to 100 or less than 50 m showed an increase of 8 %, 34 % and 63 % in the probability of having a high CACS, respectively, as compared with subjects living >200 meters away from a major road (Table 1). While there were also small associations between noise and subclinical atherosclerosis, the associations were much stronger with $PM_{2.5}$ exposure [20] (Table 1). In addition, the association between CIMT and one-year exposure to $PM_{2.5}$ was robust to the inclusion of all major known risk factors for atherosclerosis and stronger than for larger PM_{10} particles [18]. This is consistent with the greater effects of the $PM_{2.5}$ fraction on long-term cardiovascular mortality [21] and the notion that the cardiovascular effects are favored by a smaller particle size [9]. However, using a standardized exposure and analytical protocol in four European cohorts that included the European Study of Cohorts for Air Pollution Effects (ESCAPE Study) in Sweden, the Ausburg and Ruhr areas in Germany, and the Girona area in Spain, Perez et al. found that cross-sectional associations between CIMT and eight ESCAPE markers of long-term residential air pollution exposure were mostly positive but did not reach significance [22] (Table 1). A meta-analysis with 9183 individuals also showed positive but not significant associations [22]. However, a meta-analysis of eight published studies, including 18,340 participants for cross-sectional associations between CIMT and PM exposure, found a robust increase of 5 μg/m^3 $PM_{2.5}$ associated with a 1.66 % thicker CIMT (average increase of 12.1 μm) [23]. Importantly, this metanalysis included the four European cohorts where no significant associations had been found. These positive associations replicated in Asia where long-term exposures to air pollution associated with CIMT in 689 middle-aged adults in Taiwan [24].

PM exposure levels also associate with the progression of atherosclerotic disease. Kunzli et al. reported that the annual rate of CIMT progression among individuals living within 100 m of a highway was accelerated and more than twice the population mean rate of progression [25]. Wilker et al. reported data from the Normative Aging study where 380 participants were followed-up between 2004 and 2008, with two or three CIMT measurements, which showed a positive association between a one-interquartile range increase in 1-year average levels of black carbon and 1.1 % higher CMT, based on a fully adjusted model [26]. A 5-year follow-up of the MESA study also showed that among 5362 participants, with a mean annual progression of 14 μm/year, subjects exposed to 2.5 μg/m^3 exhibited a greater rate of CIMT progression over 5 years (increase of 5.0 μm/year). Importantly, reductions in $PM_{2.5}$ exposure from a fixed baseline were associated with a slower progression in the CIMT (−2.8 μm/year per each 1 μg/m^3 reduction in $PM_{2.5}$) [27], suggesting a causal link between PM exposure and atherogenesis.

Table 1 Human studies linking air pollution exposure with subclinical measures of atherosclerosis

Study	Air pollutant	Evaluation of atherosclerosis	Major findings	Reference
Kunzli et al. (2005)	$PM_{2.5}$ Ozone	CIMT	5.9 % increase in CIMT per every 10 μg $PM_{2.5}$/m^3	[15]
Hoffman et al. (2007)	$PM_{2.5}$ Distance to major road	CACS	Increased CACS with shorter distances to a major road	[19]
Diez Roux et al. (2008)	PM_{10} $PM_{2.5}$	CIMT CACS BAI	1–3 % increase in CIMT per every increase in 21 and 12.5 μg/m^3 of PM_{10} and $PM_{2.5}$ respectively	[16]
Allen et al. (2009)	PM $_{2.5}$ Distance to major road	Aortic calcification	6 % increase in the risk of aortic calcification with a 10 μg/m^3 increase in $PM_{2.5}$	[17]
Bauer et al. (2010)	$PM_{2.5}$ PM_{10} Distance to high traffic	CIMT	4.3 %, 1.7 % and 1.2 % increases in CIMT per interdecile range increases in $PM_{2.5}$ (4.2 μg/m^3), PM_{10} (6.7 μg/m^3) and distance to traffic (1939 m), respectively	[130]
Kunzli et al. (2010)	$PM_{2.5}$ Distance to highway or major road	CIMT	Greater annual progression of CIMT among individuals living < 100 m from a highway	[25]
Wilker et al. (2013)	Black Carbon	CIMT	1.1 % higher CIMT was associated with a one-interquartile range increase in 1-year average black carbon (0.26 μg/ m^3)	[26]
Adar et al. (2013)	$PM_{2.5}$	CIMT	2.5 μg/ m^3 higher $PM_{2.5}$ levels associated with greater CIMT progression over 5 years (increase of 5.0 μm/year) while greater reductions in $PM_{2.5}$ over a fixed baseline $PM_{2.5}$ associated with slowed CIMT progression	[27]
Kälsch et al. (2014)	$PM_{2.5}$ PM_{10} L_{den} L_{night}	TACS CACS	18.1 % and 3.9 % increases in burden of thoracic aortic calcification per increases of 2.4 μg/ m^3 $PM_{2.5}$ and 5 dB(A) L_{night}	[131]

Table 1 (continued)

Study	Air pollutant	Evaluation of atherosclerosis	Major findings	Reference
Perez et al. (2015)	$PM_{2.5}$ and $PM_{2.5}$abs	CIMT	Positive but not statistically significant associations between CIMT and $PM_{2.5}$, $PM_{2.5}$ abs, traffic load within 100 m of home, and traffic intensity at the nearest road. No significant inverse associations between CIMT and NO_2, NO_x, PM_{10} or PM_{coarse}	[132]
	PM_{10}			
	PM_{coarse}			
	NO_2, NO_x			
	Proximity to high traffic			
	Traffic intensity			
Su et al. (2015)	$PM_{2.5}$abs	CIMT	Average percentage increases in maximum left CIMT of 4.23 % per 1.0×10^{-5} μg/m increase in $PM_{2.5}$abs, 3.72 % per 10 μg/m^3 increase in PM_{10}, 2.81 % or 0.74 % per 20 μg/m^3 or 10 μg/m^3 increase in NO_2 or NO_x	[24]
	PM_{10}			
	NO_2			
	NO_x			

Studies are listed in chronological order based on the year of publication. BAI = Brachial artery index, CIMT = Carotid intima-media thickness, CACS = coronary artery calcification score, Lden = weighted 24 h mean of road traffic noise, Lnight = mean of night-noise (22-6 h), TACS = Thoracic aorta calcification score, $PM_{2.5}$abs = absorbance levels of $PM_{2.5}$

1.3 Particulate Matter Enhances Atherogenesis in Animal Studies

While the epidemiological studies support associations between ambient PM exposure mass and subclinical measures of atherosclerosis in humans, studies with experimental animals indicate that these associations are causal. Table 2 summarizes several of these studies using various animal models.

Exposures to air pollutants have been performed using intra-tracheal instillations (I.T.) of PM as well as inhalation of polluted air, concentrated ambient particles (CAPs) or motor vehicle emissions such as diesel exhaust (DE), gasoline exhaust (GE) and mixed vehicular emissions (MVE). Both I.T. and inhalatory studies are informative in understanding and dissecting relevant pathological events.

I.T. administration of PM_{10} or carbon black have been shown to stimulate atherogenesis in Watanabe hyperlipidemic rabbits [28, 29] and low density lipoprotein receptor deficient (LDL-R$^{-/-}$) mice [30], respectively. In those studies, atherosclerosis was assessed both in the coronary arteries [29] and in the aorta [28–30]. The effects of concentrated $PM_{2.5}$ have been evaluated with long-term inhalation exposures. Seven studies have been reported where concentrated $PM_{2.5}$ led to enhanced atherosclerosis in apolipoprotein E null (Apo E$^{-/-}$) and low-density lipoprotein receptor null (LDL-R$^{-/-}$) mice. Thus, animals were exposed to fine CAPs in suburban Sterling Forest, New York [31–35], urban Manhattan, New York [36] or

Table 2 Animal studies on the effects of air pollutants on lipids and atherosclerosis

Study	Air pollutant	Animal model	Diet	Major findings (induced by the exposure to air pollutant vs. FA controls)	Reference
Suwa et al. (2002)	I.T. PM_{10} , 2 days/week X 4 weeks	Watanabe rabbits	Chow	Increase in % lesional volume in coronary arteries and aorta	[29]
				Increase in circulating polymorphonuclear leukocytes	
				No effects on plasma lipids	
Chen and Nadziejko (2005)	Inhaled CAPs($PM_{2.5}$)	$ApoE^{-/-}$, $LDL^{-/-}$ mice and $ApoE^{-/-}$ mice	Chow	Increase in % lesional area in whole aorta in $ApoE^{-/-}$ mice but not in $ApoE^{-/-}$, $LDL^{-/-}$ mice	[31]
	6 h/day, 5 days/week X 5 months		Chow		
Sun et al. (2005)	Inhaled CAPs($PM_{2.5}$)	$ApoE^{-/-}$ mice	Chow or CED	Increase in % lesional area in cross-sections of aorta in CED-fed mice and N.S. increase in chow-fed mice	[33]
	6 h/day, 5 days/week X 6 months			Small increase in total plasma cholesterol in CED-fed mice but small decrease in chow-fed mice.	
Niwa et al. (2007)	I.T. Carbon black	LDL-$R^{-/-}$ mice	CED	Increase in % lesional area in whole aorta	[30]
	1x/week X 10 weeks			Increase in C-reactive protein	
Sun et al. (2008)	Inhaled CAPs($PM_{2.5}$)	$ApoE^{-/-}$ mice	Chow or CED	Increase in % lesional area in aorta in CED-fed mice and N.S increase in chow-red mice, assessed by ultrasound	[34]
	6 h/day, 5 days/week X 6 months			Plaques had a greater content of tissue factor protein	
Yatera et al. (2008)	I.T. PM_{10}	Watanabe rabbits	Chow	Increase in % lesional volume and % lesional area in the aorta	[28]
	2 days/w X 4 weeks			Increase in circulating monocytes 24 hours after first I.T treatment	
Araujo et al. (2008)	Inhaled CAPs($PM_{2.5}$ and UFP)	$ApoE^{-/-}$ mice	Chow	Increase in lesional area in aortic root of UFP-exposed mice N.S increase in $PM_{2.5}$-exposed mice	[45]
				Small increase in plasma total cholesterol in the $PM_{2.5}$ group	
	3x/week X 5 weeks			Increase in liver MDA in $PM_{2.5}$ and UFP-exposed mice	
				Dysfunctional plasma HDL in mice exposed to $PM_{2.5}$ and UFP, more prominent in UFP (HDL became proinflammatory)	

Castro-Soares et al. (2009)	Inhaled polluted ambient air	LDLR$^{-/-}$ mice	CED	Increase in aortic wall thickness in the CED-fed mice but not in % lesional area	[101]
			Chow	Increase in LDL susceptibility to oxidation, plasma anti-oxLDL and anti-apoB antibodies, which was more pronounced in CED-fed mice	
Ying et al. (2009)	Inhaled CAPs($PM_{2.5}$)	ApoE$^{-/-}$ mice	CED	Increase in % lesional area in cross-sections of aorta	[61]
	6 h/day, 5 days/week X 4 months			Increase in iNOS and Nitrotyrosine in the aorta	
Lund et al. (2009)	Inhaled GE	ApoE$^{-/-}$ mice	CED	Increase in aortic TBARS and MMP-9, inhibited by Tempol	[99]
	6 h/day X 1 or 7 days				
	+/– Tempol			Increase in aortic ET-1, inhibited by BQ-123 (ET_A receptor antagonist)	
	+/– BQ-123				
Campen et al. (2010)	Inhaled DE	ApoE$^{-/-}$ mice	CED	Increase in macrophage content but not % lesional area in both DE and filtered DE-exposed mice	[42]
	Inhaled DEG			Increase in aortic TBARS in both DE and filtered-DE mice	
				No effects on plasma oxLDL	
Chen et al. (2010)	Inhaled CAPs($PM_{2.5}$)	ApoE$^{-/-}$ mice	Chow	Increase in % lesional area in brachiocephalic and left common arteries by ultrasound. Increase was similar to that induced by second hand smoke with a concentration 3 times higher	[133]
	6 h/day, 5 days/week X 6 months				
Quan et al. (2010)	Inhaled CAPs($PM_{2.5}$), DE, filtered-DE, CAPs+filtered-DE	ApoE$^{-/-}$ mice	Chow	Increase in % lesional area in aorta and brachiocephalic artery, largest in CAPs or CAPs+DEG on each location, respectively	[35]
	5 h/day, 4 days/week X3 and 5 months			No effects on plasma lipids	
Kampfrath et al. (2011)	Inhaled CAPs($PM_{2.5}$)	C57BL/6 mice	Chow	Increase in superoxide production in perivascular fat and aorta, Increase in TNFα and MCP-1 in the lungs and plasma, that was Tlr4-dependent	[39]
	6 h/day, 5 days/week X 20 weeks	Nox2$^{-/-}$ mice		Increase in oxPAPC in the BALF	
		Tlr4$^{Lps\text{-}d}$ *mice*		No effects on plasma lipids	

(continued)

Table 2 (continued)

Study	Air pollutant	Animal model	Diet	Major findings (induced by the exposure to air pollutant vs. FA controls)	Reference
Bai et al. (2011)	Inhaled DE	$ApoE^{-/-}$ mice	Chow	Increase in plaque lipid content, cellularity, foam cell formation and smooth muscle cell content	[43]
	6 h/day, 5 days/week X 7 weeks			Increase in systemic lipid and DNA oxidation (8-isoprostane and 8-OH-dG in urine)	
				No effects on plasma lipids	
Lund et al. (2011)	Inhaled GE, DE and MVE	ApoE–/– mice	CED	MVE increased aortic TBARS and oxidized plasma lipoprotein, inhibited by LOX-1 receptor Ab	[100]
	+/– Ab αLOX-1 receptor			MVE led to upregulation of aortic LOX-1, ET-1 and MMP-9	
	6 h/day X 7 days				
Yin et al. (2013)	Inhaled DE	$ApoE^{-/-}$ mice	Chow	Development of prooxidative and proinflammatory HDL. No effects on cholesterol efflux capacity	[63]
				Increase in oxidized lipids in the blood (8-isoprostanes, 12-HETE, 13-HODE), BALF (12- and 15-HETEs, 13-HODE) and liver (5-HETEs)	
	6 h/day, 5 days/week X 2 weeks			Decreased plasma PON-1 activity	
				Activation of 5-lipooxygenase in the liver	
Li et al. (2013)	Inhaled reaerosolized UFP +/– D-4 F	$LDL\text{-}R^{-/-}$ mice	CED	Increase in lesional area in the aortic root	[98]
	5 h/day, 3 days/week X 10 weeks			Development of HDL antioxidant dysfunction, degree of which was associated with aortic atherosclerosis and improved by D-4 F	
				Increase in plasma oxidized lipids (5-,12- and 15-HETEs, 9- and 13-HODEs)	
				Increase plasma serum amyloid protein and TNF-α	
				No effects on plasma levels of total cholesterol, decrease in HDL and increase in triglycerides	

Chen et al. (2013)	Inhaled polluted ambient air	ApoE$^{-/-}$ mice	CED	Increase in lesional area in ascending aorta	[40]
				Increase in plasma levels of total and LDL cholesterol	
	24 h/day, 7 days/week x 2 m			Decrease in serum total antioxidant capacity	
				Increase in TNF-α in BALF and serum	
Wan et al. (2014)	Inhaled polluted ambient air	ApoE$^{-/-}$ mice	CED	Increase in lesional area in ascending aorta	[41]
				Increase in serum levels of total and LDL cholesterol	
	24 h/day, 7 days/week x 2 months			Increase in TNF-α and IL-6 in BALF and serum	
				Increase in plasma levels of visfatin	
Rao et al. (2014)		ApoE$^{-/-}$ mice		Increase in lesional area in the aortic root	[37]
		LDL-R$^{-/-}$ BMTx with CD36$^{+/+}$ vs. CD36$^{-/-}$ donor cells		Increase in 7-ketocholesterol in plasma IDL/LDL and aorta	
				BMTx with CD36$^{-/- cc}$ inhibit PM-induced atherogenesis	

Studies are shown in chronological order based on the year of publication. I.T. = Intratracheal, BALF=Bronchoalveolar lavage fluid, BMTx=bone marrow transplantation, CAPs=concentrated ambient particles, DE=diesel emissions, DEG=DE gases achieved by filtering of whole DE, GE=gasoline exhaust, MVE=mixed motor vehicle emissions (DE+GE), FA=filtered air, CED=cholesterol enriched diet, N.S.=not significant.

Columbus, Ohio [37]. In all these studies, fine CAPs had accelerating effects on atherosclerosis. Chen and Nadziejko first reported that 39–41 week-old $ApoE^{-/-}$ mice fed a chow diet and exposed to 10× ambient concentrations of $PM_{2.5}$ for 6 h per day, 5 days per week for 5 months, led to a 57 % increase in the percentage of atherosclerotic plaque area in the aortic root [38] (Table 2). Sun et al. showed that younger 6-week-old $ApoE^{-/-}$ mice, fed a chow diet and exposed to similar conditions for 6 months, displayed an upward trend in the percentage of aortic atherosclerotic plaque area (45 % increase), that was not statistically significant [33]. However, the feeding of a high fat diet, resulted in a statistically significant 58 % increase in aortic root plaque area, following exposure to $PM_{2.5}$ (Sun, Wang et al. 2005). In another study, Sun et al. reported that $ApoE^{-/-}$ mice fed a high fat diet and exposed to concentrated $PM_{2.5}$ also resulted in enhancement of the plaque area in the aortic arch as assessed by ultrasound bio-microscopy [34]. The $PM_{2.5}$ exposures also led to atherosclerotic plaques that were enriched in tissue factor, a pro-thrombotic factor that may play a causative role or may simply indicate greater atherosclerotic plaque burden. In addition, $PM_{2.5}$ exposures led to pro-oxidative effects that were NADPH oxidase and Toll Like Receptor-4-dependent (TLR-4) [39]. Interestingly, $PM_{2.5}$ exposures also led to an increase in 7-ketocholesterol (7-KCh) in the aorta, together with an increase in CD36 expression in plaque macrophages. Deletion of CD36 in bone marrow-derived cells abolished $PM_{2.5}$-mediated atherosclerosis [37].

Five to six month-exposures to concentrated $PM_{2.5}$, for a few hours per day, were required to induce atherosclerosis in all these studies. However, only 2 months were sufficient when the exposures were continuous, as shown by two studies where ApoE null mice were subjected to high concentrations of ambient polluted air in Beijing (~60 μg/m³ $PM_{2.5}$) [40, 41]. Inhalatory exposures to whole diesel exhaust emissions results in changes in plaque composition when performed during 7–8 weeks [42, 43], and enhanced atherosclerotic lesion size after 5-month exposures [35]. However, I.T. administration of diesel exhaust particles (DEP) for only 2 months was sufficient to enhance plaque size [44]. The various studies show differences in the magnitude of atherosclerotic plaque enhancement which could be due to the length of the exposures, the type of experimental diet, the PM composition and the gender and age of the mice, and thus the extent of atherosclerosis already existing at the time of exposure.

Particle size is an important determinant in PM cardiovascular toxicity. Thus, Araujo et al. reported that 5-week exposures to concentrated ultrafine particles, with PM < 0.18 μm, promoted atherosclerosis to a larger degree than $PM_{2.5}$ in ApoE null mice [45]. UFP-exposed mice developed 25 % and 55 % more aortic atherosclerosis, assessed by the mean atherosclerotic lesional area in the aortic root, compared to $PM_{2.5}$ or filtered air-exposed control mice, respectively [45], in spite of an exposure mass that was ~4 times less. However, the exposure aerosols overlapped in particle sizes since $PM_{2.5}$ included ultrafines, which makes straight comparisons very difficult given that particle mass in the $PM_{2.5}$ aerosol was mostly determined by the accumulation mode particles (0.18–2.5 μm), while the more abundant ultrafine particles contributed to less than a fifth of the $PM_{2.5}$ aerosol's total mass. It is possible that ultrafine particles are more proatherogenic likely due to their smaller size, which confer greater lung penetrability, greater number per mass and/or high content in redox cycling chemicals such as polycyclic aromatic hydrocarbons (PAHs) [9]. Indeed, removal of Volatile Organic Compounds (VOCs), rich in PAHs, resulted in decreased lesion development in ApoE null mice exposed to concentrated ultrafines [46]. Importantly, in all these studies, PM exposures also induced prooxidant and proinflammatory effects in systemic tissues, which are likely mediators in the enhanced atherosclerotic lesions that were observed in the vasculature, as it is discussed below.

2 Proatherogenic Effects Induced by Air Pollutants Are Likely Due to Systemic Prooxidant and Proinflammatory Actions

Ischemic cardiovascular and cerebrovascular diseases share a major pathogenic substrate, the development of atherosclerotic plaques and subsequent thromboembolic events. Plaques develop in the large and medium-sized elastic muscular arteries, as a result of atherosclerosis, an inflammatory process characterized by the accumulation of lipids and fibrous elements in the vascular wall [47]. Infiltrating lipids come from circulating low-density lipoprotein (LDL) particles that are retained in the vascular wall. This retention is favored by oxidative modifications and lead to activation of endothelial cells, monocyte recruitment with internalization into the vasculature, differentiation into macrophages and generation of foam cells by increased lipid uptake. Vascular infiltration of lipids and inflammatory cells further enhances oxidative stress and a vicious cycle of inflammation.

The interplay between prooxidant and antioxidant factors in the vessels may determine the degree of reactive oxygen species (ROS) generation and plaque formation [48]. Exposure to PM may shift this balance by promoting systemic pro-oxidant effects [48]. Indeed, multiple *in vitro* studies support the ability of PM to trigger and/or enhance ROS generation in vascular cells such as endothelial cells [49–53], macrophages [54–60] and possibly smooth muscle cells [34]. Experiments with animals are also supportive. Thus, long-term exposure to $PM_{2.5}$ activates NADPH oxidase via upregulation of the NADPH oxidase subunits p47phox and Rac1 [61], resulting in increased production of superoxide in monocytes, aortic tissue and perivascular fat [39]. Enhanced ROS generation leads to oxidative alteration of proteins, lipids and/or DNA. Thus, we have shown that exposure to concentrated ambient particles (CAPs) in the $PM_{2.5}$ and ultrafine particle (UFP) size ranges lead to increased hepatic lipid peroxidation that was accompanied by increased expression of Nrf2-regulated antioxidant genes in the UFP-exposed mouse livers [45, 62], while $PM_{2.5}$ exposure led to enhanced formation of 3-nitrotyrosine residues in other studies [33, 36]. In addition, PM can also induce lipid peroxidation by activating lipoxygenase pathways in systemic tissues. Indeed, we have observed that hyperlipidemic ApoE null mice exposed to diesel exhaust for 2 weeks [63], led to activation of the 5-lipooxygenase (5-LO) pathway in the liver, evidenced by upregulation of Alox-5 mRNA and protein levels and glutathione peroxidase 6 (GPx6), which resulted in increased hepatic levels of 5-hydroxyeicosatetraenoic acid (5-HETE) and malondialdehyde (MDA) [63].

Prooxidative effects can lead to proinflammatory effects in cells in culture, experimental animals and humans, as previously reviewed by Araujo [64]. These include the activation of the NF-kB [51, 52], p38 MAPK [65, 66] and ERK1/2 pathways [65] with subsequent upregulation of pro-inflammatory factors such as TNF-α, IL-8, and monocyte chemotactic protein-1 (MCP-1), [51, 52], and adhesion molecules such as VCAM [51], E-selectin, P-selectin [67] in endothelial cells. Air pollution also induces production of TNF-α, IL-6 [68–73], IL-8 [74, 75], IL1-α [76], IL1-β [77], granulocyte macrophage colony-stimulating factor (GM-CSF) [69, 78] and macrophage-inflammatory protein-2 (MIP-2) [79] in macrophages. In animals, prooxidative effects can also trigger proinflammatory cascades in the vasculature [28, 80, 81] and enhanced atherogenesis [64]. PM-activation of 5-LO is emerging as an attractive linker between prooxidative and proinflammatory effects since lipoxygenases (LO) are important contributors to lipid peroxidation, inflammation and atherosclerosis. Thus, Alox-5 is known to interact with a "scaffold-like" protein designated 5-LO-activating protein (FLAP) that facilitates transfer of arachidonic acid to 5-LO to oxygenate its 5th carbon and generate a hydroperoxide acid [5-hydroperoxyeicosatetraenoic acid (5-HPETE)], and subsequently, the unstable leukotriene (LT) A4 (LTA_4) [82–86]. 5-HPETE is immediately reduced to the 5-HETE in the cells by peroxidases such as glutathione peroxidase [87], and then 5-oxo-ETE, a proinflammatory eicosanoid [88]. LTA_4 can be converted to LTB_4 by the LTA_4 hydrolase. LTA_4 can also be converted to LTC_4, which can then be converted to LTD_4 and this into $LTDE_4$ [87]. As a group, leukotrienes show powerful proinflammatory effects in Asthma and various inflammatory disorders, including atherosclerosis. Further studies are required to evaluate the importance of 5-LO activation in PM-mediated cardiovascular effects. Importantly, studies in humans also support the association between PM exposures and systemic prooxidative and proinflammatory effects [89–94].

Although various pathways have been proposed to explain how inhalation of air pollutants can lead to all these systemic prooxidant and proinflammatory effects, we still don't know how this occurs [9, 95]. While intratracheal (I.T) administration of PM leads to obvious pulmonary inflammation, evidenced by increased total cell counts in the bronchoalveolar lavage fluid (BALF), modification of the BALF cell differential and infiltration of the lungs by inflammatory cells [29, 96, 97], we have not observed evidence of histologically-defined pulmonary inflammation in our studies with CAPs or diesel exhaust (DE) [45, 63]. However, both fine CAPs (PM < 2.5 μm) [39] and DE [63] lead to enhanced lipid peroxidation in the lungs as judged by increased levels of oxidized phospholipids [39] and oxidized fatty acids [63] in the BALF, respectively. To understand pathogenesis induced by a toxin, it is important to determine its travel path and track the effects that it triggers throughout the body. Upon inhalation, PM induces in the BALF, enhanced levels of hydroxyeicosatetraenoic acids (HETEs) and hydroxyoctadecadienoic acids (HODEs), which are oxidative products of arachidonic (C20:4) and linoleic acid (C18:2), respectively [63]. In the blood and in the liver, DE [63] and/or UFP [98] lead to enhanced plasma levels of these products, malondialdehyde (MDA) and/or 8-isoprostanes. DE exposures for 7 weeks resulted in increased levels of 8-isoprostanes in the urine as well which suggests their possible clearance by the kidneys. Altogether, PM induces lipid oxidative products in the lungs and systemic tissues that could participate in the pathogenesis and be used as tracking signals.

3 Air Pollutants Induce Lipid Oxidation: Effects on Plasma LDL and HDL

Exposures to PM may result in modest elevations of plasma lipids depending on the experimental design. Thus, Sun et al. observed that ApoE$^{-/-}$ mice exposed to fine CAPs for 6 months exhibited a small decrease in total plasma cholesterol on a chow diet but a small increase in total cholesterol when animals were fed a high-fat diet [33]. Araujo et al. noted that chow-fed ApoE$^{-/-}$ mice exposed for 5 weeks to fine but not to ultrafine CAPs showed a small increase in total plasma cholesterol without any effects on HDL cholesterol levels [45]. Likewise, Li et al. reported that inhalation of re-aerosolized UFP for 10 weeks did not affect plasma levels of total cholesterol but led to decreased HDL cholesterol and higher plasma triglyceride levels [98] in LDL-R$^{-/-}$ mice fed a high fat diet (Table 2). However, other studies have not observed any effects on the quantitative levels of plasma lipids in normolipidemic [39] or hyperlipidemic animals, either following I.T. administration [29], inhalation of CAPs [35] or inhalation of motor vehicle emissions [49]. Altogether, it appears that effects on plasma lipids are highly dependent on the genetic background, type of diet and length of exposure. In addition, the effects on total plasma cholesterol have been relatively small [33, 45] and unlikely to be a major driver of the overall pro-atherosclerotic effects. For instance, while 5-week exposure to fine CAPs did lead to a small increase in total plasma cholesterol levels, this only resulted in a trend towards larger plaques [45]. In contrast, exposure to ultrafine CAPs for a similar length of time did not influence total plasma cholesterol levels but significantly promoted atherosclerosis [45].

PM also induces qualitative effects on plasma lipoproteins that could be even more important than the effects of the levels of plasma lipids. Campen et al. [42] and Lund et al. [99] have shown that exposure of ApoE null mice to DE or GE respectively, led to increased oxidized lipids in the aorta as judged by increased levels thiobarbituric reactive substances (TBARS). Although DE exposure alone did not increase TBARS levels in the plasma, exposure to mixed vehicular emissions (MVE), consisting in a combination of gasoline and diesel exhaust, did result in enhanced plasma levels of TBARS [100], a measure of reactive aldehydes derived from the oxidation of fatty acids, suggesting a greater toxicity for MVE in comparison to DE alone. We have recently shown that ApoE$^{-/-}$ mice exposed to DE for 2 weeks led to increased plasma levels of 8-isoprostanes, 12-hydroxyeicosatetraenoic (12-HETE) and 13-hydroxyoctadecadienoic (13-HODE) acids [63]. We also found that diesel exhaust led to increased susceptibility to oxidation by air of a plasma lipoprotein fraction enriched in VLDL and LDL [63]. This is consistent with the study of Castro-

Soares et al. which indicated that 4-month exposure to ambient levels of urban air pollution in Sao Paulo, Brazil led to enhanced susceptibility of LDL to oxidation in hyperlipidemic LDL-$R^{-/-}$ mice fed a high cholesterol diet [101]. Likewise, two studies where ApoE null mice were exposed for 2 months to ambient levels of air pollution in Beijing, China, led to increased oxidative stress in the blood [40, 41], as evidenced by decreased total antioxidant capacity [40] or increased MDA [41], increased levels of oxidized LDL (oxLDL) [40, 41], serum IL-6, TNF-α [40, 41] and visfatin [40, 41], together with enhanced aortic atherosclerosis [40, 41]. PM-induced oxidized LDL has also been demonstrated by increased HETEs and HODEs in the LDL fraction of LDL-$R^{-/-}$ mice exposed to inhaled re-aerosolized UFP for 10 weeks [102]. Altogether, this indicates that ambient PM and motor vehicle emissions can increase lipid peroxidation in the plasma, resulting in LDL particles that are either more oxidized or more susceptible to oxidation than filtered air controls.

Qualitative effects on plasma lipoproteins include functional alteration of the HDL particles. There are a number of protective functions of HDL that are negatively correlated with atherosclerosis and CAD. Thus, HDL's reverse cholesterol transport function inversely associates with carotid intima-media thickness and the likelihood of angiographic coronary artery disease (CAD) [103] while HDL's anti-oxidant function is significantly impaired in subjects with acute coronary syndrome, as compared with healthy subjects or those with stable CAD [104]. In addition, pro-inflammatory HDL is predictive of the susceptibility to atherosclerosis in humans [105] and in rabbits [106]. We have shown that exposure of $ApoE^{-/-}$ mice to fine and ultrafine CAPs for 5 weeks led to the development of dysfunctional pro-inflammatory HDL [45]. The degree of HDL dysfunction was particle size-dependent since UFP exposures led to a greater degree of dysfunction than did $PM_{2.5}$ [45]. The anti-inflammatory properties of HDL were measured using an LDL-induced monocyte chemotactic assay in a co-culture of endothelial cells and smooth muscle cells. In comparison to $PM_{2.5}$, exposure to UFP not only failed to inhibit the LDL-mediated inflammatory effects but promoted more monocyte migration. The more severe degree of HDL dysfunction mimicked the bigger extent of atherosclerosis and larger Nrf2-regulated antioxidant response in the liver. We have also reported that inhalation of diesel exhaust for as short as 2 weeks not only led to the development of pro-inflammatory HDL but also induced the loss of HDL's anti-oxidant properties, turning the HDL particles from anti-oxidant into pro-oxidant lipoproteins. The anti-oxidant dysfunction correlated with markers of lipid peroxidation in the blood suggesting that it may be related to lipid peroxidation within the HDL particles [63]. Of interest, the kinetics of recovery of the anti-oxidant and anti-inflammatory capacities was different, indicating that they may be affected in a different manner. In addition, there were strong associations between hepatic and plasma levels of HETEs, HODEs and/or MDA, with the degree of HDL antioxidant/antiinflammatory dysfunction [63], and between the latter and atherosclerotic lesion formation [98], suggesting that PM-induced lipid peroxidation could lead to enhanced atherogenesis through the development of a proatherogenic plasma lipid profile consisting of qualitative modifications of the HDL and LDL.

Controlled exposure studies, as well as panel and cross-sectional studies in humans also support the association of exposure to PM with increased systemic oxidative stress, via the detection of biomarkers of oxidative alteration of proteins, lipids and/or DNA in the circulating blood or in products excreted in the urine, as reviewed by Moller and Loft [107]. For example, Liu et al. reported the association of increases in exposure to black carbon and $PM_{2.5}$ with an elevation in plasma levels of TBARS in 28 nonsmoking seniors [108]. This is in agreement with earlier reports of increased serum TBARS in association with $PM_{2.5}$ exposures [109] or after moving or living in a polluted urban location such as Mexico city [110, 111]. Measures of oxidation in circulating blood are highly relevant since they may imply the involvement of oxidatively modified plasma lipoproteins such as LDL and/or HDL. Thus, subjects exposed to biomass fuel, who were exposed to three times more particulate pollution in the kitchen than control subjects who cooked with liquefied petroleum gas, exhibited elevated oxidized LDL, platelet aggregation and higher prevalence of hypertension [112]. Interestingly, the concentration of oxLDL has been positively associated with the carbon load of airway macrophages, a marker of chonic exposure to carbon particles derived from fossil fuel burning [113], consistent with animal data where rabbits receiving I.T. PM_{10} or ApoE null mice exposed to inhaled diesel exhaust exhibited correlations between the percentage of particle-laden alveolar macrophages and the volume of atherosclerotic lesions [29] or plaque foam cell formation [43]. PM exposures have also been associated with increased blood levels of GM-CSF, IL-6, IL-1, sTNF-RII, CRP and CD40 ligand (sCD40L) [89–93], suggesting a connection between prooxidative effects in lipoproteins and proinflammatory effects in the vasculature. Thus, exposure to PM is associated with plasma lipoproteins that are functionally pro-atherogenic, consisting in LDL particles that are more oxidized and/or more susceptible to oxidation as shown in both animal and human studies, and the development of dysfunctional prooxidant and proinflammatory HDL as shown by animal studies.

4 Air Pollutants, Carbohydrate and Lipid Metabolism

The environment is increasingly being appreciated as an important risk factor in the etiology of cardiometabolic diseases such as type 2 diabetes mellitus (T2DM), obesity and liver steatosis, all of which are occurring with increasing prevalence in Western and Asian societies. Thus, persistent organic pollutants such as dioxins are known obesogens, also associated with insulin resistance, T2DM and metabolic syndrome [114]. Epidemiological and experimental studies support a causal association between exposure to air pollutants and cardiometabolic diseases. All of these diseases include a chronic inflammatory component in the vasculature and the subsequent development of atherosclerosis. This suggests that PM could also promote cardiovascular diseases via the induction of systemic metabolic effects.

4.1 Epidemiological Studies Reveal Associations Between Air Pollution and Diabetes

Several epidemiological studies provide compelling evidence for the association between exposure to ambient air pollution and incidence and/or prevalence of diabetes, diabetes-related morbidity and mortality, obesity and insulin resistance. Almost all the epidemiological studies have been conducted in North America or in Europe where exposure levels to air pollutants are relatively low. In 2008, Brook et al. reported that each 1 ppb increase in nitrogen dioxide (NO_2), a marker of traffic-related air pollutants, was associated with a significantly increased risk of developing diabetes mellitus (DM) in women attending two respiratory clinics in Hamilton and Toronto, Canada (OR = 1.04, 95 % CI = 1.00–1.08) [115]. A large cross-sectional study with 6392 participants of the Swiss cohort Study on Air Pollution and Lung and Heart Diseases (SAPALDIA) found that both PM_{10} and NO_2 were associated with the prevalence of diabetes with odds ratios of 1.40 (95 % CI = 1.17–1.67) and 1.19 (95 % CI = 1.03–1.38), respectively, per 10 μg/m^3 increase in the average home outdoor level [116]. Long-term exposure to NO_2 and nitrogen oxides (NO_x) associated significantly with impaired glucose metabolism in 363 women from the Study on the influence of Air pollution on Lung function, Inflammation and Aging (SALIA) in Germany, with odds ratio of 1.465 (95 % CI = 1.049–2.046) and 1.409 (95 % CI = 1.010–1.967), per increased interquartile range of NO_2 (14.65 μg/m^3) and NO_x (43.16 μg/m^3), respectively, although they lost statistical significance after adjustment for multiple comparisons [117]. A large ecological study assessing the relationship between $PM_{2.5}$ and diabetes prevalence in all the US counties used data from the National Diabetes Surveillance System from the Centers for Disease Control and Prevention (CDC) and annual mean $PM_{2.5}$ levels that were calculated using air quality monitors. In this study, there was a 1 % increase in diabetes prevalence for every 10 μg/m^3 increase in $PM_{2.5}$, even in counties with acceptable standards of $PM_{2.5}$ and after controlling for diabetes risk factors [118]. Follow-up prospective studies have also shown a positive association between traffic-related air pollution and incidence of T2DM. Non-diabetic women in the SALIA study in Germany were followed up to assess the 16-year (1990–2006) incidence of T2DM and the levels of complement factor C3c as a marker of subclinical inflammation [119]. The authors reported 87 new cases of diabetes during the 16-year period with hazard ratios for developing diabetes that increased by 15 % (95 % CI = 4–27) per interquartile range increase in traffic-related PM and NO_2 exposures. They also found that only women with high C3c levels at baseline had an elevated risk for T2DM whereas there was no association for women with low C3c at baseline, suggesting that air pollutants may increase the risk of T2DM by promoting subclinical inflammation. In another prospective study that included a total of 62,012 participants in Ontario, Canada, with a mean follow-up time of 8 ± 3.2 years [120], Chen et al. reported that the adjusted hazard ratio for incident diabetes was 1.11 (95 % CI = 1.02–1.21) for every 10 μg/m^3 increase of $PM_{2.5}$, after adjusting for all covariates and comorbid conditions. Likewise, Coogan et al. reported a incidence rate ratio (IRR) for diabetes of 1.63 (95 % CI = 0.78–3.44) with $PM_{2.5}$ and 1.25 (95 % CI 1.07–1.46) with NOx, for every 10 μg/m^3 increase in each pollutant, in a cohort of 4204

African American women who lived in Los Angeles at baseline in 1995, followed-up over 10 years. [121]. Two prospective cohorts, the Nurses' Health Study (NHS) and the Health Professionals Follow-Up Study (HPFS), were used to investigate the relationship between chronic exposure to fine (<2.5 μm), quasi fine (10–2.5 μm) and coarse (>10 μm) particulate matter with incident T2DM [122]. The fully-adjusted hazards ratio for inter quartile range increases in $PM_{2.5}$, $PM_{10\text{-}2.5}$ and PM_{10} were not significant at 1.07 (95% CI=0.92–1.24); 1.06 (95% CI=0.94–1.20); and 1.04 (95% CI=0.93–1.16), respectively. However, there was a significantly strong association with DM for NHS women living <50 m versus >200 m from the nearest road, HR=1.14 (95% CI=1.03–1.27) thus suggesting an association between traffic-related air pollutants and incident DM among women.

4.2 Animal Studies Suggest Mechanisms How Air Pollution Induces Metabolic Effects

The increasing number of epidemiological studies that has suggested associations between low levels of air pollution and the propensity for developing T2DM prompted studies on animal models to investigate the potential biological mechanisms mediating these effects. Sun et al. reported one of the first studies describing the effects of air pollution on diabetes in 6-week old C57BL/6 mice fed a high-fat diet, exposed to concentrated $PM_{2.5}$ for 24 weeks. $PM_{2.5}$-exposed animals displayed elevated fasting glucose, insulin and Homeostatic model assessment (HOMA)-Insulin resistance (IR) [123]. These effects were accompanied by vascular endothelial dysfunction, increased levels of circulating inflammatory biomarkers, adipose tissue inflammation and increased visceral adiposity. Ensuing studies have uncovered a multitude of systemic effects induced by air pollutants on various organs, extending from the endothelium to adipose tissue, liver and intestines. To investigate the effects of early-life inhalation exposure to fine PM, 3-week old C57BL/6 mice were exposed to concentrated $PM_{2.5}$ for a duration of 10 weeks. $PM_{2.5}$ induced whole-body insulin resistance, increased abdominal fat with inflammation and defective vasomotor responses to insulin, characteristic of insulin resistance [124]. Exposure of C57BL/6 mice to $PM_{2.5}$ for 10 months also lead to pronounced insulin resistance and glucose intolerance [125], accompanied by reduced Akt phosphorylation in the liver, macrophage infiltration and inflammation in the lungs and adipose tissue. $PM_{2.5}$-exposed mice also exhibited decreased mitochondrial count and size, together with reduced uncoupling protein 1 (UCP1) protein in the adipose tissue indicating mitochondrial dysfunction. It seems that NADPH oxidases may play a critical role in $PM_{2.5}$-induced development of T2DM/IR since p47phox null mice, lacking the p47phox cytosolic subunit, which is critical for the function of NADPH oxidase displayed similar glucose tolerance, HOMA-IR indexes, visceral and adipose fat accumulation in both filtered air and $PM_{2.5}$ groups, which indicates a significant role for oxidative stress in $PM_{2.5}$-mediated metabolic abnormalities.

Interestingly, PM systemic effects involve carbohydrate and lipid metabolism concurrently. Thus, C57BL/6 mice exposed to concentrated $PM_{2.5}$ for 10 weeks exhibited not only resulted in glucose intolerance and impaired glycogen storage in the liver likely due to insulin resistance but also, increased plasma tryglycerides, larger content of cholesterol and triglycerides in the liver together with increased fat accumulation [126]. Staining for collagen deposition and macrophage activation indicated increased fibrosis and inflammation in the $PM_{2.5}$ group, consistent with a non-alcoholic steatohepatitis (NASH) phenotype [126]. Likewise, C57BL/6 mice fed a high fat diet and exposed to $PM_{2.5}$ for 17 weeks displayed insulin resistance and hepatic lipid accumulation as well [127]. This was likely due to increased lipogenesis as $PM_{2.5}$-exposed mice showed increased mRNA expression of SREBP-1 in the liver, a key transcription factor involved in the activation of lipogenic genes. Thus, important lipid synthetic enzymes such as acetyl-CoA carboxylase 2 (ACC2), fatty acid synthase (FAS), and diacylglycerol acyl transferase (DGAT2) were significantly upregulated in the liver of PM exposed mice in a CC-chemokine receptor-2 (CCR2)-dependent manner since CCR2 deficient mice exposed to the same protocol displayed reduced SREBP-1-dependent gene activation and hepatic triglyceride accumulation [127].

Exposures to PM via I.T. administration also result in the induction of systemic metabolic effects. Thus, diabetic obese db/db mice that received intratracheal DEP every 2 weeks for 18 weeks displayed increased hepatic levels of hexanoyl-lysine (HEL) adduct, indicative of increased oxidative stress together with enhanced liver steatosis [128]. Interestingly, I.T. administration of PM_{10} to BALB/c mice has also been shown to affect lipid metabolism in the liver resulting in increased levels of phospholipids, specifically phosphatidylcholine, as well as composition of total fatty acids with a more pronounced level of unsaturated fatty acids, in particular docosahexaenoic acid (DHA), and an increase of unsaturation index [129], which suggests that PM effects in lipid metabolism may lead and/or couple with systemic inflammatory effects.

5 Perspectives

Multiple epidemiological studies strongly support the associations of exposure to air pollutants and adverse health effects. For a long time, the focus was placed in the triggering and/or exacerbation of respiratory diseases such as bronchospastic disorders including asthma and chronic pulmonary obstructive disease, allergic disorders or lung cancer. However, work over the last 2 decades has put in manifest the development of cardiovascular and cerebrovascular diseases, mostly of ischemic nature, together with a variety of systemic metabolic effects resulting in insulin resistance, type 2 diabetes mellitus, liver steatosis and obesity. It appears that the ability of air pollutants to enhance ROS generation and promote oxidative stress in the lungs and systemic tissues underlie their potential to induce proinflammatory, metabolic and vascular effects. Although different pathways have been proposed to explain how their inhalation results in the induction of various systemic effects, it is still not clear how this occurs. Air pollutant effects on lipid peroxidation, leading to alteration of plasma lipoproteins, and the generation of a functionally proatherogenic plasma lipid profile, may explain at least in part, PM effects on the vasculature and the development of atherosclerosis. At the same time, air pollutant effects on lipid and carbohydrate metabolism may also predispose for the infiltration of lipids and inflammatory cells in the vasculature, with the subsequent formation of atherosclerotic lesions. Future studies are required to elucidate the precise mechanism(s) how pulmonary effects are translated into the systemic tissues, the identification of biomarkers that can provide early signals for the induction of deleterious vascular effects, and a better understanding of the potential interaction of air pollutant exposures with predisposing conditions and/or factors that may confer genetic susceptibility. Prevention of air pollution toxicity should largely be focused on reducing the levels of air pollution and avoidance of exposure. However, pharmacological developments aimed at reducing air pollution effects may be necessary, especially for populations that are predisposed, at higher risk or unable to avoid significant exposures.

Acknowledgments Writing of this article was supported by the National Institute of Environmental Health Sciences, National Institutes of Health (R56 ES016959 to Jesus A. Araujo).

References

1. Brook RD, Franklin B, Cascio W, Hong Y, Howard G, Lipsett M, Luepker R, Mittleman M, Samet J, Smith Jr SC, Tager I, Expert Panel on Population and Prevention Science of the American Heart Association. Air pollution and cardiovascular disease: a statement for healthcare professionals from the Expert Panel on Population and Prevention Science of the American Heart Association. Circulation. 2004;109(21):2655–71.
2. Bhatnagar A. Environmental cardiology: studying mechanistic links between pollution and heart disease. Circ Res. 2006;99(7):692–705.
3. Brook RD, Rajagopalan S, Pope CA III, Brook JR, Bhatnagar A, Diez-Roux AV, Holguin F, Hong Y, Luepker RV, Mittleman MA, Peters A, Siscovick D, Smith SC Jr, Whitsel L, Kaufman JD, Council on the Kidney in Cardiovascular Disease, and Council on Nutrition, Physical Activity and Metabolism on Behalf of the American Heart Association Council on Epidemiology and Prevention. Particulate matter air pollution and cardiovascular disease. An update to the scientific statement from the American Heart Association. Circulation 2010;121:2331–78.
4. Pope 3rd CA, Burnett RT, Thurston GD, Thun MJ, Calle EE, Krewski D, Godleski JJ. Cardiovascular mortality and long-term exposure to particulate air pollution: epidemiological evidence of general pathophysiological pathways of disease. Circulation. 2004;109(1):71–7.
5. World Health Organization. Burden of disease from Ambient Air Pollution for 2012. Geneva: Public Health Social and Environmental Determinants of Health Department, World Health Organization; 2014.
6. Hoyert DL, Xu J. Deaths: preliminary data for 2011. Natl Vital Stat Rep. 2012;61(6):1–52.
7. World Health Organization. World Health Statistics. Geneva: World Health Organization; 2013. p. 1–172.

8. Newby DE, Mannucci PM, Tell GS, Baccarelli AA, Brook RD, Donaldson K, Forastiere F, Franchini M, Franco OH, Graham I, Hoek G, Hoffmann B, Hoylaerts MF, Kunzli N, Mills N, Pekkanen J, Peters A, Piepoli MF, Rajagopalan S, Storey RF, ESC Working Group on Thrombosis, European Association for Cardiovascular Prevention and Rehabilitation; ESC Heart Failure Association. Expert position paper on air pollution and cardiovascular disease. Eur Heart J. 2015;36(2):83–93b.
9. Araujo JA, Nel AE. Particulate matter and atherosclerosis: role of particle size, composition and oxidative stress. Part Fibre Toxicol. 2009;6:24. 2761850.
10. Lim SS, Vos T, Flaxman AD, Danaei G, Shibuya K, Adair-Rohani H, Amann M, Anderson HR, Andrews KG, Aryee M, Atkinson C, Bacchus LJ, Bahalim AN, Balakrishnan K, Balmes J, Barker-Collo S, Baxter A, Bell ML, Blore JD, Blyth F, Bonner C, Borges G, Bourne R, Boussinesq M, Brauer M, Brooks P, Bruce NG, Brunekreef B, Bryan-Hancock C, Bucello C, Buchbinder R, Bull F, Burnett RT, Byers TE, Calabria B, Carapetis J, Carnahan E, Chafe Z, Charlson F, Chen H, Chen JS, Cheng AT, Child JC, Cohen A, Colson KE, Cowie BC, Darby S, Darling S, Davis A, Degenhardt L, Dentener F, Des Jarlais DC, Devries K, Dherani M, Ding EL, Dorsey ER, Driscoll T, Edmond K, Ali SE, Engell RE, Erwin PJ, Fahimi S, Falder G, Farzadfar F, Ferrari A, Finucane MM, Flaxman S, Fowkes FG, Freedman G, Freeman MK, Gakidou E, Ghosh S, Giovannucci E, Gmel G, Graham K, Grainger R, Grant B, Gunnell D, Gutierrez HR, Hall W, Hoek HW, Hogan A, Hosgood HD 3rd, Hoy D, Hu H, Hubbell BJ, Hutchings SJ, Ibeanusi SE, Jacklyn GL, Jasrasaria R, Jonas JB, Kan H, Kanis JA, Kassebaum N, Kawakami N, Khang YH, Khatibzadeh S, Khoo JP, Kok C, Laden F, Lalloo R, Lan Q, Lathlean T, Leasher JL, Leigh J, Li Y, Lin JK, Lipshultz SE, London S, Lozano R, Lu Y, Mak J, Malekzadeh R, Mallinger L, Marcenes W, March L, Marks R, Martin R, McGale P, McGrath J, Mehta S, Mensah GA, Merriman TR, Micha R, Michaud C, Mishra V, Mohd Hanafiah K, Mokdad AA, Morawska L, Mozaffarian D, Murphy T, Naghavi M, Neal B, Nelson PK, Nolla JM, Norman R, Olives C, Omer SB, Orchard J, Osborne R, Ostro B, Page A, Pandey KD, Parry CD, Passmore E, Patra J, Pearce N, Pelizzari PM, Petzold M, Phillips MR, Pope D, Pope CA 3rd, Powles J, Rao M, Razavi H, Rehfuess EA, Rehm JT, Ritz B, Rivara FP, Roberts T, Robinson C, Rodriguez-Portales JA, Romieu I, Room R, Rosenfeld LC, Roy A, Rushton L, Salomon JA, Sampson U, Sanchez-Riera L, Sanman E, Sapkota A, Seedat S, Shi P, Shield K, Shivakoti R, Singh GM, Sleet DA, Smith E, Smith KR, Stapelberg NJ, Steenland K, Stöckl H, Stovner LJ, Straif K, Straney L, Thurston GD, Tran JH, Van Dingenen R, van Donkelaar A, Veerman JL, Vijayakumar L, Weintraub R, Weissman MM, White RA, Whiteford H, Wiersma ST, Wilkinson JD, Williams HC, Williams W, Wilson N, Woolf AD, Yip P, Zielinski JM, Lopez AD, Murray CJ, Ezzati M, AlMazroa MA, Memish ZA. A comparative risk assessment of burden of disease and injury attributable to 67 risk factors and risk factor clusters in 21 regions, 1990–2010: a systematic analysis for the Global Burden of Disease Study 2010. Lancet 2012;380(9859):2224–60.
11. U.S. EPA. Air quality criteria for particulate matter (Final Report, Oct 2004). Washington, DC: U.S. Environmental Protection Agency; 2004. EPA 600/P-99/002aF-bF.
12. Mustafic H, Jabre P, Caussin C, Murad MH, Escolano S, Tafflet M, Perier MC, Marijon E, Vernerey D, Empana JP, Jouven X. Main air pollutants and myocardial infarction: a systematic review and meta-analysis. JAMA. 2012;307(7):713–21.
13. Shah ASV, Langrish JP, Nair H, McAllister DA, Hunter AL, Donaldson K, Newby DE, Mills NL. Global association of air pollution and heart failure: a systematic review and meta-analysis. 2013. The Lancet.
14. Miller KA, Siscovick DS, Sheppard L, Shepherd K, Sullivan JH, Anderson GL, Kaufman JD. Long-term exposure to air pollution and incidence of cardiovascular events in women. N Engl J Med. 2007;356(5):447–58.
15. Kunzli N, Jerrett M, Mack WJ, Beckerman B, LaBree L, Gilliland F, Thomas D, Peters J, Hodis HN. Ambient air pollution and atherosclerosis in Los Angeles. Environ Health Perspect. 2005;113(2):201–6. 1277865.
16. Diez Roux AV, Auchincloss AH, Franklin TG, Raghunathan T, Barr RG, Kaufman J, Astor B, Keeler J. Long-term exposure to ambient particulate matter and prevalence of subclinical atherosclerosis in the Multi-Ethnic Study of Atherosclerosis. Am J Epidemiol. 2008;167(6): 667–75.

17. Allen RW, Criqui MH, Diez Roux AV, Allison M, Shea S, Detrano R, Sheppard L, Wong ND, Stukovsky KH, Kaufman JD. Fine particulate matter air pollution, proximity to traffic, and aortic atherosclerosis. Epidemiology. 2009;20(2):254–64.
18. Bauer M, Moebus S, Mohlenkamp S, Dragano N, Nonnemacher M, Fuchsluger M, Kessler C, Jakobs H, Memmesheimer M, Erbel R, Jockel KH, Hoffmann B, H.N.R.S.I. Group. Urban particulate matter air pollution is associated with subclinical atherosclerosis: results from the HNR (Heinz Nixdorf Recall) study. J Am Coll Cardiol. 2010;56(22):1803–8.
19. Hoffmann B, Moebus S, Mohlenkamp S, Stang A, Lehmann N, Dragano N, Schmermund A, Memmesheimer M, Mann K, Erbel R, Jockel KH. Residential exposure to traffic is associated with coronary atherosclerosis. Circulation. 2007;116(5):489–96.
20. Kalsch H, Hennig F, Moebus S, Mohlenkamp S, Dragano N, Jakobs H, Memmesheimer M, Erbel R, Jockel KH, Hoffmann B, G. Heinz Nixdorf Recall Study Investigative. Are air pollution and traffic noise independently associated with atherosclerosis: the Heinz Nixdorf Recall Study. Eur Heart J. 2014;35(13):853–60.
21. Pope 3rd CA, Burnett RT, Thun MJ, Calle EE, Krewski D, Ito K, Thurston GD. Lung cancer, cardiopulmonary mortality, and long-term exposure to fine particulate air pollution. JAMA. 2002;287(9):1132–41.
22. Perez L, Wolf K, Hennig F, Penell J, Basagaña X, Foraster M, Aguilera I, Agis D, Beelen R, Brunekreef B, Cyrys J, Fuks KB, Adam M, Baldassarre D, Cirach M, Elosua R, Dratva J, Hampel R, Koenig W, Marrugat J, de Faire U, Pershagen G, Probst-Hensch NM, de Nazelle A, Nieuwenhuijsen MJ, Rathmann W, Rivera M, Seissler J, Schindler C, Thiery J, Hoffmann B, Peters A, Künzli N. Air pollution and atherosclerosis: a cross-sectional analysis of four European cohort studies in the ESCAPE study. Environ Health Perspect. 2015;123(6):597–605, 4455580.
23. Provost EB, Madhloum N, Int Panis L, De Boever P, Nawrot TS. Carotid intima-media thickness, a marker of subclinical atherosclerosis, and particulate air pollution exposure: the meta-analytical evidence. PLoS One. 2015;10(5):e0127014. 4430520
24. Su TC, Hwang JJ, Shen YC, Chan CC. Carotid intima-media thickness and long-term exposure to traffic-related air pollution in middle-aged residents of Taiwan: a cross-sectional study. Environ Health Perspect. 2015;123(8):773–8. 4529007.
25. Kunzli N, Jerrett M, Garcia-Esteban R, Basagana X, Beckermann B, Gilliland F, Medina M, Peters J, Hodis HN, Mack WJ. Ambient air pollution and the progression of atherosclerosis in adults. PLoS One. 2010;5(2), e9096. 2817007.
26. Wilker EH, Mittleman MA, Coull BA, Gryparis A, Bots ML, Schwartz J, Sparrow D. Long-term exposure to black carbon and carotid intima-media thickness: the normative aging study. Environ Health Perspect. 2013;121(9):1061–7. 3764069.
27. Adar SD, Sheppard L, Vedal S, Polak JF, Sampson PD, Diez Roux AV, Budoff M, Jacobs Jr DR, Barr RG, Watson K, Kaufman JD. Fine particulate air pollution and the progression of carotid intima-medial thickness: a prospective cohort study from the multi-ethnic study of atherosclerosis and air pollution. PLoS Med. 2013;10(4), e1001430. 3637008.
28. Yatera K, Hsieh J, Hogg JC, Tranfield E, Suzuki H, Shih CH, Behzad AR, Vincent R, van Eeden SF. Particulate matter air pollution exposure promotes recruitment of monocytes into atherosclerotic plaques. Am J Physiol Heart Circ Physiol. 2008;294(2):H944–53.
29. Suwa T, Hogg JC, Quinlan KB, Ohgami A, Vincent R, van Eeden SF. Particulate air pollution induces progression of atherosclerosis. J Am Coll Cardiol. 2002;39(6):935–42.
30. Niwa Y, Hiura Y, Murayama T, Yokode M, Iwai N. Nano-sized carbon black exposure exacerbates atherosclerosis in LDL-receptor knockout mice. Circ J. 2007;71(7):1157–61.
31. Chen LC, Nadziejko C. Effects of subchronic exposures to concentrated ambient particles (CAPs) in mice. V. CAPs exacerbate aortic plaque development in hyperlipidemic mice. Inhal Toxicol. 2005;17(4-5):217–24.
32. Pan B, Ma Y, Ren H, He Y, Wang Y, Lv X, Liu D, Ji L, Yu B, Wang Y, Chen YE, Pennathur S, Smith JD, Liu G, Zheng L. Diabetic HDL is dysfunctional in stimulating endothelial cell migration and proliferation due to down regulation of SR-BI expression. PLoS One. 2012;7(11), e48530. 3487724.
33. Sun Q, Wang A, Jin X, Natanzon A, Duquaine D, Brook RD, Aguinaldo JG, Fayad ZA, Fuster V, Lippmann M, Chen LC, Rajagopalan S. Long-term air pollution exposure and acceleration of atherosclerosis and vascular inflammation in an animal model. JAMA. 2005;294(23):3003–10.

34. Sun Q, Yue P, Kirk RI, Wang A, Moatti D, Jin X, Lu B, Schecter AD, Lippmann M, Gordon T, Chen LC, Rajagopalan S. Ambient air particulate matter exposure and tissue factor expression in atherosclerosis. Inhal Toxicol. 2008;20(2):127–37.
35. Quan C, Sun Q, Lippmann M, Chen LC. Comparative effects of inhaled diesel exhaust and ambient fine particles on inflammation, atherosclerosis, and vascular dysfunction. Inhal Toxicol. 2010;22(9):738–53. 3073494.
36. Ying Z, Yue P, Xu X, Zhong M, Sun Q, Mikolaj M, Wang A, Brook RD, Chen LC, Rajagopalan S. Air pollution and cardiac remodeling: a role for RhoA/Rho-kinase. Am J Physiol Heart Circ Physiol. 2009;296(5):H1540–50.
37. Rao X, Zhong J, Maiseyeu A, Gopalakrishnan B, Villamena FA, Chen LC, Harkema JR, Sun Q, Rajagopalan S. CD36-dependent 7-ketocholesterol accumulation in macrophages mediates progression of atherosclerosis in response to chronic air pollution exposure. Circ Res. 2014;115(9):770–80. 4275116.
38. Gunnison A, Chen LC. Effects of subchronic exposures to concentrated ambient particles (CAPs) in mice. VI. Gene expression in heart and lung tissue. Inhal Toxicol. 2005;17(4-5): 225–33.
39. Kampfrath T, Maiseyeu A, Ying Z, Shah Z, Deiuliis JA, Xu X, Kherada N, Brook RD, Reddy KM, Padture NP, Parthasarathy S, Chen LC, Moffatt-Bruce S, Sun Q, Morawietz H, Rajagopalan S. Chronic fine particulate matter exposure induces systemic vascular dysfunction via NADPH oxidase and TLR4 pathways. Circ Res. 2011;108(6):716–26. 3085907.
40. Chen T, Jia G, Wei Y, Li J. Beijing ambient particle exposure accelerates atherosclerosis in ApoE knockout mice. Toxicol Lett. 2013;223(2):146–53.
41. Wan Q, Cui X, Shao J, Zhou F, Jia Y, Sun X, Zhao X, Chen Y, Diao J, Zhang L. Beijing ambient particle exposure accelerates atherosclerosis in ApoE knockout mice by upregulating visfatin expression. Cell Stress Chaperones. 2014;19(5):715–24. 4147068.
42. Campen MJ, Lund AK, Knuckles TL, Conklin DJ, Bishop B, Young D, Seilkop S, Seagrave J, Reed MD, McDonald JD. Inhaled diesel emissions alter atherosclerotic plaque composition in ApoE(-/-) mice. Toxicol Appl Pharmacol. 2010;242(3):310–7.
43. Bai N, Kido T, Suzuki H, Yang G, Kavanagh TJ, Kaufman JD, Rosenfeld ME, van Breemen C, Eeden SF. Changes in atherosclerotic plaques induced by inhalation of diesel exhaust. Atherosclerosis. 2011;216(2):299–306.
44. Miller MR, McLean SG, Duffin R, Lawal AO, Araujo JA, Shaw CA, Mills NL, Donaldson K, Newby DE, Hadoke PW. Diesel exhaust particulate increases the size and complexity of lesions in atherosclerotic mice. Part Fibre Toxicol. 2013;10:61. 3907045.
45. Araujo JA, Barajas B, Kleinman M, Wang X, Bennett BJ, Gong KW, Navab M, Harkema J, Sioutas C, Lusis AJ, Nel AE. Ambient particulate pollutants in the ultrafine range promote early atherosclerosis and systemic oxidative stress. Circ Res. 2008;102(5):589–96. 3014059.
46. Keebaugh AJ, Sioutas C, Pakbin P, Schauer JJ, Mendez LB, Kleinman MT. Is atherosclerotic disease associated with organic components of ambient fine particles? Sci Total Environ. 2015;533:69–75.
47. Lusis AJ. Atherosclerosis. Nature. 2000;407(6801):233–41. 2826222.
48. Araujo JA. Nrf2 and the promotion of atherosclerosis: lessons to be learned. Clin Lipidol. 2011;7(2):123–6.
49. Bai Y, Suzuki AK, Sagai M. The cytotoxic effects of diesel exhaust particles on human pulmonary artery endothelial cells in vitro: role of active oxygen species. Free Radic Biol Med. 2001;30(5):555–62.
50. Hirano S, Furuyama A, Koike E, Kobayashi T. Oxidative-stress potency of organic extracts of diesel exhaust and urban fine particles in rat heart microvessel endothelial cells. Toxicology. 2003;187(2-3):161–70.
51. Li R, Ning Z, Majumdar R, Cui J, Takabe W, Jen N, Sioutas C, Hsiai T. Ultrafine particles from diesel vehicle emissions at different driving cycles induce differential vascular pro-inflammatory responses: implication of chemical components and NF-kappaB signaling. Part Fibre Toxicol. 2010;7:6.

52. Montiel-Davalos A, Ibarra-Sanchez Mde J, Ventura-Gallegos JL, Alfaro-Moreno E, Lopez-Marure R. Oxidative stress and apoptosis are induced in human endothelial cells exposed to urban particulate matter. Toxicol In Vitro. 2010;24(1):135–41.
53. Li Z, Hyseni X, Carter JD, Soukup JM, Dailey LA, Huang YC. Pollutant particles enhanced H2O2 production from NAD(P)H oxidase and mitochondria in human pulmonary artery endothelial cells. Am J Physiol Cell Physiol. 2006;291(2):C357–65.
54. Soukup JM, Ghio AJ, Becker S. Soluble components of Utah Valley particulate pollution alter alveolar macrophage function in vivo and in vitro. Inhal Toxicol. 2000;12(5):401–14.
55. Hiura TS, Kaszubowski MP, Li N, Nel AE. Chemicals in diesel exhaust particles generate reactive oxygen radicals and induce apoptosis in macrophages. J Immunol. 1999;163(10):5582–91.
56. Li N, Alam J, Venkatesan MI, Eiguren-Fernandez A, Schmitz D, Di Stefano E, Slaughter N, Killeen E, Wang X, Huang A, Wang M, Miguel AH, Cho A, Sioutas C, Nel AE. Nrf2 is a key transcription factor that regulates antioxidant defense in macrophages and epithelial cells: protecting against the proinflammatory and oxidizing effects of diesel exhaust chemicals. J Immunol. 2004;173(5):3467–81.
57. Ohyama M, Otake T, Adachi S, Kobayashi T, Morinaga K. A comparison of the production of reactive oxygen species by suspended particulate matter and diesel exhaust particles with macrophages. Inhal Toxicol. 2007;19 Suppl 1:157–60.
58. Lee CC, Kang JJ. Extract of motorcycle exhaust particles induced macrophages apoptosis by calcium-dependent manner. Chem Res Toxicol. 2002;15(12):1534–42.
59. Chio C-P, Chen S-C, Chiang K-C, Chou W-C, Liao C-M. Oxidative stress risk analysis for exposure to diesel exhaust particle-induced reactive oxygen species. Sci Total Environ. 2007;387(1-3):113–27.
60. Goldsmith CA, Imrich A, Danaee H, Ning YY, Kobzik L. Analysis of air pollution particulate-mediated oxidant stress in alveolar macrophages. J Toxicol Environ Health A. 1998;54(7): 529–45.
61. Ying Z, Kampfrath T, Thurston G, Farrar B, Lippmann M, Wang A, Sun Q, Chen LC, Rajagopalan S. Ambient particulates alter vascular function through induction of reactive oxygen and nitrogen species. Toxicol Sci. 2009;111(1):80–8. 2726294.
62. Gong KW, Zhao W, Li N, Barajas B, Kleinman M, Sioutas C, Horvath S, Lusis AJ, Nel A, Araujo JA. Air-pollutant chemicals and oxidized lipids exhibit genome-wide synergistic effects on endothelial cells. Genome Biol. 2007;8(7):R149. PMC2323217.
63. Yin F, Lawal A, Ricks J, Fox JR, Larson T, Navab M, Fogelman AM, Rosenfeld ME, Araujo JA. Diesel exhaust induces systemic lipid peroxidation and development of dysfunctional pro-oxidant and pro-inflammatory high-density lipoprotein. Arterioscler Thromb Vasc Biol. 2013;33(6):1153–61.
64. Araujo J. Particulate air pollution, systemic oxidative stress, inflammation, and atherosclerosis. Air Qual Atmos Health. 2011;4(1):79–93.
65. Mo Y, Wan R, Chien S, Tollerud DJ, Zhang Q. Activation of endothelial cells after exposure to ambient ultrafine particles: the role of NADPH oxidase. Toxicol Appl Pharmacol. 2009;236(2):183–93.
66. Sumanasekera WK, Ivanova MM, Johnston BJ, Dougherty SM, Sumanasekera GU, Myers SR, Ali Y, Kizu R, Klinge CM. Rapid effects of diesel exhaust particulate extracts on intracellular signaling in human endothelial cells. Toxicol Lett. 2007;174(1-3):61–73.
67. Montiel-Davalos A, Alfaro-Moreno E, Lopez-Marure R. PM2.5 and PM10 induce the expression of adhesion molecules and the adhesion of monocytic cells to human umbilical vein endothelial cells. Inhal Toxicol. 2007;19 Suppl 1:91–8.
68. Alfaro-Moreno E, Martinez L, Garcia-Cuellar C, Bonner JC, Murray JC, Rosas I, Rosales SP, Osornio-Vargas AR. Biologic effects induced in vitro by PM10 from three different zones of Mexico City. Environ Health Perspect. 2002;110(7):715–20.
69. Fujii T, Hayashi S, Hogg JC, Mukae H, Suwa T, Goto Y, Vincent R, van Eeden SF. Interaction of alveolar macrophages and airway epithelial cells following exposure to particulate matter produces mediators that stimulate the bone marrow. Am J Respir Cell Mol Biol. 2002;27(1):34–41.

70. Osornio-Vargas AR, Bonner JC, Alfaro-Moreno E, Martinez L, Garcia-Cuellar C, Ponce-de-Leon Rosales S, Miranda J, Rosas I. Proinflammatory and cytotoxic effects of Mexico City air pollution particulate matter in vitro are dependent on particle size and composition. Environ Health Perspect. 2003;111(10):1289–93.
71. Pozzi R, De Berardis B, Paoletti L, Guastadisegni C. Inflammatory mediators induced by coarse (PM2.5-10) and fine (PM2.5) urban air particles in RAW 264.7 cells. Toxicology. 2003;183(1-3):243–54.
72. Becker S, Mundandhara S, Devlin RB, Madden M. Regulation of cytokine production in human alveolar macrophages and airway epithelial cells in response to ambient air pollution particles: further mechanistic studies. Toxicol Appl Pharmacol. 2005;207(2 Suppl):269–75.
73. Amakawa K, Terashima T, Matsuzaki T, Matsumaru A, Sagai M, Yamaguchi K. Suppressive effects of diesel exhaust particles on cytokine release from human and murine alveolar macrophages. Exp Lung Res. 2003;29(3):149–64.
74. Becker S, Dailey L, Soukup JM, Silbajoris R, Devlin RB. TLR-2 is involved in airway epithelial cell response to air pollution particles. Toxicol Appl Pharmacol. 2005;203(1):45–52.
75. Monn C, Becker S. Cytotoxicity and induction of proinflammatory cytokines from human monocytes exposed to fine (PM2.5) and coarse particles (PM10-2.5) in outdoor and indoor air. Toxicol Appl Pharmacol. 1999;155(3):245–52.
76. Brown DM, Donaldson K, Stone V. Effects of PM10 in human peripheral blood monocytes and J774 macrophages. Respir Res. 2004;5:29.
77. Jimenez LA, Drost EM, Gilmour PS, Rahman I, Antonicelli F, Ritchie H, MacNee W, Donaldson K. PM(10)-exposed macrophages stimulate a proinflammatory response in lung epithelial cells via TNF-alpha. Am J Physiol Lung Cell Mol Physiol. 2002;282(2):L237–48.
78. Ishii H, Hayashi S, Hogg JC, Fujii T, Goto Y, Sakamoto N, Mukae H, Vincent R, van Eeden SF. Alveolar macrophage-epithelial cell interaction following exposure to atmospheric particles induces the release of mediators involved in monocyte mobilization and recruitment. Respir Res. 2005;6:87.
79. Imrich A, Ning Y, Lawrence J, Coull B, Gitin E, Knutson M, Kobzik L. Alveolar macrophage cytokine response to air pollution particles: oxidant mechanisms. Toxicol Appl Pharmacol. 2007;218(3):256–64.
80. Tamagawa E, Bai N, Morimoto K, Gray C, Mui T, Yatera K, Zhang X, Xing L, Li Y, Laher I, Sin DD, Man SF, van Eeden SF. Particulate matter exposure induces persistent lung inflammation and endothelial dysfunction. Am J Physiol Lung Cell Mol Physiol. 2008;295(1):L79–85. 2494798.
81. Mutlu GM, Green D, Bellmeyer A, Baker CM, Burgess Z, Rajamannan N, Christman JW, Foiles N, Kamp DW, Ghio AJ, Chandel NS, Dean DA, Sznajder JI, Budinger GR. Ambient particulate matter accelerates coagulation via an IL-6-dependent pathway. J Clin Invest. 2007;117(10):2952–61. 1978421.
82. Brash AR. Lipoxygenases: occurrence, functions, catalysis, and acquisition of substrate. J Biol Chem. 1999;274(34):23679–82.
83. Funk CD. Prostaglandins and leukotrienes: advances in eicosanoid biology. Science. 2001;294(5548):1871–5.
84. Kuhn H, Thiele BJ. The diversity of the lipoxygenase family. Many sequence data but little information on biological significance. FEBS Lett. 1999;449(1):7–11.
85. Yoshimoto T, Takahashi Y. Arachidonate 12-lipoxygenases. Prostaglandins Other Lipid Mediat. 2002;68–69:245–62.
86. Yamamoto S. Mammalian lipoxygenases: molecular structures and functions. Biochim Biophys Acta. 1992;1128(2-3):117–31.
87. Poeckel D, Funk CD. The 5-lipoxygenase/leukotriene pathway in preclinical models of cardiovascular disease. Cardiovasc Res. 2010;86(2):243–53.
88. Powell WS, Rokach J. Biochemistry, biology and chemistry of the 5-lipoxygenase product 5-oxo-ETE. Prog Lipid Res. 2005;44(2-3):154–83.
89. van Eeden SF, Tan WC, Suwa T, Mukae H, Terashima T, Fujii T, Qui D, Vincent R, Hogg JC. Cytokines involved in the systemic inflammatory response induced by exposure to particulate matter air pollutants (PM(10)). Am J Respir Crit Care Med. 2001;164(5):826–30.

90. Delfino RJ, Staimer N, Tjoa T, Polidori A, Arhami M, Gillen DL, Kleinman MT, Vaziri ND, Longhurst J, Zaldivar F, Sioutas C. Circulating biomarkers of inflammation, antioxidant activity, and platelet activation are associated with primary combustion aerosols in subjects with coronary artery disease. Environ Health Perspect. 2008;116(7):898–906. 2453158.
91. Ruckerl R, Phipps RP, Schneider A, Frampton M, Cyrys J, Oberdorster G, Wichmann HE, Peters A. Ultrafine particles and platelet activation in patients with coronary heart disease—results from a prospective panel study. Part Fibre Toxicol. 2007;4:1. 1790903.
92. Chuang KJ, Chan CC, Su TC, Lee CT, Tang CS. The effect of urban air pollution on inflammation, oxidative stress, coagulation, and autonomic dysfunction in young adults. Am J Respir Crit Care Med. 2007;176(4):370–6.
93. Dubowsky SD, Suh H, Schwartz J, Coull BA, Gold DR. Diabetes, obesity, and hypertension may enhance associations between air pollution and markers of systemic inflammation. Environ Health Perspect. 2006;114(7):992–8. 1513328.
94. Moller P, Loft S. Oxidative damage to DNA and lipids as biomarkers of exposure to air pollution. Environ Health Perspect. 2010;118(8):1126–36. PMC2920082.
95. Brook RD, Rajagopalan S, Pope 3rd CA, Brook JR, Bhatnagar A, Diez-Roux AV, Holguin F, Hong Y, Luepker RV, Mittleman MA, Peters A, Siscovick D, Smith Jr SC, Whitsel L, Kaufman JD, E. American Heart Association Council on, C.o.t.K.i.C.D. Prevention, P.A. Council on Nutrition, and Metabolism. Particulate matter air pollution and cardiovascular disease: an update to the scientific statement from the American Heart Association. Circulation. 2010;121(21):2331–78.
96. Kooter IM, Pennings JLA, Opperhuizen A, Cassee FR. Gene expression pattern in spontaneously hypertensive rats exposed to urban particulate matter (EHC-93). Inhal Toxicol. 2005;17(1):53–65.
97. Wise H, Balharry D, Reynolds LJ, Sexton K, Richards RJ. Conventional and toxicogenomic assessment of the acute pulmonary damage induced by the instillation of Cardiff PM10 into the rat lung. Sci Total Environ. 2006;360(1-3):60–7.
98. Li R, Navab M, Pakbin P, Ning Z, Navab K, Hough G, Morgan TE, Finch CE, Araujo JA, Fogelman AM, Sioutas C, Hsiai T. Ambient ultrafine particles alter lipid metabolism and HDL anti-oxidant capacity in LDLR-null mice. J Lipid Res. 2013;54(6):1608–15. 3646462.
99. Lund AK, Lucero J, Lucas S, Madden MC, McDonald JD, Seagrave JC, Knuckles TL, Campen MJ. Vehicular emissions induce vascular MMP-9 expression and activity associated with endothelin-1-mediated pathways. Arterioscler Thromb Vasc Biol. 2009;29(4):511–7.
100. Lund AK, Lucero J, Harman M, Madden MC, McDonald JD, Seagrave JC, Campen MJ. The oxidized low-density lipoprotein receptor mediates vascular effects of inhaled vehicle emissions. Am J Respir Crit Care Med. 2011;184(1):82–91.
101. Soares SR, Carvalho-Oliveira R, Ramos-Sanchez E, Catanozi S, da Silva LF, Mauad T, Gidlund M, Goto H, Garcia ML. Air pollution and antibodies against modified lipoproteins are associated with atherosclerosis and vascular remodeling in hyperlipemic mice. Atherosclerosis. 2009;207(2):368–73.
102. Li N, Sioutas C, Cho A, Schmitz D, Misra C, Sempf J, Wang M, Oberley T, Froines J, Nel A. Ultrafine particulate pollutants induce oxidative stress and mitochondrial damage. Environ Health Perspect. 2003;111(4):455–60.
103. Khera AV, Cuchel M, de la Llera-Moya M, Rodrigues A, Burke MF, Jafri K, French BC, Phillips JA, Mucksavage ML, Wilensky RL, Mohler ER, Rothblat GH, Rader DJ. Cholesterol efflux capacity, high-density lipoprotein function, and atherosclerosis. N Engl J Med. 2011;364(2):127–35. 3030449.
104. Patel PJ, Khera AV, Jafri K, Wilensky RL, Rader DJ. The anti-oxidative capacity of high-density lipoprotein is reduced in acute coronary syndrome but not in stable coronary artery disease. J Am Coll Cardiol. 2011;58(20):2068–75.
105. Ansell BJ, Navab M, Hama S, Kamranpour N, Fonarow G, Hough G, Rahmani S, Mottahedeh R, Dave R, Reddy ST, Fogelman AM. Inflammatory/antiinflammatory properties of high-density lipoprotein distinguish patients from control subjects better than high-density lipoprotein cholesterol levels and are favorably affected by simvastatin treatment. Circulation. 2003;108(22):2751–6.

106. Van Lenten BJ, Wagner AC, Navab M, Anantharamaiah GM, Hama S, Reddy ST, Fogelman AM. Lipoprotein inflammatory properties and serum amyloid A levels but not cholesterol levels predict lesion area in cholesterol-fed rabbits. J Lipid Res. 2007;48(11):2344–53.
107. Moller P, Loft S. Oxidative damage to DNA and lipids as biomarkers of exposure to air pollution. 2010. Environ Health Perspect.
108. Liu L, Ruddy T, Dalipaj M, Poon R, Szyszkowicz M, You H, Dales RE, Wheeler AJ. Effects of indoor, outdoor, and personal exposure to particulate air pollution on cardiovascular physiology and systemic mediators in seniors. J Occup Environ Med. 2009;51(9):1088–98.
109. Sorensen M, Daneshvar B, Hansen M, Dragsted LO, Hertel O, Knudsen L, Loft S. Personal PM2.5 exposure and markers of oxidative stress in blood. Environ Health Perspect. 2003;111(2):161–6.
110. Medina-Navarro R, Lifshitz A, Wacher N, Hicks JJ. Changes in human serum antioxidant capacity and peroxidation after four months of exposure to air pollutants. Arch Med Res. 1997;28(2):205–8.
111. Sanchez-Rodriguez MA, Retana-Ugalde R, Ruiz-Ramos M, Munoz-Sanchez JL, Vargas-Guadarrama LA, Mendoza-Nunez VM. Efficient antioxidant capacity against lipid peroxide levels in healthy elderly of Mexico City. Environ Res. 2005;97(3):322–9.
112. Dutta A, Mukherjee B, Das D, Banerjee A, Ray MR. Hypertension with elevated levels of oxidized low-density lipoprotein and anticardiolipin antibody in the circulation of premenopausal Indian women chronically exposed to biomass smoke during cooking. Indoor Air. 2011;21(2):165–76.
113. Jacobs L, Emmerechts J, Hoylaerts MF, Mathieu C, Hoet PH, Nemery B, Nawrot TS. Traffic air pollution and oxidized LDL. PLoS One. 2011;6(1), e16200. 3023773.
114. Taylor KW, Novak RF, Anderson HA, Birnbaum LS, Blystone C, Devito M, Jacobs D, Kohrle J, Lee DH, Rylander L, Rignell-Hydbom A, Tornero-Velez R, Turyk ME, Boyles AL, Thayer KA, Lind L. Evaluation of the association between persistent organic pollutants (POPs) and diabetes in epidemiological studies: a national toxicology program workshop review. Environ Health Perspect. 2013;121(7):774–83. 3701910.
115. Brook RD, Jerrett M, Brook JR, Bard RL, Finkelstein MM. The relationship between diabetes mellitus and traffic-related air pollution. J Occup Environ Med. 2008;50(1):32–8.
116. Eze IC, Schaffner E, Fischer E, Schikowski T, Adam M, Imboden M, Tsai M, Carballo D, von Eckardstein A, Kunzli N, Schindler C, Probst-Hensch N. Long-term air pollution exposure and diabetes in a population-based Swiss cohort. Environ Int. 2014;70:95–105.
117. Teichert T, Vossoughi M, Vierkotter A, Sugiri D, Schikowski T, Schulte T, Roden M, Luckhaus C, Herder C, Kramer U. Association between traffic-related air pollution, subclinical inflammation and impaired glucose metabolism: results from the SALIA study. PLoS One. 2013;8(12), e83042. 3858363.
118. Pearson JF, Bachireddy C, Shyamprasad S, Goldfine AB, Brownstein JS. Association between fine particulate matter and diabetes prevalence in the U.S. Diabetes Care. 2010;33(10):2196–201. 2945160.
119. Kramer U, Herder C, Sugiri D, Strassburger K, Schikowski T, Ranft U, Rathmann W. Traffic-related air pollution and incident type 2 diabetes: results from the SALIA cohort study. Environ Health Perspect. 2010;118(9):1273–9. 2944089.
120. Chen H, Burnett RT, Kwong JC, Villeneuve PJ, Goldberg MS, Brook RD, van Donkelaar A, Jerrett M, Martin RV, Brook JR, Copes R. Risk of incident diabetes in relation to long-term exposure to fine particulate matter in Ontario. Canada Environ Health Perspect. 2013;121(7):804–10. 3701997.
121. Coogan PF, White LF, Jerrett M, Brook RD, Su JG, Seto E, Burnett R, Palmer JR, Rosenberg L. Air pollution and incidence of hypertension and diabetes mellitus in black women living in Los Angeles. Circulation. 2012;125(6):767–72. 3326581.
122. Puett RC, Hart JE, Schwartz J, Hu FB, Liese AD, Laden F. Are particulate matter exposures associated with risk of type 2 diabetes? Environ Health Perspect. 2011;119(3):384–9. 3060003.
123. Sun Q, Yue P, Deiuliis JA, Lumeng CN, Kampfrath T, Mikolaj MB, Cai Y, Ostrowski MC, Lu B, Parthasarathy S, Brook RD, Moffatt-Bruce SD, Chen LC, Rajagopalan S. Ambient air pol-

lution exaggerates adipose inflammation and insulin resistance in a mouse model of diet-induced obesity. Circulation. 2009;119(4):538–46. 3845676.
124. Xu X, Yavar Z, Verdin M, Ying Z, Mihai G, Kampfrath T, Wang A, Zhong M, Lippmann M, Chen LC, Rajagopalan S, Sun Q. Effect of early particulate air pollution exposure on obesity in mice: role of p47phox. Arterioscler Thromb Vasc Biol. 2010;30(12):2518–27. 3065931.
125. Xu X, Liu C, Xu Z, Tzan K, Zhong M, Wang A, Lippmann M, Chen LC, Rajagopalan S, Sun Q. Long-term exposure to ambient fine particulate pollution induces insulin resistance and mitochondrial alteration in adipose tissue. Toxicol Sci. 2011;124(1):88–98. 3196653.
126. Zheng Z, Xu X, Zhang X, Wang A, Zhang C, Huttemann M, Grossman LI, Chen LC, Rajagopalan S, Sun Q, Zhang K. Exposure to ambient particulate matter induces a NASH-like phenotype and impairs hepatic glucose metabolism in an animal model. J Hepatol. 2013;58(1):148–54. 3527686.
127. Liu C, Xu X, Bai Y, Wang TY, Rao X, Wang A, Sun L, Ying Z, Gushchina L, Maiseyeu A, Morishita M, Sun Q, Harkema JR, Rajagopalan S. Air pollution-mediated susceptibility to inflammation and insulin resistance: influence of CCR2 pathways in mice. Environ Health Perspect. 2014;122(1):17–26. 3888572.
128. Tomaru M, Takano H, Inoue K, Yanagisawa R, Osakabe N, Yasuda A, Shimada A, Kato Y, Uematsu H. Pulmonary exposure to diesel exhaust particles enhances fatty change of the liver in obese diabetic mice. Int J Mol Med. 2007;19(1):17–22.
129. Rizzo AM, Corsetto PA, Farina F, Montorfano G, Pani G, Battaglia C, Sancini G, Palestini P. Repeated intratracheal instillation of PM10 induces lipid reshaping in lung parenchyma and in extra-pulmonary tissues. PLoS One. 2014;9(9), e106855. 4178018.
130. Bauer M, et al. Urban particulate matter air pollution is associated with subclinical atherosclerosis: results from the HNR (Heinz Nixdorf Recall) study. J Am Coll Cardiol. 2010;56(22): 1803–8.
131. Kalsch H, et al. Coronary artery calcification outperforms thoracic aortic calcification for the prediction of myocardial infarction and all-cause mortality: the Heinz Nixdorf Recall Study. Eur J Prev Cardiol. 2014;21(9):1163–70.
132. Perez L, et al. Air pollution and atherosclerosis: a cross-sectional analysis of four European cohort studies in the ESCAPE study. Environ Health Perspect. 2015;123(6):597–605.
133. Chen LC, et al. Atherosclerosis lesion progression during inhalation exposure to environmental tobacco smoke: a comparison to concentrated ambient air fine particles exposure. Inhal Toxicol. 2010;22(6):449–59.

Oxidative Stress and Alzheimer's Disease

Sandeep Kumar Singh, Rudy Castellani, and George Perry

1 Introduction

The human brain, which constitutes only 2 % of body weight, consumes approximate 20 % of total oxygen supplied by the respiratory system [1]. Neurons, the basic functional unit of the brain, are susceptible to oxidative damage because of a high metabolic rate compared to other cells [2]. The oxidation of proteins, lipids and nucleic acids in neuronal cells is a common pathological feature of AD [3]. Polyunsaturated fatty acids (PUFAs) are more abundant in neurons and can interact with reactive oxygen species (ROS), leading to a self-propagating cascade of lipid peroxidation and molecular destruction [4]. Furthermore, glutathione, an essential antioxidant, is present in very low amounts in neurons [5]. Therefore, neurons are highly susceptible to oxidative stress.

Oxidative stress implies a situation where reactive oxygen species (ROS), like superoxide (O-2), hydroxyl (OH.), free radicals and its products are in excess compared to the antioxidant defense system. Oxygen free radicals and ROS are a common product of different biochemical processes in aerobic cellular metabolism [6, 7]. Free radicals are very reactive and unstable molecules with unpaired electrons in outer orbit. Superoxide (O2-.), hydroxyl (OH.), nitric monoxide (NO), hydrogen

S.K. Singh
Department of Biochemistry, Faculty of Science, Banaras Hindu University, Varanasi, UP 221 005, India

R. Castellani
Department of Pathology, School of Medicine, University of Maryland, Baltimore, MD, USA

G. Perry, Ph.D. (✉)
College of Sciences, The University of Texas at San Antonio, One UTSA Circle, San Antonio, TX 78249, USA
e-mail: george.perry@utsa.edu

S.C. Bondy, A. Campbell (eds.), *Inflammation, Aging, and Oxidative Stress*, Oxidative Stress in Applied Basic Research and Clinical Practice, DOI 10.1007/978-3-319-33486-8_10 peroxide (H_2O_2), and peroxynitrite ($ONOO^-$) are well known free radicals and ROS in aerobic cellular metabolism. Hydrogen peroxide and peroxynitrite do not readily cause toxicity to biomolecules; these must undergo further reactions to generate free radicals. For instance, hydrogen peroxide may degrade into highly reactive hydroxyl radical via Fenton chemistry or a Haber-Weiss reaction [8, 9]. Overproduction of free radicals can cause extreme damage to biomolecules like proteins, unsaturated fatty acids and mitochondrial DNA, inflammation, aging, tissue damage and subsequent cellular apoptosis [10]. Apart from their high reactive toxicity, free radicals have biological activities such as signal transduction, defense against invading pathogens, gene transcription, and regulation of soluble guanylate cyclase activity. The aerobic biological antioxidant defense system coevolved with oxidative free radicals to counteract [11]. Antioxidants are scavengers of ROS and its precursors. The biological antioxidant defense system comprises two groups; enzymatic and non-enzymatic antioxidants. Enzymatic antioxidants are catalase, superoxide dismutase (SOD), glutathione reductase and peroxidase. and other supporting enzymes. Non-enzymatic antioxidants like ascorbic acid, vitamin-A, vitamin-E, carotenoids and polyphenolic compounds, are derived from dietary sources which directly scavenge free radicals; in some cases our system itself synthesizes chelating agents to trap redox metals, preventing free radical production [12]. The brain is one of the famously metabolically active organs in human body, whose sole carbon energy source is glucose. It accounts for 2 % of total body mass and consumes 20 % of total oxygen in a resting individual. Therefore it is highly prone to electron leakage in oxidative phosphorylation and reaction of redox active metals like Cu and Fe with molecular oxygen [13] which cause subsequent reactive oxygen species (ROS) generation. Over production of these highly reactive species cause abnormal protein aggregation and leads to neurodegeneration. Apart from several other genetic and environmental factors, oxidative stress is the leading factor in most neurodegenerative disorders such as Alzheimer's disease and Parkinson's disease.

2 Amyloid-β (Aβ) Induced Toxicity and Oxidative Stress

Aβ is produced via proteolytic processing of Amyloid-β precursor protein (AβPP) by two membrane-bound proteases, β and γ-secretase, which are also known as β-site AβPP cleaving enzyme1(BACE-1), and a multi-protein complex consisting of presenilin (PS), nicastrin (NCT), anterior pharynx defective-1(APH-1), and presenilin enhancer protein-2 (PEN-2) respectively [14, 15]. β-secretase acts at N-terminal end of AβPP and produces a C-terminal fragment of 99 amino acid which is later cleaved transmembranely by γ-secretase and produces Aβ-peptides [14, 16]. After AβPP processing through β and γ-secretases, there occurs a generation of peptides of variable lengths, among them the 42-amino acid form of Aβ ($A\beta_{42}$), more toxic than the more abundantly produced 40-amino acid form of Aβ (Aβ40), possibly because of its faster self-aggregation into oligomers [14, 16, 17]. The above process which results in Aβ production is called the amyloidogenic pathway. Cleavage of AβPP by α-secretase instead of β-secretase produces a non-toxic small fragment, sAβPPα, which is required for normal function of neurons. This process is called non-amyloidogenic pathway of AβPP processing [18]. Increased generation and/or decreased clearance of Aβ-peptides leads to Aβ accumulation outside the neurons, which stimulates diverse cell signaling pathways, eventually resulting in synaptic degeneration, neuronal loss, and decline in cognitive function [16, 17, 19–21].

Currently, some researchers are focusing on elucidating the relationship between oxidative stress and Aβ toxicity [22]. The level of hydrogen peroxide and lipid peroxidase was found significantly increased after Aβ treatment *in-vitro* [23]. Further evidence from *in-vivo* studies, using AD transgenic mouse model carrying mutants of APP and PS-1, showed increased nitric oxide production and hydrogen peroxide as well as elevated oxidative modifications of lipids and proteins associated with age-linked Aβ accumulation, confirming that Aβ promotes oxidative stress [24–28]. It has been demonstrated consistently in AD cell and animal models that natural antioxidants, such as EGb 761, curcumin, and green tea catechins, can exert neuroprotective functions by attenuating Aβ-induced ROS generation and neuronal apoptosis [29–32].

3 Tau Pathology and Oxidative Stress

Another hallmark of AD pathology is hyperphosphorylated tau protein, a major component of neurofibrillary tangles (NFT) that correlates with neurodegeneration and cognitive decline [33]. In normal conditions, tau binds with tubulin and maintains cell integrity, but in AD, tau hyperphosphorylates and promotes microtubule assembly, resulting in formation of neurofibrillary tangles [34]. There is involvement of a number of protein kinases and protein phosphatases in the abnormal phosphorylation of tau, including glycogen synthase kinase-3 beta (GSK-3 beta), mitogen-activated protein kinase (MAPK), calcium-calmodulin kinase, cyclin-dependent kinase 5, and protein kinase C [35]. It has been suggested that the accumulation of Aβ may appear before the tau pathology and that Aβ aggregates may be one of a cascade of molecular events leading to tau hyperphosphorylation [36–38]. On the other hand, it was reported that overexpression of tau inhibited kinesin-dependent transport of neurofilaments, peroxisomes, and Golgi-derived vesicles into neurites, causing transport discrepancies in primary neuronal cells including the AβPP trafficking [39]. In particular, the transport of AβPP into axons and dendrites was blocked, resulting in its accumulation in the cell body [39].

There is less evidence that oxidative stress is interlinked with tau pathology. It has also been shown that the cells overexpressing tau protein have increased susceptibility to oxidative stress, perhaps due to the reduction of peroxisomes [39]. In a *drosophila* model of human tauopathies expressing a disease related mutant form of human tau (tauR406W), reduction of gene dosage of mitochondrial sod2 or thioredoxin reductase (trxr) enhanced tau-induced neurodegenerative histological abnormalities and neuronal apoptosis [40]. In contrast, overexpression of these antioxidant enzymes or treatment with vitamin E attenuated tau-induced neuronal cell death [40]. Moreover, in cortical neurons derived from a transgenic rat model expressing a human truncated variant form of tau protein, it was observed that the levels of ROS were elevated when compared to control non-transgenic neurons, while antioxidants such as vitamin C significantly abolished the ROS elevation [41, 42]. These observations clearly suggest that tau-induced neurotoxicity is at least partially mediated by oxidative stress [40]. The linkage between tau pathology and oxidative stress was further elucidated in P301S and P301L transgenic mouse models carrying the human tau gene with P301S or P301L mutations, which exhibit an aggregation and deposition of hyperphosphorylated tau and develop NFT [43]. Mitochondrial dysfunction together with reduced NADH-ubiquinone oxidoreductase activity was found in P301L tau transgenic mice, which was associated with increased ROS production, impaired mitochondrial respiration, and ATP synthesis in aged animals [44]. Similarly, the brains of P301S transgenic mice showed signs of raised oxidative stress comprising elevated protein carbonyl levels in mitochondria cortex and changes in the activity and content of mitochondrial enzymes involved in ROS formation and energy metabolism, suggesting that oxidative stress and mitochondrial dysfunction might play an important role in tau pathology [45]. Consistently, administration of P301S mice with coenzyme Q10, an antioxidant and a key component of the electron transport chain, significantly increased complex I activity and reduced lipid peroxidation while improving survival and behavioural deficits of the mice [46]. Furthermore, the junction of Aβ and tau pathologies on mitochondria dysfunction was clearly established in a triple transgenic mouse model [pR5/APP/PS2], which exhibits both Aβ and tau pathologic features of the disease in animals' brains [47]. Proteomic studies of samples from the triple AD mice brain demonstrated a substantial deregulation of 24 proteins, of which one third were mitochondrial proteins mostly related to complexes I and IV of the oxidative phosphorylation system [48]. Notably, deregulation of mitochondrial complex I and complex IV was shown to be tau and Aβ dependent respectively [48]. The effects of Aβ and tau on mitochondrial function were found to be synergistic and age-associated, resulting in reduced mitochondrial respiratory capacity and the decline of ATP synthesis, which led to synaptic loss and neuronal cell death [48].

Growing evidence has also shown that oxidative stress may have an important role in the hyperphosphorylation and polymerization of tau. Oxidation of fatty acids, found to be higher in AD brains, was reported to facilitate the polymerization of tau, and thus might serve as a link between oxidative stress and tau pathology in AD [49]. In Tg2576 AD transgenic mice, reduction of cytoplasmic SOD1 or deficiency in mitochondrial SOD2 [50] induced tau phosphorylation, suggesting that ROS may play a crucial role in tau hyperphosphorylation [51]. Oxidative stress activates p38 MAPK which is further responsible for phosphorylating tau protein *in-vitro* [52]. In hippocampal and cortical brain regions of AD patients, activated p38 is found exclusively localized to NFT and co-immunoprecipitated with PHF-tau, suggesting it might be involved in the phosphorylation of tau *in-vivo* [53]. Thus, p38 may play a role as an important candidate which links the phosphorylation of tau with increased oxidative stress in case of AD.

4 Mitochondrial Dysfunction and Alzheimer's Disease

Mitochondria are very important organelles of the human body. They play a very important role in normal cellular metabolism and energy production. There is ample literature supporting a relation between oxidative damage and mitochondrial dysfunction in relation to the pathogenesis of AD. Oxidative damage occurs early in the brain of AD patients, before the onset of significant Aβ induced pathology [54]. Oxidative damage also leads deposition of Aβ in transgenic AβPP mutant mice [55], with changes in gene expression related to mitochondrial metabolism and apoptosis taking place even earlier in neuronal cells undergoing oxidative damage [56]. Several results from experimental studies provide insight for oxidative damage and mitochondrial dysfunction in relation to AD. In fetal guinea pig neurons, intracellular Aβ level was found significantly increased after hydrogen peroxide (H_2O_2) treatment [57]. In a transgenic AβPP-mutant mouse, hemizygous deficiency of the mitochondrial antioxidant enzyme MnSOD markedly increased brain Aβ levels and their deposition [58]. In another transgenic AβPP-mutant mouse, elevated β-secretase activity and hence increased Aβ levels were found by the action of energy metabolism inhibitors (insulin, 2-deoxyglucose, 3-nitropropionic acid and kainic acid) [59].

Several pathways related to oxidative stress, mitochondrial dysfunction, and AD pathology have recently been uncovered. Oxidative stress may activate different signalling pathways that alter the processing of AβPP or tau. For example, activation of glycogen synthase kinase increases tremendous tau hyperphosphorylation [60], and oxidative stress up-regulates the expression of β-secretase through activation of c-Jun amino-terminal kinase and p38 mitogen-activated protein kinase (MAPK) [61]. There is some evidence that mtDNA may play a very important role in the mitochondrial dysfunction seen in AD. When AD patient mtDNA is transferred into mtDNA-deficient cell lines, the resulting 'cybrids' reproduce the respiratory enzyme deficiency seen in the brain and other tissues in AD, suggesting that the defect is carried at least in part by mtDNA abnormalities [62]. However, it has been a challenge to identify AD-specific mtDNA mutations. 145 AD patients and 128 controls were studied and no significant association was found with mitochondrial haplogroup or with inherited mtDNA mutations [63]. Also there was no relationship with acquired mtDNA mutations when a coding region (for CO1) was examined [64]. However, in the same way that promoter appeared more subtle to damage than coding regions in nuclear genes15, the mtDNA control region showed an elevation in acquired mutations in AD [65]. AD brains had on average a 63 % increase in heteroplasmic mtDNA control-region mutations, and in individuals older than 80 years there was a 130 % increase in mutations. These mutations preferentially suppressed mitochondrial transcription and replication and altered known mtDNA regulatory elements. Lastly, several recent studies support that many proteins implicated in AD pathogenesis have direct physical association with mitochondrial proteins or mitochondria directly. AβPP has a dual endoplasmic reticulum/mitochondrial-targeting sequence, and in transgenic mice overexpressing AβPP and transfected cells, it clogged the mitochondrial protein importation machinery, causing mitochondrial dysfunction and diminished energy metabolism [66]. Aβ binds to a mitochondrial-matrix protein called Aβ-binding alcohol dehydrogenase (ABAD) [67]. A decoy peptide blocks the interaction of Aβ and ABAD and suppresses free radical generation in neurons and Aβ-induced apoptosis. Conversely, ABAD overexpressed transgenic AβPP-mutant mice promoting neuronal oxidative stress and loss of memory. Other groups have also found that interaction of Aβ with mitochondria inhibits cytochrome oxidase activity and increases ROS generation [27, 68]. In addition, Aβ also hinders α-ketoglutarate dehydrogenase activity in isolated mitochondria [69], causes shortage of α- cytochrome oxidase activities [70] and ketoglutarate dehydrogenase [71], and has previously been found in the brain and other tissues of AD patients. Different components of γ-secretase complex such as Presenilin have also been localized to mitochondria, where they form an active γ-secretase complex [72].

5 Metal Homeostasis and Oxidative Stress in Alzheimer's Disease

In most biological systems, redox-active transition metals such as copper and iron are required in very trace amounts for normal metabolism and functioning. The excess deposition of these metals as well as metalloproteins in the body can produce higher amounts of free metal ions inside the body. These ions are needed in trace amounts to catalyse "redox cycling" and mediate oxidative stress reaction [73, 74]. Redox cycling is the process whereby cellular reductants such as vitamin C or thiols reduce the oxidized transition metal ions. This superoxide radical is chemically unreactive, but works as the reductant of oxidized metal ions, which are available to generate hydroxyl radical ($OH^{\cdot}$) from H_2O_2 through the Haber–Weiss reaction. The $OH^{\cdot}$ reacts with all biomacromolecules at diffusion-controlled rates, i.e., acts away from its site of production. It is normally generated by the Fenton reaction between reduced transition metals ions and H_2O_2. Under normal conditions, the antioxidant cascade is responsible for preventing ROS damage. This antioxidant protection is provided by two different methods: (a) converting superoxide to O^{2-} and H_2O_2 through mitochondrial manganese superoxide dismutase (MnSOD) and cytosolic copper–zinc superoxide dismutase (CuZnSOD), and (b) elimination of by-products of oxygen reduction by oxidase (i.e. monoamine oxidase) by enzymes (catalases and peroxidases). In AD, a link between metals and oxidative stress modifications is well established [75–77] .

6 Conclusions

In summary, evidence exists that oxidative stress is inextricably linked with several major pathological processes in AD including Aβ-induced neurotoxicity, tau pathology, mitochondria dysfunction, and metal dyshomeostasis. Abundant ROS may be generated due to mitochondrial dysfunction and/or aberrant accumulation of transition metals, possibly caused by a combination of abnormal Aβ accumulation and tau pathology, eventually resulting in oxidative stress. Oxidative stress, which facilitates the neurotoxicity induced by abnormal Aβ aggregation and hyperphosphorylation of tau proteins, may augment Aβ aggregation as well as facilitate tau phosphorylation and polymerization, further enhancing a variety of neurotoxic events including ROS production, thus promoting the initiation and progression of AD. Under controlled conditions, oxidative stress can readily induce apoptosis and antioxidants can defend and extend a cell's lifespan under pressure. On the other hand, in complex disease conditions, it is difficult to show the relationship of oxidative stress to AD. The studies are unclear, complex, and open-ended, with narrow conclusions being promoted. Oxidative stress is an important factor contributing to the development of AD. Scavenging of ROS or inhibition of their formation may delay the onset or slow the progression of AD through various mechanisms including, but not limited to, inhibiting ions of Aβ production and aggregation, reduction of oxidative stress-mediated neuronal toxicity, decrease of tau hyperphosphorylation and polymerization, and restoration of metal homeostasis and mitochondria dysfunction. Therefore, AD prevention or treatment using antioxidant therapeutic approaches may be one way to target a number of different molecular events related to oxidative stress and AD pathogenesis.

Acknowledgements Support provided by the Semmes Foundation and by a grant from the National Institute on Minority Health and Health Disparities (G12MD007591) from the National Institutes of Health.

References

1. Sokoloff L. Energetics of functional activation in neural tissues. Neurochem Res. 1999;24:321–9.
2. Tholey G, Ledig M. Neuronal and astrocytic plasticity: metabolic aspects. Ann Med Intern (Paris). 1990;141 Suppl 1:13–8.
3. Pratico D. Oxidative stress hypothesis in Alzheimer's disease: a reappraisal. Trends Pharmacol Sci. 2008;29:609–15.
4. Nunomura A, Castellani RJ, Zhu X, Moreira PI, Perry G, Smith MA. Involvement of oxidative stress in Alzheimer disease. J Neuropathol Exp Neurol. 2006;65:631–41.
5. Pocernich C, Butterfield D. Elevation of glutathione as a therapeutic strategy in Alzheimer disease. Biochim Biophys Acta. 2012;1822(5):625–30.
6. Halliwell B. Free radicals, antioxidants, and human disease: curiosity, cause, or consequence? Lancet. 1994;344:721–4.
7. Poulson HE, Prieme H, Loft S. Role of oxidative DNA damage in cancer initiation and promotion. Eur J Cancer Prev. 1998;7:9–16.
8. Markesbery WR. Oxidative stress hypothesis in Alzheimer's disease. Free Radic Biol Med. 1997;23:134–47.
9. Liochev SI. The mechanism of "Fenton-like" reactions and their importance for biological systems. A biologist's view. Met Ions Biol Syst. 1999;36(1999):1–39.
10. Salganik RI. The benefits and hazards of antioxidants: controlling apoptosis and other protective mechanisms in cancer patients and the human population. J Am Coll Nutr. 2001;20(suppl):464S–72.
11. Yu BP. Cellular defences against damage from reactive oxygen species. Biol Rev. 1994;74:139–62.

12. Gilgun-Sherki Y, Melamed E, Offen D. Oxidative stress induced-neurodegenerative diseases: the need for antioxidants that penetrate the blood brain barrier. Neuropharmacology. 2001;40:959–75.
13. Halliwell B, Gutterer JM. Free radicals in biology and medicine. Oxford: Oxford University Press; 1999.
14. Querfurth HW, LaFerla FM. Alzheimer's disease. N Engl J Med. 2010;362:329–44.
15. Steiner H. Uncovering gamma-secretase. Curr Alzheimer Res. 2004;1:175–81.
16. Walsh DM, Selkoe DJ. Abeta oligomers-a decade of discovery. J Neurochem. 2007;101:1172–84.
17. Yankner BA, Lu T. Amyloid beta-protein toxicity and the pathogenesis of Alzheimer disease. J Biol Chem. 2009;284:4755–9.
18. Esler WP, Wolfe MS. A portrait of Alzheimer secretases—new features and familiar faces. Science. 2001;293:1449–54.
19. Roberson ED, Mucke L. 100 years and counting: prospects for defeating Alzheimer's disease. Science. 2006;314:781–4.
20. Maccioni R, Maccioni RB, Munoz JP, Barbeito L. The molecular bases of Alzheimer's disease and other neurodegenerative disorders. Arch Med Res. 2001;32:367–438.
21. Hsiao K, Chapman P, Nilsen S, Eckman C, Harigaya Y, Younkin S, Yang F, Cole G. Correlative memory deficits, Abeta elevation, and amyloid plaques in transgenic mice. Science. 1996;274: 99–102.
22. Mattson MP. Cellular actions of beta-amyloid precursor protein and its soluble and fibrillogenic derivatives. Physiol Rev. 1997;77:1081–132.
23. Behl C, Davis JB, Lesley R, Schubert D. Hydrogen peroxide mediates amyloid beta protein toxicity. Cell. 1994;77:817–27.
24. Matsuoka Y, Picciano M, La Francois J, Duff K. Fibrillar beta-amyloid evokes oxidative damage in a transgenic mouse model of Alzheimer's disease. Neuroscience. 2001;104:609–13.
25. Smith MA, Hirai K, Hsiao K, Pappolla MA, Harris PL, Siedlak SL, Tabaton M, Perry G. Amyloid-beta deposition in Alzheimer transgenic mice is associated with oxidative stress. J Neurochem. 1998;70:2212–5.
26. Mohmmad Abdul H, Sultana R, Keller JN, St Clair DK, Markesbery WR, Butterfield DA. Mutations in amyloid precursor protein and presenilin-1 genes increase the basal oxidative stress in murine neuron al cells and lead to increased sensitivity to oxidative stress mediated by amyloid beta -peptide (1-42), HO and kainic acid: implications for Alzheimer's disease. J Neurochem. 2006;96:1322–35.
27. Manczak M, Anekonda TS, Henson E, Park BS, Quinn J, Reddy PH. Mitochondria are a direct site of A beta accumulation in Alzheimer's disease neurons: implications for free radical generation and oxidative damage in disease progression. Hum Mol Genet. 2006;15:1437–49.
28. Apelt J, Bigl M, Wunderlich P, Schliebs R. Aging-related increase in oxidative stress correlates with developmental pattern of beta-secretase activity and beta-amyloid plaque formation in transgenic Tg2576 mice with Alzheimer- like pathology. Int J Dev Neurosci. 2004;22:475–84.
29. Cole GM, Teter B, Frautschy SA. Neuroprotective effects of curcumin. Adv Exp Med Biol. 2007;595:197–212.
30. Mandel SA, Amit T, Kalfon L, Reznichenko L, Youdim MB. Targeting multiple neurodegenerative diseases etiologies with multimodal -acting green tea catechins. J Nutr. 2008;138: 1578S–83.
31. Zhao Y, Zhao B. Natural antioxidants in prevention and management of Alzheimer's disease. Front Biosci. 2012;4:794–808.
32. Smith JV, Luo Y. Studies on mo lecular mechanisms of Ginkgo biloba extract. Appl Microbiol Biotechnol. 2004;64:465–72.
33. Goedert M, Spillantini MG. A century of Alzheimer's disease. Science. 2006;314:777–81.
34. Iqbal K, Liu F, Gong CX, Alonso-Adel C, Grundke-Iqbal I. Mechanisms of tau -induced neurodegeneration. Acta Neuropathol. 2009;118:53–69.
35. Billingsley ML, Kincaid RL. Regulated phosphorylation and dephosphorylation of tau protein: effects on microtubule interaction, intracellular trafficking and neurodegeneration. Biochem J. 1997;323(Pt 3):577–91.

36. Zheng WH, Bastianetto S, Mennicken F, Ma W, Kar S. Amyloid beta peptide induces tau phosphorylation and loss of cholinergic neurons in rat primary septal cultures. Neuroscience. 2002;115:201–11.
37. Lewis J, Dickson DW, Lin WL, Chisholm L, Corral A, Jones G, Yen SH, Sahara N, Skipper L, Yager D, Eckman C, Hardy J, Hutton M, McGowan E. Enhanced neurofibrillary degeneration in transgenic mice expressing mutant tau and APP. Science. 2001;293:1487–91.
38. Gotz J, Chen F, van Dorpe J, Nitsch RM. Formation of neurofibrillary tang les in P3011 tau transgenic mice induced by Abeta 42 fibrils. Science. 2001;293:1491–5.
39. Stamer K, Vogel R, Thies E, Mandelkow E, Mandelkow EM. Tau blocks traffic of organelles, neurofilaments, and APP vesicles in neurons and enhances oxidative stress. J Cell Biol. 2002;156:1051–63.
40. Dias-Santagata D, Fulga TA, Duttaroy A, Feany MB. Oxidative stress mediates tau -induced neurodegeneration in Drosophila. J Clin Invest. 2007;117:236–45.
41. Cente M, Filipcik P, Pevalova M, Novak M. Expression of a truncated tau protein induces oxidative stress in a rodent model of tauopathy. Eur J Neurosci. 2006;24:1085–90.
42. Cente M, Filipcik P, Mandakova S, Zilka N, Krajciova G, Novak M. Expression of a truncated human tau protein induces aqueous -phase free radicals in a rat model of tauopathy: implications for targeted antioxidative therapy. J Alzheimers Dis. 2009;17:913–20.
43. Yoshiyama Y, Higuchi M, Zhang B, Huang SM, Iwata N, Saido TC, Maeda J, Suhara T, Trojanowski JQ, Lee VM. Synapse loss and microglial activation precede tangles in a P301S tauopathy mouse model. Neuron. 2007;53:337–51.
44. David DC, Hauptmann S, Scherping I, Schuessel K, Keil U, Rizzu P, Ravid R, Drose S, Brandt U, Muller WE, Eckert A, Gotz J. Proteomic and functional analyses reveal a mitochondrial dysfunction in P301L tau transgenic mice. J Biol Chem. 2005;280:23802–14.
45. Dumont M, Stack C, Elipenahli C, Jainuddin S, Gerges M, Starkova NN, Yang L, Starkov AA, Beal F. Behavioral deficit, oxidative stress, and mitochondrial dysfunction precede tau pathology in P301S transgenic mice. FASEB J. 2011;25:4063–72.
46. Elipenahli C, Stack C, Jainuddin S, Gerges M, Yang L, Starkov A, Beal MF, Dumont M. Behavioral improvement after chronic administration of coenzyme Q10 in P301S transgenic mice. J Alzheimers Dis. 2011;28:173–82.
47. Eckert A, Schulz KL, Rhein V, Gotz J. Convergence of amyloid-beta and tau pathologies on mitochondria in vivo. Mol Neurobiol. 2010;41:107–14.
48. Rhein V, Song X, Wiesner A, Ittner LM, Baysang G, Meier F, Ozmen L, Bluethmann H, Drose S, Brandt U, Savaskan E, Czech C, Gotz J, Eckert A. Amyloid-beta and tau synergistically impair the oxidative phosphorylation system in triple transgenic Alzheimer's disease mice. Proc Natl Acad Sci U S A. 2009;106:20057–62.
49. Gamblin TC, King ME, Kuret J, Berry RW, Binder LI. Oxidative regulation of fatty acid-induced tau polymerization. Biochemistry. 2000;39:14203–10.
50. Melov S, Adlard PA, Morten K, Johnson F, Golden TR, Hinerfeld D, Schilling B, Mavros C, Masters CL, Volitakis I, Li QX, Laughton K, Hubbard A, Cherny RA, Gibson B, Bush AI. Mitochondrial oxidative stress causes hyperphosphorylation of tau. PLoS One. 2007;2, e536.
51. Murakami K, Murata N, Noda Y, Tahara S, Kaneko T, Kinoshita N, Hatsuta H, Murayama S, Barnham KJ, Irie K, Shirasawa T, Shimizu T. SOD1 (copper/zinc superoxide dismutase) deficiency drives amyloid beta protein oligomerization and memory loss in mouse model of Alzheimer disease. J Biol Chem. 2011;286:44557–68.
52. Goedert M, Hasegawa M, Jakes R, Lawler S, Cuenda A, Cohen P. Phosphorylation of microtubule-associated protein tau by stress-activated protein kinases. FEBS Lett. 1997;409:57–62.
53. Zhu X, Rottkamp CA, Boux H, Takeda A, Perry G, Smith MA. Activation of p38 kinase links tau phosphorylation, oxidative stress, and cell cycle- related events in Alzheimer disease. J Neuropathol Exp Neurol. 2000;59:880–8.
54. Nunomura A, et al. Oxidative damage is the earliest event in Alzheimer disease. J Neuropathol Exp Neurol. 2001;60:759–67.
55. Pratico D, Uryu K, Leight S, Trojanoswki JQ, Lee VM. Increased lipid peroxidation precedes amyloid plaque formation in an animal model of Alzheimer amyloidosis. J Neurosci. 2001;21:4183–7.

56. Reddy PH, et al. Gene expression profiles of transcripts in amyloid precursor protein transgenic mice: up-regulation of mitochondrial metabolism and apoptotic genes is an early cellular change in Alzheimer's disease. Hum Mol Genet. 2004;13:1225–40.
57. Ohyagi Y, et al. Selective increase in cellular Aβ42 is related to apoptosis but not necrosis. Neuroreport. 2000;11:167–71.
58. Li F, et al. Increased plaque burden in brains of APP mutant MnSOD heterozygous knockout mice. J Neurochem. 2004;89:1308–12.
59. Velliquette RA, O'Connor T, Vassar R. Energy inhibition elevates β-secretase levels and activity and is potentially amyloidogenic in APP transgenic mice: possible early events in Alzheimer's disease pathogenesis. J Neurosci. 2005;25:10874–83.
60. Lovell MA, Xiong S, Xie C, Davies P, Markesbery WR. Induction of hyperphosphorylated tau in primary rat cortical neuron cultures mediated by oxidative stress and glycogen synthase kinase-3. J Alzheimers Dis. 2004;6:659–71.
61. Tamagno E, et al. β-Site APP cleaving enzyme up-regulation induced by 4-hydroxynonenal is mediated by stress-activated protein kinases pathways. J Neurochem. 2005;92:628–36.
62. Swerdlow RH, et al. Cybrids in Alzheimer's disease: a cellular model of the disease? Neurology. 1997;49:918–25.
63. Elson JL, et al. Does the mitochondrial genome play a role in the etiology of Alzheimer's disease? Hum Genet. 2006;119:241–54.
64. Lin MT, Simon DK, Ahn CH, Kim LM, Beal MF. High aggregate burden of somatic mtDNA point mutations in aging and Alzheimer's disease brain. Hum Mol Genet. 2002;11:133–45.
65. Coskun PE, Beal MF, Wallace DC. Alzheimer's brains harbor somatic mtDNA control-region mutations that suppress mitochondrial transcription and replication. Proc Natl Acad Sci U S A. 2004;101:10726–31.
66. Anandatheerthavarada HK, Biswas G, Robin MA, Avadhani NG. Mitochondrial targeting and a novel trans-membrane arrest of Alzheimer's amyloid precursor protein impairs mitochondrial function in neuronal cells. J Cell Biol. 2003;161:41–54.
67. Lustbader JW, et al. ABAD directly links Aβto mitochondrial toxicity in Alzheimer's disease. Science. 2004;304:448–52.
68. Crouch PJ, et al. Copper-dependent inhibition of human cytochrome coxidase by a dimeric conformer of amyloid-β1-42. J Neurosci. 2005;25:672–9.
69. Casley CS, Canevari L, Land JM, Clark JB, Sharpe MA. β-amyloid inhibits integrated mitochondrial respiration and key enzyme activities. J Neurochem. 2002;80:91–100.
70. Parker WD, Filley CM, Parks JK. Cytochrome oxidase deficiency in Alzheimer's disease. Neurology. 1990;40:1302–3.
71. Gibson GE, et al. Reduced activities of thiamine-dependent enzymes in the brains and peripheral tissues of patients with Alzheimer's disease. Arch Neurol. 1988;45:836–40.
72. Hansson CA, et al. Nicastrin, presenilin, APH-1, and PEN-2 form active γ-secretase complexes in mitochondria. J Biol Chem. 2004;279:51654–60.
73. Nunomura A, et al. Neuronal RNA oxidation in Alzheimer's disease and Down's syndrome. Ann N Y Acad Sci. 1999;893:362–4.
74. Sayre LM, Perry G, Smith MA. Redox metals and neurodegenerative disease. Curr Opin Chem Biol. 1999;3(2):220–5.
75. Smith MA, et al. Increased iron and free radical generation in preclinical Alzheimer's disease and mild cognitive impairment. J Alzheimers Dis. 2010;19(1):363–72.
76. Liu G, et al. Metal chelators coupled with nanoparticles as potential therapeutic agents for Alzheimer's disease. J Nanoneurosci. 2009;1(1):42–55.
77. Moreira PI, et al. Alzheimer's disease and the role of free radicals in the pathogenesis of the disease. CNS Neurol Disord Drug Targets. 2008;7(1):3–10.

Reactive Oxygen Species and Protein Oxidation in Neurodegenerative Disease

Edward H. Sharman

1 Introduction

As humans age, they become increasingly susceptible to a number of neurological diseases. In contrast to birth defects arising from specific genetic mutations, most neurodegenerative diseases and age-related cognitive decline are not associated with a single mutation of high penetrance, and symptoms typically develop slowly starting in middle age or later. Thus environmental and nutritive factors, in addition to differences in genetic background, are likely to be important causes of these conditions.

An association between an increased presence of reactive oxygen species (ROS) and neurodegenerative diseases has been acknowledged for quite some time. More generally, it formed the basis for the free radical theory of aging [1]. In this view, ROS were judged to be the unavoidable and deleterious byproducts of aerobic metabolism that slowly damaged biomolecules and overwhelmed endogenous antioxidant repair mechanisms, resulting in gradual loss of biological function, increased incidence of degenerative disease, and ultimately death. With respect to neurodegeneration in particular, as the amounts of these byproducts increase with aging, the efficiency of the associated triage processes is reduced, allowing buildup of the toxic proteinaceous inclusions and aggregates that are universally associated with degenerative disease.

As a corollary, administration of substances or genetic manipulation that could either quench free radical ROS or raise the activity of antioxidant enzymes were anticipated to slow or reverse the course of degenerative diseases and lengthen healthy lifespan. Despite a very substantial amount of both animal and clinical research directed along these lines, only modest or nonexistent benefits have been demonstrated up to this point, suggesting that the mere occurrence of increased ROS is not directly causative of degeneration. At most, since not all toxic aggregates

E.H. Sharman, Ph.D. (✉)
Department of Neurology, University of California, Irvine, Irvine, CA 92697, USA
e-mail: esharman@uci.edu

S.C. Bondy, A. Campbell (eds.), *Inflammation, Aging, and Oxidative Stress*,
Oxidative Stress in Applied Basic Research and Clinical Practice,
DOI 10.1007/978-3-319-33486-8_11

are oxidized, increased ROS and protein oxidation should be considered as contributors to, rather than the driving force of degenerative pathology.

Indeed, just as evidence was accumulating that there was little to gain by the direct suppression of free radical levels, it became increasingly clear that both ROS and the sites of oxidation on biomolecules—particularly proteins—were integral parts of signaling pathways key to maintaining cellular homeostasis. Thus the focus of research has shifted to understanding how these pathways become dysfunctional during the induction and progression of neurodegeneration and during the 'normal' aging process itself.

2 Characteristics of Neurodegenerative Diseases Associated with Aging

While heritable genetic mutations have been identified as the cause of many neurological diseases, the focus of this chapter is on age-associated 'sporadic' forms of high-incidence diseases not associated with any specific genetic defect. These include most cases of Alzheimer's disease, Parkinson's disease and non-familial cases of ALS.

1. Alzheimer's disease (AD) is the most common neurodegenerative disorder—accounting for 60–70 % of dementia cases—and is the third leading cause of death in the United States. As of 2010, it was estimated that 26 million people worldwide were living with AD and by 2050 the number of people with this disorder may reach 100 million [2]. AD is characterized by progressive cognitive deficits that typically manifest slowly and gradually worsen with time. Later, control of critical bodily functions is lost, so that following diagnosis, the average life expectancy of patients is 3–9 years. The single most important risk factor for AD is aging; otherwise, nearly all (~95 %) of AD cases are sporadic and of unknown etiology. Histologically, AD is associated with the increased deposition of amyloid-β-containing plaques and neurofibrillary tangles [3].
2. Parkinson's disease (PD) is a degenerative disorder affecting the central nervous system. Its symptoms, particularly early in its course, reflect a decline in motor function and consist of resting tremors, muscle rigidity, akinesia, and bradykinesia. Later, dementia, depression and sleep difficulties may develop. As the second most common neurodegenerative disorder after Alzheimer's disease, most cases occur after the age of 50 years of age and it affects more than 1 % of the population over 60. Nearly all cases (90 %) are sporadic, while about 10 % are inherited [4]. Sufficient loss of dopaminergic neurons in the midbrain substantia nigra results in the motor symptoms of PD. Accompanying the loss of these neurons is the formation of dysfunctional aggregates of α-synuclein, a neuroprotective protein highly expressed in CNS and normally involved in proper mitochondrial function and synaptic vesicle formation. These aggregates comprise the bulk of Lewy bodies, the hallmark lesions of PD located in the substantia nigra and putamen—regions involved in motor control and learning [5].

3. Amyotrophic lateral sclerosis (ALS; otherwise known as motor neuron disease (MND) and sometimes referred to as Lou Gehrig's disease) is a fatal disorder involving the degeneration and death of lower motor neurons in the ventral (anterior) horn of the spinal cord and upper motor neurons in the brainstem and motor cortex. Ensuing symptoms include overactive tendon reflexes and clonus resulting from upper motor neuron dysfunction, and progressive muscular weakness and fasciculation in the hands or legs and/or slurred speech or dysphagia consequent to lower motor neuron dysfunction [6]. In approximately 10 % of cases the cause of disease can be attributed to a specific genetic mutation inherited in a family. Regardless of family heritability, several hundred mutations in ~40 genes are associated with ALS, yet only some 15 % of cases can be associated with a specific genetic mutation regardless of family heritability [7]. The remaining 85 % of cases are apparently sporadic, at present without identifiable genetic influence. The only clearly identifiable risk factor is age, while a number of environmental risk factors—including dietary fat, strenuous exercise, trauma, mercury, selenium, agriculture, and smoking—have been implicated, but not confirmed, in studies typically involving small populations in specific localities, or larger studies that could not be replicated [8, 9].

3 ROS, Protein Oxidation, and Oxidative Signaling

Classical ROS include superoxide ($O_2^{\cdot -}$), hydrogen peroxide (H_2O_2), nitric oxide ($NO^{\cdot}$) and hydroxyl $OH^{\cdot}$. Of these, $O_2^{\cdot -}$, $NO^{\cdot}$ and H_2O_2 are less reactive and are thus able to interact at greater distances, and often react at specific sites on target biomolecules. Sites so modified on target proteins can induce well-defined changes in protein functionality, thus constituting one component of a signaling pathway that communicates cellular redox status or other information.

$NO^{\cdot}$ has well-known roles as a signaling molecule. $NO^{\cdot}$ is produced under tight control enzymatically by the several isoforms of nitric oxide synthase, or alternatively by reduction of NO_2^- or NO_3^-, and is involved in numerous signaling pathways in the CNS and elsewhere (for reviews, see [10] and [11]). Targets include cysteine thiols, iron in heme and iron-sulfur prosthetic groups, and the phenolic hydroxyl groups of tyrosine residues. While $NO^{\cdot}$ reacts directly with heme iron, it must first react with superoxide or another radical species before cysteine thiols or phenolic hydroxyl groups can be nitrated [12]. While production of nitrotyrosines is more likely indicative of excessive levels of $NO^{\cdot}$ and nitrosative stress, there are a number of indications that nitrosylation of cysteine thiols is utilized in signaling pathways under normal physiological conditions (reviewed in [13]). First off, thiol nitrosylation is found to be selective rather than indiscriminate. Selectivity due to source adjacency can occur on proteins associated with—and thus close to—nitric oxide synthase (NOS), a key site of $NO^{\cdot}$ production. Thiols are also more readily nitrosylated—and thus preferentially modified—when found within motifs containing amino acids with acidic or basic side chains that facilitate the thiol deprotonation involved in nitrosylation. Finally, hydrophobicity facilitates nitrosylation, with the consequence that thiols of membrane-associated proteins are more readily attacked by $NO^{\cdot}$.

In contrast, $OH^{\cdot}$ is so highly reactive that it is thought typically to oxidize the first biomolecular site it encounters, and so is thought unlikely to be a useful signaling molecule.

The high reactivity of some ROS, particularly hydroxyl and peroxynitrite, has led to the idea that they oxidize the first target they encounter, and consequently that the locations of their oxidative lesions are random. The assumption embedded in this 'random oxidative damage' model is that the sites where these ROS are generated are also random. However, large quantities of ROS are generated by membrane-bound enzymes at well-defined organellar and cellular sites, suggesting that rather than being generated randomly, most ROS are produced at well defined locations. Furthermore, as techniques have been developed that enable identification of the specific sites of oxidation, it is becoming increasingly clear that specific sites on a limited set of molecules are much more frequently oxidized than others. Some of these sites can be linked readily to signaling pathways found to be sensitive to cellular redox status. Consistent with this is that the observed sites of oxidation are more reactive, are more exposed to ROS, and/or are on molecules nearby where ROS are generated, so that less random oxidation actually occurs than might be expected. Nevertheless, organisms possess both effective processes for repairing nearly all ROS-induced damage, and the ability to degrade damaged molecules that cannot be repaired.

Hypochlorite (OCl^-) is a reactive species generated by myeloperoxidase in activated microglia; since such cells and associated chronic inflammation are found in most neurodegenerative diseases, this species of ROS may be of some significance.

Proteins can be modified oxidatively not only by ROS and $NO^{\cdot}$, but by the end-products of lipid peroxidation and the cyclooxygenase pathway, by cysteine thiol oxidation to form disulfide adducts with glutathione, and by chlorine and bromine. The most abundant of these oxidative modifications are protein carbonylation, 3-nitrotyrosine, binding of HNE, and glutathionylation.

ROS, particularly H_2O_2, have often been used as reagents for inducing oxidative stress in vitro. However, other evidence points to increases in their in vivo production as being under regulatory control, and not necessarily as a stochastic causative agent of stress. Regulated production of superoxide is critical to mitochondrial functioning. In mitochondria, eleven potential sources of $O_2^{\cdot -}$ have been identified and once generated superoxide is rapidly converted to H_2O_2 by superoxide dismutase. So long as the $O_2^{\cdot -}$ and H_2O_2 concentrations are low, these species are essential not only for normal function, but also for communication between mitochondria and the rest of the cell; only at higher concentrations do these species become toxic [14]. Mitochondrial H_2O_2 production is modulated by $NO^{\cdot}$ levels, mediated by interaction of $NO^{\cdot}$ with cytochrome oxidase [15]. Just as for mitochondria, different cellular ROS concentrations may have specific biological consequences. For example, the basal cytosolic concentration of H_2O_2 is controlled at a steady state of ~10 nM under normal conditions; at 1 μM, it provides proliferative signaling in human breast cancer cells, and induces apoptosis and necrosis at levels of 3 μM and above [15]. As a more specific case, in cardiomyocyte slices, increased amounts of $O_2^{\cdot-}$ are produced dose-dependently by aldosterone, mediated by the mineralocorticoid receptor, the mitochondrial respiratory complex, and NADPH oxidase [16].

Oxidation of protein cysteine thiols by H_2O_2 is an important redox signaling pathway and one in which only thiols with low pK_as in specific locations of redox sensitive proteins are susceptible to attack; in this pathway, varying levels of H_2O_2 are likely involved [17]. Initial oxidation of cysteine thiols typically results in sulfenic acids (–SOH) and changes to protein conformation and activity; further oxidation produces sulfinic ($–SO_2H$) acids. Enzymes exist to reduce both sulfenic and sulfinic forms, so these modifications are effectively reversible and can be used for signaling. Sulfonic acids ($–SO_3H$) are generated by more drastic oxidation, but no enzymes to reduce these have been found to date [17].

Lipids constitute a substantial part of cellular membranes. Polyunsaturated fatty acid (PUFA) components of membrane lipids are readily susceptible to oxidation, while the saturated and mono-unsaturated components are much less so. Even so, not all oxidative species are capable of oxidizing PUFAs. Notably $NO^{\cdot}$ and $O_2^{\cdot-}$ are unreactive, while the protonated form of superoxide, $HO_2^{\cdot}$, is capable of PUFA oxidation. Initiation of lipid oxidation appears to involve reaction of labile iron and H_2O_2 to produce $HO^{\cdot}$, which then abstracts a hydrogen atom or adds across a double bond, in either case generating a carbon-centered radical. Such radicals then most often initiate a chain reaction by reacting with molecular oxygen to produce a peroxyl radical, which then abstracts hydrogen from a neighboring lipid PUFA, thus propagating the chain reaction by producing another carbon radical, etc. [12]. The chain reaction can be broken by reaction with antioxidants such as α-tocopherol, or other free radicals such as $NO^{\cdot}$. As uncovered in studies of purified compounds, fatty acid peroxyls can react with additional oxygen molecules by several different routes that result in carbon chain scission or cyclization. Scission results in the generation of a variety of reactive shorter-chained unsaturated aldehydes or ketones [18]. One such aldehyde, 4-hydroxy-*trans*-2-nonenal (HNE), is the oxidative scission product of ω-6 fatty acids and is produced in a wide variety of systems subjected to oxidative stress. 4-hydroxy-*trans*-2-hexenal (HHE) is the analogous scission product from ω-3 fatty acids. Other reactive aldehydes that occur in neurodegeneration and aging include malondialdehyde (MDA), glyoxal, methyl glyoxal, acrolein, and 4-oxo-2-nonenal (ONE), a close, more-oxidized relative of HNE.

If carbon chain cyclization rather than scission occurs, compounds with several ring structures including isofurans and F_2 isoprostanes are produced. Human cerebrospinal fluid (CSF) levels of F_2 isoprostanes increase with age [19]. Increased levels are also found in CSF of AD patients [19] and plasma of PD patients [20], but levels in CSF of ALS patients were unchanged [21]. Moreover, increased levels also are associated with poorer performance on at least one test of cognitive ability in non-demented adults [22], suggesting that higher oxidative levels in CNS may not only be associated with neurodegenerative diseases but may also impact cognition generally and be associated with the reduced cognitive ability associated with normal aging.

Under normal intracellular oxygen tension, isoprostanes are favored, but as local oxygen tension increases isoprostane production is suppressed and isofuran generation predominates. Notably, increased levels of isofurans, but not isoprostanes, are measured in the substantia nigra of Parkinson's patients, which could be ascribed to an elevated oxygen tension resulting from mitochondrial dysfunction in this disease; unsurprisingly, levels of both are normal in substantia nigral tissues of AD patients [23].

Levels of several products of lipid peroxidation and their protein adducts are increased in relevant tissues of patients suffering from a number of neurodegenerative diseases, compared to those of individuals of similar age (Table 1). Levels of free HNE are increased in brains of Alzheimer's disease patients [24], as are levels of acrolein [31]. HNE reacts with cysteine, histidine or lysine residues to form protein adducts that can cause changes in protein conformation and function [46]. HNE protein adducts are increased in brains of patients afflicted with Alzheimer's disease [24, 27], Parkinson's disease [26], and ALS [28].

Table 1 Regions showing increased levels of protein oxidative products in human aging and neurodegenerative disease, versus healthy controls

	Neurodegenerative condition/affected region			
Oxidative product	Normal cognitive aging, ≥60 year *vs.* 20 year	AD	ALS	PD
Free HNE		Amygdala, hippocampus, parahippocampal gyrus [24]	Serum, CSF [25]	
HNE adduct	Oculomotor nucleus [26]	Hippocampus, frontal cortex [27]	Ventral horn motor neurons [28]	SN [26]
4-HHE adduct		Hippocampus [29]	Spinal cord [30]	
Acrolein adduct		Cortex [31]		SN αSYN [32]
F_2 isoprostanes	CSF [19]	Hippocampus [29]	CSF nc [21]	Plasma [20]
Protein carbonyl	Frontal pole [33]	Hippocampus, inferior parietal lobule [34]	Lumbar spinal cord [35]	SN [36]
Protein 3-nitrotyrosine	Hippocampus [37]	Hippocampus [27]	Spinal cord motor neurons [38]	SN [39]
Protein 3-chlorotyrosine	Ebc, ffa's [40]	Hippocampus [41]		
AGEs	Collagen [42]	Hippocampal plaques, tangles [43]	Spinal cord motor neurons [44]	SN [45]

AGEs advanced glycation end-products, *Ebc* exhaled breath condensate, *ffa's* free amino acids, *HNE* 4-hydroxy-*trans*-2-nonenal, *HHE* hydroxyhexenal, *nc* no change, *SN* substantia nigra

HNE adducts may play a particularly crucial role in the development of ALS. The motor neurons that are decimated by this disease are especially vulnerable to the excitotoxic stress elicited by excessive concentrations of synaptic glutamate. Neighboring astrocytes normally act to maintain healthy levels of this neurotransmitter by secreting its transporter EAAT2 into the synaptic cleft. However, EAAT2 is one of the proteins targeted by HNE [28], and levels of this transporter are drastically reduced in ALS spinal cord tissue [47] leading to glutamate excitotoxicity and motor neuron death.

Protein adducts of HHE, the unsaturated aldehyde analogous to HNE that results from oxidation of ω-3 fatty acids, are also elevated in hippocampus of AD patients [29] and spinal cord tissue of ALS patients [30].

Adducts of acrolein, an aldehyde more reactive than either HNE or HHE, are also found in brain (Table 1). These occur in hippocampal tissue of AD patients [31] and an acrolein adduct of α-synuclein has been identified in neurons in substantia nigra of PD patients [32]. Such adducts of α-synuclein may well be importantly involved in the mechanism of PD since (1) α-synuclein aggregates are a PD hallmark, (2) α-synuclein plays a role in protecting neurons from oxidative stress and in preventing PUFA oxidation, and (3) oxidative protein adducts often inactivate the protein's function [5].

When activated, microglia in the brain increase levels of myeloperoxidase, an enzyme capable of generating HOCl, which then reacts with proteins to produce 3-chlorotyrosine residues. Hippocampal chlorotyrosine levels are elevated some 3-fold in the brains of AD patients compared with normal similarly-aged subjects [41].

A preliminary indication that the reversal of protein carbonylation may be mediated enzymatically is the observation that levels of carbonylated proteins are lowered when β-mercaptoethanol—a thiol chemically incapable of carbonyl reduction—is added to cell or tissue homogenates; moreover, in muscle cell homogenates this effect is abrogated by knockdown of glutaredoxin 1 by siRNA transfection [48].

4 Redox Signaling Proteins Important in Neurodegenerative Diseases and Aging

Surveys of ROS adducts to specific proteins have been conducted, resulting in identification of a relatively small number of affected proteins in neurological tissues of either human patients in the case of AD [46], or of mouse models of ALS or PD [46, 49]. Many of the adducts present in mild cognitive impairment (MCI), a disease precursor to Alzheimer's, also occur in AD; as degeneration progresses from MCI to AD, many more adduct proteins appear once AD has been diagnosed (Table 2). Significantly, adducts of many proteins are found in more than one disease, and nearly all proteins fall into classes that are notably dysfunctional during neurodegeneration (Table 2). The actual or putative involvement of each of these proteins in neurodegenerative disease has been critically discussed [46].

Table 2 Oxidation of specific proteins is detected in affected CNS tissues from Alzheimer's disease patients, or in Parkinson's disease (PD) and amyotrophic lateral sclerosis (ALS) model mice

		Neurodegenerative condition			
Function	Protein	Amnestic MCI	AD	PD	ALS
Energy dysfunction	α-ATP synthase		✓		
	β-ATP synthase				NO[a]
	creatine kinase BB		✓		
	α-enolase (Eno1)	✓	✓	✓	HNE, NO[a]
	Triosephosphate isomerase		✓		
	Phosphoglycerate mutase 1		✓		
	Pyruvate kinase M2	✓			
	Lactate dehydrogenase			✓	
Excitotoxicity	Glutamine synthase	✓	✓		
Proteasomal dysfunction	Ubiquitin carboxy-terminal hydrolase 1		✓		
Neuritic abnormalities	Dihydropyrimidinase-related protein 2 (CRMP2)		✓		HNE
Phosphorylation regulation	Mitogen-activated protein kinase 1	✓			
	Peptidyl prolyl *cis-trans* isomerase 1	✓	✓		
Synaptic abnormalities	γ-N-ethylmaleimide-sensitive factor attachment protein (γ-SNAP)		✓		
	Syntaxin-binding protein 1	✓			
pH maintenance	Carbonic anhydrase 2	✓	✓	✓	
Protein quality control	Heat shock protein 70	✓			HNE
	Heat shock cognate 71		✓		NO[a]
	Park7 (DJ-1)			C106-SOH	

✓ carbonylated protein, *HNE* HNE protein adduct, *NO* nitrotyrosine-containing protein, *MCI* mild cognitive impairment, a precursor condition to AD. Data from [46], except ALS NO data labeled [a]are from [49]

The number of redox signaling proteins continues to grow; several ones recently identified are worthy of discussion.

DJ-1 (Park7) is a multifunctional protein that was originally identified as an oncogene [50]. Mutations in DJ-1 are associated with early onset PD. In addition to its involvement in chaperone activity, transcriptional regulation, and oncogenesis, DJ-1 has shown protection against oxidative stress, involving at least three different functions. In the first, it acts as an atypical peroxidase involved in scavenging mitochondrial H_2O_2, while in the second, it interacts with and activates SOD1 [50]. In the third, it acts as a deglycase to reverse the protein glycation lesions generated by the lipid peroxide product glyoxal [51]. Moreover, knockout of DJ-1 in the SOD1^{G93A} mutant mouse model of ALS increases disease severity and shortens survival [52].

Cysteine C106 in DJ-1 is located in a pocket lined with polar residues that makes its thiol moiety especially susceptible to oxidation by H_2O_2, concentrations of which are elevated in the dopaminergic neurons at risk of loss in PD. Oxidation of this thiol is necessary for DJ-1 to be translocated to mitochondria, where it must be located to carry out many of its neuroprotective functions [53]. One likely protective function is the prevention of α-synuclein fibril formation and only the doubly oxidized (sulfinic acid) form of DJ-1 is capable of carrying this out [54].

Oxidative resistance-1 (OXR1) is an intriguing signaling protein that lacks the enzymatic ability to directly quench ROS, but rather suppresses complement and STAT3 activation in response to oxidative stress [55]. As a possible defensive response to the increased oxidative stress associated with ALS, OXR1 levels are increased in the spinal cords of late-stage ALS patients, and in presymptomatic ALS-mutant mice; its cysteine 753 is uniquely oxidized by H_2O_2, providing a possible mechanism for its oxidative stress response [56]. Notably, when OXR1 is overexpressed in the $SOD1^{G93A}$ mouse model of ALS, it lengthens survival, improves motor deficits, and delays pathology [55]. Moreover, its effects extend to models of ALS involving ALS-linked mutations in FUS and TDP43; in cellular systems Oxr1 overexpression corrects mitochondrial defects, cytoplasmic mis-localization, protein aggregation, and splicing changes associated with the ALS-linked mutations in these proteins [57]. This is in contrast to the ineffectiveness of a number of antioxidant enzymes that directly quench ROS.

Nitrosylation of the metabolic enzyme glyceraldehyde-3-phosphate dehydrogenase (GAPDH) can act as an apoptotic switch in response to neuronal excitotoxicity. On the one hand, nitrosylation to form SNO-GAPDH abolishes catalytic activity and can result in initiation of apoptosis by facilitating binding to Siah1, an apoptotic ligase. Counter to this, nitrosylation of the associated protein GOSPEL (GAPDH's competitor of Siah Protein Enhances Life) induces competitive binding to GAPDH and thus inhibits formation of the pro-apoptotic SNO-GAPDH-Siah complex [58]. The neuroprotective power of abrogating this complex is indicated by the effects of R-(–)-deprenyl. At subnanomolar concentrations, this monoamine oxidase inhibitor blocks formation of the SNO-GAPDH-Siah complex [59]; this may be the mechanism by which it slows the progression of early PD [60], and lengthens the lifespan of rats [61].

The prominence of increased amounts of oxidative products and protein adducts in aging and neurodegenerative diseases (Table 1) underscores the need to better understand the functioning of agents that normally scavenge them, and how these agents become dysfunctional in neurodegenerative disease. A recent study serves as an example.

As noted above, high levels of HNE and HNE protein adducts are observed in brains of AD patients, as well as in those of AD mouse models. Aldehyde dehydrogenase 2 (ALDH2) detoxifies reactive aldehydes including HNE. This enzyme occurs at increased levels in the temporal cortex and putamen of AD brains, in an apparent attempt to cope with the increased levels of oxidants such as HNE. Transgenic mice lacking ALDH2 exhibit many of the behavioral symptoms of AD as they age and thus have recently have been proposed as a model for AD [62].

5 Protein Quality Control and Neurodegenerative Diseases

The action of enzymes to reverse or repair oxidative modifications to biomolecules can only go so far to maintain homeostasis. Degradative processes, carried out by proteasomes or via autophagy, are critical for recycling biomolecules including oxidized proteins that would otherwise accumulate with lethal results. Indeed these processes appear to be gradually overwhelmed in many neurodegenerative diseases, resulting in the appearance of aberrant protein aggregates—frequently substantially oxidized—that are the distinguishing hallmarks of these diseases.

When young, neurons appear to have substantial proteasomal processing capacity over and above normal requirements: only 20 % of proteasomes at 190 nM concentration in hippocampal neurons from 2-day-old rats were observed to be actively processing substrate proteins *in situ*, indicating that 80 % may be available to respond to the occurrence of stress conditions [63]. The new technique of cryoelectrotomography was used to obtain these results. It is reasonable to suppose that this substantial reserve in degradative capacity may be dissipated with aging or the occurrence of neurodegenerative disease. Of interest in the future would be to use this or other comparable methods to measure changes in proteasomal numbers and proportions of proteasomes actively performing degradation in aged and/or diseased cells. Observing over time increased numbers of proteasomes in neuronal cells—and saturation of actively processing organelles—prior to or after onset of symptoms would provide mechanistic support for the 'incremental damage accumulation' hypothesis of neurodegenerative disease.

Conclusion

ROS generation is an inescapable consequence of living in an aerobic environment and utilizing aerobic metabolism. ROS do react with and cause damage to biomolecules. Products of oxidized lipids and protein adducts increase in the nervous system with age, and are prominent characteristics of virtually all neurodegenerative diseases. High levels of many of the same protein oxidative adducts are found in the most affected tissues of a number of neurodegenerative diseases, pointing to the likelihood that they all share a common inability to completely repair the same types of protein oxidative stress.

While aerobic organisms have developed effective, but apparently imperfect, repair mechanisms to cope with this damage, they appear to tolerate and indeed endeavor to maintain low levels of ROS. To do so they must sense ROS levels in order to regulate them and to modulate the damage response as oxidative levels change. Regulation and maintenance of optimal repair processes is of particular importance to the nervous system due to its high levels of ROS generation and extremely low rates of cell turnover. ROS sensing appears to be accomplished by enhancing the reactivity of redox-sensitive groups in proteins often located close to sources of ROS. The degree of oxidation of a number of these proteins is increased in neurodegenerative disease. The functions of many of these same proteins are critical to processes that become dysfunctional during disease progression, thus linking mechanisms of disease progression with sensing of aberrantly high oxidant levels.

Ever since the connection between increased oxidative levels and neurodegenerative disease was established, numerous attempts have been made to treat these diseases by suppressing ROS levels with antioxidants, essentially without success. One hopes that improved understanding of ROS sensing pathways, and of their ability to modulate processes essential to healthy nervous system functioning, will provide the tools to develop effective treatments for these presently incurable conditions.

References

1. Harman D. Aging: a theory based on free radical and radiation chemistry. J Gerontol. 1956;11(3):298–300.
2. Rodriguez-Rodriguez A, Martel-Monagas L, Lopez-Rodriguez A. Enhancing the communication flow between Alzheimer patients, caregivers, and neuropsychologists. Adv Exp Med Biol. 2010;680:601–7.
3. Querfurth HW, LaFerla FM. Alzheimer's disease. N Engl J Med. 2010;362:329–44.
4. Milani P, Ambrosi G, Gammoh O, Blandini F, Cereda C. SOD1 and DJ-1 converge at Nrf2 pathway: a clue for antioxidant therapeutic potential in neurodegeneration. Oxid Med Cell Longev. 2013;2013:836760.
5. Perluigi M, Coccia R, Butterfield DA. 4-Hydroxy-2-nonenal, a reactive product of lipid peroxidation, and neurodegenerative diseases: a toxic combination illuminated by redox proteomics studies. Antioxid Redox Signal. 2012;17:1590–609.
6. Rowland LP, Shneider NA. Amyotrophic lateral sclerosis. N Engl J Med. 2001;344: 1688–700.
7. Singleton AB, Traynor BJ. Genetics. For complex disease genetics, collaboration drives progress. Science. 2015;347(6229):1422–3.
8. Trojsi F, Monsurrò MR, Tedeschi G. Exposure to environmental toxicants and pathogenesis of amyotrophic lateral sclerosis: state of the art and research perspectives. Int J Mol Sci. 2013;14:15286–311.
9. D'Amico E, Factor-Litvak P, Santella RM, Mitsumoto H. Clinical perspective on oxidative stress in sporadic amyotrophic lateral sclerosis. Free Radic Biol Med. 2013;65:509–27.
10. Cossenza M, Socodato R, Portugal CC, Domith ICL, Gladulich LFH, Encarnação TG, Calaza KC, Mendonça HR, Campello-Costa P, Paes-de-Carvalho R. Nitric oxide in the nervous system: biochemical, developmental, and neurobiological aspects. Vitam Horm. 2014;96: 79–125.
11. Shahani N, Sawa A. Nitric oxide signaling and nitrosative stress in neurons: role for S-nitrosylation. Antioxid Redox Signal. 2011;14:1493–504.
12. Halliwell B, Gutteridge JMC. Free radicals in biology and medicine. 4th ed. Oxford, UK: Oxford University Press; 2007. p. 241.
13. Nakamura T, Tu S, Akhtar MW, Sunico CR, Okamoto S-I, Lipton SA. Aberrant protein s-nitrosylation in neurodegenerative diseases. Neuron. 2013;78:596–614.
14. Mailloux RJ. Teaching the fundamentals of electron transfer reactions in mitochondria and the production and detection of reactive oxygen species. Redox Biol. 2015;4:381–98.
15. Boveris A, Cadenas E. Mitochondrial production of hydrogen peroxide regulation by nitric oxide and the role of ubisemiquinone. IUBMB Life. 2000;50:245–50.

16. Nolly MB, Caldiz CI, Yeves AM, Villa-Abrille MC, Morgan PE, Amado Mondaca N, Portiansky EL, Chiappe de Cingolani GE, Cingolani HE, Ennis IL. The signaling pathway for aldosterone-induced mitochondrial production of superoxide anion in the myocardium. J Mol Cell Cardiol. 2014;67:60–8.
17. Reczek CR, Chandel NS. ROS-dependent signal transduction. Curr Opin Cell Biol. 2014;33:8–13.
18. Spickett CM. The lipid peroxidation product 4-hydroxy-2-nonenal: advances in chemistry and analysis. Redox Biol. 2013;1:145–52.
19. Montine TJ, Peskind ER, Quinn JF, Wilson AM, Montine KS, Galasko D. Increased cerebrospinal fluid F2-isoprostanes are associated with aging and latent Alzheimer's disease as identified by biomarkers. Neuromolecular Med. 2010;13:37–43.
20. Seet RCS, Lee C-YJ, Lim ECH, Tan JJH, Quek AML, Chong W-L, Looi W-F, Huang S-H, Wang H, Chan Y-H, Halliwell B. Oxidative damage in Parkinson disease: measurement using accurate biomarkers. Free Radic Biol Med. 2009;48:560–6.
21. Montine TJ, Beal MF, Cudkowicz ME, O'Donnell H, Margolin RA, McFarland L, Bachrach AF, Zackert WE, Roberts LJ, Morrow JD. Increased CSF F2-isoprostane concentration in probable AD. Neurology. 1999;52:562–5.
22. Li G, Millard SP, Peskind ER, Zhang J, Yu CE, Leverenz JB, Mayer C, Shofer JS, Raskind MA, Quinn JF, Galasko DR, Montine TJ. Cross-sectional and longitudinal relationships between cerebrospinal fluid biomarkers and cognitive function in people without cognitive impairment from across the adult life span. JAMA Neurol. 2014;71:742–51.
23. Fessel JP, Hulette C, Powell S, Roberts LJ, Zhang J. Isofurans, but not F2-isoprostanes, are increased in the substantia nigra of patients with Parkinson's disease and with dementia with Lewy body disease. J Neurochem. 2003;85:645–50.
24. Markesbery WR, Lovell MA. Four-hydroxynonenal, a product of lipid peroxidation, is increased in the brain in Alzheimer's disease. Neurobiol Aging. 1998;19:33–6.
25. Simpson EP, Henry YK, Henkel JS, Smith RG, Appel SH. Increased lipid peroxidation in sera of ALS patients: a potential biomarker of disease burden. Neurology. 2004;62:1758–65.
26. Yoritaka A, Hattori N, Uchida K, Tanaka M, Stadtman ER, Mizuno Y. Immunohistochemical detection of 4-hydroxynonenal protein adducts in Parkinson disease. Proc Natl Acad Sci U S A. 1996;93:2696–701.
27. Smith MA, Richey Harris PL, Sayre LM, Beckman JS, Perry G. Widespread peroxynitrite-mediated damage in Alzheimer's disease. J Neurosci. 1997;17:2653–7.
28. Pedersen WA, Fu W, Keller JN, Markesbery WR, Appel S, Smith RG, Kasarskis E, Mattson MP. Protein modification by the lipid peroxidation product 4-hydroxynonenal in the spinal cords of amyotrophic lateral sclerosis patients. Ann Neurol. 1998;44:819–24.
29. Bradley MA, Xiong-Fister S, Markesbery WR, Lovell MA. Elevated 4-hydroxyhexenal in Alzheimer's disease (AD) progression. Neurobiol Aging. 2012;33:1034–44.
30. Shibata N, Yamada S, Uchida K, Hirano A, Sakoda S, Fujimura H, Sasaki S, Iwata M, Toi S, Kawaguchi M, Yamamoto T, Kobayashi M. Accumulation of protein-bound 4-hydroxy-2-hexenal in spinal cords from patients with sporadic amyotrophic lateral sclerosis. Brain Res. 2004;1019:170–7.
31. Calingasan NY, Uchida K, Gibson GE. Protein-bound acrolein: a novel marker of oxidative stress in Alzheimer's disease. J Neurochem. 1999;72:751–6.
32. Shamoto-Nagai M, Maruyama W, Hashizume Y, Yoshida M, Osawa T, Riederer P, Naoi M. In parkinsonian substantia nigra, a-synuclein is modified by acrolein, a lipid-peroxidation product, and accumulates in the dopamine neurons with inhibition of proteasome activity. J Neural Transm. 2007;114:1559–67.
33. Smith CD, Carney JM, Starke-Reed PE, Oliver CN, Stadtman ER, Floyd RA, Markesbery WR. Excess brain protein oxidation and enzyme dysfunction in normal aging and in Alzheimer disease. Proc Natl Acad Sci U S A. 1991;88:10540–3.
34. Hensley K, Hall N, Subramaniam R, Cole P, Harris M, Aksenov M, Aksenova M, Gabbita SP, Wu JF, Carney JM. Brain regional correspondence between Alzheimer's disease histopathology and biomarkers of protein oxidation. J Neurochem. 1995;65:2146–56.

35. Shaw PJ, Ince PG, Falkous G, Mantle D. Oxidative damage to protein in sporadic motor neuron disease spinal cord. Ann Neurol. 1995;38:691–5.
36. Floor E, Wetzel MG. Increased protein oxidation in human substantia nigra pars compacta in comparison with basal ganglia and prefrontal cortex measured with an improved dinitrophenylhydrazine assay. J Neurochem. 1998;70:268–75.
37. Venkateshappa C, Harish G, Mahadevan A, Srinivas Bharath MM, Shankar SK. Elevated oxidative stress and decreased antioxidant function in the human hippocampus and frontal cortex with increasing age: implications for neurodegeneration in Alzheimer's disease. Neurochem Res. 2012;37:1601–14.
38. Abe K, Pan LH, Watanabe M, Kato T, Itoyama Y. Induction of nitrotyrosine-like immunoreactivity in the lower motor neuron of amyotrophic lateral sclerosis. Neurosci Lett. 1995;199:152–4.
39. Good PF, Hsu A, Werner P, Perl DP, Olanow CW. Protein nitration in Parkinson's disease. J Neuropathol Exp Neurol. 1998;57:338–42.
40. Syslová K, Böhmová A, Mikoška M, Kuzma M, Pelclová D, Kačer P. Multimarker screening of oxidative stress in aging. Oxid Med Cell Longev. 2014;2014:562860.
41. Green PS, Mendez AJ, Jacob JS, Crowley JR, Growdon W, Hyman BT, Heinecke JW. Neuronal expression of myeloperoxidase is increased in Alzheimer's disease. J Neurochem. 2004;90(3): 724–33.
42. Sell DR, Lane MA, Johnson WA, Masoro EJ, Mock OB, Reiser KM, Fogarty JF, Cutler RG, Ingram DK, Roth GS, Monnier VM. Longevity and the genetic determination of collagen glycoxidation kinetics in mammalian senescence. Proc Natl Acad Sci U S A. 1996;93:485–90.
43. Smith MA, Taneda S, Richey PL, Miyata S, Yan SD, Stern D, Sayre LM, Monnier VM, Perry G. Advanced Maillard reaction end products are associated with Alzheimer disease pathology. Proc Natl Acad Sci U S A. 1994;91:5710–4.
44. Chou SM, Wang HS, Taniguchi A, Bucala R. Advanced glycation endproducts in neurofilament conglomeration of motoneurons in familial and sporadic amyotrophic lateral sclerosis. Mol Med. 1998;4:324–32.
45. Castellani R, Smith MA, Richey PL, Perry G. Glycoxidation and oxidative stress in Parkinson disease and diffuse Lewy body disease. Brain Res. 1996;737:195–200.
46. Butterfield DA, Perluigi M, Reed T, Muharib T, Hughes CP, Robinson RAS, Sultana R. Redox proteomics in selected neurodegenerative disorders: from its infancy to future applications. Antioxid Redox Signal. 2012;17:1610–55.
47. Rothstein JD, Van Kammen M, Levey AI, Martin LJ, Kuncl RW. Selective loss of glial glutamate transporter GLT-1 in amyotrophic lateral sclerosis. Ann Neurol. 1995;38:73–84.
48. Wong C-M, Marcocci L, Das D, Wang X, Luo H, Zungu-Edmondson M, Suzuki YJ. Mechanism of protein decarbonylation. Free Radic Biol Med. 2013;65:1126–33.
49. Casoni F, Basso M, Massignan T, Gianazza E, Cheroni C, Salmona M, Bendotti C, Bonetto V. Protein nitration in a mouse model of familial amyotrophic lateral sclerosis: possible multifunctional role in the pathogenesis. J Biol Chem. 2005;280:16295–304.
50. Abdullah R, Basak I, Patil KS, Alves G, Larsen JP, Møller SG. Parkinson's disease and age: the obvious but largely unexplored link. Exp Gerontol. 2015;68:33–8.
51. Richarme G, Mihoub M, Dairou J, Bui LC, Leger T, Lamouri A. Parkinsonism-associated protein DJ-1/Park7 is a major protein deglycase that repairs methylglyoxal- and glyoxal-glycated cysteine, arginine, and lysine residues. J Biol Chem. 2015;290:1885–97.
52. Lev N, Barhum Y, Lotan I, Steiner I, Offen D. DJ-1 knockout augments disease severity and shortens survival in a mouse model of ALS. PLoS One. 2015;10, e0117190.
53. Madian AG, Hindupur J, Hulleman JD, Diaz-Maldonado N, Mishra VR, Guigard E, Kay CM, Rochet J-C, Regnier FE. Effect of single amino acid substitution on oxidative modifications of the Parkinson's disease-related protein, DJ-1. Mol Cell Proteomics. 2012;11:M111.010892.
54. Zhou W, Zhu M, Wilson M, Petsko G, Fink AL. The oxidation state of DJ-1 regulates its chaperone activity toward α-synuclein. J Mol Biol. 2006;356:1036–48.
55. Liu KX, Edwards B, Lee S, Finelli MJ, Davies B, Davies KE, Oliver PL. Neuron-specific antioxidant OXR1 extends survival of a mouse model of amyotrophic lateral sclerosis. Brain. 2015;138:1167–81.

56. Oliver PL, Finelli MJ, Edwards B, Bitoun E, Butts DL, Becker EBE, Cheeseman MT, Davies B, Davies KE. Oxr1 is essential for protection against oxidative stress-induced neurodegeneration. PLoS Genet. 2011;7, e1002338.
57. Finelli MJ, Liu KX, Wu Y, Oliver PL, Davies KE. Oxr1 improves pathogenic cellular features of ALS-associated FUS and TDP-43 mutations. Hum Mol Genet. 2015;24(12):3529–44.
58. Nakamura T, Prikhodko OA, Pirie E, Nagar S, Akhtar MW, Oh CK, McKercher SR, Ambasudhan R, Okamoto S, Lipton SA. Aberrant protein S-nitrosylation contributes to the pathophysiology of neurodegenerative diseases. Neurobiol Dis. 2015;84:99–108.
59. Hara MR, Thomas B, Cascio MB, Bae B-I, Hester LD, Dawson VL, Dawson TM, Sawa A, Snyder SH. Neuroprotection by pharmacologic blockade of the GAPDH death cascade. Proc Natl Acad Sci U S A. 2006;103:3887–9.
60. The Parkinson Study Group. Effects of tocopherol and deprenyl on the progression of disability in early Parkinson's disease. N Engl J Med. 1993;328:176–83.
61. Kitani K, Kanai S, Miyasaka K, Carrillo MC, Ivy GO. The necessity of having a proper dose of (–)deprenyl (D) to prolong the life spans of rats explains discrepancies among different studies in the past. Ann N Y Acad Sci. 2006;1067:375–82.
62. D'Souza Y, Elharram A, Soon-Shiong R, Andrew RD, Bennett BM. Characterization of Aldh2 (-/-) mice as an age-related model of cognitive impairment and Alzheimer's disease. Mol Brain. 2015;8:27.
63. Asano S, Fukuda Y, Beck F, Aufderheide A, Förster F, Danev R. Baumeister Wolfgang. Proteasomes. a molecular census of 26S proteasomes in intact neurons. Science. 2015;347: 439–42.

Oxidative Stress Is a Driver of Normal and Pathological Ovarian Aging

Ulrike Luderer

1 Normal Ovarian Development and Ovarian Aging

Primordial germ cells (PGCs), the germline stem cells, are formed in the yolk sac and migrate via the hindgut to the urogenital ridge on the mesonephros during embryonic development. In female embryos PGCs are called oogonia, and these mitotically self-renewing cells are present only during prenatal life. At gestational day (GD) 13.5 in mice or gestational week 11 in humans, oogonia initiate meiosis and then arrest in the diplotene stage of the first meiotic prophase until meiosis resumes at the time of ovulation [1–3]. Once they enter meiosis, the ovarian germ cells are called oocytes, and they initially exist in syncytial clones, called oocyte nests or cysts, which are derived from a single PGC. During the first postnatal week in mice and gestational week 18–19 in humans, oocyte nest breakdown and formation of primordial follicles occurs [2–4]. During this process about one third of oocytes are lost via apoptosis [5].

Ovarian follicles are the functional unit of the mammalian ovary. They consist of oocytes surrounded by and communicating via gap junctions with specialized somatic cells called granulosa cells. Primordial follicles, the least mature folllicles, contain only a single partial layer of flattened granulosa cells. Primordial follicles are continuously recruited into the growing pool throughout life by paracrine

U. Luderer (✉)
Division of Occupational and Environmental Medicine, Department of Medicine, University of California Irvine, Irvine, CA, USA

Department of Developmental and Cell Biology, University of California Irvine, Irvine, CA, USA

Program in Public Health, University of California Irvine, Irvine, CA, USA
e-mail: uluderer@uci.edu

S.C. Bondy, A. Campbell (eds.), *Inflammation, Aging, and Oxidative Stress*, Oxidative Stress in Applied Basic Research and Clinical Practice,
DOI 10.1007/978-3-319-33486-8_12

signals [6–8]. During recruitment, granulosa cells differentiate, assuming a cuboidal morphology, and proliferate. Follicles with a single layer of cuboidal granulosa cells are called primary follicles and those with more than one layer of granulosa cells are called secondary follicles. As the follicle continues to grow, fluid-filled vesicles appear in the granulosa cell layer and these eventually coalesce to form a single vesicle, the antrum, and these are called antral follicles. In larger follicles a basement membrane separates the nonvascularized granulosa cell layers from the outer cell layers, the vascularized theca cells. It is generally accepted that women and other female mammals are born with a finite number of germ cells that continuously declines throughout life due to ovulation and an apoptotic process called follicular atresia. In recent years this dogma has been challenged, with some studies providing evidence for germline stem cells and oogenesis in the adult ovary [9–11]. Nonetheless, aging in mammalian females is characterized by a decline in ovarian follicle numbers, culminating in reproductive senescence, which in most species occurs long before death [12–17]. Only about one percent of follicles mature to ovulation, the rest undergo atresia, which in the postnatal ovary occurs predominantly at the early antral stage of follicular development [18].

In addition to declining oocyte numbers, ovarian aging is characterized by decreased oocyte quality. Ovulated oocytes from older compared to younger females have increased aneuploidy, shorter telomeres, decreased mitochondrial fraction, and increased endoplasmic reticulum and Golgi fractions [19–25]. Double strand DNA breaks are increased in primordial and antral follicle oocytes from older women and older mice compared to those of younger females, while expression of DNA repair genes is decreased [26]. Consistent with decreased oocyte quality, older women undergoing assisted reproduction have higher percentages of aneuploid and mosaic blastocysts, decreased pregnancy rates and increased pre- and post-implantation mortality compared to younger women [21, 27].

2 Oxidative Stress and Normal Ovarian Aging

Ovarian aging is accelerated compared to aging of other tissues and organs. Complete cessation of ovarian function or menopause occurs at 50 ± 4 years, typically several decades before death [28]. Normal aging in women is associated with increased markers of systemic oxidative stress, such as urinary F_2-isoprostanes, which are products of free-radical catalyzed lipid peroxidation of arachidonic acid. 15-F_{2t}-isoprostane (15-F_{2t}-IsoP) is one of the most common F_2-IsoPs and converts to 2,3-dinor-5,6-dihydro-15-F_{2t}-IsoP (15-F_{2t}-IsoP-M), which may be a better biomarker of oxidative stress because it is not subject to autoxidation and is not produced by the kidneys [29]. These authors showed that urinary 15-F_{2t}-IsoP-M increased with age, menopausal status, body mass index (BMI), and smoking in middle aged and older women [29]. The association of systemic biomarkers of oxidative stress with menopausal status suggests that oxidative stress may play a role in ovarian aging, and several studies have attempted to examine this more directly.

Multiple studies have examined the hypothesis that oxidative stress is associated with ovarian aging in women undergoing assisted reproduction (ART). Conflicting results have been obtained in studies that attempted to measure reactive oxygen species (ROS) or oxidative damage in follicular fluid, but many of these studies utilized assays that are not appropriate for cell-free fluids or have poor specificity, sensitivity and/or reproducibility (e.g. luminol chemiluminescence for ROS or malondialdehyde for lipid peroxidation) ([30, 31], reviewed in [32]). Studies that utilized more robust methods have found evidence for increased oxidative damage in follicular fluid of older compared to younger women. Wiener-Megnazi et al utilized a thermochemiluminescence assay that measures the oxidation of molecules in biological fluids to unstable electrically excited species such as carbonyls and observed increasing oxidative activity in follicular fluid with increasing age of the woman [33]. Moreover, increased oxidative activity was associated with decreased numbers of oocytes ovulated and fertilized eggs progressing to cleaved embryos. Using another sensitive method, measurement of loss of reduced thiols, Tatone and coworkers demonstrated decreased levels of reduced thiols in follicular fluid of older women [32]. The latter group also reported that enzymatic activities of antioxidant enzymes in follicular fluid differed in older compared to younger women, with decreased activities of glutathione-S-transferase (GST) and catalase and increased activity of superoxide dismutase (SOD)[34].

Other studies have measured similar endpoints in oocytes and cumulus cells (specialized granulosa cells surrounding the oocyte) of women undergoing ART. Protein levels of cytosolic SOD1 and mitochondrial SOD2 and enzymatic SOD activity decreased with age in cumulus cells, and higher SOD activity was associated with greater likelihood of successful ART outcome [35]. Similarly in another study, cumulus cells from older compared to younger women had decreased mRNA and protein expression of catalase, *SOD1*, and *SOD2* [36]. In contrast to rodent studies described below, single oocyte transcriptomics did not show a strong difference in the expression of oxidative stress or other stress response genes in oocytes from older compared to younger women [37]. This lack of observed differences may have to do with the study design in which the young group (women<=35 years, with a mean age of 31.1) did not differ greatly in age from the "old" group (37–39 years, with mean of 37.8).

Multiple lines of evidence from rodent models support a role for ROS and oxidative damage in ovarian aging. Oxidative lipid, protein, and DNA damage, detected by 4-hydroxynonenal (4-HNE), nitrotyrosine (NTY), and 8-hydroxy-2′-deoxyguanosine (8-OH-dG) immunostaining, respectively, increased in ovarian follicles, and the accumulation of a product of protein and lipid oxidation, lipofuscin [38], increased in the ovaries of mice with increasing age [39]. Ovarian expression of the mitochondrial antioxidant genes peroxiredoxin 3 and thioredoxin 2 (*Txn2*), and the cytosolic antioxidants glutaredoxin 1 (*Glrx1*) and glutathione-S-transferase mu 2 (*Gstm2*), decreased with increasing age in mice, suggesting that declining antioxidant defenses may be contributing to oxidative damage in the aging ovary [39]. In another study, vitamin E levels were increased and glutathione reductase (GSR) enzymatic activity was decreased in ovaries of 8–9 month old rats compared to 26 day old prepubertal rats, while catalase and GPX activities, and TBARS were unchanged [40]. Decreased expression of the glutathione (GSH)-related genes *Glrx1*, glutamate cysteine ligase modifier subunit (*Gclm*), *Gsr*, GLRX-related protein, as well as of *Sod1*, and *Txn1* was measured in oocytes from aged compared to young mice [41, 42]. GSH concentrations and the ratio of GSH to GSSG were decreased in oocytes from aged compared to young mice [25]. 4-HNE was recently shown to covalently modify the succinate dehydrogenase protein of complex II of the mitochondrial electron transport chain in oocytes, increasing ROS production and lipid peroxidation, causing loss of mitochondrial membrane potential, and inducing apoptosis [43]. Consistent with decreased antioxidant capacity, oocytes from 11 to 12 month old mice cultured with 250 μM hydrogen peroxide had decreased survival compared to oocytes from 1 month old mice [26].

If oxidative stress drives ovarian aging, then antioxidant supplementation should slow ovarian aging. However, few studies have attempted to test this hypothesis. In a model of ovarian aging induced by repeated superovulation cycles in rapid succession, ovarian levels of oxidative protein damage (protein carbonyls) and DNA damage (8-OH-dG) increased, while protein levels of SOD1 and SOD2, number of oocytes ovulated, the percentage of oocytes with normal mitochondrial distribution, and the percentage of fertilized oocytes progressing to the blastocyst stage decreased with successive superovulation cycles [44]. All of these effects were partially mitigated by supplementation with L-carnitine, but not by supplementation with the GSH precursor *N*-acetylcysteine or ascorbate, consistent with a key role of mitochondrial dysfunction [44]. Administration of pharmacological doses of the antioxidant vitamins C (10 g/kg diet) and E (0.6 g/kg diet) to female mice from the age of weaning or from the age of 32 weeks prevented the age-related declines in numbers of oocytes ovulated with hormonal stimulation and in the percentage of oocytes with normal distributions of chromosomes on the metaphase II plate during meiosis II [24]. Despite these beneficial effects on oocyte quantity and quality, this dietary regimen actually decreased the litter frequency, litter size, and total number of offspring [45]. Supplementation of mice with 0.1 mM NAC in drinking water from 2 months of age for up to 12 months delayed the age-related decline in litter size and partially mitigated the age-related increased percentage of fragmented and lysed oocytes and decrease in ovarian telomere length [46]. Supplementation with 1 mM NAC was less protective for all of these endpoints. The same group also examined the ability of supplementation with 30 mg/L resveratrol in the drinking water from 2 to 3 months of age for 6 or 12 months on ovarian aging [47]. They observed partial mitigation of the age-related declines in litter size, ovarian follicle numbers, percentage of oocytes with normal meiotic spindles and chromosome alignment, ovarian expression of Sirtuin 1 (*Sirt1*), and ovarian telomere length in the resveratrol-treated mice, with the beneficial effects more pronounced after 12 months of treatment [47]. These beneficial ovarian effects of resveratrol may be mediated in part by modulation of antioxidant pathways and improvement of mitochondrial function, as well as by its well-known ability to activate SIRT1 and increase telomerase activity [48, 49]. In vitro supplementation with 1 mM melatonin during oocyte maturation was able to mitigate the ROS increase, caspase activation, and oocyte fragmentation that characterize postovulatory oocyte aging, and improve IVF success measured as development to the blastocyst stage [50].

3 Oxidative Stress and Pathological Ovarian Aging, Evidence from Genetically Modified Mouse Models

Premature ovarian failure (POF) or primary ovarian insufficiency (POI) is clinically defined as cessation of ovarian function prior to the age of 40 [51, 52]. POF occurs in 1 % of all women, but the cause is unknown in the large majority of cases. Chromosomal abnormalities, including disorders of the X chromosome, are thought to be responsible for about 10 % of cases, while single gene mutations and multigenic variants are thought to be responsible for another 10–15 % of cases [53]. Menopause before the age of 45 years is often called early menopause, and recent genomewide association studies have identified a few common genetic variants that are associated with early menopause [54]. Other known causes of POF include ionizing radiation to the pelvis and treatment with antineoplastic agents for cancer [55–60]. Oxidative stress has been implicated in the causal pathways by which these and other toxic agents cause POF, and genetically modified mouse models further implicate oxidative stress in the pathophysiology of POF.

If oxidative stress drives ovarian aging then mice deficient in antioxidant genes might be expected to have accelerated ovarian aging and POF. Although knockout mouse models of many of the major antioxidant genes have been generated in recent years, detailed analyses of ovarian aging have been conducted in only a few of these.

Mice null for the modifier subunit of glutamate cysteine ligase (GCLM), the rate-limiting enzyme in the synthesis of GSH, have globally decreased GSH levels, but survive to old age in the absence of exposure to exogenous stressors. Compared to *Gclm+/+* mice, total ovarian GSH concentrations are decreased by about 75 %, ovarian GSH:GSSG ratios are decreased by about half, and the GSH:GSSG Nernst potential is oxidized in *Gclm−/−* mice, consistent with chronic ovarian oxidative stress [61]. Total GSH concentrations in the oocytes are decreased by more than 80 % in *Gclm−/−* mice [62]. Consistent with chronic oxidative stress, *Gclm−/−* ovaries have increased oxidative lipid damage, detected by 4-HNE immunostaining, and oxidative protein damage, detected by NTY immunostaining, compared to *Gclm+/+* ovaries [61]. *Gclm−/−* mice have similar numbers of oocytes and ovarian follicles at birth and weaning as *Gclm+/+* mice, but the subsequent age-related decline in ovarian follicle numbers is accelerated in *Gclm−/−* mice [61]. *Gclm−/−* mice also show evidence of poor oocyte quality, with delayed development of the male pronucleus, low rates of development to the blastocyst stage, and fewer implantations of embryos derived from oocytes of *Gclm−/−* females whether fertilized in vivo or in vitro [62]. The accelerated decline in ovarian follicles in *Gclm−/−*

mice is driven by accelerated decline in the primordial follicles, but there is no evidence for increased apoptosis of primordial follicles. Instead, there is evidence for increased activation of primordial follicles into the growing pool, followed by increased apoptosis of the recruited follicles at secondary and antral stages of development [61]. These in vivo data from GSH deficient mice are supported by a study using cultured neonatal (PND 4) rat ovaries, which are enriched in primordial follicles, and biochemical depletion of GSH using buthionine sulfoximine (BSO), a competitive inhibitor of *Gclm*. Treatment of cultured ovaries for 4 days with BSO significantly decreases both primordial and primary follicle numbers [63].

The transcription factor nuclear factor-erythroid 2-related factor 2 (*Nrf2*) regulates the transcription of multiple antioxidant genes, including GSH synthesis genes, various GSTs, *Sod2*, and NAD(P)H quinone oxidoreductase [64–66]. *Nrf2−/−* mice have decreased ovarian GCLM and GCLC protein levels and modestly decreased ovarian GSH concentrations [67]. Similar to *Gclm−/−* mice, ovarian follicle numbers early in life do not differ between *Nrf2−/−* and *Nrf2+/+* mice, but by middle age, the numbers of primordial follicles are significantly decreased in *Nrf2−/−* compared to *Nrf2+/+* ovaries [67]. *Nrf2−/−* mice also have smaller litters than *Nrf2+/+* mice, suggesting decreased oocyte quality, but the cause of the small litter size has not been further elucidated [68].

Oxidative stress may play a role in POF caused by deletion of several other genes that are important in ovarian follicular development. The forkhead box transcription factor FOXL2 is highly expressed in pre-granulosa cells during ovarian development and in granulosa cells throughout reproductive life and is necessary for normal ovarian development. *Foxl2−/−* mice have premature activation of primordial follicles without subsequent formation of primary follicles due to failure of granulosa cell differentiation, massive follicular atresia, and POF [69, 70]. Humans with germline mutations in *FOXL2* have Blepharophimosis Ptosis Epicanthus Inversus Syndrome, in which craniofacial abnormalities occur in association with POF [71]. *FOXL2* promoter activity and mRNA and protein expression in KGN human granulosa cells increase rapidly in response to oxidative stress (150 μM H_2O_2 for 2 h) [72]. This is partially due to transactivation by FOXL2 of its own promoter, as well as post-translational modifications that increase FOXL2 activity [72]. Collectively, these result in increased expression of oxidative stress response genes that are targets of FOXL2, including *SOD2* and immediate early response protein 3 (IER3) [72, 73] and increased cellular concentrations of GSH [74]. FOXL2 also participates in its own eventual downregulation by transactivating *SIRT1*, which decreases FOXL2 activity via deacetylation [72]. Several of the disease-causing mutations of *FOXL2* are predicted to impair activation of the stress response by FOXL2 [72], suggesting that oxidative stress may play a central role in POF caused by FOXL2 mutations or deletion. Another forkhead box transcription factor, *Foxo3a,* is localized to oocytes in the ovary, and its deletion also causes premature primordial follicle activation and POF [75, 76]. In mouse oocytes, *Sirt1, Foxo3a,* and *Sod2* expression increase in response to oxidative stress (25 μM H_2O_2 for 30 min), and inhibition of SIRT1 prevents the oxidative stress-induced increase in *Sod2* expression [77], suggesting a role for oxidative stress in POF caused by deletion of *Foxo3a*.

4 Evidence for a Role of Oxidative Stress in Pathological Ovarian Aging from Toxicological Models

4.1 Ionizing Radiation

The ovary is highly sensitive to ionizing radiation, causing temporary cessation of ovarian cycles and in many cases POF [55–57, 78, 79]. The ED_{50} for oocyte destruction by γ-radiation in young women is estimated to be less than 2 Gy [55, 56]. Although ionizing radiation directly ionizes cellular macromolecules such as DNA, the majority of ionizing radiation–induced damage is thought to be indirect, due to ROS generation from the ionization of water [80–82]. Gamma radiation destroys follicles at all stages of development [78, 79, 83]; however, primordial follicles appear to be most sensitive [84]. Degeneration of primordial and primary follicles was observed as early as 2 h after γ-irradiation of postnatal day (PND) 21 mice with a dose of 8.3 Gray [78, 83].

Interestingly, oocytes are relatively resistant to γ-radiation prior to their arrest in the diplotene stage of the first meiotic prophase, which occurs at GD 18.5–19.5 in the mouse [85–87]. Using a late fetal, GD 18.5 mouse ovary culture model, which is enriched in primordial follicles, irradiation with 0.5 Gy γ-rays at day 2 of culture (the equivalent of PND 1) resulted in phosphorylation of the Trp53 (also known as p53) homolog, Trp63 (also known as p63), and activation of caspases 9 and 3 in oocytes by 6 h post-irradiation, with few follicles remaining by 7 days after irradiation [85]. *Trp63−/−* ovaries were protected against γ-irradiation-induced oocyte destruction [85]. Exposure of wild type C57BL/6 mice at PND 5 to 0.45 or 4.5 Gy γ-rays led to increases protein and mRNA expression of the BH3-only proapoptotic BCL2 family members PUMA and NOXA in primordial follicle oocytes, while ovarian PUMA and NOXA are not induced after irradiation of *Trp63−/−* mice [88]. Complete destruction of primordial follicles occurs after exposure to 0.45 or 4.5 Gy γ-radiation at PND 5 in wild type, *p53−/−*, and *NOXA−/−* C57BL/6 mice, while PUMA−/− and *PUMA−/−;NOXA−/−* double knockout mice are protected, despite induction of equivalent levels of double strand DNA breaks [88].

Several lines of evidence point to oxidative stress as a driver of γ-radiation-induced follicle depletion. Pretreatment of mice with 100 μg of the antioxidant melatonin significantly protected against 8.3 Gy γ-radiation-induced primordial follicle destruction at all time points, while 10 μg was less protective; beneficial effects of melatonin were less consistent for primary and larger growing follicles [83]. Rapid, sustained increases in ROS occurred in human COV434 granulosa cells within 30 min after 1 or 5 Gy γ-irradiation, followed by apoptotic death at 6 h [89]. Stable overexpression of one or both subunits of the rate-limiting enzyme in GSH synthesis, *Gclc* and *Gclm*, increased GSH synthesis, prevented the radiation-induced rise in ROS, and prevented apoptotic death of the cells [89].

Compared to low linear energy transfer (LET) γ-radiation, neutrons and charged particles typical of space radiation have high LET and are generally thought to be more damaging to cells than γ-radiation at equivalent doses, but few studies have examined their effects on ovarian follicles. Exposure of PND 7 C57BL/6 N mice to

1 Gy of neutrons at energies of 0.525 or 2.13 MeV or 1 Gy γ-rays, induces apoptosis in 78, 66, and 43 % of oocytes, respectively, and in granulosa cells of 53, 18, and 29 % of follicles, respectively [90]. Exposure of 3 month old C57BL/6 J mice to 0.05, 0.3 or 0.5 Gy charged iron particles (LET = 179 keV/μm) causes nonsignificant 11 % and statistically significant 68 % and 79 % declines, respectively, in primordial follicle numbers at 1 week after irradiation compared to controls; by 8 weeks after irradiation, primordial follicle numbers are depleted by 57 % and 99 % at the 0.05 and 0.5 Gy doses, respectively [91]. Primary follicle numbers are similarly decreased, while secondary and antral follicles are decreased to a lesser extent. The decline in follicle numbers is preceded by increases in oocyte and/or granulosa cell double strand DNA breaks (phosphorylated H2AFX), as well as 4-HNE, NTY, PUMA, and activated caspase 3 immunostaining at 6 h after irradiation [91]. Dietary supplementation with the thiol antioxidant α-lipoic acid (150 mg/kg diet) significantly mitigates the effects of 0.5 Gy charged iron on all of these endpoints, consistent with oxidative stress playing a mechanistic role in the depletion of ovarian follicles by charged iron particles [91].

4.2 *Antineoplastic Agents*

Anticancer treatment with alkylating agents such as cyclophosphamide (CP) or anthracyclines such as doxorubicin can cause temporary loss of menstrual cycles and POF [60, 92–95]. Both of these classes of drugs exert their therapeutic effects by causing DNA damage and resultant death of cancer cells, but the same mechanism is also responsible for many of their adverse side effects. These drugs destroy follicles at all stages of development in humans and experimental animals [96–101]. Mice are more sensitive than rats to the primordial follicle toxicity of CP [100, 102], but rats are highly sensitive to destruction of secondary and antral follicles by CP [103, 104].

CP requires metabolic activation to exert its anticancer activity, undergoing oxidation by cytochrome P450 enzymes to 4-hydroxycyclophosphamide, which undergoes ring-opening to aldophosphamide, which spontaneously decomposes to phosphoramide mustard (PM), which can in turn decompose to chloroethylaziridine (CEZ) [105–107]. PM is thought to be the active metabolite of CP, both in terms of its anticancer activity and its ovarian toxicity [99, 106]. Treatment of cultured neonatal mouse and rat ovaries with PM depletes primordial follicles at concentrations ≥3 and 30 μM, respectively [108, 109]. The primordial follicle toxicity of PM is partially due to the volatile metabolite CEZ [63]. Depletion of primordial follicles by PM does not involve caspase activation [108]. However, immunostaining for phosphorylated histone H2AX, a marker of double-stranded DNA breaks, is increased in oocytes of primordial follicles at concentrations that cause follicle depletion and at time points prior to the onset of follicle degeneration, suggesting that the DNA strand breaks initiate follicular destruction [109]. Double-strand DNA breaks can be caused by oxidative DNA lesions [110]. Consistent with a role for oxidative stress in CP-induced follicle destruction, supplementation with GEE partially rescues the depletion of primordial, primary, and secondary follicles by

60 μM PM in cultured neonatal rat ovaries [63]. Surprisingly, depletion of GSH with BSO pretreatment does not exacerbate the destruction of primordial and primary follicles by PM treatment [63]. It is possible that the concentration of PM used maximally depletes cellular GSH so that addition of BSO does not result in any further GSH depletion, but GSH concentrations were not measured in the study by Madden and coworkers.

As noted above, CP treatment also destroys secondary and antral follicles. A single i.p. injection of 50 mg/kg CP in adult rats significantly increases activation of caspases 3 and 9 and TUNEL in granulosa cells of secondary and antral follicles 24 h later [104, 111]. These data suggest that destruction of these larger follicles involves caspase-dependent apoptosis of granulosa cells, in contrast to destruction of small follicles by PM described above. The role of oxidative stress in CP-induced granulosa cell apoptosis was studied using a human granulosa cell tumor line, COV434 cells, which possess many characteristics of normal granulosa cells, and a preactivated form of CP, 4-hydroperoxycyclophosphamide (4HC), which spontaneously decomposes in solution to 4-hydroxycyclophosphamide [106, 112]. Treatment with 4HC (1–50 μM) rapidly and concentration-dependently decreases GSH concentrations and increases ROS (dichlorofluorescein fluorescence), followed at 12 and 24 h by oxidative DNA damage (8-OHdG immunofluorescence), activation of caspase 3, and DNA fragmentation (TUNEL) [113]. Supplementation with GEE or with the antioxidants ascorbate or dithiothreitol protects against the induction of apoptosis by 4HC, while depletion of GSH with BSO enhances the induction of apoptosis by 4HC [113]. These data show that an early rise in ROS initiates the induction of apoptosis in granulosa cells by 4HC.

Treatment with doxorubicin causes dose-dependent increases in primordial follicle double strand DNA breaks (phosphorylated histone H2AX) and caspase 3 activation in cultured human ovary cortical fragments (1–100 μg/mL) and in human cortical fragments xenografted into SCID mice and in the native ovaries of the mice (10 mg/kg dose) [114]. Mice null for eEF2 kinase, which phosphorylates eukaryotic elongation factor 2 (eEF2) have persistent ovarian follicles and corpora lutea well into advanced age, when wild type ovaries are devoid of follicles [115]. Moreover, eEF2K−/− oocytes were resistant to doxorubicin-induced apoptosis and this was suggested to be due to downregulation of the endoplasmic reticulum response to oxidative stress induced by doxorubicin [115] .

Overall, the studies summarized above provide strong evidence that depletion of ovarian follicles by CP and doxorubicin is caused by oxidative stress-induced DNA damage and apoptosis induction in granulosa cells and oocytes of follicles at all stages of development.

4.3 Polycyclic Aromatic Hydrocarbons (PAHs)

PAHs are ubiquitous environmental pollutants formed during the incomplete combustion of organic materials such as wood, fossil fuels, tobacco, and foods [116]. Major routes of exposure to PAHs include consumption of grilled, roasted, or smoked foods, smoking or exposure to environmental tobacco smoke, and exposure to particulate matter air pollution [116–120]. Metabolic activation of PAHs generates reactive metabolites that can form DNA adducts, as well as metabolites that can undergo redox cycling, generating ROS and leading to oxidative DNA damage [121].

Epidemiological studies show that women who smoke have decreased fecundity [122–125] and earlier onset of menopause [28, 126, 127] than women who do not smoke. Exposure to cigarette smoke depletes primordial follicles in mice [128]. The ovotoxicity of cigarette smoke is likely due to its high PAH content. Experimental studies show that the PAHs benzo[a]pyrene (BaP), 9,10-dimethyl-1,2-benzanthracene (DMBA), and 3-methylcholanthrene (3-MC) destroy primordial and primary follicles in peripubertal mice and rats after single high doses [129–131] or repeated lower doses [132]. Primordial follicles in human ovarian explants are also sensitive to destruction by DMBA [133]. More mature secondary and antral follicles are also sensitive to destruction by PAHs [132, 134]. Culture of large antral rat follicles with DMBA causes concentration-dependent increases in ROS (dichlorofluorescein and dihydrorhodamine fluorescence) at 12 h, followed by increased BAX protein at 24 h, and caspase 3 activation and DNA fragmentation at 48 h [135]. Biochemical depletion of GSH using buthionine sulfoximine enhances these effects of DMBA, and supplementation with the cell permeable GSH analog GSH ethyl ester (GEE) is protective, suggesting that the rise in ROS initiates granulosa cell apoptosis caused by exposure of antral follicles to DMBA [135].

In addition to causing ovotoxicity after ovarian development is complete, in vivo and in vitro studies show that the developing ovary is highly sensitive to PAH exposure. Oral treatment of pregnant mice with 10, 40, or 160 mg/kg BaP daily from GD 7–16 causes dose-dependent declines in the fertility of their F1 female offspring, with complete infertility and nearly complete lack of ovarian follicles on histology in the highest dose group [136]. A more recent study utilizing mice deficient in GSH due to lack of *Gclm* provides evidence that these developmental ovarian effects of BaP are mediated by oxidative stress. Oral treatment of pregnant *Gclm+/−* dams with 2 or 10 mg/kg/day BaP from GD 7–16 dose-dependently depleted germ cells, manifested as decreased primordial and growing follicle numbers and decreased fertility, in their F1 female offspring, and the effects were significantly more severe in *Gclm−/−* F1 females than in their *Gclm+/+* littermates [137].

In mice, the neonatal period is a critical developmental window when primordial follicles are forming from oocyte nests [4]. Culture of neonatal (PND 4) rat ovaries with DMBA concentration- and time-dependently depletes primordial and small primary follicles, with the former being more sensitive (ED50 about 250 nM) [138]. Using the same experimental system with 1 μM DMBA, GSTP protein and mRNA expression were increased prior to any observable decrease in follicle numbers [139]. Single exposures to 12.5 or 75 nM DMBA, which did not decrease primordial or primary follicle numbers, significantly increased mRNA expression of *Gstm, Sod1, Sod2,* as well as *Gstp* [140]. Taken together, the early upregulation of antioxidant gene expression in response to exposure to DMBA is consistent with a response to oxidative stress induction by DMBA.

DMBA treatment of neonatal mice (1 mg/kg/day for 7 days) or cultured neonatal mouse ovaries (50 nM) increased activation of primordial follicles (PCNA, phosphorylated Akt immunostaining), while increasing apoptosis in primary and secondary follicles (activation of caspases 2 and 3 and TUNEL) [141]. ROS have been implicated in phosphorylation of AKT [142–144]. In addition, upregulation of genes in the methionine metabolism pathway was interpreted by the authors as suggesting a role for DMBA-induced oxidative stress in the neonatal ovarian toxicity of DMBA. The same group subsequently reported similar results for the PAHs BaP and 3-MC [145, 146]. In addition, they reported that neonatal treatment with BaP (1.5 or 3.0 mg/kg/day for 7 days) leads to increased mitochondrial $O_2{-}\bullet$ generation measured using the fluorophore MitoSOX, increased lipid peroxidation measured using the fluorophore BODIBY, and decreased in vitro sperm binding and sperm fusion in oocytes ovulated at 6 weeks of age [145]. The latter result shows that a transient neonatal exposure to BaP causes persistent oxidative stress and decreased oocyte quality later in life.

4.4 Chromium (Cr)

Cr is a heavy metal that exists in several oxidation states. CrVI is a known carcinogen, which is rapidly reduced to the more stable CrIII in biological systems by GSH, ascorbic acid, and cysteine with concomitant generation of reactive Cr species and ROS [147]. Cr is used in numerous industrial processes including electroplating, leather tanning, steel production, and wood preservative manufacturing [148]. Exposure to the general population occurs due to contamination of drinking water supplies with Cr from industrial and natural sources in many parts of the world, including the United States [149]. During lactation, CrIII is excreted in breast milk, exposing the nursing offspring to CrIII.

Adult, lactational, and prenatal exposure to CrVI in drinking water all cause follicular atresia and deplete ovarian follicles in rats [150–153]. Administration of 50, 100, or 200 mg/L potassium dichromate (CrVI) via drinking water from the day of birth until weaning at PND 21 dose-dependently decreases primordial and primary follicle numbers in the ovaries of F1 offspring at PND 21, 45, and 65 and decreases secondary and antral follicle numbers at PND 21 and 45, with recovery of secondary and antral follicle numbers at PND 65 [150, 153]. Co-administration of 500 mg/L ascorbic acid in drinking water [153] or oral gavage with 500 mg/kg/day ascorbic acid [150] are protective against CrVI-induced follicle depletion and apoptosis. Plasma and ovarian hydrogen peroxide and lipid peroxidation product concentrations increase, enzymatic activities of GPX, GSR, SOD, and catalase, and concentrations of ascorbic acid decrease with CrVI dose, and ascorbic acid supplementation partially prevents these effects at PND 21, 45, and 65 [150]. In a similar study, treatment of lactating dams with 50 mg/L CrVI during the same window increased immunostaining for apoptosis markers, hydrogen peroxide, and lipid peroxidation products, and decreased enzymatic activity of antioxidant enzymes catalase, SOD, GPX, and GSR in ovaries of F1 females at PND 25, and co-treatment with the free radical scavenger edaravone mitigated these effects [154]. These results show that lactational plus direct exposure of the pups to CrVI in the drinking water induces systemic and ovarian oxidative stress and support a mechanistic role for oxidative stress in ovarian follicle depletion by CrVI during this developmental window.

Treatment of pregnant rat dams with 25 mg/L CrVI via the drinking water from GD 9.5 to 14.5 results in decreased germ cell numbers and increased percentages of germ cells remaining in nests at GD 15.5 and 17.5 and at PND 1, 4, and 25, and induces accelerated declines in fertility of the F1 female offspring [151, 155]. The decline in germ cell numbers is associated with increased ovarian immunostaining for proapoptotic markers TUNEL, cleaved caspase 3, BAX, and p53/p27 and decreased immunostaining for the antiapoptotic marker XIAP [151, 155]. Culture of GD 13.5 fetal rat ovaries for up to 12 days with 0.1 mg/L CrVI similarly leads to germ cell apoptosis, as well as inhibition of germ cell nest breakdown and primordial follicle formation [156]. In addition, products of lipid peroxidation increase and enzymatic activities of SOD, catalase, and GPX decrease in the culture media during the 12 days of culture in ovaries treated with CrVI compared to control ovaries, supporting a role for oxidative stress in the effects of CrVI on germ cell numbers and follicle formation during the fetal-neonatal window of development [156].

These ovary culture and in vivo studies are supported by studies of the effects of CrVI on apoptosis, cell proliferation, and oxidative stress in primary cultures of granulosa and theca cells from prepubertal rats and spontaneously immortalized granulosa cells (SIGC). Culture of rat granulosa cells with 10 μM (520 mg/L) potassium dichromate for 12 and 24 h induces apoptosis via the mitochondrial apoptotic pathway, induces cell cycle arrest, and inhibits proliferation, and these effects are largely prevented by co-treatment with ascorbic acid [157, 158]. Culture of primary granulosa cells, theca cells, and SIGCs with 10 μM (520 mg/L) CrVI for 12 or 24 h decreases intracellular ascorbic acid concentrations, increases hydrogen peroxide and lipid peroxidation product concentrations, decreases mRNA expression of the antioxidant genes *Sod1, Sod2, catalase, Glrx1, Gsr, Gstm1, Gstm2, Gsta4, Txn1, Txn2, Txnrd2,* and *Prdx3*, and enzymatic activity of GPX, GSR, and GST, and all of these effects are largely prevented by ascorbic acid pretreatment [150].

Overall, this body of work shows that prenatal or lactational exposure of rat dams to CrVI depletes ovarian germ cells and/or follicles in the F1 offspring by increasing ROS generation and decreasing antioxidant mRNA expression and enzymatic activities, resulting in oxidative stress, which induces apoptosis, inhibits proliferation, interferes with oocyte nest breakdown, and eventually results in premature ovarian failure. The ability of the antioxidant vitamin ascorbic acid and the free radical scavenger edavarone to prevent these effects further supports oxidative stress as the driver of CrVI-induced premature ovarian failure.

4.5 *Methoxychlor*

Methoxychlor is an organochlorine insecticide that was widely used as a replacement for DDT because it is more rapidly metabolized and less persistent. Although concern about its toxicity led to the banning of methoxychlor in the United States and European Union in 2002 [159], it is still widely used in other countries.

In vivo dosing of adult female mice with methoxychlor daily for 20 days at 32 mg/kg/day, but not lower doses, does not affect primordial, primary, or secondary follicle numbers, but induces atresia of antral follicles and significantly decreases the number of antral follicles in the ovaries [160]. The same methoxychlor dosing regimen increases ovarian hydrogen peroxide concentrations, oxidative DNA damage (8-OHdG immunostaining), and oxidative protein damage (NTY immunostaining), and decreased ovarian enzymatic activity and mRNA expression of the antioxidant enzymes GPX, catalase, and SOD1 [161]. In isolated, cultured mouse antral follicles, methoxychlor induces atresia and inhibits growth in a concentration- (1–100 μg/mL) and time- (0–96 h) dependent manner [162]. Expression of the antioxidant genes *Gpx1* and catalase increases at 48 h of exposure before atresia is evident, and declines, along with expression of *Sod2* by 96 h when the follicles display prominent signs of atresia [162]. Consistent with a role for oxidative stress in mediating the induction of atresia by methoxychlor in antral follicles, co-treatment of follicles with 1, 5 or 10 mM NAC prevents all of these aforementioned effects [162].

In mice, the neonatal period is a critical developmental window when primordial follicles are forming from oocyte nests [4]. The effects of methoxychlor on the ovary have been studied during this critical window in neonatal mice dosed with 50 or 100 mg/kg/day for 7 days [163]. The proportion of primordial follicles is decreased and the proportions of transitional and primary follicles are increased following treatment with the higher dose only [163]. These changes in follicle numbers are associated with upregulation of genes in pathways known to be involved in primordial follicle recruitment, in particular the phosphatidylinositol-3-kinase (PI3K)/AKT pathway [163]. Moreover, culture of neonatal mouse ovaries with methoxychlor for 96 h increases oxidative DNA damage [163], and ROS are known to increase AKT signaling [143, 144, 164]. These data suggest that methoxychlor accelerates recruitment of primordial follicles into the growing pool and that this involves upregulation of PI3K/AKT signaling by ROS. The same study also showed that neonatal in vivo dosing with methoxychlor dose-dependently decreases oocyte quality at 6 weeks of age, measured by decreased in vitro sperm binding and sperm fusion with zona pellucida-free oocytes and that incubation of superovulated oocytes with an ovotoxic methoxychlor metabolite increases lipid peroxidation measured by decreased ratio of red:green BODIPY fluorescence [163]. The latter results suggest that neonatal methoxychlor causes long-lasting oxidative damage to the oocyte plasma membrane that adversely affects sperm binding and fusion.

4.6 *Phthalates*

Phthalates are commonly used as plasticizers to impart flexibility and in cosmetics and other personal care products. Phthalates are diesters of *o*-phthalic acid containing hydrocarbon side chains of various lengths. Phthalates are metabolized to monoesters, which are generally thought to be the toxicologically active metabolites.

High doses (1 or 2 g/kg) of di(2-ethylhexyl) phthalate (DEHP) or dibutyl phthalate (DBP) in vivo decrease serum concentrations of 17β-estradiol, disrupt estrous cycling, and prevent ovulation in adult female rats [165]. In cultured rat granulosa cells DEHP and its monoester metabolite, mono(2-ethylhexyl) phthalate (MEHP, 50 μM) decrease mRNA and protein expression of aromatase, which is required to convert testosterone to 17β-estradiol [166]. Suppression of aromatase by MEHP occurs via activation of peroxisome proliferator activated receptors γ and α (PPARγ and PPARα), which are nuclear receptors and transcription factors [166]. Interestingly, other phthalates and their metabolites do not suppress aromatase expression, but do share another mechanism with DEHP/MEHP by which serum 17β-estradiol concentrations are decreased: they induce hepatic 17β-hydroxysteroid dehydrogenase Type IV activity, which metabolizes 17β-estradiol to its less active metabolite estrone [165]. MEHP induction of 17β-hydroxysteroid dehydrogenase Type IV appears to be mediated by activation of PPARα, but not PPARγ [166].

More recent studies have examined the ovarian effects of lower doses of DEHP and MEHP in mice. In vivo oral treatment of adult mice with 2 μg/kg/day to 750 mg/kg/day DEHP for 10 or 30 days decreased the percentage of primordial follicles and increased the percentage of primary follicles in a non-monotonic fashion, with the greatest alterations in follicle percentages observed at the 20 and 200 mg/kg doses at 10 days and the 200 μg/kg and 20 mg/kg doses after 30 days of dosing [167]. These effects suggest increased recruitment of primordial follicles into the growing pool, and this conclusion is supported by upregulation of PI3K/AKT signaling in the whole ovaries and in primordial and primary follicles [167]. Both DEHP (≥1.0 μg/mL, 2.6 μM) and MEHP (≥0.1 μg/mL, 0.4 μM) concentration- and time-dependently inhibit growth and induce atresia in cultured mouse antral follicles [168–170]. Growth inhibition is preceded by increased generation of ROS (dichlorofluorescein fluorescence), and co-treatment with NAC prevents the increase in ROS and growth inhibition, strongly supporting oxidative stress as an initiator of the adverse effects of DEHP and MEHP on antral follicle growth and survival [168, 169]. Although the role of oxidative stress in mediating the effects of in vivo DEHP treatment on primordial follicle recruitment have not been directly examined, activation of AKT signaling by ROS has been demonstrated in other systems. Taken together these studies in mice suggest that DEHP/MEHP increase primordial follicle recruitment into the growing pool and induce atresia in follicles at the antral stage of development.

5 Evidence From Maternal Adverse Nutrition Models

In recent years, a growing number of studies have shown that maternal under- or over-nutrition during pregnancy and/or lactation have myriad adverse developmental effects on the offspring, but relatively few studies have examined the effects on the ovary.

5.1 Maternal Under-Nutrition

Maternal protein restriction (10 % casein) during gestation and lactation reportedly did not affect F1 offspring secondary and antral follicle numbers at 21 days of age, while gestational or lactational only restriction significantly decreased both secondary and antral follicle numbers compared to females fed normal, 20 % casein diets (primordial and primary follicle counts were not reported) [171]. Maternal protein restriction (8 %) during lactation in another study reportedly decreased primordial and primary follicle numbers, and non-significantly decreased secondary and antral follicle numbers at 3 months of age compared to 23 % protein diet controls [172]. Pregnant Wistar rats were fed low (8 %) protein diet or normal, isocaloric 20 % protein diet during gestation only; pups were culled to 4 females and cross-fostered to normal calorie dams on PND 3, while control female offspring were also cross-fostered to control dams. Low protein F1 females had increased ovarian and oviductal lipid peroxidation, increased mtDNA copy number, and decreased ovarian and oviductal telomere length, and these were associated with lower primordial follicle density at 6 months of age and lower AMH (a serum marker of ovarian reserve that declines with age) at 3 months that was similar to control AMH at 6 months [173]. Overall, these studies suggest that maternal protein restriction during gestation and/or lactation decreases the ovarian reserve of primordial follicles and subsequent follicular development and induces ovarian oxidative stress, which may lead to more rapid onset of ovarian senescence.

5.2 Maternal Over-Nutrition During Gestation

Feeding 6 weeks old female mice a high fat diet with 22 % fat (0.15 % cholesterol), 19 % protein, and 49.5 % carbohydrates for 4 weeks causes lipid accumulation, oxidative and endoplasmic reticulum stress, and mitochondrial damage in the oocytes compared to mice fed a matched diet with 6 % fat and 64.7 % carbohydrate [174]. These effects are associated with significantly increased rates of anovulation and decreased rate of fertilization in the high fat diet fed mice [174]. The high fat diet fed mice also gained more weight and had higher serum glucose and cholesterol concentrations than mice fed the control diet.

Alms1 mutant mice ("Blobby" mice) develop obesity and insulin resistance on normal chow diets by 14 weeks of age due to hyperphagia. They have decreased ovulation rates in response to exogenous gonadotropins due to failure to release oocytes from preovulatory follicles. ER stress inhibitors salubrinal or BGP-15 rescue this decreased ovulation and also increase ovulation rate in normal mice [175]. The oocyte-cumulus complexes of these mice have increased expression of ER stress markers *Atf4*, *Hspa1a* and *1b*, increased lipid accumulation, decreased secretion of the ER product PTX3 (essential for ovulation), increased oocyte meiotic spindle abnormalities, decreased mitochondrial membrane potential, increased autophagy, and decreased percentages of normal ovulated oocytes and of early embryonic development of normal oocytes, and all of these effects were mitigated by salubrinal or BGP-15 treatment [175]. Fetuses from obese Blobby mothers transferred into uteri of normal mice were heavier and had decreased mtDNA copy numbers in liver and kidney and increased mtDNA mutations and these were also mitigated by salubrinal or BGP-15 treatment of mothers [175].

The data from these two models of maternal over-nutrition with increased weight gain, glucose intolerance and hyperlipidemia provide strong evidence that oxidative stress causes decreased oocyte quality and diminished fertility. It would be interesting for future studies to examine the impact of these diets on the age-related decline in ovarian follicle numbers in both the F0 and F1 generations.

6 Conclusions

As women have increasingly joined the work force in the past half century, delaying child-bearing until after the age of 35 has become more common. This has made understanding the normal process of ovarian aging, and especially mechanisms underlying accelerated ovarian aging, more urgent. Accelerated ovarian aging and premature ovarian failure also increase the risk for cardiovascular disease, osteoporosis, and Alzheimer's disease. Therefore, improved understanding of the pathophysiology of these conditions and developing interventions is important for overall health. This chapter has reviewed the accumulating evidence that oxidative stress is an important driver of both normal and pathological ovarian aging. While the cause of premature ovarian failure is not determined in the majority of clinical cases, research in animal models clearly demonstrates that diverse chemical, physical, and nutritional exposures are capable of accelerating the age-related decline in ovarian follicles and decreased oocyte quality that characterize premature ovarian failure. Moreover, multiple lines of research now demonstrate the importance of the prenatal and early postnatal environment on subsequent ovarian aging. Importantly, various antioxidants have demonstrated efficacy in preventing or mitigating the adverse ovarian effects of chemical toxicants, ionizing radiation, and adverse maternal nutrition in animal models. Public health interventions to reduce exposures that accelerate ovarian aging and chemoprevention studies in women at increased risk of premature ovarian failure are clearly warranted.

Acknowledgments This work was supported in part by grants from the National Institutes of Health (ES020454) and the National Aeronautic and Space Administration (NNX14AC50G).

References

1. Yao HH. The pathway to femaleness: current knowledge on embryonic development of the ovary. Mol Cell Endocrinol. 2005;230:87–93.
2. Pepling ME, Sundman EA, Patterson NL, Gephardt GW, Medico LJ, Wilson KJ. Differences in oocyte development and estradiol sensitivity among mouse strains. Reproduction. 2010;139:349–57.
3. Pepling ME. From primordial germ cell to primordial follicle: mammalian female germ cell development. Genesis. 2006;44:622–32.
4. Pepling ME. Follicular assembly: mechanisms of action. Reproduction. 2012;143:139–49.
5. Myers M, Morgan FH, Liew SH, Zerafa N, Gamage TU, Sarraj M, Cook M, Kapic I, Sutherland A, Scott CL, Strasser A, Findlay JK, Kerr JB, Hutt KJ. PUMA regulates germ cell loss and primordial follicle endowment in mice. Reproduction. 2014;148:211–9.
6. Zheng W, Nagaraju G, Liu Z, Liu K. Functional roles of the phosphatidylinositol 3-kinases (PI3Ks) signaling in the mammalian ovary. Mol Cell Endocrinol. 2012;356:24–30.
7. Parrott JA, Skinner MK. Kit-ligand/stem cell factor induces primordial follicle development and initiates folliculogenesis. Endocrinology. 1999;140:4262–71.
8. Durlinger ALL, Kramer P, Karels B, de Jong FH, Uilenbroek JTJ, Grootegoed JA, Themmen APN. Control of primordial follicle recruitment by anti-müllerian hormone in the mouse ovary. Endocrinology. 1999;140:5789–96.
9. White YAR, Woods DC, Takai Y, Ishihara O, Seki H, Tilly JL. Oocyte formation by mitotically active germ cells purified from ovaries of reproductive Age women. Nat Med. 2012. doi:10.1038/nm.2669.
10. Johnson J, Canning J, Kaneko T, Pru JK, Tilly JL. Germline stem cells and follicular renewal in the postnatal mammalian ovary. Nature. 2004;428:145–50.
11. Johnson J, Bagley J, Skaznik-Wikiel M, Lee H-J, Adams GB, Nikura Y, Tschudy KS, Tilly JC, Cortes ML, Forkert R, Spitzer T, Iacomini J, Scadden DT, Tilly JL. Oocyte generation in adult mammalian ovaries by putative germ cells in bone marrow and peripheral blood. Cell. 2005;122:303–15.
12. Gosden RG, Laing SC, Felicio LS, Nelson JF, Finch CE. Imminent oocyte exhaustion and reduced follicular recruitment mark the transition to acyclicity in aging C57BL/6J mice. Biol Reprod. 1983;28:255–60.
13. Faddy MJ, Gosden RG, Gougeon A, Richardson SJ, Nelson JF. Accelerated disappearance of ovarian follicles in Mid-life: implications for forecasting menopause. Hum Reprod. 1992;7:1342–6.
14. Miller PB, Charleston JS, Battaglia DE, Klein NA, Soules MR. Morphometric analysis of primordial follicle number in pigtailed monkey ovaries: symmetry and relationship with Age. Biol Reprod. 1999;61:553–6.
15. Mandl AM, Shelton M. A quantitative study of oocytes in young and Old nulliparous laboratory rats. J Endocrinol. 1959;18:444–50.
16. Jones KP, Walker LC, Anderson D, Lacreuse A, Robson SL, Hawkes K. Depletion of ovarian follicles with Age in chimpanzees: similarities to humans. Biol Reprod. 2007;77:247–51.
17. Cloutier CT, Coxworth JE, Hawkes K. Age-related decline in ovarian follicle stocks differ between chimpanzees (*Pan troglodytes*) and humans. Biol Reprod. 2015;37:10.
18. Hirshfield AN. Size-frequency analysis of atresia in cycling rats. Biol Reprod. 1988;38:1181–8.
19. Keefe DL, Liu L. Telomeres and reproductive aging. Reprod Fertil Dev. 2009;21:10–4.
20. Ottolenghi C, Uda M, Hamatani T, Crisponi L, Garcia J-E, Ko M, Pilia G, Sforza C, Schlessinger D, Forabosco A. Aging of oocyte, ovary, and human reproduction. Ann N Y Acad Sci. 2004;1034:117–31.
21. Broekmans FJ, Knauff EAH, te Velde ER, Macklon NS, Fauser BC. Female reproductive ageing: current knowledge and future trends. Trends Endocrinol Metab. 2007;18:58–65.

22. Eichenlaub-Ritter U, Vogt E, Yin H, Gosden R. Spindles, mitochondria and redox potential in ageing oocytes. Reprod Biomed Online. 2003;8:45–58.
23. Tarín JJ, Peréz-Albalá S, Cano A. Cellular and morphological traits of oocytes retrieved from aging mice after exogenous ovarian stimulation. Biol Reprod. 2001;65:141–50.
24. Tarín JJ, Pérez-Albalá S, Cano A. Oral antioxidants counteract the negative effects of female aging on oocyte quantity and quality in the mouse. Mol Reprod Dev. 2002;61:385–97.
25. Tarín JJ. Potential effects of Age-associated oxidative stress on mammalian oocytes/embryos. Mol Hum Reprod. 1996;2:717–24.
26. Titus S, Li F, Stobezki R, Akula K, Unsal E, Jeong K, Dickler M, Robson M, Moy F, Goswami S, Oktay K. Impairment of BRCA1-related DNA double-strand break repair leads to ovarian aging in mice and humans. Sci Transl Med. 2013;5:ra21.
27. Liu J, Wang W, Sun X, Liu L, Jin H, Li M, Witz C, Williams D, Griffith J, Skorupski J, Haddad G, Gill J. DNA microarray reveals that high proportions of human blastocysts from women of advanced maternal Age Are aneuploid and mosaic. Biol Reprod. 2012;87(148):141–9.
28. van Noord PAH, Boersma H, Dubas JS, te Velde ER, Dorland M. Age at natural menopause in a population-based screening cohort: the role of menarche, fecundity, and lifestyle factors. Fertil Steril. 1997;68:95–102.
29. Dorjgochoo T, Gao Y-T, Chow W-H, X-o S, Yang G, Cai Q, Rothman N, Cai H, Li H, Deng X, Shrubsole MJ, Murff H, Milne G, Zheng W, Dai Q. Obesity, Age, and oxidative stress in middle-aged, and older women. Antioxid Redox Signal. 2011;14:2453–60.
30. Das S, Chattopadhyay R, Ghosh S, Ghosh S, Goswami SK, Chakravarty BN, Chaudhury K. Reactive oxygen species level in follicular fluid—Embryo quality marker in IVF? Hum Reprod. 2006;21:2403–7.
31. Attaran M, Pasqualatto EB, Falcone T, Goldberg JM, Miller KF, Agarwal A, Sharma RK. The effect of follicular fluid reactive oxygen species on the outcome of in vitro fertilization. Int J Fertil. 2000;45:314–20.
32. Tatone C, Amicarelli F, Carbone MC, Monteleone P, Caserta D, Marci R, Artini PG, Piomboni P, Focarelli R. Cellular and molecular aspects of ovarian follicle ageing. Hum Reprod Update. 2008;14:131–42.
33. Wiener-Megnazi Z, Vardi L, Lissak A, Shnizer S, Reznick AZ, Ishai D, Lahav-Baratz S, Shiloh H, Koifman M, Dirnfeld M. Oxidative stress indices in follicular fluid as measured by the thermochemi-luminescence assay correlate with outcome parameters in in vitro fertilization. Fertil Steril. 2004;82 Suppl 3:1171–6.
34. Carbone MC, Tatone C, Monache SD, Marci R, Caserta D, Colonna R, Amicarelli F. Antioxidant enzymatic defences in human follicular fluid: characterization and Age-dependent changes. Mol Hum Reprod. 2003;9:639–43.
35. Matos L, Stevenson D, Gomes F, Silva-Carvalho JL, Almeida H. Superoxide DIsmutase expression in human cumulus oophorus cells. Mol Hum Reprod. 2009;15:411–9.
36. Tatone C, Carbone MC, Falone S, Aimola P, Giardinelli A, Caserta D, Marci R, Pandolfi A, Ragnelli AM, Amicarelli F. Age-dependent changes in the expression of superoxide dismutases and catalase Are associated with Ultrastructural modifications in human granulosa cells. Mol Reprod Dev. 2006;12:655–60.
37. Grøndahl ML, Yding Andersen C, Bogstad J, Nielsen FC, Meinertz H, Borup R. Gene expression profiles of single human mature oocytes in relation to Age. Hum Reprod. 2010;25:957–68.
38. Reeg S, Grune T. Protein oxidation in aging: does It play a role in aging progression? Antioxid Redox Signal. 2015;23:239–55.
39. Lim J, Luderer U. Oxidative damage increases and antioxidant gene expression decreases with aging in the mouse ovary. Biol Reprod. 2011;84:775–82.
40. Yeh J, Bowman MJ, Browne RW, Chen N. Reproductive aging results in a reconfigured ovarian antioxidant defense profile in rats. Fertil Steril. 2005;84 Suppl 2:1109–13.
41. Pan H, Ma P, Zhu W, Schultz RM. Age-associated increase in aneuploidy and changes in gene expression in mouse eggs. Dev Biol. 2008;316:397–407.

42. Hamatani T, Falco G, Carter MG, Akutsu H, Stagg CA, Sharov AA, Dudekula DB, VanBuren V, Ko MSH. Age-associated alteration of gene expression patterns in mouse oocytes. Hum Mol Genet. 2004;13:2263–78.
43. Lord T, Martin JH, Aitken RJ. Accumulation of electrophilic aldehydes during postovulatory aging of mouse oocytes causes reduced fertility, oxidative stress, and apoptosis. Biol Reprod. 2015;92:33.
44. Miyamoto K, Sato EF, Kasahara E, Jikumara M, Hiramoto K, Tabata H, Katsuragi M, Odo S, Utsumi K, Inoue M. Effect of oxidative stress during repeated ovulation on the structure and functions of the ovary, oocytes, and their mitochondria. Free Radic Biol Med. 2010;49:674–81.
45. Tarín JJ, Pérez-Albalá S, Pertusa JF, Cano A. Oral administration of pharmacological doses of vitamins C and E reduces reproductive fitness and impairs the ovarian and uterine functions of female mice. Theriogenology. 2002;57:1539–50.
46. Liu J, Liu M, Ye X, Liu K, Huang J, Wang L, Ji G, Liu N, Tang X, Baltz JM, Keefe DL, Liu L. Delay in oocyte aging in mice by the antioxidant *N*-acetyl-L-cysteine (NAC). Hum Reprod. 2012;27:1411–20.
47. Liu M, Yin Y, Ye X, Zeng M, Zhao Q, Keefe DL, Liu L. Resveratrol protects against Age-associated infertility in mice. Reproduction. 2013;28:707–17.
48. Pirola L, Fröjdö S. Resveratrol: One molecule, many targets. IUBMB Life. 2008;60:323–32.
49. Baur JA, Pearson KJ, Price NL, Jamieson HA, Lerin C, Kalra A, Prabhu VV, Allard JS, Lopez-Lluch G, Lewis K, Pistell PJ, Poosala S, Becker KG, Boss O, Gwinn D, Wang M, Ramaswamy S, Fishbein KW, Spencer RG, Lakatta EG, Le Couteur D, Shaw RJ, Navas P, Puigserver P, Ingram DK, de Cabo R, Sinclair DA. Resveratrol improves health and survival of mice on a high-calorie diet. Nature. 2006;444:337–42.
50. Lord T, Nixon B, Jones KT, Aitken JR. Melatonin prevents postovulatory oocyte aging in the mouse and extends the window for optimal fertilization in vitro. Biol Reprod. 2013;88(67):61–9.
51. Rebar RW. Mechanisms of premature menopause. Endocrinol Metab Clin North Am. 2005;34:923–33.
52. Nelson LM. Primary ovarian insufficiency. N Engl J Med. 2009;360:606–14.
53. Qin Y, Jiao X, Simpson JL, Chen Z-J. Genetics of primary ovarian insufficiency: New developments and opportunities. Hum Reprod Update. 2015;21(6):787–808.
54. Murray A, Bennett CE, Perry JRB, Weedon MN, Jacobs PA, Morris DH, Orr N, Schoemaker MJ, Jones M, Ashworth A, Swerdlow AJ, ReproGen Consortium. Common genetic variants Are significant risk factors for early menopause: results from the breakthrough generations study. Hum Mol Genet. 2011;20(1):186–92.
55. Wallace WHB, Thomson AB, Kelsey TW. The radiosensitivity of the human oocyte. Hum Reprod. 2003;18:117–21.
56. Wallace WHB, Shalet SM, Hendry JH, Morris-Jone PH, Gattamaneni HR. Ovarian failure following abdominal irradiation in childhood: the radiosensitivity of the human oocyte. Br J Radiol. 1989;62:995–8.
57. Lo Presti A, Ruvulo G, Gancitano RA, Cittadini E. Ovarian function following radiation and chemotherapy for cancer. Eur J Obstet Gynecol Reprod Biol. 2004;113:S33–40.
58. Molina JR, Barton DL, Loprinzi CL. Chemotherapy-induced ovarian failure: manifestations and management. Drug Saf. 2005;28:401–16.
59. Green DM, Kawashima T, Stovall M, Leisenring W, Sklar CA, Mertens AC, Donaldson SS, Byrne J, Robison LL. Fertility of female survivors of childhood cancer: a report from the childhood cancer survivor study. J Clin Oncol. 2009;27:2677–85.
60. Chemaitilly W, Mertens AC, Mitby P, Whitton J, Stovall M, Yasui Y, Robison LL, Sklar CA. Acute ovarian failure in the childhood cancer survivor study. J Clin Endocrinol Metab. 2006;91:1723–8.
61. Lim J, Nakamura BN, Mohar I, Kavanagh TJ, Luderer U. Glutamate cysteine ligase modifier subunit (*gclm*) null mice have increased ovarian oxidative stress and accelerated Age-related ovarian failure. Endocrinology. 2015;156(9):3329–43.

62. Nakamura BN, Fielder TJ, Hoang YD, Lim J, McConnachie LA, Kavanagh TJ, Luderer U. Lack of maternal glutamate cysteine ligase modifier subunit (*gclm*) decreases oocyte glutathione concentrations and disrupts Preimplantation development in mice. Endocrinology. 2011;152:2806–15.
63. Madden JA, Hoyer PB, Devine PJ, Keating AF. Involvement of a volatile metabolite during phosphoramide mustard-induced ovotoxicity. Toxicol Appl Pharmacol. 2014;277:1–7.
64. Zhu H, Itoh K, Yamamoto M, Zweier JL, Li Y. Role of Nrf2 signaling in regulation of antioxidants and phase 2 enzymes in cardiac fibroblasts: protection against reactive oxygen and nitrogen species-induced cell injury. FEBS Lett. 2005;579:3029–36.
65. Ramos-Gomez M, Dolan PM, Itoh K, Yamamoto M, Kensler TW. Interactive effects of nrf2 genotype and Oltipraz on benzo[a]pyrene DNA adducts and tumor yield in mice. Carcinogenesis. 2003;24:461–7.
66. Chan JY, Kwong M. Impaired expression of glutathione synthetic enzyme genes in mice with targeted deletion of the Nrf2 basic-leucine zipper protein. Biochim Biophys Acta. 2000;1517:19–26.
67. Lim J, Ortiz L, Nakamura BN, Hoang YD, Banuelos J, Flores VN, Chan JY, Luderer U. Effects of deletion of the transcription factor *Nrf2* and benzo[a]pyrene treatment on ovarian follicles and ovarian surface epithelial cells in mice. Reprod Toxicol. 2015;58:24–32.
68. Hu X, Roberts JR, Apopa PL, Kan YW, Ma Q. Accelerated ovarian failure induced by 4-vinyl cyclohexene diepoxide in Nrf2 null mice. Mol Cell Biol. 2006;26:940–54.
69. Uda M, Ottolenghi C, Crisponi L, Garcia J-E, Deiana M, Kimber W, Forabosco A, Cao A, Schlessinger D, Pilia G. *Foxl2* disruption causes mouse ovarian failure by pervasive blockage of follicle development. Hum Mol Genet. 2004;13:1171–81.
70. Schmidt D, Ovitt CE, Anlag K, Fehsenfeld S, Gredsted L, Treier A-C, Treier M. The murine winged-helix transcription factor Foxl2 is required for granulosa cell differentiation and ovary maintenance. Development. 2004;131:933–42.
71. Benayoun BA, Dipietromaria A, Bazin C, Veitia RA. FOXL2: at the crossroads of female sex determination and ovarian function. In: Forkhead transcription factors: vital elements in biology and medicine. Landes Bioscience and Springer; 2009. pp. 207–226.
72. Benayoun BA, Batista F, Auer J, Dipietromaria A, L'Hote D, De Baere E, Veitia RA. Positive and negative feedback regulates the transcription factor FOXL2 in response to cell stress: evidence for a regulatory imbalance induced by disease-causing mutations. Hum Mol Genet. 2009;18:632–44.
73. Batista F, Vaiman D, Dausset J, Fellous M, Veitia RA. Potential targets of FOXL2, a transcription factor involved in craniofacial and follicular development, identified by transcriptomics. Proc Natl Acad Sci U S A. 2007;104:3330–5.
74. Benayoun BA, Georges AB, L'Hote D, Andersson N, Dipietromaria A, Todeschini A-L, Caburet S, Bazin C, Anttonen M, Veitia RA. Transcription factor FOXL2 protects granulosa cells from stress and delays cell cycle: role of its regulation by the SIRT1 deacetylase. Hum Mol Genet. 2011;20:1673–86.
75. Sui X-X, Luo L-L, Xu J-J, Fu Y-C. Evidence that FOXO3a is involved in oocyte apoptosis in the neonatal Rat ovary. Biochem Cell Biol. 2010;88:621–8.
76. Castrillon DH, Miao L, Kollipara R, Horner JW, DePinho RA. Suppression of ovarian follicle activation in mice by the transcription factor Foxo3a. Science. 2003;301:215–8.
77. Di Emidio G, Falone S, Vitti M, D'Alessandro AM, Vento M, Di Pietro C, Amicarelli F, Tatone C. SIRT1 signalling protects mouse oocytes against oxidative stress and is deregulated during aging. Hum Reprod. 2014;29:2006–17.
78. Lee CJ, Park HH, Do BR, Yoon YD, Kim JK. Natural and radiation-induced degeneration of the primordial and primary follicles in the mouse ovary. Anim Reprod Sci. 2000;59:109–17.
79. Lee CJ, Yoon Y-D. ϒ-radiation-induced follicular degeneration in the prepubertal mouse ovary. Mutat Res. 2005;578:247–55.
80. Ward JF. The complexity of DNA damage: relevance to biological consequences. Int J Radiat Biol. 1994;66:427–32.

81. Ward JF. DNA damage produced by ionizing radiation in mammalian cells: identities, mechanisms of formation, and reparability. Prog Nucleic Acid Res Mol Biol. 1988;35:95–125.
82. Riley PA. Free radicals in biology: oxidative stress and the effects of ionizing radiation. Int J Radiat Biol. 1994;65:27–33.
83. Kim JK, Lee CJ. Effect of exogenous melatonin on the ovarian follicles in ϒ-irradiated mouse. Mutat Res. 2000;449:33–9.
84. Kujjo LL, Ronningen R, Ross P, Pereira RJG, Rodriguez R, Beyhan Z, Goissis MD, Baumann T, Kagawa W, Camsari C, Smith GW, Kurumizaka H, Yokoyama S, Cibelli JB, Perez GI. RAD51 plays a crucial role in halting cell death program induced by ionizing radiation in bovine oocytes. Biol Reprod. 2012;86:76.
85. Livera G, Petre-Lazar B, Guerquin M-J, Trautmann E, Coffigny H, Habert R. p63 null mutation protects mouse oocytes from radio-induced apoptosis. Reproduction. 2008;135:3–12.
86. Guerquin M-J, Duquenne C, Coffigny H, Rouiller-Fabre V, Lambrot R, Bakalska M, Frydman R, Habert R, Livera G. Sex-specific differences in fetal germ cell apoptosis induced by ionizing radiation. Hum Reprod. 2009;24:670–8.
87. Hanoux V, Pairault C, Bakalska M, Habert R, Livera G. Caspase-2 involvement during ionizing radiation-induced oocyte death in the mouse ovary. Cell Death Differ. 2007;14:671–81.
88. Kerr JB, Hutt KJ, Michalak EM, Cook M, Vandenberg CJ, Liew SH, Bouillet P, Mills A, Scott CL, Findlay JK, Strasser A. DNA damage-induced primordial follicle oocyte apoptosis and loss of fertility require TAp63-mediated induction of *puma* and *noxa*. Mol Cell. 2012;48:343–52.
89. Cortés-Wanstreet MM, Giedzinski E, Limoli CL, Luderer U. Overexpression of glutamate cysteine ligase protects human COV434 granulosa tumor cells against oxidative and γ-radiation-induced cell death. Mutagenesis. 2009;24:211–24.
90. Nitta Y, Hoshi M. Relationship between oocyte apoptosis and ovarian tumors induced by high and Low LET radiations in mice. Int J Radiat Biol. 2003;79:241–50.
91. Mishra B, Ortiz L, Luderer U. Charged Iron Particles, Typical of Space Radiation, Destroy Ovarian Follicles. Hum. Reprod. 2016; epub ahread of print. doi:10.1093/humrep/dew126.
92. Byrne J. Long-term genetic and reproductive effects of ionizing radiation and chemotherapeutic agents on cancer patients and their offspring. Teratology. 1999;59:210–5.
93. Howell S, Shalet S. Gonadal damage from chemotherapy and radiotherapy. Endocrinol Metab Clin North Am. 1998;27:927–43.
94. Meirow D, Nugent D. The effects of radiotherapy and chemotherapy on female reproduction. Hum Reprod Update. 2001;7:535–43.
95. Nicosia SV, Matus-Riley M, Meadows AT. Gonadal effects of cancer therapy in girls. Cancer. 1985;55:2364–72.
96. Davis BJ, Maronpot RR. Chemically associated toxicity and carcinogenicity of the ovary. In: Huff J, Boyd J, Barrett JC, editors. Cellular and molecular mechanisms of hormonal carcinogenesis: environmental influences, in. New York: Wiley-Liss; 1996. p. 285–308.
97. Jarrell J, Young Lai EV, Barr R, McMahon A, Belbeck L, O'Connell G. Ovarian toxicity of cyclophosphamide alone and in combination with ovarian irradiation in the Rat. Cancer Res. 1987;47:2340–3.
98. Kumar R, Biggart JD, McEvoy J, McGeown MG. Cyclophosphamide and reproductive function. Lancet. 1972;299:1212–4.
99. Plowchalk DR, Mattison DR. Phosphoramide mustard is responsible for the ovarian toxicity of cyclophosphamide. Toxicol Appl Pharmacol. 1991;107:472–81.
100. Shiromizu K, Thorgeirsson SS, Mattison DR. Effect of cyclophosphamide on oocyte and follicle number in Sprague-Dawley rats, C57BL/6N and DBA/2N mice. Pediatr Pharmacol. 1984;4:213–21.
101. Warne GL, Fairley KF, Hobbs JB, Martin FIR. Cyclophosphamide-induced ovarian failure. N Engl J Med. 1973;289:1159–62.
102. Meirow D, Lewis H, Nugent D, Epstein M. Subclinical depletion of primordial follicular reserve in mice treated with cyclophosphamide: clinical importance and proposed accurate investigative tool. Hum Reprod. 1999;14:1903–7.

103. Davis BJ, Heindel JJ. Ovarian toxicants: multiple mechanisms of action. In: Korach KS, editor. Reproductive and developmental toxicology. New York: Marcel Dekker, Inc; 1998. p. 373–95.
104. Lopez SG, Luderer U. Effects of cyclophosphamide and buthionine sulfoximine on ovarian glutathione and apoptosis. Free Radic Biol Med. 2004;36:1366–77.
105. Dirven HAAM, van Ommen B, van Bladeren PJ. Involvement of human glutathione S-transferase isoenzymes in the conjugation of cyclophosphamide metabolites with glutathione. Cancer Res. 1994;54:6215–20.
106. Gamcsik MP, Dolan ME, Andersson BS, Murray D. Mechanisms of resistance to the toxicity of cyclophosphamide. Curr Pharm Des. 1999;5:587–605.
107. Chang TK, Weber GF, Crespi CL, Waxman DJ. Differential activation of cyclophosphamide and ifosphamide by cytochromes P-450 2B and 3A in human liver microsomes. Cancer Res. 1993;53:5629–37.
108. Desmeules P, Devine PJ. Characterizing the ovotoxicity of cyclophosphamide metabolites on cultured mouse ovaries. Toxicol Sci. 2006;90:500–9.
109. Petrillo SK, Desmeules P, Truong T-Q, Devine PJ. Detection of DNA damage in oocytes of small ovarian follicles following phosphoramide mustard exposures of cultured rodent ovaries *in vitro*. Toxicol Appl Pharmacol. 2011;253:84–102.
110. Lloyd DR, Carmichael PL, Phillips DH. Comparison of the formation of 8-hydroxy-2'-deoxyguanosine and single- and double-strand breaks in DNA mediated by Fenton reactions. Chem Res Toxicol. 1998;11:420–7.
111. Devine PJ, Perreault SD, Luderer U. Roles of reactive oxygen species and antioxidants in ovarian toxicity. Biol Reprod. 2012;86:1–10.
112. Flowers J, Ludeman SM, Gamcsik MP, Colvin OM, Shao K-L, Boal JH, Springer JB, Adams DJ. Evidence for a role of chloroethylaziridine in the cytotoxicity of cyclophosphamide. Cancer Chemother Pharmacol. 2000;45:335–44.
113. Tsai-Turton M, Luong BT, Tan Y, Luderer U. Cyclophosphamide-induced apoptosis in COV434 human granulosa cells involves oxidative stress and glutathione depletion. Toxicol Sci. 2007;98:216–30.
114. Soleimani R, Heytens E, Darzynkiewicz Z, Oktay K. Mechanisms of chemotherapy-induced human ovarian aging: double strand DNA breaks and microvascular compromise. Aging (Milano). 2011;3:1–12.
115. Chu H-P, Liao Y, Novak JS, Hu Z, Merkin JJ, Shymkiv Y, Braeckman BP, Dorovkov MV, Nguyen A, Clifford PM, Nagele RG, Harrison DE, Ellis RE, Ryazanov AG. Germline quality control: eEF2K stands guard to eliminate defective oocytes. Dev Cell. 2014;28:561–72.
116. ATSDR. Toxicological profile for polycyclic aromatic hydrocarbons. Atlanta, GA: US Department of Health and Human Services, Public Health Service, Agency for Toxic Substances and Disease Registry; 1995.
117. Shopland DR, Burns DM, Benowitz NL, Amacher RH, editors. Risks associated with smoking cigarettes with low machine-measured yields of tar and nicotine. U.S. Department of Health and Human Services, Public Health Service, National Institutes of Health, National Cancer Institute; 2001.
118. Lu X, Kambe F, Cao X, Kozaki Y, Kaji T, Ishii T, Seo H. 3β-hydroxysteroid-Δ24 reductase is a hydrogen peroxide-scavenger, protecting cells from oxidative stress-induced apoptosis. Endocrinology. 2008;149:3267–73.
119. Lodovici M, Akpan V, Evangelisti C, Dolara P. Sidestream tobacco smoke as the main predictor of exposure to polycyclic aromatic hydrocarbons. J Appl Toxicol. 2004;24:277–81.
120. Menzie CA, Potocki BB, Santodonato J. Ambient concentrations and exposure to carcinogenic PAHs in the environment. Environ Sci Technol. 1992;26:1278–84.
121. Xue W, Warshawsky D. Metabolic activation of polycyclic aromatic hydrocarbon and heterocyclic aromatic hydrocarbons and DNA damage: a review. Toxicol Appl Pharmacol. 2005;206:73–93.
122. Jensen TK, Henriksen TB, Hjollund NH, Scheike T, Kolstad H, Giwercman A, Ernst E, Bonde JP, Skakkebaek NE, Olsen J. Adult and prenatal exposures to tobacco smoke as risk indicators of fertility among 430 Danish couples. Am J Epidemiol. 1998;148:992–7.

123. Baird DD, Wilcox AJ. Cigarette smoking associated with delayed conception. JAMA. 1985;253:2979–83.
124. Alderete E, Eskenazi B, Sholtz R. Effect of cigarette smoking and coffee drinking on time to conception. Epidemiology. 1995;6:403–8.
125. Hassan MAM, Killick SR. Negative lifestyle is associated with a significant reduction in fecundity. Fertil Steril. 2004;81:384–92.
126. Mattison DR, Plowchalk DR, Meadows MJ, Miller MM, Malek A, London S. The effect of smoking on oogenesis, fertilization, and implantation. Semin Reprod Endocrinol. 1989;7:291–304.
127. Harlow BL, Signorello LB. Factors associated with early menopause. Maturitas. 2000;35:3–9.
128. Tuttle AM, Stampfli M, Foster WG. Cigarette smoke causes follicle loss in mice ovaries at concentrations representative of human exposure. Hum Reprod. 2009;24:1452–9.
129. Mattison DR. Difference in sensitivity of Rat and mouse primordial oocytes to destruction by polycyclic aromatic hydrocarbons. Chem Biol Interact. 1979;28:133–7.
130. Mattison DR, Thorgeirsson SS. Ovarian aryl hydrocarbon hydroxylase activity and primordial oocyte toxicity of polycyclic aromatic hydrocarbons in mice. Cancer Res. 1979;39:3471–5.
131. Mattison DR, White NB, Nightingale MR. The effect of benzo(a)pyrene on fertility, primordial oocyte number, and ovarian response to pregnant Mare's serum gonadotropin. Pediatr Pharmacol. 1980;1:143–51.
132. Borman SM, Christian PJ, Sipes IG, Hoyer PB. Ovotoxicity in female fischer rats and B6 mice induced by Low-dose exposure to three polycyclic aromatic hydrocarbons: comparison through calculation of an ovotoxic index. Toxicol Appl Pharmacol. 2000;167:191–8.
133. Matikainen T, Perez GI, Jurisicova A, Pru JK, Schlezinger JJ, Ryu H-Y, Laine J, Sakai T, Korsmeyer SJ, Casper RF, Sherr DH, Tillly JL. Aromatic hydrocarbon receptor-driven Bax gene expression is required for premature ovarian failure caused by biohazardous environmental chemicals. Nat Genet. 2001;28:355–60.
134. Mattison DR. Morphology of oocyte and follicle destruction by polycyclic aromatic hydrocarbons in mice. Toxicol Appl Pharmacol. 1980;53:249–59.
135. Tsai-Turton M, Nakamura BN, Luderer U. Induction of apoptosis by 9,10-dimethyl-1,2-benzanthracene (DMBA) in cultured preovulatory Rat follicles is preceded by a rise in reactive oxygen species and is prevented by glutathione. Biol Reprod. 2007;77:442–51.
136. MacKenzie KM, Angevine DM. Infertility in mice exposed in utero to benzo(a)pyrene. Biol Reprod. 1981;24:183–91.
137. Lim J, Lawson GW, Nakamura BN, Ortiz L, Hur JA, Kavanagh TJ, Luderer U. Glutathione-deficient mice have increased sensitivity to transplacental benzo[a]pyrene-induced premature ovarian failure and ovarian tumorigenesis. Cancer Res. 2013;73:908–17.
138. Igawa Y, Keating AF, Rajapaksa KS, Sipes IG, Hoyer PB. Evaluation of ovotoxicity induced by 7,12-dimethylbenz[a]anthracene and its 3,4-diol metabolite utilizing a Rat *in vitro* ovarian culture system. Toxicol Appl Pharmacol. 2009;234:361–9.
139. Bhattacharya P, Keating AF. Protective role for ovarian glutathione *S*-transferase isoform Pi during 7,12-dimethylbenz[a]anthracene-induced ovotoxicity. Toxicol Appl Pharmacol. 2012;260:201–8.
140. Madden JA, Hoyer PB, Devine PJ, Keating AF. Acute 7,12-dimethylbenz[a]anthracene exposure causes differential concentration-dependent follicle depletion and gene expression in neonatal Rat ovaries. Toxicol Appl Pharmacol. 2014;276:179–87.
141. Sobinoff AP, Mahony M, Nixon B, Roman SD, McLaughlin EA. Understanding the villain: DMBA-induced preantral ovotoxicity involves selective follicular destruction and primordial follicle activation through PI3K/Akt and mTOR signaling. Toxicol Sci. 2011;123:563–75.
142. Zhang X, Tang N, Hadden TJ, Rishi AK. Akt, foxo and regulation of apoptosis. Biochim Biophys Acta. 1813;2011:1978–86.
143. LaHair MM, Howe CJ, Rodriguez-Mora O, McCubrey JA, Franklin RA. Molecular pathways leading to oxidative stress-induced phosphorylation of Akt. Antiox Redox Signal. 2006;8:1749–56.

144. King SM, Quartuccio SM, Vanderhyden BC, Burdette JE. Early transformative changes in normal ovarian surface epithelium induced by oxidative stress require Akt upregulation, DNA damage and epithelial-stromal interaction. Carcinogenesis. 2013;34:1125–33.
145. Sobinoff AP, Pye V, Nixon B, Roman SD, McLaughlin EA. Jumping the Gun: smoking constituent BaP causes premature primordial follicle activation and impairs oocyte fusibility through oxidative stress. Toxicol Appl Pharmacol. 2012;260:70–80.
146. Sobinoff AP, Nixon B, Roman SD, McLaughlin EA. Staying alive: PI3K pathway promotes primordial follicle activation and survival in response to 3MC-induced ovotoxicity. Toxicol Sci. 2012;128:258–71.
147. O'Brien TJ, Ceryak S, Patierno SR. Complexities of chromium carcinogenesis: role of cellular response, repair, and recovery mechanisms. Mutat Res. 2003;533:3–36.
148. Keegan GM, Learmonth ID, Case CP. A systematic comparison of the actual, potential, and theoretical health effects of cobalt and chromium exposures from industry and surgical implants. Crit Rev Toxicol. 2008;38:645–74.
149. Sutton R. Chromium-6 in U.S. Tap Water. Environmental Working Group; 2011.
150. Stanley JA, Sivakumar KK, Nithy TK, Arosh JA, Hoyer PB, Burghardt RC, Banu SK. Postnatal exposure to chromium through Mother's milk accelerates follicular atresia in F1 offspring through increased oxidative stress and depletion of antioxidant enzymes. Free Radic Biol Med. 2013;61:179–96.
151. Sivakumar KK, Stanley JA, Arosh JA, Pepling ME, Burghardt RC, Banu SK. Prenatal exposure to chromium induces early reproductive senescence by increasing germ cell apoptosis and advancing germ cell cyst breakdown in the F1 offspring. Dev Biol. 2014;388:22–34.
152. Murthy RC, Junaid M, Saxena DK. Ovarian dysfunction in mice following chromium (VI) exposure. Toxicol Lett. 1996;89:147–54.
153. Banu SK, Samuel JB, Arosh JA, Burghardt RC, Aruldhas MM. Lactational exposure to hexavalent chromium delays puberty by impairing ovarian development, steroidogenesis, and pituitary hormone synthesis in developing Wistar rats. Toxicol Appl Pharmacol. 2008;232:180–9.
154. Stanley JA, Sivakumar KK, Arosh JA, Burghardt RC, Banu SK. Edavarone mitigates hexavalent chromium-induced oxidative stress and depletion of antioxidant enzymes while estrogen restores antioxidant enzymes in the Rat ovary in F1 offspring. Biol Reprod. 2014;91:12.
155. Banu SK, Stanley JA, Sivakumar KK, Arosh JA, Barhoumi R, Burghardt RC. Identifying a novel role for X-prolyl aminopeptidase (*xpnpep*) 2 in CrVI-induced adverse effects on germ cell nest breakdown and follicle development in rats. Biol Reprod. 2015;92:67.
156. Stanley JA, Arosh JA, Burghardt RC, Banu SK. A fetal whole ovarian culture model for the evaluation of CrVI-induced developmental toxicity during germ cell nest breakdown. Toxicol Appl Pharmacol. 2015;289(1):58–69. doi:10.1016/j.taap.2015.09.002.
157. Stanley JA, Lee J, Nithy TK, Arosh JA, Burghardt RC, Banu SK. Chromium-VI arrests cell cycle and decreases granulosa cell proliferation by down-regulating cyclin-dependent kinases (CDK) and Up-regulating CDK-inhibitors. Reprod Toxicol. 2011;32:112–23.
158. Banu SK, Stanley JA, Lee J, Stephen SD, Arosh JA, Hoyer PB, Burghardt RC. Hexavalent chromium-induced apoptosis of granulosa cells involves selective Sub-cellular translocation of Bcl-2 members, ERK1/2 and p53. Toxicol Appl Pharmacol. 2011;251:253–66.
159. USEPA. Update of human health ambient water quality criteria: methoxychlor 72-43-5. Washington, DC: US Environmental Protection Agency, Office of Science and Technology, Office of Water; 2015.
160. Borgeest C, Symonds DA, Mayer LP, Hoyer PB, Flaws JA. Methoxychlor May cause ovarian follicular atresia and proliferation of the ovarian surface epithelium in the mouse. Toxicol Sci. 2002;68:473–8.
161. Gupta RK, Schuh RA, Fiskum G, Flaws JA. Methoxychlor causes mitochondrial dysfunction and oxidative damage in the mouse ovary. Toxicol Appl Pharmacol. 2006;216:436–45.
162. Gupta RK, Miller KP, Babus JK, Flaws JA. Methoxychlor inhibits growth and induces atresia of antral follicles through an oxidative stress pathway. Toxicol Sci. 2006;93:382–9.

163. Sobinoff AP, Pye V, Nixon B, Roman SD, McLaughlin EA. Adding insult to injury: effects of xenobiotic-induced preantral ovotoxicity on ovarian development and oocyte fusibility. Toxicol Sci. 2010;118:653–66.
164. Sedding DG. FoxO transcription factors in oxidative stress response and ageing—a New fork on the Way to longevity? Biol Chem. 2008;389:279–83.
165. Lovekamp-Swan T, Davis BJ. Mechanisms of phthalate ester toxicity in the female reproductive system. Environ Health Perspect. 2003;111:139–45.
166. Lovekamp-Swan T, Jetten AM, Davis BJ. Dual activation of PPARα and PPARγ by mono-(2-ethylhexyl) phthalate in Rat ovarian granulosa cells. Mol Cell Endocrinol. 2003;201:133–41.
167. Hannon PR, Peretz J, Flaws JA. Daily exposure to Di(2-ethylhexyl) phthalate alters estrous cyclicity and accelerates primordial follicle recruitment potentially Via dysregulation of the phosphatidylinositol 3-kinase signaling pathway in adult mice. Endocrinology. 2014. doi:10.1095/biolreprod.114.119032.
168. Wang W, Craig ZR, Basavarajappa MS, Hafner KS, Flaws JA. Mono (2-ethylhexyl) phthalate induces oxidative stress and inhibits growth of mouse antral follicles. Biol Reprod. 2012;87:152.
169. Wang W, Craig ZR, Basavarajappa MS, Gupta RK, Flaws JA. Di(2-ethylhexyl) phthalate inhibits growth of mouse ovarian antral follicles through an oxidative stress pathway. Toxicol Appl Pharmacol. 2012;258:288–95.
170. Hannon PR, Brannick KE, Wang W, Gupta RK, Flaws JA. Di(2-ethylhexyl) phthalate inhibits antral follicle growth, induces atresia, and inhibits steroid hormone production in cultured antral follicles. Toxicol Appl Pharmacol. 2015;284:42–53.
171. Guzmán C, García-Becerra R, Aguilar-Medina MA, Méndex I, Merchant-Larios H, Zambrano E. Maternal protein restriction during pregnancy and/or lactation negatively affects follicular ovarian development and steroidogenesis in the prepubertal Rat offspring. Arch Med Res. 2014;45:294–300.
172. da Silva FT, de Bittencourt BF, Sampaio FJB, da Fonte Ramos C. Maternal malnutrition durnig lactation affects folliculogenesis, gonadotropins, and leptin receptors in adult rats. Nutrition. 2009;26:1000–7.
173. Aiken CE, Tarry-Adkins JL, Ozanne SE. Suboptimal nutrition *in utero* causes DNA damage and accelerated aging of the female reproductive tract. FASEB J. 2013;27:3959–65.
174. Wu LL-Y, Dunning KR, Yang X, Russell DL, Lane M, Norman RJ, Robker RL. High-Fat diet causes lipotoxicity responses in cumulus-oocyte complexes and decreased fertilization rates. Endocrinology. 2010;151:5438–45.
175. Wu LL, Russell DL, Wong SL, Chen M, Tsai T-S, St. John JC, Norman RJ, Febbraio MA, Carroll J, Robker RL. Mitochondrial dysfunction in oocytes of obese mothers: transmission to offspring and reversal by pharmacological endoplasmic reticulum stress inhibitors. Development. 2015;142:681–91.

Part III
Retardation of Cellular Aging

Anti-Oxidants, the Aging Brain and Age-Related Brain Disorders

Michel Baudry and Bernard Malfroy-Camine

1 Introduction

Oxidative stress affects all cells, tissues and organs, and the brain is no exception. It has long been recognized, through post mortem analysis of rodent, primate and human brains, that aged brains from all mammalian species exhibit hallmarks of oxidative stress, including lipid peroxidation and nucleic acid and protein oxidation. In addition, a number of studies have shown that oxidative damage to the brain increases with age. Accumulation of cellular damage over time due to oxidative stress may lead to impairments in higher brain function, including memory and cognition. An age-associated decline in learning and memory has been well documented in humans [1]. While in its mild form, age-related decline in memory function is not life-threatening, it becomes much more dramatic in the pathological form exhibited in patients with Alzheimer's disease. In addition, individuals presenting with mild cognitive impairment have a greater chance of developing Alzheimer's disease compared to the general population [2]. Such age-related loss of memory is not unique to humans, and is present in a variety of mammals, including rats and mice [3, 4]. While there is still no consensus regarding the nature of the biological process(es) that underlie(s) age-related decline in cognitive function, it has been frequently proposed that age-related accumulation of oxidative damage in neurons is the main culprit. In addition, oxidative stress has now been firmly implicated in

M. Baudry (✉)
Graduate College of Biomedical Sciences, Western University of Health Sciences, Pomona, CA, USA
e-mail: mbaudry@westernu.edu

B. Malfroy-Camine
MindSet Rx, Inc., Arlington, MA, USA

S.C. Bondy, A. Campbell (eds.), *Inflammation, Aging, and Oxidative Stress*, Oxidative Stress in Applied Basic Research and Clinical Practice,
DOI 10.1007/978-3-319-33486-8_13

various chronic neurodegenerative diseases, many of which associated with aging, including age-related mild cognitive impairment, Alzheimer's disease, Parkinson's disease, Huntington's disease, Friedreich's ataxia, ataxia telangiectasia, and prion-related disease, as all these diseases are associated with mitochondrial dysfunction [5]. Oxidative stress has also been shown to play a critical role in acute neurodegeneration, such as in stroke or traumatic brain injury [6]. Finally, oxidative stress may also be involved in the etiology of certain psychiatric diseases, namely schizophrenia [7] and autism [8]. As many excellent reviews have been devoted to the topic of age-related oxidative stress and neurodegenerative disorders, this review will first focus on the use of antioxidants to reduce oxidative stress associated with normal aging. We will then discuss how the same strategies could apply to the treatment of age-related disorders where enhanced oxidative stress has also been implicated. Finally, we will discuss potential reasons for the failure of the majority of clinical trials attempting to use antioxidants for the treatment of age-related neurodegenerative disorders, and propose the idea that the use of small catalytic scavengers could offer a promising alternative to that of small antioxidant molecules.

2 Age-Associated Increase in Oxidative Stress and Effects of Antioxidants

Aging is associated with increased free radical levels and damage associated with oxidative stress in mammalian brain, including lipid peroxidation, protein oxidation, and oxidized nucleic acids [9–14]. The reason for this age-related increase in oxidative stress in the brains of aged mammals is still unclear, although mitochondria preparations from the brains of aged rodents exhibit a significant age-dependent increase in superoxide and hydrogen peroxide production [14, 15] and several studies have reported age-associated decreases in superoxide dismutase activity and catalase activity in rat brain [16]. Supplementing diets of aging mammals with antioxidants or free radical scavengers has generally been shown to attenuate age-related cognitive decline and oxidative stress [17–19]. Such findings support the idea that free radical accumulation might indeed impair memory function. The decline in learning and memory aged mice exhibit when tested with the spatial swim maze has been correlated with an increase in protein carbonyl content in cerebral cortex [4]. Lipid peroxidation is also significantly much higher in hippocampus and inferior parietal lobule of elderly individuals who exhibit mild cognitive impairment [20]. Aged rats that perform just as well as young rats in spatial learning tasks do not show the increased levels of nucleic acid and protein oxidation in hippocampus that aged cognitively impaired rats do. Furthermore, age-dependent cognitive decline in rodents can be attenuated by treating them with antioxidants and free radical scavengers, such as N-tert-butyl-alpha-phenylnitrone [17], melatonin [18] and alpha-lipoic acid [19]. More recent studies have further evaluated the effects of several classes of antioxidants on cognitive processes in aging. Antioxidants have

Table 1 Classification of antioxidants

Different classes of antioxidants
• Enzymatic antioxidants: superoxide dismutase (SOD), catalase (CAT), and glutathione peroxidase (GSHpx) and glutathione reductase (GR)
• Small molecule scavengers: Vitamin E, Vitamin C, carotenoids, flavonoids, glutathioned, DHEA, CoQ10, resveratrol, etc.
• Large protein antioxidants: albumin, etc.
• Small catalytic scavengers: Mn-TBAP, EUKn

generally been categorized into several different classes (Table 1). Enzymatic antioxidants are the typical enzymes that eliminate the oxygen free radicals (Superoxide dismutase, SOD) hydrogen peroxide (Catalase, CAT), and the enzymes producing and regenerating glutathione (Glutathione peroxidase (GSHpx) and Glutathione Reductase (RH). A large number of small molecule scavengers, either naturally occurring as well as synthetic, have been shown to eliminate reactive oxygen species (ROS) in a stoichiometric manner, which means that large amount of these molecules are required to decrease oxidative stress under conditions of increased ROS production. In addition, a small number of large proteins, such as albumin, are capable of scavenging ROS. Finally, several small catalytic scavengers of ROS and nitrogen oxygen species (NOS) have been evaluated for their effects on cognitive decline in aging.

As previously mentioned, Vitamin C has been repeatedly tested on age-related cognitive decline based on its antioxidant properties. In general, animal studies have all found that Vitamin C can reduce age-related cognitive deficits [21]. These results have prompted numerous clinical trials to test the effects of Vitamin C in cognitive decline associated with Alzheimer's disease or other conditions presenting cognitive impairment (see [22] and [23] for recent reviews). It is clear that in humans, the results have not been as conclusive as in animal studies, and, while it is possible that Vitamin C deficiency contributes to age-related cognitive decline, the effects of dietary intake of large doses of Vitamin C remain questionable.

Similarly Vitamin E is considered to be a powerful antioxidant, and has been directly or indirectly, through its presence in numerous dietary elements, tested for its beneficial effects on age-related cognitive decline. Like for Vitamin C, the conclusions from many studies is that it is not clear that Vitamin E has a clear significant effect on age-related cognitive decline and age-related dementia, including MCI and AD [24].

While selenium (Se) is not directly an antioxidant, it is an essential element of selenoproteins, in particular GSHpx, which are involved in ROS production and elimination. A recent French study, "Etude du Vieillissement Arteriel (EVA)" consisted in a 9-year longitudinal study to determine the relationship between cognitive decline, vascular disease and oxidative stress in the elderly [25]. The results suggested that decrease in plasma Se levels was associated with cognitive decline independently of vascular factors. The results of the clinical trial combining Vitamin E and Se (PRE-ADVISE trial) for the treatment of dementia are not yet available and more data are needed to support the beneficial effects of Se alone or in combination with other antioxidants on age-related decline in cognitive function.

Over the past many years we have been developing a class of compounds, usually referred to as Salen-Mn complexes, which have SOD-like [26] and catalase-like [27] activity. Previous work done using SOD2 knock-out mice provided indirect evidence that salen-manganese compounds such as EUK-8, EUK-134, or EUK-189 are able to cross the blood brain barrier and to be mito-protective [28]. Based on structure-activity studies [27], we designed another, cyclized analog, EUK-207 [29]. EUK-207 was found to have a significantly longer half-life than EUK-189 in mice, and was amongst the most effective compounds in extending the lifespan of SOD2 knock-out mice [27]. We used two of these Salen-Mn complexes, EUK-189 and EUK-207, to study the role of oxidative stress in normal aging of the brain [29]. As they age, mice develop deficits in learning and cognitive function which can be monitored by using fear conditioning. We found that there was a dramatic decrease in cognitive function from 8 to 11 months of age. Chronic treatment with EUK-189 and EUK-207 initiated at 8 months of age almost fully prevented this cognitive deficit. We also found that between 8 and 11 months of age the brain showed a significant increase in markers for oxidative stress, including lipid peroxidation and protein and nucleic acid oxidation. Chronic treatment with EUK-189 and EUK-207 significantly decreased all three markers for oxidative stress. Interestingly, there was a significant negative correlation between cognitive function and lipid peroxidation.

These data were confirmed and expanded in another study using older mice [30]. A 6 months treatment with low doses of EUK-189 and EUK-207 (approximately 15 and 150 μg/kg/day through Alzed minipumps implanted subcutaneously) dramatically reduced the cognitive deficit that occurs from 16 to 20 or 23 months of age and significantly decreased oxidative stress. In addition, again we found significant negative correlations between cognitive function and oxidative stress (Fig. 1).

Earlier work with a carboxyfullerene SOD mimetic showed a dramatic decrease in age-dependent learning and memory deficits and oxidative stress in mice [31]. The multiple catalytic activities of EUK-189 and EUK-207 might prove to be even more beneficial, and could account for the very low efficacious doses, as compared to that of the carboxyfullerene tested. Thus, a dose of the EUK compounds of about 15 μg/kg/day was as potent as a dose of 10 mg/kg/day of carboxyfullerene. In addition, it is important to note that the carboxyfullerene is only an SOD mimetic whereas the EUK compounds tested can also protect against damage caused by hydrogen peroxide, and through catalase-like mechanisms [27, 32], reactive nitrogen species.

These data fully support the notion that accumulation of oxidative stress-related damage plays a key role in brain aging, and in particular in age-related decline in cognitive function.

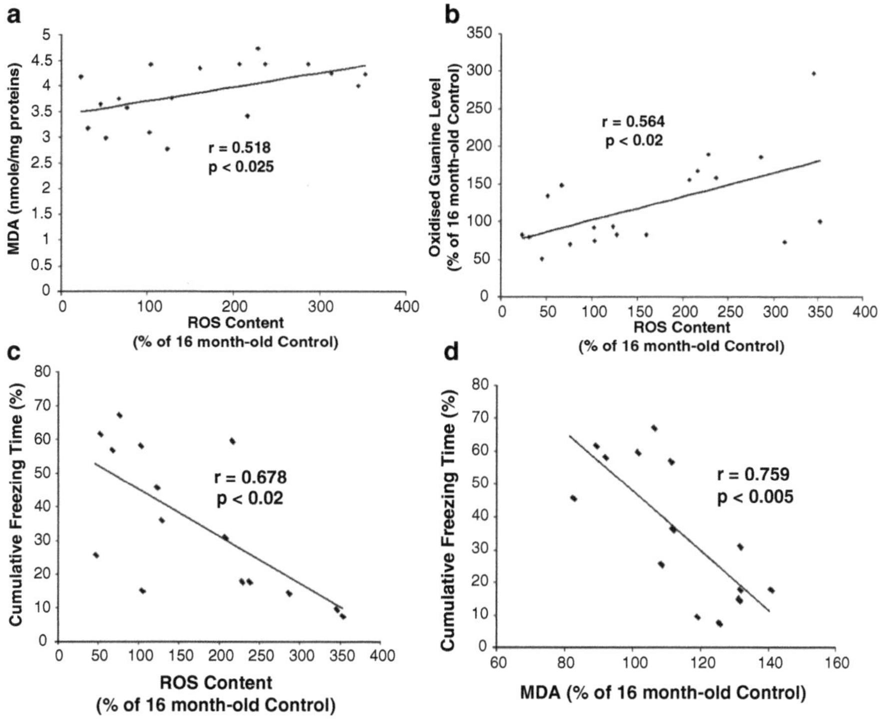

Fig. 1 Multiple correlations between markers of oxidative stress and learning performance in 24-month old mice. (**a**) Correlation between ROS content and lipid peroxidation. (**b**) Correlation between ROS content and oxidized DNA. (**c**) Correlation between ROS content and learning performance in the fear conditioning paradigm. (**d**) Correlation between lipid peroxidation and learning performance in the fear conditioning paradigm. Markers of oxidative stress in hippocampus of 24-month old mice (ROS content, Oxidized guanine, and MDA levels) were expressed as percent of their means values measured in 16-month old mice

3 Role of Oxidative Stress in Age-Related Disorders

The mechanisms linking age-related increase in oxidative stress are intimately related to those associated with neurodegeneration (Fig. 2). Under both conditions, the first step appears to be mitochondrial dysfunction, resulting in the increased production of ROS and reactive nitrogen species (RNS). The resulting oxidative stress then produces synaptic dysfunction and neuronal damage through pathways that are not yet completely elucidated. The background in which these events take place, such as age, genetic mutations, environmental factors, imposes additional constraints and determines the particular features of the neuronal damage. In this section, we will focus on three particular neurodegenerative disorders for which a role of oxidative stress and therefore the effects of antioxidants have been specifically determined.

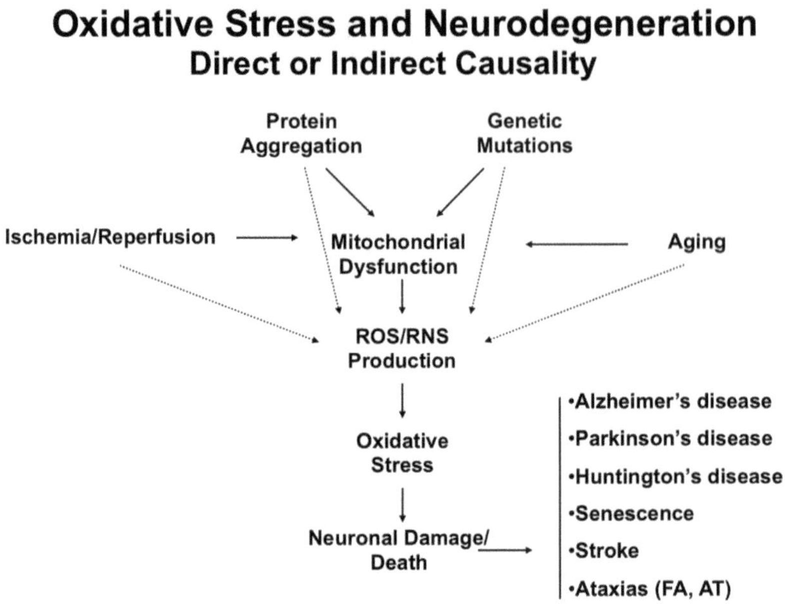

Fig. 2 Potential links between oxidative stress and neurodegeneration. A number of risk factors (aging, genetic mutations, ischemia, etc.) are postulated to produce mitochondrial dysfunction resulting in increased ROS/RNS formation. In turn, oxidative stress results in neuronal damage/death associated with many neurodegenerative disorders

3.1 Alzheimer's Disease

Alzheimer's disease (AD) is a progressive neurodegenerative disorder characterized by memory loss and cognitive deficits that ultimately lead to severe dementia [33]. AD is the most prominent form of dementia found in the elderly and the number of Americans inflicted with the disease is expected to reach nearly 15 million over the next several decades [34, 35]. AD pathology is characterized by the presence of senile plaques made up of aggregated extracellular amyloid β (Aβ) peptides and intracellular neurofibrillary tangles mainly composed of phosphorylated tau protein [33]. In addition, AD also results in synapse loss and neuronal death in hippocampus, entorhinal cortex, basal forebrain, and neocortical association cortices [36].

Over the past decade, a number of studies have revealed that oxidative stress might play a critical part in AD. Protein oxidation [37–39], lipid peroxidation [40–42], and DNA and RNA oxidation [43–46] are significantly elevated in AD brains. In addition, the Aβ peptide may be responsible for AD-associated oxidative stress, as it induces production of reactive oxygen species (ROS) in both neuronal [47, 48] and astrocyte cell cultures [47]. Protein oxidation is also significantly elevated in cultured hippocampal neurons incubated with a number of different Aβ peptides (see [49] for a review), and treating rat synaptic plasma membranes with $A\beta_{1-40}$

increased lipid peroxidation [50]. Oxidative stress could also potentially be responsible for the neurodegeneration observed in AD because ROS production accompanies Aβ-induced neuronal apoptosis and the antioxidants α-tocopherol and N-acetylcysteine inhibit Aβ-induced neuronal apoptosis [51]. Consistent with these studies, early on we had also found that the catalytic antioxidant EUK-8 could prevent Aβ-induced lipid peroxidation and neuronal death in organotypic hippocampal slice cultures [52].

While there is strong evidence linking AD with oxidative stress, whether or not oxidative stress is the initiator or a product and mediator of its pathogenesis remains unclear. Mitochondrial abnormalities have been shown to precede the development of neurofibrillary tangles in AD [53, 54], and in mouse models of AD [55]. Moreover, recent reports indicate that superoxide might play a causal role in several manifestations of AD [56, 57]. Part of the difficulty to test the role of oxidative stress in AD originates from the lack of quantitative information regarding the levels of free radicals that are continuously generated under physiological as well as pathological conditions, the low levels of antioxidants that can be administered, and their limited brain penetration [58]. Because signs of oxidative stress appear before the development of neurofibrillary tangles in AD and in mouse models of AD, it has been suggested that oxidative stress contributes to the initiation of the disease. In addition, recent evidence indicate that the amyloid ß peptides directly interact with mitochondria [59, 60], and it has also been proposed that continuous mitochondrial dysfunction and increased ROS production participate in the progression of AD pathology [61, 62]. In support of this mechanism, several reports have indicated that antioxidants ameliorate mitochondrial function [63–67]. However, very few studies have convincingly demonstrated the usefulness of antioxidant treatment in mouse models of AD or in human AD [58, 68–70].

Thus, while a relationship between AD and brain oxidative stress is clear, whether or not oxidative stress initiates AD pathogenesis or is a product/mediator of AD remains an open question. We recently published the results of a study designed to address this question [71]. As EUK-207 had previously been shown to significantly reduce age-associated cognitive impairment and oxidative stress in middle-aged and aged wild-type mice [29, 30], we used this compound as a tool to elucidate the relationship between oxidative stress and AD. In the first study, chronic treatment with EUK-207 was initiated at 4 months of age before the appearance of cognitive deficits and AD pathology in 3xTg-AD mice, and was maintained until cognitive deficits became apparent in vehicle-treated 3xTg-AD mice. Chronic EUK-207 treatment significantly reduced intraneuronal accumulation of ß-amyloid peptide, as well as tau and hyperphosphorlated tau levels in hippocampus and amygdala of 9 month-old 3xTg-AD mice. In addition, EUK-207 treatment reversed the increased levels of brain lipid peroxidation and oxidized guanine in hippocampus and amygdala of 3xTg-AD mice to values close to those found in wild-type animals. EUK-207 also reversed deficits in performance of 9 month-old 3xTg-AD mice in both the context and cue fear conditioning tests. The treatment also significantly decreased brain $A\beta_{1-42}$ levels. In addition, our results revealed high correlations between learning and memory performance and levels of lipid peroxidation and between levels of detergent-soluble $Aß_{1-42}$ and lipid peroxidation. Our results therefore clearly demonstrated that oxidative stress is a critical mediator in the development of AD-like pathology and cognitive impairment.

In the second study, chronic treatment with EUK-207 initiated at 9 months of age and continued for 3 months significantly reduced brain levels of lipid peroxidation assessed in 12 month-old mice, whether Non-Tg or 3xTg-AD. Interestingly, the levels of brain lipid peroxidation determined at 12 month of age in the 3xTg-AD mice were significantly higher than the values determined at 9 month, and EUK-207 treatment reversed them to those found in the 9 month-old vehicle-treated Non-Tg mice. These results indicated that this regimen of EUK-207 treatment almost completely prevented the increased in lipid peroxidation taking place in the 3xTg-AD between 9 and 12 months. Similarly, EUK-207 treatment significantly reduced the levels of DNA oxidation in various brain structures in both Non-Tg and 3xTg-AD. We particularly analyzed these changes in hippocampus and amygdala, as both structures are implicated in fear conditioning. Although it was difficult to compare the levels of oxidized DNA observed in the two studies, it is worth noting that the percent increase between the 3xTg-AD and the Non-Tg was higher in the second study, suggesting that the increase in DNA oxidation follows a similar pattern to that of lipid peroxidation. These results clearly indicate that EUK-207 treatment was effective in reducing oxidative stress in both Non-Tg and in 3xTg-AD. As in the first study, this reduction in oxidative load was associated with decreased intraneuronal accumulation of ß-amyloid peptide, as well as tau and hyperphosphorylated tau levels in hippocampus and amygdala of 12 month-old 3xTg-AD mice, as compared to vehicle-treated 3xTg-AD mice. From previous published studies with these mice, accumulation of ß-amyloid peptide and hyperphosphorylated tau is a continuous process once it is initiated, and older animals exhibit larger increases than younger ones. While it is difficult to assess the absolute degree of changes in ß-amyloid peptide and hyperphosphorylated tau in our two studies, it is interesting to note that the effect of EUK-207 appeared to be somewhat larger at 12 month than at 9 month, suggesting that the treatment does result in a slowing down of the accumulation of both ß-amyloid peptide and hyperphosphorylated tau. It is thus tempting to conclude that continuous production of reactive oxygen species is causally related to accumulation of ß-amyloid peptide and hyperphosphorylated tau. One study reported that ß-amyloid accumulation increased between 9 and 12 months of age in the same mouse model of AD [72], strongly supporting the idea that the development of this pathology is indeed a continuous process in this mouse model. Finally, EUK-207 treatment almost completely reversed the deficits in performance of 12 month-old 3xTg-AD mice in both the context and cue fear conditioning tests to values close to those found in vehicle-treated Non-Tg mice. Our results therefore clearly demonstrated that oxidative stress is a critical mediator in the development of cognitive impairment and suggested that the Aβ-amyloid plaques and hyperphosphorylated tau tangles, which are the hallmarks of AD, might be, at least in part, a consequence of oxidative stress rather than its direct cause.

Studies performed in the Tg2576 mouse model of AD also support our results. Like the 3xTg-AD mice, the Tg2576 mice develop cognitive impairments and

Aß-associated pathology in an age-dependent manner. In addition, these mice as well as AD patients develop cataract due to aggregation of Aβ in the lens [73, 74]. Chronic treatment with the catalytic antioxidant EUK-189 was found to significantly decrease lens opacification that occurs over time in these mice [74]. Overexpressing the mitochondrial form of superoxide dismutase in Tg2756 mice protects against learning and memory impairments observed in the Morris water maze and in contextual and cued fear conditioning, and significantly reduces the number of Aß plaques in brain [57]. In addition, supplementing the diets of Tg2576 mice with the antioxidant Vitamin E beginning at 4 months of age significantly reduced lipid peroxidation, levels of $A\beta_{1-40}$ and $A\beta_{1-42}$, and the number of amyloid plaques in the brain of 13 month-old Tg2576 mice [75]. However, the relationship between oxidative stress and the development of AD-associated tau pathology was not evaluated in these two reports because Tg2576 mice do not express mutant tau protein and therefore do not exhibit any neurofibrillary alterations.

EUK-207 treatment in 3xTg-AD mice reversed brain oxidative stress to levels that were similar to those found in Non-Tg mice. On the other hand, levels of Aβ peptides, tau, and hyperphosphorylated tau, while significantly reduced compared to those in 9 or 12 month-old vehicle-treated 3xTg-AD mice, were still elevated. This suggests that accumulation of these proteins is only partially due to oxidative stress, and that other factors might contribute to their accumulation. Nevertheless, although intraneuronal Aβ and hyperphosphorylated tau were still present in hippocampus and amygdala of EUK-207-treated mice, they still performed as well as the Non-Tg vehicle mice in both the context and cue test. These results suggest that Aβ and hyperphosphorylated tau can reach significant levels within neurons before synaptic function and plasticity is impaired.

3.2 Parkinson's Disease and Other Neurodegenerative Disorders

Parkinson's disease (PD) is a common neurodegenerative disorder that is pathologically characterized by the selective degeneration of dopaminergic (DAergic) neurons of the substantia nigra [76, 77]. A growing body of evidence suggests that oxidative stress induced by reactive oxygen species (ROS) is involved in this selective nigral cell degeneration [78–82]. Postmortem studies of PD patients have provided evidence for chemical changes that are indicative of oxidative stress in the substantia nigra. Such changes include increased levels of lipid peroxidation, protein oxidation, 3-nitrotyrosine formation, DNA oxidation and breaks, and decreased levels of ROS scavenging enzymes such as glutathione peroxidase (GSHpx) and catalase [83–86].

A number of neurotoxins that selectively damage DAergic neurons have been used to create animal models of PD. Injections of either 1-methyl-4-phenylpyridinium (MPP^+), the active metabolite of 1-methyl-4-phenyl-1,2,3,6-tetrahydropyridine (MPTP), or 6-hydroxydopamine (6-OHDA) have been shown to produce some of the biochemical and pathological changes that occur in PD [87, 88]. MPP^+ and 6-OHDA are selectively taken up by the plasma membrane dopamine (DA) transporters and subsequently accumulate within mitochondria. Both neurotoxins are presumed to increase the formation of ROS, e.g. hydrogen peroxide (H_2O_2), superoxide ($O_2^{\bullet-}$), and hydroxyl radicals ($^{\bullet}OH$). However, there are still a number of unanswered questions regarding the similarities and differences between the modes of actions of these two DAergic neuron toxins.

The free radical nitric oxide ($NO^{\bullet}$) readily interacts with $O_2^{\bullet-}$ to produce peroxynitrite ($ONOO^-$), a potent oxidant and nitrating agent [89, 90]. One consequence of increased $ONOO^-$ formation is protein nitration on tyrosine residues, and analysis of protein nitration has thus been used as a marker of oxidative stress under various experimental conditions [91–94]. $ONOO^-$ has been proposed as a mediator of nigrostriatal damage in PD [95–98], an idea supported by its effects on DA synthesis in mouse striatum [99].

We showed that the catalytic antioxidant EUK-134 completely blocked TH nitration and the subsequent decreases in DA uptake, number of THir neurons, and morphological alterations of DAergic neurons produced by MPP+ and 6-OHDA, suggesting a potential therapeutic role for EUK-134 in the treatment of PD [100].

Environmental factors such as the herbicide paraquat and neonatal iron exposure are known risk factors for sporadic PD. In mice degeneration of the nigrostriatal pathway can be induced by both paraquat and iron, yielding a model for PD. The catalytic antioxidant EUK-189 was found to significantly protect dopaminergic neurons, both in in vitro experiments and in vivo [101].

Since 1997, several new animal models for Parkinson's disease have been developed, in part based on the discovery that certain protein aggregations and gene mutations lead to the development of PD in human patients [102]. These include a-synucleopathies, linked to abnormal accumulation of a-synuclein, and mutations in PTEN-induced kinase 1 (PINK1), parkin, DJ-1 and leucine-rich repeat kinase 2 ($LRRK_2$). Interestingly, several of these mutations result in mitochondrial dysfunction/mitophagy and ROS accumulation has been shown to be correlated to the neurodegeneration associated with PD [103–105]. The potential links between oxidative stress and PD-related neurodegeneration, several studies have attempted to prevent or to treat PD patients with a variety of antioxidants, and in particular, the so-called mitochondria-targeted antioxidants. While these compounds have shown promising effects in experimental models of PD, the results from clinical studies have not been conclusive [106]. Other treatments included Vitamin E in conjunction with MAO inhibitors, such as selegiline or rasigiline. Again, clinical results were not overwhelmingly convincing [107].

3.3 *Amyotrophic Lateral Sclerosis (ALS)*

Another neurodegenerative disease in which mitochondrial dysfunction has been implicated is amyotrophic lateral sclerosis (ALS) [108]. The existing experimental model for the disease uses mice expressing certain sod1 mutations identified as causative in some forms of inherited ALS. Mutations in Cu, Zn superoxide dismutase (SOD1) cause a subset of amyotrophic lateral sclerosis (ALS). Well-documented evidence indicates that a gain of toxic activity by the mutant SOD1 kills motoneurons [109]. The nature of this toxic activity is unknown but an enhanced reactive oxidative species (ROS) generation has been suspected [109]. Mutant SOD1 can produce enhanced oxidative damage to cells. Expression of mutant SOD1 causes an elevated hydroxyl radical production [110, 111] and increasing levels of markers for oxidative damage in proteins, nucleic acids, and lipids in ALS patients and in mice [112–114]. In cultured cells, mutant SOD1 causes oxidative damage and neuronal death that is inhibited by antioxidative agents, suggesting that oxidative damage plays a role in causing this neuronal death [115–117]. However, whether such a role exists in vivo, and whether antioxidative treatments can both reduce oxidative damage and slow disease progression has been unclear. The ALS mice exhibit symptoms analogous to those of ALS patients, motor weakness progressing to total paralysis and death. Jung et al. [118] treated an ALS mouse line, G93A "low expressors", with EUK-8 and EUK-134 (30 mg/kg, ip) from the age of 60 days until their death. Mean survival of the G93A mice was increased by 17 and 23 days in groups treated with EUK-8 and EUK-134, respectively. Survival after disease onset, as detected by an abrupt loss of muscle strength, was increased by 22 % and 68 %, respectively. In spinal cord samples, collected at death, indicators of oxidative stress, malonyldialdehyde and protein carbonyl content, were decreased in treated groups. In addition, treated groups had decreased staining for protein tyrosine-nitration in the motorneurons.

3.4 Other Neurodegenerative Diseases

Oxidative stress has been implicated in several other neurodegenerative diseases. These diseases include Friedreich's ataxia (FA), a genetic disease due to a decreased synthesis of mitochondrial frataxin caused by impaired gene transcription. While low frataxin affects several organs, including most importantly the heart, it also leads to a debilitating ataxia. Frataxin is involved in iron homeostasis in mitochondria, which could conceivably lead to increased oxidative stress through the Fenton reaction that releases the highly damaging hydroxyl radicals from the interaction of free iron with hydrogen peroxide. Links between oxidative stress and FA have been reviewed in Armstrong et al. [119]. However, while the rationale for using antioxidants to prevent heart hypertrophy and ataxia that develops in FA is strong, results with stoichiometric antioxidants had little success [120].

Ataxia Telangiectasia (AT) is a rare autosomal recessive disorder linked to mutations in the ATM gene. ATM is a protein kinase that functions as an intracellular redox sensor [121], thus suggesting a critical involvement of oxidative stress in the many manifestations of this disease, including ataxia. Chronic treatment of ATM null mice with EUK-189 improved locomotor activity and decreased lipid peroxidation in the brain [122]. Interestingly, overexpression of mitochondrial catalase in ATM null mice improved a number of the phenotypes associated with AT [123].

Transmissible spongiform encephalopathies, also known as prion diseases, affect both humans (e.g. Creutzfeldt-Jakob disease) and animals (bovine spongiform encephalopathy). The pathogenesis of these diseases involves the conversion of the prion protein (PrPc) into an insoluble isoform (PrPres). Post mortem analysis of the brain demonstrates clear-cut signs of oxidative stress. In a mouse model for prion disease EUK-189 demonstrated modest but significant efficacy as shown by an increased survival and a decrease in markers for oxidative stress in the brain [124].

4 Potential Therapeutic Applications of Small Catalytic Scavengers in Age-Related Neurodegenerative Disorders

A number of molecules claimed to have antioxidant properties have been tested in animal models for neurodegenerative diseases such as stroke, ALS, Alzheimer's disease, Friedreich's ataxia, Huntington's disease and Parkinson's disease, and were found to be highly protective. They include in particular lazaroids, ascorbate, N-acetyl cysteine (NAC), PBN, idebenone and MitoQ. It would therefore seem that such antioxidant molecules should be beneficial for treating age-associated decline in learning and memory and neurodegenerative diseases in humans. However, in general efficacy in animal models consistently failed to translate into efficacy in human clinical trials and to-date no antioxidant therapeutic has been approved for cognitive impairment or neurodegenerative diseases. We conclude this review by a discussion of potential reasons explaining these failures, and by a suggestion to move forward in further exploring the potential of antioxidants in these diseases.

There are several possible explanations for such disappointing human data. The first one is that most compounds touted as antioxidants target only oxygen free radicals. While oxygen free radicals undoubtedly play a role in oxidative stress, they might not be the primary cause for oxidative damage. Indeed there is evidence pointing to hydrogen peroxide playing a more significant role. For example overexpression of mitochondrial catalase in mice was found to increase lifespan [125]. Thus, oxygen free radical-targeting antioxidants might not be the most appropriate compounds to alleviate the deleterious effects of oxidative stress in human disease.

A second explanation lies in the fact that antioxidants such as ascorbate, NAC, PBN, idebenone or MitoQ are themselves oxidized as they exert their antioxidant activities. In other words they can act once and are either inactivated or need to be recycled to continue to be efficacious, while oxidative stress is a continuously ongoing process, particularly in chronic diseases. Thus, they may be seen as stoichiometric antioxidants. It has been estimated that approximately 4 % of the oxygen that humans breathe is transformed into ROS. This translates into approximately 25–30 g a day. To inactivate even a small fraction of this amount would require daily intake of tens of grams of antioxidants, which have molecular weights much higher than that of ROS, even assuming 100 % of bioavailability.

Finally, it is possible that the level of oxidative and nitrosative stress in the human brain is higher than in mice. While it is extremely difficult to study this in a quantitative way, it can be noted that the human brain, which accounts for an average 2 % of body weight, consumes 20 % of the oxygen we breathe. It is thus possible that in order to significantly decrease oxidative and nitrosative stress in the human brain, even higher doses of antioxidants might be needed. For these reasons one may consider that the hypothesis of the involvement of oxidative stress in aging and age-related diseases in humans has not been properly tested yet.

In 1991 we proposed that, in order to be optimally effective, antioxidants had to act catalytically on mediators for oxidative stress. Indeed, in an early work we reported that a prototype catalytic antioxidant formed by complexing manganese to desferrioxamine was efficacious in preventing kainic acid neurodegeneration in mice [126]. We subsequently discovered that a class of compounds usually referred to as salen-Mn complexes, had catalytic antioxidant properties, being able to efficiently dismute oxygen free radicals, and to hydrolyze hydrogen peroxide into oxygen and water. These compounds have proven to be efficacious in a number of animal models for oxidative stress-related diseases, including neurodegenerative diseases as described above. Their ability to catalytically inactivate both oxygen free radicals and hydrogen peroxide, and RNS [32] most likely explains their versatility and potency. For example, as described above, doses as low as 15 μg/kg/day were sufficient to prevent age-related cognitive impairment in mice [30].

No human data on their efficacy after systemic administration is available yet. However, interestingly, topical application of EUK-134 in humans was shown to potently prevent UV-induced skin lipid peroxidation even when applied after irradiation, under conditions where the stoichiometric antioxidant α-tocopherol was ineffective [127]. We propose that such compounds, which can be described as broad-spectrum catalytic antioxidants, might prove efficacious in humans in a number of diseases associated with oxidative stress, including age-related neurodegenerative diseases.

References

1. Davis HP, Small SA, Stern Y, Mayeux R, Feldstein SN, Keller FR. Acquisition, recall, and forgetting of verbal information in long-term memory by young, middle-aged, and elderly individuals. Cortex. 2003;39:1063–91.
2. Petersen RC, Doody R, Kurz A, Mohs RC, Morris JC, Rabins PV, et al. Current concepts in mild cognitive impairment. Arch Neurol. 2001;58:1985–92.
3. Barnes CA, Markowska AL, Ingram DK, Kametani H, Spangler EL, Lemken VJ, et al. Acetyl-1-carnitine. 2: Effects on learning and memory performance of aged rats in simple and complex mazes. Neurobiol Aging. 1990;11:499–506.
4. Forster MJ, Dubey A, Dawson KM, Stutts WA, Lal H, Sohal RS. Age-related losses of cognitive function and motor skills in mice are associated with oxidative protein damage in the brain. Proc Natl Acad Sci U S A. 1996;93:4765–9.
5. Federico A, Cardaioli E, Da Pozzo P, Formichi P, Gallus GN, Radi E. Mitochondria, oxidative stress and neurodegeneration. J Neurol Sci. 2012;322:254–62.

6. Ansari MA, Roberts KN, Scheff SW. A time course of contusion-induced oxidative stress and synaptic proteins in cortex in a rat model of TBI. J Neurotrauma. 2008;25:513–26.
7. Behrens MM, Sejnowski TJ. Does schizophrenia arise from oxidative dysregulation of parvalbumin-interneurons in the developing cortex? Neuropharmacology. 2009;57:193–200.
8. Villagonzalo KA, Dodd S, Dean O, Gray K, Tonge B, Berk M. Oxidative pathways as a drug target for the treatment of autism. Expert Opin Ther Targets. 2010;14:1301–10.
9. Calabrese V, Scapagnini G, Ravagna A, Colombrita C, Spadaro F, Butterfield DA, et al. Increased expression of heat shock proteins in rat brain during aging: relationship with mitochondrial function and glutathione redox state. Mech Ageing Dev. 2004;125:325–35.
10. Cini M, Moretti A. Studies on lipid peroxidation and protein oxidation in the aging brain. Neurobiol Aging. 1995;16:53–7.
11. Hamilton ML, Van Remmen H, Drake JA, Yang H, Guo ZM, Kewitt K, et al. Does oxidative damage to DNA increase with age? Proc Natl Acad Sci U S A. 2001;98:10469–74.
12. O'Donnell E, Lynch MA. Dietary antioxidant supplementation reverses age-related neuronal changes. Neurobiol Aging. 1998;19:461–7.
13. Siqueira IR, Fochesatto C, de Andrade A, Santos M, Hagen M, Bello-Klein A, et al. Total antioxidant capacity is impaired in different structures from aged rat brain. Int J Dev Neurosci. 2005;23:663–71.
14. Sohal RS, Ku HH, Agarwal S, Forster MJ, Lal H. Oxidative damage, mitochondrial oxidant generation and antioxidant defenses during aging and in response to food restriction in the mouse. Mech Ageing Dev. 1994;74:121–33.
15. Sawada M, Carlson JC. Changes in superoxide radical and lipid peroxide formation in the brain, heart and liver during the lifetime of the rat. Mech Ageing Dev. 1987;41:125–37.
16. Rao G, Xia E, Richardson A. Effect of age on the expression of antioxidant enzymes in male Fischer F344 rats. Mech Ageing Dev. 1990;53:49–60.
17. Carney JM, Starke-Reed PE, Oliver CN, Landum RW, Cheng MS, Wu JF, et al. Reversal of age-related increase in brain protein oxidation, decrease in enzyme activity, and loss in temporal and spatial memory by chronic administration of the spin-trapping compound N-tert-butyl-alpha-phenylnitrone. Proc Natl Acad Sci U S A. 1991;88:3633–6.
18. Raghavendra V, Kulkarni SK. Possible antioxidant mechanism in melatonin reversal of aging and chronic ethanol-induced amnesia in plus-maze and passive avoidance memory tasks. Free Radic Biol Med. 2001;30:595–602.
19. Stoll S, Rostock A, Bartsch R, Korn E, Meichelbock A, Muller WE. The potent free radical scavenger alpha-lipoic acid improves cognition in rodents. Ann N Y Acad Sci. 1994;717:122–8.
20. Butterfield DA, Reed T, Perluigi M, De Marco C, Coccia R, Cini C, et al. Elevated protein-bound levels of the lipid peroxidation product, 4-hydroxy-2-nonenal, in brain from persons with mild cognitive impairment. Neurosci Lett. 2006;397:170–3.
21. Shahidi S, Komaki A, Mahmoodi M, Atrvash N, Ghodrati M. Ascorbic acid supplementation could affect passive avoidance learning and memory in rat. Brain Res Bull. 2008;76:109–13.
22. Hansen SN, Tveden-Nyborg P, Lykkesfeldt J. Does vitamin C deficiency affect cognitive development and function? Nutrients. 2014;6:3818–46.
23. Harrison FE, Bowman GL, Polidori MC. Ascorbic acid and the brain: rationale for the use against cognitive decline. Nutrients. 2014;6:1752–81.
24. Joshi YB, Pratico D. Vitamin E in aging, dementia, and Alzheimer's disease. Biofactors. 2012;38:90–7.
25. Berr C, Arnaud J, Akbaraly TN. Selenium and cognitive impairment: a brief-review based on results from the EVA study. Biofactors. 2012;38:139–44.
26. Baudry M, Etienne S, Bruce A, Palucki M, Jacobsen E, Malfroy B. Salen-manganese complexes are superoxide dismutase-mimics. Biochem Biophys Res Commun. 1993; 192:964–8.
27. Doctrow SR, Huffman K, Marcus CB, Tocco G, Malfroy E, Adinolfi CA, et al. Salen-manganese complexes as catalytic scavengers of hydrogen peroxide and cytoprotective agents: structure-activity relationship studies. J Med Chem. 2002;45:4549–58.

28. Melov S, Doctrow SR, Schneider JA, Haberson J, Patel M, Coskun PE, et al. Lifespan extension and rescue of spongiform encephalopathy in superoxide dismutase 2 nullizygous mice treated with superoxide dismutase-catalase mimetics. J Neurosci. 2001;21:8348–53.
29. Liu R, Liu IY, Bi X, Thompson RF, Doctrow SR, Malfroy B, et al. Reversal of age-related learning deficits and brain oxidative stress in mice with superoxide dismutase/catalase mimetics. Proc Natl Acad Sci U S A. 2003;100:8526–31.
30. Clausen A, Doctrow S, Baudry M. Prevention of cognitive deficits and brain oxidative stress with superoxide dismutase/catalase mimetics in aged mice. Neurobiol Aging. 2010;31:425–33.
31. Quick KL, Ali SS, Arch R, Xiong C, Wozniak D, Dugan LL. A carboxyfullerene SOD mimetic improves cognition and extends the lifespan of mice. Neurobiol Aging. 2008;29:117–28.
32. Sharpe MA, Ollosson R, Stewart VC, Clark JB. Oxidation of nitric oxide by oxomanganese-salen complexes: a new mechanism for cellular protection by superoxide dismutase/catalase mimetics. Biochem J. 2002;366:97–107.
33. Selkoe DJ. Alzheimer's disease: genes, proteins, and therapy. Physiol Rev. 2001;81:741–66.
34. Katzman R, Saitoh T. Advances in Alzheimer's disease. FASEB J. 1991;5:278–86.
35. Salmon DP, Thomas RG, Pay MM, Booth A, Hofstetter CR, Thal LJ, et al. Alzheimer's disease can be accurately diagnosed in very mildly impaired individuals. Neurology. 2002;59:1022–8.
36. DeKosky ST, Scheff SW, Styren SD. Structural correlates of cognition in dementia: quantification and assessment of synapse change. Neurodegeneration. 1996;5:417–21.
37. Good PF, Werner P, Hsu A, Olanow CW, Perl DP. Evidence of neuronal oxidative damage in Alzheimer's disease. Am J Pathol. 1996;149:21–8.
38. Butterfield DA, Hensley K, Cole P, Subramaniam R, Aksenov M, Aksenova M, et al. Oxidatively induced structural alteration of glutamine synthetase assessed by analysis of spin label incorporation kinetics: relevance to Alzheimer's disease. J Neurochem. 1997;68:2451–7.
39. Smith MA, Harris PL, Sayre LM, Perry G. Iron accumulation in Alzheimer disease is a source of redox-generated free radicals. Proc Natl Acad Sci U S A. 1997;94:9866–8.
40. Sayre LM, Zelasko DA, Harris PL, Perry G, Salomon RG, Smith MA. 4-Hydroxynonenal-derived advanced lipid peroxidation end products are increased in Alzheimer's disease. J Neurochem. 1997;68:2092–7.
41. Butterfield DA, Drake J, Pocernich C, Castegna A. Evidence of oxidative damage in Alzheimer's disease brain: central role for amyloid beta-peptide. Trends Mol Med. 2001;7:548–54.
42. Butterfield DA, Castegna A, Lauderback CM, Drake J. Evidence that amyloid beta-peptide-induced lipid peroxidation and its sequelae in Alzheimer's disease brain contribute to neuronal death. Neurobiol Aging. 2002;23:655–64.
43. Gabbita SP, Lovell MA, Markesbery WR. Increased nuclear DNA oxidation in the brain in Alzheimer's disease. J Neurochem. 1998;71:2034–40.
44. Nunomura A, Perry G, Pappolla MA, Wade R, Hirai K, Chiba S, et al. RNA oxidation is a prominent feature of vulnerable neurons in Alzheimer's disease. J Neurosci. 1999;19:1959–64.
45. Lovell MA, Markesbery WR. Ratio of 8-hydroxyguanine in intact DNA to free 8-hydroxyguanine is increased in Alzheimer disease ventricular cerebrospinal fluid. Arch Neurol. 2001;58:392–6.
46. Nakanishi H, Wu Z. Microglia-aging: roles of microglial lysosome—and mitochondria-derived reactive oxygen species in brain aging. Behav Brain Res. 2009;201:1–7.
47. Harris ME, Carney JM, Cole PS, Hensley K, Howard BJ, Martin L, et al. beta-Amyloid peptide-derived, oxygen-dependent free radicals inhibit glutamate uptake in cultured astrocytes: implications for Alzheimer's disease. Neuroreport. 1995;6:1875–9.
48. Yatin SM, Varadarajan S, Link CD, Butterfield DA. In vitro and in vivo oxidative stress associated with Alzheimer's amyloid beta-peptide (1–42). Neurobiol Aging. 1999;20:325–30. discussion 39–42.
49. Varadarajan S, Yatin S, Aksenova M, Butterfield DA. Review: Alzheimer's amyloid beta-peptide-associated free radical oxidative stress and neurotoxicity. J Struct Biol. 2000;130:184–208.

50. Avdulov NA, Chochina SV, Igbavboa U, O'Hare EO, Schroeder F, Cleary JP, et al. Amyloid beta-peptides increase annular and bulk fluidity and induce lipid peroxidation in brain synaptic plasma membranes. J Neurochem. 1997;68:2086–91.
51. Tamagno E, Parola M, Guglielmotto M, Santoro G, Bardini P, Marra L, et al. Multiple signaling events in amyloid beta-induced, oxidative stress-dependent neuronal apoptosis. Free Radic Biol Med. 2003;35:45–58.
52. Bruce AJ, Malfroy B, Baudry M. beta-Amyloid toxicity in organotypic hippocampal cultures: protection by EUK-8, a synthetic catalytic free radical scavenger. Proc Natl Acad Sci U S A. 1996;93:2312–6.
53. de la Monte SM, Neely TR, Cannon J, Wands JR. Oxidative stress and hypoxia-like injury cause Alzheimer-type molecular abnormalities in central nervous system neurons. Cell Mol Life Sci. 2000;57:1471–81.
54. Zhu X, Perry G, Moreira PI, Aliev G, Cash AD, Hirai K, et al. Mitochondrial abnormalities and oxidative imbalance in Alzheimer disease. J Alzheimers Dis. 2006;9:147–53.
55. Yao J, Irwin RW, Zhao L, Nilsen J, Hamilton RT, Brinton RD. Mitochondrial bioenergetic deficit precedes Alzheimer's pathology in female mouse model of Alzheimer's disease. Proc Natl Acad Sci U S A. 2009;106:14670–5.
56. Ma T, Hoeffer CA, Wong H, Massaad CA, Zhou P, Iadecola C, et al. Amyloid beta-induced impairments in hippocampal synaptic plasticity are rescued by decreasing mitochondrial superoxide. J Neurosci. 2011;31:5589–95.
57. Massaad CA, Washington TM, Pautler RG, Klann E. Overexpression of SOD-2 reduces hippocampal superoxide and prevents memory deficits in a mouse model of Alzheimer's disease. Proc Natl Acad Sci U S A. 2009;106:13576–81.
58. Lee HP, Zhu X, Casadesus G, Castellani RJ, Nunomura A, Smith MA, et al. Antioxidant approaches for the treatment of Alzheimer's disease. Expert Rev Neurother. 2010;10:1201–8.
59. Borger E, Aitken L, Muirhead KE, Allen ZE, Ainge JA, Conway SJ, et al. Mitochondrial beta-amyloid in Alzheimer's disease. Biochem Soc Trans. 2011;39:868–73.
60. Tillement L, Lecanu L, Papadopoulos V. Further evidence on mitochondrial targeting of beta-amyloid and specificity of beta-amyloid-induced mitotoxicity in neurons. Neurodegener Dis. 2011;8:331–44.
61. Mancuso M, Orsucci D, LoGerfo A, Calsolaro V, Siciliano G. Clinical features and pathogenesis of Alzheimer's disease: involvement of mitochondria and mitochondrial DNA. Adv Exp Med Biol. 2010;685:34–44.
62. Muller WE, Eckert A, Kurz C, Eckert GP, Leuner K. Mitochondrial dysfunction: common final pathway in brain aging and Alzheimer's disease—therapeutic aspects. Mol Neurobiol. 2010;41:159–71.
63. Dragicevic N, Smith A, Lin X, Yuan F, Copes N, Delic V, et al. Green tea epigallocatechin-3-gallate (EGCG) and other flavonoids reduce Alzheimer's amyloid-induced mitochondrial dysfunction. J Alzheimers Dis. 2011;26:507–21.
64. Manczak M, Mao P, Calkins MJ, Cornea A, Reddy AP, Murphy MP, et al. Mitochondria-targeted antioxidants protect against amyloid-beta toxicity in Alzheimer's disease neurons. J Alzheimers Dis. 2010;20 Suppl 2:S609–31.
65. Picone P, Bondi ML, Montana G, Bruno A, Pitarresi G, Giammona G, et al. Ferulic acid inhibits oxidative stress and cell death induced by Ab oligomers: improved delivery by solid lipid nanoparticles. Free Radic Res. 2009;43:1133–45.
66. Rhein V, Giese M, Baysang G, Meier F, Rao S, Schulz KL, et al. Ginkgo biloba extract ameliorates oxidative phosphorylation performance and rescues abeta-induced failure. PLoS One. 2010;5, e12359.
67. Simpkins JW, Yi KD, Yang SH, Dykens JA. Mitochondrial mechanisms of estrogen neuroprotection. Biochim Biophys Acta. 2010;1800:1113–20.
68. Bonda DJ, Lee HP, Lee HG, Friedlich AL, Perry G, Zhu X, et al. Novel therapeutics for Alzheimer's disease: an update. Curr Opin Drug Discov Devel. 2010;13:235–46.
69. Darvesh AS, Carroll RT, Bishayee A, Geldenhuys WJ, Van der Schyf CJ. Oxidative stress and Alzheimer's disease: dietary polyphenols as potential therapeutic agents. Expert Rev Neurother. 2010;10:729–45.

70. Wang J, Ho L, Zhao W, Ono K, Rosensweig C, Chen L, et al. Grape-derived polyphenolics prevent Abeta oligomerization and attenuate cognitive deterioration in a mouse model of Alzheimer's disease. J Neurosci. 2008;28:6388–92.
71. Clausen A, Xu X, Bi X, Baudry M. Effects of the superoxide dismutase/catalase mimetic EUK-207 in a mouse model of Alzheimer's disease: protection against and interruption of progression of amyloid and tau pathology and cognitive decline. J Alzheimers Dis. 2012;30:183–208.
72. Hirai K, Aliev G, Nunomura A, Fujioka H, Russell RL, Atwood CS, et al. Mitochondrial abnormalities in Alzheimer's disease. J Neurosci. 2001;21:3017–23.
73. Goldstein LE, Muffat JA, Cherny RA, Moir RD, Ericsson MH, Huang X, et al. Cytosolic beta-amyloid deposition and supranuclear cataracts in lenses from people with Alzheimer's disease. Lancet. 2003;361:1258–65.
74. Melov S, Wolf N, Strozyk D, Doctrow SR, Bush AI. Mice transgenic for Alzheimer disease beta-amyloid develop lens cataracts that are rescued by antioxidant treatment. Free Radic Biol Med. 2005;38:258–61.
75. Sung S, Yao Y, Uryu K, Yang H, Lee VM, Trojanowski JQ, et al. Early vitamin E supplementation in young but not aged mice reduces Abeta levels and amyloid deposition in a transgenic model of Alzheimer's disease. FASEB J. 2004;18:323–5.
76. Agid Y, Javoy-Agid F, Ruberg M. Biochemistry of Parkinson's disease. In: Marsden CD, Fahn S, editors. Movement Disorders. Stoneham, MA: Butterworth & Co; 1987. p. 166–230.
77. Forno S. Pathology of Parkinson's Disease. In: Marsden CD, Fahn S, editors. Movement Disorders Neurology 2. London: Butterworth Scientific; 1992. p. 21–40.
78. Bandmann O, Vaughan J, Holmans P, Marsden CD, Wood NW. Association of slow acetylator genotype for N-acetyltransferase 2 with familial Parkinson's disease. Lancet. 1997;350:1136–9.
79. De Erausquin GA, Costa E, Hanbauer I. Calcium homeostasis, free radical formation, and trophic factor dependence mechanisms in Parkinson's disease. Pharmacol Rev. 1994;46:467–82.
80. Dexter DT, Wells FR, Lees AJ, Agid F, Agid Y, Jenner P, et al. Increased nigral iron content and alterations in other metal ions occurring in brain in Parkinson's disease. J Neurochem. 1989;52:1830–6.
81. Goeth ME, Freyberger A. P. R. Oxidative stress: a role in the pathogenesis of Parkinson's disease. J Neural Transmission. 1990;29:241–9.
82. Youdim MB, Lavie L. Selective MAO-A and B inhibitors, radical scavengers and nitric oxide synthase inhibitors in Parkinson's disease. Life Sci. 1994;55:2077–82.
83. Jenner P. Oxidative mechanism in nigral cell death in Parkinson's disease. Mov Disord. 1998;13:24–34.
84. Keller JN, Mattson MP. Roles of lipid peroxidation in modulation of cellular signaling pathways, cell dysfunction, and death in the nervous system. Rev Neurosci. 1998;9:105–16.
85. Mylonas C, Kouretas D. Lipid peroxidation and tissue damage. In Vivo. 1999;13:295–309.
86. Prezedborski S, Jackson-Lewis V. Mechanisms of MPTP toxicity. Movement Disord. 1998;13:35–8.
87. Bloem BR, Irwin I, Buruma OJ, Haan J, Roos RA, Tetrud JW, et al. The MPTP model: versatile contributions to the treatment of idiopathic Parkinson's disease. J Neurol Sci. 1990;97:273–93.
88. Burns RS, Chiueh CC, Markey SP, Ebert MH, Jacobowitz DM, Kopin IJ. A primate model of parkinsonism: selective destruction of dopaminergic neurons in the pars compacta of the substantia nigra by N-methyl-4-phenyl-1,2,3,6-tetrahydropyridine. Proc Natl Acad Sci U S A. 1983;80:4546–50.
89. Beckman JS, Koppenol WH. Nitric oxide, superoxide, and peroxynitrite: the good, the bad, and ugly. Am J Physiol. 1996;271:C1424–37.
90. Beckman JS. Oxidative damage and tyrosine nitration from peroxynitrite. Chem Res Toxicol. 1996;9:836–44.
91. Di Stasi AM, Mallozzi C, Macchia G, Petrucci TC, Minetti M. Peroxynitrite induces tryosine nitration and modulates tyrosine phosphorylation of synaptic proteins. J Neurochem. 1999;73:727–35.

92. Fontana M, Pecci L, Macone A, Cavallini D. Antioxidant properties of the decarboxylated dimer of aminoethylcysteine ketimine: assessment of its ability to scavenge peroxynitrite. Free Radic Res. 1998;29:435–40.
93. Mesenge C, Charriaut-Marlangue C, Verrecchia C, Allix M, Boulu RR, Plotkine M. Reduction of tyrosine nitration after N(omega)-nitro-L-arginine-methylester treatment of mice with traumatic brain injury. Eur J Pharmacol. 1998;353:53–7.
94. Wu W, Chen Y, Hazen SL. Eosinophil peroxidase nitrates protein tyrosyl residues. Implications for oxidative damage by nitrating intermediates in eosinophilic inflammatory disorders. J Biol Chem. 1999;274:25933–44.
95. Good PF, Hsu A, Werner P, Perl DP, Olanow CW. Protein nitration in Parkinson's disease. J Neuropathol Exp Neurol. 1998;57:338–42.
96. Hantraye P, Brouillet E, Ferrante R, Palfi S, Dolan R, Matthews RT, et al. Inhibition of neuronal nitric oxide synthase prevents MPTP-induced parkinsonism in baboons. Nat Med. 1996;2:1017–21.
97. Przedborski S, Jackson-Lewis V, Yokoyama R, Shibata T, Dawson VL, Dawson TM. Role of neuronal nitric oxide in 1-methyl-4-phenyl-1,2,3,6-tetrahydropyridine (MPTP)-induced dopaminergic neurotoxicity. Proc Natl Acad Sci U S A. 1996;93:4565–71.
98. Schulz JB, Matthews RT, Muqit MM, Browne SE, Beal MF. Inhibition of neuronal nitric oxide synthase by 7-nitroindazole protects against MPTP-induced neurotoxicity in mice. J Neurochem. 1995;64:936–9.
99. Ara J, Przedborski S, Naini AB, Jackson-Lewis V, Trifiletti RR, Horwitz J, et al. Inactivation of tyrosine hydroxylase by nitration following exposure to peroxynitrite and 1-methyl-4-phenyl-1,2,3,6-tetrahydropyridine (MPTP). Proc Natl Acad Sci U S A. 1998;95:7659–63.
100. Pong K, Doctrow SR, Baudry M. Prevention of 1-methyl-4-phenylpyridinium- and 6-hydroxydopamine-induced nitration of tyrosine hydroxylase and neurotoxicity by EUK-134, a superoxide dismutase and catalase mimetic, in cultured dopaminergic neurons. Brain Res. 2000;881:182–9.
101. Peng J, Stevenson FF, Doctrow SR, Andersen JK. Superoxide dismutase/catalase mimetics are neuroprotective against selective paraquat-mediated dopaminergic neuron death in the substantial nigra: implications for Parkinson disease. J Biol Chem. 2005;280:29194–8.
102. Cookson MR, Bandmann O. Parkinson's disease: insights from pathways. Hum Mol Genet. 2010;19:R21–7.
103. Schildknecht S, Gerding HR, Karreman C, Drescher M, Lashuel HA, Outeiro TF, et al. Oxidative and nitrative alpha-synuclein modifications and proteostatic stress: implications for disease mechanisms and interventions in synucleinopathies. J Neurochem. 2013;125:491–511.
104. Zuo L, Motherwell MS. The impact of reactive oxygen species and genetic mitochondrial mutations in Parkinson's disease. Gene. 2013;532:18–23.
105. Gaki GS, Papavassiliou AG. Oxidative stress-induced signaling pathways implicated in the pathogenesis of Parkinson's disease. Neuromolecular Med. 2014;16:217–30.
106. Jin H, Kanthasamy A, Ghosh A, Anantharam V, Kalyanaraman B, Kanthasamy AG. Mitochondria-targeted antioxidants for treatment of Parkinson's disease: preclinical and clinical outcomes. Biochim Biophys Acta. 1842;2014:1282–94.
107. AlDakheel A, Kalia LV, Lang AE. Pathogenesis-targeted, disease-modifying therapies in Parkinson disease. Neurotherapeutics. 2014;11:6–23.
108. Jung C, Higgins CM, Xu Z. Mitochondrial electron transport chain complex dysfunction in a transgenic mouse model for amyotrophic lateral sclerosis. J Neurochem. 2002;83:535–45.
109. Xu Z. Mechanism and treatment of motoneuron degeneration in ALS: what have SOD1 mutants told us? Amyotroph Lateral Scler Other Motor Neuron Disord. 2000;1:225–34.
110. Bogdanov MB, Ramos LE, Xu Z, Beal MF. Elevated "hydroxyl radical" generation in vivo in an animal model of amyotrophic lateral sclerosis. J Neurochem. 1998;71:1321–4.
111. Liu R, Althaus JS, Ellerbrock BR, Becker DA, Gurney ME. Enhanced oxygen radical production in a transgenic mouse model of familial amyotrophic lateral sclerosis. Ann Neurol. 1998;44:763–70.

112. Andrus PK, Fleck TJ, Gurney ME, Hall ED. Protein oxidative damage in a transgenic mouse model of familial amyotrophic lateral sclerosis. J Neurochem. 1998;71:2041–8.
113. Ferrante RJ, Browne SE, Shinobu LA, Bowling AC, Baik MJ, MacGarvey U, et al. Evidence of increased oxidative damage in both sporadic and familial amyotrophic lateral sclerosis. J Neurochem. 1997;69:2064–74.
114. Hall ED, Andrus PK, Oostveen JA, Fleck TJ, Gurney ME. Relationship of oxygen radical-induced lipid peroxidative damage to disease onset and progression in a transgenic model of familial ALS. J Neurosci Res. 1998;53:66–77.
115. Dugan LL, Turetsky DM, Du C, Lobner D, Wheeler M, Almli CR, et al. Carboxyfullerenes as neuroprotective agents. Proc Natl Acad Sci U S A. 1997;94:9434–9.
116. Ghadge GD, Lee JP, Bindokas VP, Jordan J, Ma L, Miller RJ, et al. Mutant superoxide dismutase-1-linked familial amyotrophic lateral sclerosis: molecular mechanisms of neuronal death and protection. J Neurosci. 1997;17:8756–66.
117. Kruman II, Pedersen WA, Springer JE, Mattson MP. ALS-linked Cu/Zn-SOD mutation increases vulnerability of motor neurons to excitotoxicity by a mechanism involving increased oxidative stress and perturbed calcium homeostasis. Exp Neurol. 1999;160:28–39.
118. Jung C, Rong Y, Doctrow S, Baudry M, Malfroy B, Xu Z. Synthetic superoxide dismutase/catalase mimetics reduce oxidative stress and prolong survival in a mouse amyotrophic lateral sclerosis model. Neurosci Lett. 2001;304:157–60.
119. Armstrong JS, Khdour O, Hecht SM. Does oxidative stress contribute to the pathology of Friedreich's ataxia? A radical question. FASEB J. 2010;24:2152–63.
120. Pandolfo M. Friedreich ataxia. Arch Neurol. 2008;65:1296–303.
121. Ditch S, Paull TT. The ATM protein kinase and cellular redox signaling: beyond the DNA damage response. Trends Biochem Sci. 2012;37:15–22.
122. Browne SE, Roberts 2nd LJ, Dennery PA, Doctrow SR, Beal MF, Barlow C, et al. Treatment with a catalytic antioxidant corrects the neurobehavioral defect in ataxia-telangiectasia mice. Free Radic Biol Med. 2004;36:938–42.
123. D'Souza AD, Parish IA, Krause DS, Kaech SM, Shadel GS. Reducing mitochondrial ROS improves disease-related pathology in a mouse model of ataxia-telangiectasia. Mol Ther. 2013;21:42–8.
124. Brazier MW, Doctrow SR, Masters CL, Collins SJ. A manganese-superoxide dismutase/catalase mimetic extends survival in a mouse model of human prion disease. Free Radic Biol Med. 2008;45:184–92.
125. Schriner SE, Linford NJ, Martin GM, Treuting P, Ogburn CE, Emond M, Coskun PE, Ladiges W, Wolf N, VanRemmenn H, Wallace DC, Rabinovitch PS. Extesnsion of murine life spam by overexpression of catalse targeted to mitochondria. Science. 2005;308:1909–11.
126. Bruce A, Najm I, Malfroy B, Baudry M. Effects of desferrioxamine-manganese, a superoxide dismutase-mimic, on kainate-induced pathology in rat brain. Neurodegeneration. 1992;1:265–71.
127. Declercq L, Sente I, Hellemans L, Corstjens H, Maes D. Use of the synthetic superoxide dismutase/catalase mimetic EUK-134 to compensate for seasonal antioxidant deficiency by reducing pre-existing lipid peroxides at the human skin surface. Int J Cosmet Sci. 2004;26:255–63.
128. Nicolle MM, Gonzalez J, Sugaya K, Baskerville KA, Bryan D, Lund K, et al. Signatures of hippocampal oxidative stress in aged spatial learning-impaired rodents. Neuroscience. 2001;107:415–31.

Antioxidants and Redox-Based Therapeutics in Parkinson's Disease

Dilshan S. Harischandra, Huajun Jin, Anamitra Ghosh, Vellareddy Anantharam, Arthi Kanthasamy, and Anumantha G. Kanthasamy

1 Parkinson's Disease

Parkinson's disease (PD) is a progressive neurodegenerative disorder affecting more than one million individuals over the age of 60 within the United States [1]. According to one recent article, the number of new cases increased by about 50,000 annually [2]. Although PD is an age-related disorder affecting nearly 3 % of people over 60 years and 4–5 % of those over age 85, nearly 10 % of PD patients are under 40 years of age [3]. Epidemiological studies suggest that sporadic PD cases (90 %) are predominantly late onset, whereas the remainder (10 %) is characterized by early onset occurring mainly in familial clusters [3, 4]. Familial or early onset PD has been linked with mutations in several genes such as *parkin*, *ubiquitin C-terminal hydrolase L1*, *α-synuclein*, *leucine-rich repeat kinase 2* (*LRRK2*), *PINK-1* or *DJ1* [5–8]. The cause of sporadic or non-familial PD is not known, but several reports suggest environmental toxins, genetic factors, mitochondrial dysfunction, apoptosis, oxidative stress and neuroinflammation to be among the possible factors behind PD's neurodegeneration [9–12]. Among the environmental toxins implicated in the pathogenesis and progression of the disease, the list includes infectious agents, pesticides, herbicides and heavy metals [3]. Recent investigations have focused on inflammation and oxidative stress as the central players in the pathogenesis of PD.

In both idiopathic and genetic cases of PD, oxidative stress appears to be the common underlying mechanism contributing to the cascade leading to selective neurodegeneration in substantia nigra (SN) neurons and their terminals in the striatum.

D.S. Harischandra • H. Jin • A. Ghosh • V. Anantharam • A. Kanthasamy • A.G. Kanthasamy (✉)
Parkinson's Disorder Research Laboratory, Iowa Center for Advanced Neurotoxicology, Department of Biomedical Sciences, Iowa State University, Ames, IA 50011, USA
e-mail: dilshan@iastate.edu; egb761@iastate.edu; Anamitra.Ghosh@vai.org; anantram@iastate.edu; arthik@iastate.edu; akanthas@iastate.edu

S.C. Bondy, A. Campbell (eds.), *Inflammation, Aging, and Oxidative Stress*, Oxidative Stress in Applied Basic Research and Clinical Practice,
DOI 10.1007/978-3-319-33486-8_14

An imbalance between reactive oxygen species (ROS) generation and elimination mechanisms, such as impaired cellular antioxidant machinery, contributes to the pathogenesis of PD and other neurodegenerative disorders. The resulting oxidative stress is intimately linked to the other aspects of the degenerative process such as mitochondrial dysfunction, inflammation, protein misfolding and DNA damage.

2 Oxidative Stress in PD

Oxidative stress in PD is supported by post-mortem studies on the SN of PD patients showing increased levels of oxidized lipids [13], proteins, nucleic acids [14, 15] and impaired antioxidant mechanisms such as a reduced glutathione (GSH) and oxidized glutathione (GSSG) ratio [16]. Therefore, oxidative stress appears to play a major role in the cascade of biological changes culminating in dopaminergic cell death even though the precise mechanisms involving oxidative stress-mediated nigral cell degeneration in PD are not clear. However, accumulating evidence indicates that dopamine (DA), a neurotransmitter under physiological conditions, may also serve as a neurotoxin and thereby participate in the neurodegenerative process. The mechanism of dopamine neurotoxicity is strongly linked to oxidative metabolism. Under physiological conditions, dopamine can be oxidized enzymatically through monoamine oxidases (MAO) to dihydroxyphenylacetic acid (DOPAC) and subsequently methylated by catechol-O-methyltransferase (COMT) to homovanillic acid (HVA), or from dopamine to 3-methoxytyramine (3-MT) via COMT and further oxidized to HVA through MOA. During this MAO-mediated DA turnover process, hydrogen peroxide (H_2O_2) is produced as a byproduct of DA deamination, a process serving as an inherent source of oxidative stress in the nigrostriatal system. Dopamine can also be non-enzymatically oxidized by O_2 yielding quinones and H_2O_2. These quinones also undergo intramolecular cyclization immediately followed by cascading oxidative reactions ending in the formation of a black, insoluble polymeric pigment known as neuromelanin [17, 18]. Neuromelanin renders dopaminergic neurons more susceptible to auto-oxidation through quinone modification of dopamine, which increases basal levels of oxidative stress in SN [17]. What is becoming clear is that degrading dopamine either enzymatically or non-enzymatically generates H_2O_2, which is easily converted through the Fenton's reaction to highly toxic hydroxyl radicals (•OH) when in the presence of the high levels of ferrous iron (Fe^{2+}) normally found in the SN [19].

3 Hydroxyl Radicals, Superoxide and Hydrogen Peroxide in PD

As mentioned above, DA oxidation plays an important role in generating hydroxyl radical in the central nervous system, including the degeneration of dopaminergic neurons. Dopamine, like many catecholamines (dihydroquinones, QH2), can easily be oxidized by O_2 under physiological conditions. During this oxidation process, both semiquinones (•QH) and quinones (Q) are generated, resulting in •OH (via Fenton's reaction), the most toxic free radical in living cells [20]. The resulting DA quinones also exert further neurotoxicity by covalently binding to cellular nucleophiles such as GSH and protein cysteinyl residues, which normally function as antioxidants important for cell survival [21, 22]. Moreover, DA quinones bind and modify several proteins implicated in PD pathophysiology such as α-synuclein, DJ-1 and parkin [23–25]. However, among the various types of oxidative damage in cellular macromolecules, damage to nucleic acids is particularly hazardous as it could alter the genetic information. Among the five nucleobases—uracil, thymine, cytosine, adenine and guanine - guanine is most susceptible to nucleic acid oxidation through hydroxyl radicals [26, 27]. Hydroxyl radical-mediated lesioning of the DNA strand produces 8-hydroxyguanosine (8OHG), the most studied oxidized DNA product. Moreover, DNA damage in PD also appears to be at the level of 8OHG and 8-hydroxyl-2-deoxyguanosine (8-OHdG) as elevated 8OHG and reduced 8-OHdG have been observed in the SN and cerebrospinal fluid (CSF) of PD patients [15, 28].

Although the human brain comprises only 2 % of the total body weight, it is especially prone to oxidative stress as it receives 15 % of the cardiac output and 20 % of total O_2 consumption of the body, making it highly metabolically active tissue that critically relies on oxidative phosphorylation to meet energy demands. Oxidative phosphorylation also produces potentially damaging radicals such as the superoxide anion O_2^- as a result of a one-electron reduction of O_2. Superoxide occurs widely in nature through a variety of enzymatic processes including xanthine oxidase and NADPH oxidase, a multimeric enzyme that generates both O^{2-} and H_2O_2 [29]. Superoxide has the capacity to damage components of the electron transport chain and other cellular constituents. Superoxides are also produced at microsomal membranes, with electron transport systems dependent on NADH or NADPH, via detoxification of toxic compounds and the catalyzed oxidation of fatty acids [30]. NADPH oxidase (also known as PHOX) is a membrane-bound enzyme that contributes to the production of O_2^- from O_2 in microglial cells leading to dopaminergic neuron damage [31, 32]. NADPH oxidase is a multimeric enzyme composed of plasma membrane bound gp91phox and p22phox subunits and cytosolic p40phox, p47phox and p67phox subunits. Upon activation, the cytosolic subunits undergo phosphorylation and translocate to the membrane, where together with small G proteins they associate with the membrane-bound subunits. The assembled and active enzyme complex then catalyzes the transfer of a single electron from NADPH to O_2 to release superoxide. Moreover, O_2^- is generated as a normal byproduct of the mitochondrial electron transfer chain. Depending on the availability of substrates and cofactors, O_2^- can react as a one-electron oxidant, oxidizing hydroquinones to semiquinone radicals, ascorbate, or epinephrine with the concomitant production of H_2O_2 or a one-electron reductant, e.g., quinones or peroxides in the presence of transition metals [30]. The flux of O_2^- is a function of the concentration of potential electron donors, the local concentration of O_2 and the second-order rate constants for the reactions between them. Two modes of operation by isolated mitochondria result in significant O_2^- production, predominantly from complex I, when (1) the mitochondria are not making ATP and consequently have a high Δp (proton motive force) and a reduced CoQ (coenzyme Q) pool, and when (2) a high NADH/NAD^+ ratio exists in the mitochondrial matrix. For mitochondria that are actively making ATP, and consequently have a lower Δp and NADH/NAD^+ ratio, the extent of O_2^- production is far lower.

Given the ability of mitochondria to produce superoxides and hamper a neuron's ability to produce ATP, which subsequently lead to apoptosis, several toxin-based models are employed to study PD and related molecular mechanisms. For instance, to mimic oxidative stress mechanisms in PD, researchers use rotenone, a complex I inhibitor, as well as other chemical inhibitors of electron flow that act further downstream in the electron transport chain because they increase ROS production and subsequent mitochondria-dependent apoptosis. Rotenone binds to the ubiquinone binding site of complex I and disrupts the electron transfer between the terminal iron-sulfur (FeS) cluster N2 and ubiquinone [33]. This process interferes with NADH's ability to produce ATP and pass electrons to CoQ, creating excess electrons within the mitochondrial matrix [34]. This complex I inhibition causes electrons to react with O_2 prematurely, incompletely reducing it to superoxide radicals instead of water. Therefore, rotenone-induced oxidative stress activates a downstream apoptotic cascade in dopaminergic cells, which helps explain the observed systemic reduction in complex I activity and oxidative stress in PD brains [35–37]. Another highly lipophilic, selective neurotoxicant similar to rotenone is 1-methyl-4-phenyl-1,2,3,6-tetrahydropyridine (MPTP), which induces clinical features very similar to human PD [38]. In the brain, MPTP is quickly metabolized to 1-methyl-4-phenyl-2,3-dihydropyridinium ($MPDP^+$) via monoamine oxidase B (MAO-B) in astroglial cells and serotonergic cells [39, 40]. $MPDP^+$ is an unstable molecule that undergoes spontaneous oxidation to its active metabolite 1-methyl-4-phenylpyridinium (MPP^+). MPP^+ is selectively taken up by dopaminergic neurons via dopaminergic transporter (DAT) where it exerts its neurotoxicity by inhibiting mitochondrial complex I, thereby leading to ATP reduction and superoxide generation [41]. Oxidopamine or 6-hydroxydopamine (6-OHDA) is another synthetic neurotoxicant that selectively destroys dopaminergic neurons by generating ROS such as superoxide radicals [42, 43]. Like MPP^+, 6-OHDA enters neurons via DAT [44]. It activates cell death pathways by generating intracellular free radicals and mitochondrial inhibition [45]. Therefore, 6-OHDA, like DA, could generate hydroxyl radicals and superoxide radicals by the deamination process via MAO or autooxidation, and iron-catalyzed via the Fenton reaction, thus further strengthening the free radical hypothesis of PD. However, 6-OHDA's exact mechanism of ROS production and neurotoxicity remains unclear.

4 Alkoxy Radicals ($RO^•$) and Peroxy Radicals ($ROO^•$) in PD

The most favorable biological substrates for peroxidation are the polyunsaturated fatty acid (PUFA) components of cell and subcellular membranes. Lipid peroxides result from the addition of double bonds or hydrogen abstraction in the presence of oxygen. Since PUFAs are more sensitive than saturated fatty acids, it is apparent that the activated methylene (RH) bridge represents a critical target site. The double bond adjacent to a methylene group weakens the methylene C-H bond, thereby rendering the hydrogen more susceptible to abstraction [46]. Like many radical reactions, lipid peroxidation is a multi-step process with initiation, propagation and termination. At the initiation step, ROS such as hydroperoxyl radicals or hydroxyl radicals react with PUFA to produce unstable fatty acid radicals that continue reacting with O_2 to produce unstable, intermediate fatty acid peroxy radicals. These fatty acid peroxy radicals and fatty acid radicals undergo chain reactions that produce organic hydroperoxides, which in turn can remove hydrogen from another PUFA [47]. This chain reaction is termed propagation, implying that one initiating hit can result in the conversion of numerous PUFAs to lipid hydroperoxides. Since most biological membranes are composed of PUFA, lipid peroxidation is considered the main molecular mechanism underlying oxidative damage to cell structures and in toxicity-induced cell death. The end products of lipid peroxidation are reactive aldehydes such as 4-hydroxy-trans-2-nonenal (4-HNE), 4-oxo-trans-2-nonenal (4-ONE), malondialdehyde (MDA), acrolein, isoprostanes, and isofurans [48, 49]. These markers are derived from arachidonic acid (ARA), which is released from neural membrane glycerophospholipids through the activation of cytosolic phospholipase A_2 ($cPLA_2$), an enzyme coupled with NMDA receptors through a G protein-independent mechanism [50, 51].

The primary end product of lipid peroxidation, 4-HNE, is a highly reactive lipophilic α,β-alkenal that forms stable adducts with nucleophilic groups on proteins such as thiols and amines [52], and it chemically modifies cellular macromolecules and DNA. Moreover, 4-HNE shows time- and dose-dependent activation of caspase-8, caspase-9 and caspase-3 as well as apoptotic cell death accompanied by DNA fragmentation [53]. Mechanistically, 4-HNE reduces glutathione (GSH) inhibition [54] of the NFkB signaling pathway [55], disinhibits mitochondrial complexes I and II, and it deactivates p53 [56] and poly-(ADP-ribose) polymerase (PARP) [57]. The increased levels of 4-HNE immunopositive neurons in the brain tissue and cerebrospinal fluid of PD patients indicate not only a pathophysiological role for oxidative stress in these diseases, but also a role for 4-HNE in neuronal apoptosis [53, 58, 59].

5 Nitric Oxide (NO) in PD

Nitric oxide (NO) is another potential source of oxidative stress. NO is produced by nitric oxide synthase (NOS) through converting L-arginine to L-citrulline utilizing NADPH oxidase and O_2 as cofactors [60, 61]. There are three isoforms of NOS: neuronal NOS (nNOS), inducible NOS (iNOS) and endothelial NOS (eNOS), of which nNOS is expressed in several neuronal subtypes except dopaminergic neurons. In contrast to nNOS, iNOS is not normally expressed in the brain; however, under pathological conditions, iNOS can be induced. Activated glial cells produce iNOS, which leads to increased production of NO. Indeed, elevated iNOS levels mediated by CD23 have been reported in the SN of patients with PD [62]. MPTP administration in mice also produces glial cell-mediated increases in iNOS expression and NO production [63]. Consequently, mice lacking the iNOS gene are less susceptible to MPTP-induced losses of SN DA neurons [64]. The MPTP-induced striatal dopamine depletion, however, remains intact in iNOS null mice, as does MPTP-induced microglial activation [63, 65]. Although poorly reactive, NO and O_2^- free radicals can combine to form the highly reactive nitrogen species peroxynitrite ($ONOO^-$), which can cause oxidative damage to various proteins such as tyrosine hydroxylase (TH) and α-synuclein [66, 67]. Iron content increases in the SN of PD patients and in animal models of the disease [68]. Through a superoxide-driven Fenton's reaction between hydrogen peroxide and the free ferrous iron catalyst, a substantial amount of highly reactive hydroxyl radicals (OH) can be produced. Reactive astrocytes produce myeloperoxidase (MPO), which oxidizes non-reactive nitrites (NO_2^-) that contribute to protein nitrosylation [69]. MPO is also implicated in the production of the non-radical oxidant hypochlorous acid (HOCl), which can damage macromolecules directly [70]. Altogether, an inflammatory, oxidative environment can be produced by activated glial cells in the SN region.

6 Antioxidants as Therapeutics for PD

PD is a multifactorial disease wherein glial activation, inflammation, oxidative stress and mitochondrial dysfunction play central roles in dopaminergic neurodegeneration, specifically in the nigrostriatum. Increasing efforts are being devoted to searching for neuroprotective agents that will protect against the irreversible loss of neurons. Administration of a dopamine agonist or levodopa has been widely used to treat PD symptoms, but does not alter disease pathogenesis. Dopaminergic neuroprotection in animal models of PD has been demonstrated with various substances including glial cell line-derived neurotrophic factor (GDNF), brain-derived neurotrophic factor (BDNF) and TGF-β. Additionally, various anti-inflammatory agents, such as NSAIDs, COX inhibitors, statins, pioglitazone and minocycline, have been used in different animal models of PD. However, most of these compounds failed in either preclinical trials or in human phase I trials due to their inability to cross the blood-brain barrier or to limited bioavailability. Moreover, they also cause side effects and toxicity in animals. Hence, developing successful neuroprotective therapeutic approaches to halt progression of PD requires a better understanding of the disease mechanism.

7 Vitamin Antioxidant Therapy

Oxidative stress is closely linked to other components of the degenerative process, such as mitochondrial dysfunction, excitotoxicity, nitric oxide toxicity and inflammation. That is, neuronal injury and cell death in both acute and chronic pathological conditions can result from oxidative damage, for example, through superoxide (O_2^-), hydroxyl (OH^-), peroxyl (RO_2^-), hydrogen peroxide (H_2O_2), and peroxynitrite ($ONOO^-$). Therefore, various vitamin antioxidants have been tested for their efficacy as scavengers of oxygen radicals and their potential as neuroprotective agents. Naturally, dietary sources supply many antioxidants. Vitamins C and E, β-carotene and coenzyme Q are the best known dietary antioxidants, of which Vitamin E is present in vegetable oils and found abundantly in wheat germ [71]. This fat soluble vitamin is absorbed in the gut and carried in the plasma by lipoproteins. Of eight natural isomeric forms of vitamin E, α-tocopherol is the most common and potent isomer. Being lipid soluble, vitamin E can effectively prevent lipid peroxidation of plasma membranes [71]. In the MPTP mouse model of PD, vitamin E inhibited the iron accumulation and thus reversed the MPTP-induced increase in oxidized glutathione (GSSG) and lipid peroxidation levels in brain tissues [72]. Moreover, in the 6-OHDA-induced rat model of PD, vitamin E significantly attenuated the effects of 6-OHDA on GSH and SOD in most brain regions [73], indicating that vitamin antioxidants may serve as potential therapeutic agents in retarding the progression of neurodegeneration. However, epidemiological evidence regarding the associations between antioxidant vitamin intake and PD is limited and inconsistent. Observational data from humans suggest that the combined administration of high-dose α-tocopherol (vitamin E) and ascorbate (vitamin C) supplementation slows the progression of PD [74] and that the dietary intake of vitamin E and β-carotene lowers the risk of developing PD [75]. In contrast, results from double-blind, randomized controlled trials found vitamin E to have no benefits in PD patients [76, 77].

8 Other Plant-Based Antioxidants

Plants contain a wide variety of endogenous, free radical-scavenging antioxidants such as phenolic compounds (e.g., phenolic acids, flavonoids, quinones, coumarins, lignans, stilbenes, tannins), nitrogen compounds (e.g., alkaloids, amines, betalains), terpenoids (including carotenoids) and some other endogenous metabolites rich in antioxidant activity. Many of these have shown protective effects against oxidative-induced neuronal death [71]. Although consumer demand for phytotherapeutic agents is growing, they need scientific validation before plant-derived extracts gain wider acceptance and use.

Apocynin (4-hydroxy-3-methoxyacetophenone) is a non-toxic plant-derived molecule that has been well-studied in cell culture and animals models of PD in our lab and elsewhere [78, 79]. Apocynin can effectively block NADPH oxidase and reduce ROS generation during neuronal injury or stress. Recently, we demonstrated that the apocynin dimer diapocynin is also neuroprotective and anti-neuroinflammatory in the MPTP animal model as well as in the progressively degenerative $LRRK2_{R1441G}$ transgenic mouse model [80, 81]. Importantly, we were able to demonstrate that diapocynin crosses the blood brain barrier, which is one of the main limitations for antioxidant therapies. Upon reaching the midbrain of MPTP-treated mice, it attenuates the nigral activation of microglial and astroglial cells, inhibits the proinflammatory molecule iNOS and the production of NADPH oxidase-mediated superoxide formation and decreases oxidative stress, thereby protecting the nigrostriatum and improving neurobehavioral performance, suggesting its potential as a therapeutic candidate for clinical trials of human PD patients. In these studies, both apocynin and diapocynin were orally administered at 300 mg/kg body weight. Although these high doses were not toxic to animals, there is a need for more efficacious apocynin analogs that will translate into human clinical trials.

In recent years resveratrol has gained much attention as a therapeutic for prevention and treatment of neurodegeneration disorders. Resveratrol is present in a variety of vegetables, fruits, grains, teas, and wines. It is protective against a number of cardiovascular and neurodegenerative diseases and cancer. Although the mechanisms behind resveratrol's health benefits have not yet been clearly elucidated, a number of studies have reported on its antioxidant, anti-inflammatory, and metal-chelating properties [82, 83], as well as its ability to activate Sirtuin 1 (SIRT1) and vitagenes, which can prevent the deleterious effects triggered by oxidative stress [84]. In fact, SIRT1 activation by resveratrol is gaining importance in the development of innovative treatment strategies for stroke and other neurodegenerative disorders [84].

Quercetin, found abundantly in vegetables and fruits, is another natural antioxidant flavonoid capable of protecting cells against oxidative damage, and thus has therapeutic potential for the prevention and treatment of cardiovascular disease, cancer, and neurodegenerative disease. Importantly, there is now compelling evidence of its neuroprotective role in various neurodegenerative diseases [85–88]. In nature, quercetin mainly occurs as glycosides, ethers, and to a lesser extent, sulfates. When tested in PC12 cells in a cell culture model of AD, a glycoside form of quercetin, quercetin-3'-glucoside, reduced H_2O_2-induced ROS generation and also protected against Aβ-induced cell death [89]. Isoquercitrin, another glycoside form of quercetin, was neuroprotective against 6-OHDA-induced neurotoxicity in a PC12 cell model of PD [90]. When the 6-OHDA-treated PC12 cells were pre-treated with isoquercitrin, the levels of ROS-scavenging enzymes (SOD, catalase, and GPx) increased and lipid peroxidation decreased. Similarly, quercetin treatment reduced protein carbonyl content and lipid hydroperoxide (LPO) levels in the striatum of 6-OHDA-treated rats [91].

9 Mitochondria-Targeted Antioxidant Therapy

Mitochondrial oxidative stress, mitochondrial DNA deletions, altered mitochondrial morphology and mitochondrial interactions with pathogenic proteins increase oxidative damage leading to dopaminergic neurodegeneration in PD. Therefore, therapeutic approaches targeting mitochondrial dysfunction and related oxidative stress hold great promise as potential cures for PD. MPTP and other complex-I inhibitors such as rotenone, maneb, paraquat, fenzaquin and trichloroethylene result in the loss of nigral dopaminergic neurons in mouse models of PD, implicating mitochondrial dysfunction in PD pathogenesis. Moreover, reduced complex-I activity and an increased susceptibility to MPP^+, the toxic metabolite of MPTP, were also observed in mitochondrial DNA from PD patients, clearly demonstrating the mtDNA-encoded defects in PD. Based on all the evidence, it could be inferred that intervening in one or more of these processes could alleviate the harmful effects of mitochondrial dysfunction. During the past decade, numerous antioxidant analogs have been developed to specifically target mitochondria and have been shown to improve mitochondrial function in experimental models of PD. To target small-molecule antioxidants to mitochondria, two general strategies have so far been shown to be safe and effective in pre-clinical studies: conjugations to lipophilic cations or incorporation into mitochondria-targeted peptides. Since lipophilic cations can easily pass through the lipid bilayers of plasma membranes and the mitochondrial inner membrane, they accumulate in the mitochondrial matrix in response to the large mitochondrial membrane potential (from outer positive to inner negative) [92]. The best characterized and most widely used lipophilic cation for conjugating small molecules is triphenylphosphonium (TPP), which has traditionally been used to determine mitochondrial inner membrane potential. Using TPP chemistry, Murphy and colleagues [92] developed a series of orally bioavailable mitochondria-targeted antioxidants, including MitoQ$_{10}$, MitoVitE and MitoTEMPOL [93].

Mito-Q$_{10}$ (Mito-quinone), the most studied mitochondria-targeted antioxidant, protects dopaminergic neurons from 6-OHDA in a cell model [94] and from MPTP-induced toxicity in a mouse model of PD [95]. Mito-Q$_{10}$ consists of TPP covalently attached to the ubiquinone moiety of Coenzyme Q (CoQ$_{10}$) through a ten-carbon alkyl chain. CoQ$_{10}$ is a component of the electron transport chain enabling cellular respiration and it works as a strong endogenous antioxidant. Like its parent CoQ$_{10,}$ MitoQ continually scavenges peroxyl, peroxynitrite and superoxide, thereby protecting mitochondria against lipid peroxidation. MitoVitE is a TPP-conjugated mitochondria-targeted antioxidant, which by coupling the antioxidant phenolic moiety of α-tocopherol, gets taken up by mitochondria about 80 times more effectively than vitamin E itself and thus affords better protection against oxidative damage [96]. MitoVitE also reduces H_2O_2-induced caspase activity and can prevent cell death in fibroblasts in patients with Friedrich ataxia, an inherited nervous system disease associated with decreased frataxin and increased iron-catalyzed oxidative damage [97]. In a study targeting cerebellar granule cells, MitoVitE diminished the ethanol-induced accumulation of intracellular oxidants and counteracted the suppression of not only glutathione peroxidase/glutathione reductase functions, but also the protein expression of γ-glutamylcysteine synthetase and total cellular glutathione levels [98]. MitoTEMPOL is another TPP derivative, but one with the stable piperidine nitroxide radical TEMPOL (4-hydroxy-2,2,6,6-tetramethylpiperidine-1-oxy). MitoTEMPOL also acts as a cytosolic SOD mimetic, converting superoxide molecules into water, and is able to detoxify ferrous iron by oxidizing it to ferric iron. Although MitoTEMPOL has not yet been tested in experimental models of PD, it reduced protein oxidation and mitochondrial and cytosolic ROS production in rat models of breast cancer [99] and diabetes [100], respectively. In an LRRK2^{R1441G} mouse model of PD, the novel mitochondria-targeted antioxidant MitoApo developed with apocynin, a plant-derived antioxidant and NADPH oxidase inhibitor, markedly improved coordinated motor skills and olfactory function [101]. The authors also showed that the presence of a highly lipophilic and delocalized cationic moiety in MitoApo-C_{11} makes it more cell-permeable and bioavailable [101]. In our own MitoApo studies, we have observed significant neuroprotection against MPP^+-induced loss of dopaminergic neurons in primary mesencephalic culture wherein MitoApo reduced glial cell-mediated inflammatory reactions. Moreover, administration of MitoApo in mice protects dopaminergic neurons and terminals from MPTP toxicity by reducing inflammatory reactions and oxidative stress (unpublished data). MitoPBN is a TPP derivative of phenoxy-butyl-nitrone. The spin trap PBN was chosen based on PBN's well-known reactivity with carbon-centered radicals [102]. MitoPBN is rapidly taken up by mitochondria and can block the oxygen-induced activation of uncoupled proteins [102].

Another major alternative approach to targeting antioxidants to mitochondria is through the use of small positively charged peptides call Szeto-Schiller (SS)-peptides [103]. SS-peptides contain an aromatic cationic sequence that facilitates the delivery of small molecules directly to mitochondria where they localize in the inner mitochondrial membrane with an approximately 1000–5000 fold accumulation [103, 104]. These SS-peptides can scavenge H_2O_2 and peroxynitrite and inhibit lipid peroxidation. Their antioxidant action can be attributed to the tyrosine or dimethyltyrosine residue [105]. By reducing mitochondrial ROS, these peptides inhibit mitochondrial permeability transition and cytochrome *c* release, thus preventing oxidant-induced neuronal apoptosis. Among the SS-peptides recently developed, SS-31 (D-Arg-(2′6′-dimethyltyrosine)-Lys-Phe-NH_2) and SS-20 (Phe-D-Arg-Phe-Lys-NH_2) have been studied most and both comprise a dimethyltyrosine residue, which reacts with a variety of free radicals and inhibits lipid peroxidation [106]. Studies with isolated mitochondria showed that both SS-31 and SS-20 prevented MPP^+-induced inhibition of oxygen consumption and ATP production and mitochondrial swelling, indicating their protective effect in cell culture models of PD [107]. Furthermore, SS-31 exhibited complete dose-dependent protection against the MPTP-induced loss of dopamine and its metabolites in the striatum, as well as against the loss of tyrosine hydroxylase immunoreactive neurons in the SN. These findings provide strong evidence that these neuroprotective peptides, which target both mitochondrial dysfunction and oxidative damage, are a promising approach for the treatment of PD [107].

10 Conclusions

PD is a complex, multifactorial disease condition strongly influenced by environmental factors. Exposure to different environmental conditions including pesticides, heavy metals, solvents (trichloroethylene), polychlorinated biphenols (PCBs) and repeated head injury increases the risk of developing sporadic PD later in life. Although the exact mechanisms underlying neurodegeneration in PD is not well understood, substantial evidence has implicated mitochondrial dysfunction and oxidative damage as important components of PD pathogenesis. Since the brain is particularly vulnerable to the effects of ROS due to its high demand for oxygen and its abundance of highly peroxidisable substrates, mitochondria-targeted interventions have emerged as a tool for modulating oxidative stress in the prevention and treatment of PD. As described above, a series of mitochondria-targeted antioxidants have been developed over the past few years showing great results in *in vitro* and *in vivo* models of PD. Despite their efficacy in animal studies, similar outcomes for these novel antioxidant therapies have not been borne out in clinical studies of neurodegenerative diseases. Therefore, research on effective strategies targeting mitochondria with bioactive molecules capable of penetrating the blood brain barrier is essential. Moreover, increasing the innate cellular antioxidant defense through other mitochondrial drug targets may be as important. PGC-1α, a master regulator of mitochondrial biogenesis, and Nrf2, a natural antioxidant and inflammation fighter, are possible therapeutic targets for PD, with important roles in the function and survival of dopaminergic neurons in the SN. However, at present, antioxidants and mitochondria-targeted therapeutics have seen very limited success in the prevention or treatment of PD, and randomized clinical trials in humans, as well as animal studies, are urgently needed to identify and understand the effects of mitochondria-targeted therapeutics in the treatment of PD.

Acknowledgements This chapter was supported by the National Institutes of Health grants NS039958, NS088206, ES10586 and NS78247. The W. Eugene and Linda Lloyd Endowed Chair to AGK and the Dean Endowed Professorship to AK is also acknowledged. We also thank Mr. Gary Zenitsky for assistance in preparing this manuscript.

References

1. Jankovic J, Stacy M. Medical management of levodopa-associated motor complications in patients with Parkinson's disease. CNS Drugs. 2007;21(8):677–92.
2. Dauer W, Przedborski S. Parkinson's disease: mechanisms and models. Neuron. 2003;39(6):889–909.
3. Mizuno Y, et al. Parkin and Parkinson's disease. Curr Opin Neurol. 2001;14(4):477–82.
4. Tanner CM. Is the cause of Parkinson's disease environmental or hereditary? Evidence from twin studies. Adv Neurol. 2003;91:133–42.
5. Gwinn-Hardy K. Genetics of parkinsonism. Mov Disord. 2002;17(4):645–56.
6. Tang B, et al. Association of PINK1 and DJ-1 confers digenic inheritance of early-onset Parkinson's disease. Hum Mol Genet. 2006;15(11):1816–25.
7. Weng YH, et al. PINK1 mutation in Taiwanese early-onset parkinsonism : clinical, genetic, and dopamine transporter studies. J Neurol. 2007;254(10):1347–55.
8. Bonifati V, et al. Re: LRRK2 mutation analysis in Parkinson disease families with evidence of linkage to PARK8. Neurology. 2008;70(24):2348. author reply 2348–9.
9. Schapira AH. Evidence for mitochondrial dysfunction in Parkinson's disease—a critical appraisal. Mov Disord. 1994;9(2):125–38.
10. Ben-Shachar D, Zuk R, Glinka Y. Dopamine neurotoxicity: inhibition of mitochondrial respiration. J Neurochem. 1995;64(2):718–23.

11. Hoehn MM, Yahr MD. Parkinsonism: onset, progression, and mortality. 1967. Neurology. 1998;50(2):318. and 16 pages following.
12. McGeer PL, Yasojima K, McGeer EG. Inflammation in Parkinson's disease. Adv Neurol. 2001;86:83–9.
13. Bosco DA, et al. Elevated levels of oxidized cholesterol metabolites in Lewy body disease brains accelerate alpha-synuclein fibrilization. Nat Chem Biol. 2006;2(5):249–53.
14. Nakabeppu Y, et al. Oxidative damage in nucleic acids and Parkinson's disease. J Neurosci Res. 2007;85(5):919–34.
15. Zhang J, et al. Parkinson's disease is associated with oxidative damage to cytoplasmic DNA and RNA in substantia nigra neurons. Am J Pathol. 1999;154(5):1423–9.
16. Sian J, et al. Alterations in glutathione levels in Parkinson's disease and other neurodegenerative disorders affecting basal ganglia. Ann Neurol. 1994;36(3):348–55.
17. Graham DG. Oxidative pathways for catecholamines in the genesis of neuromelanin and cytotoxic quinones. Mol Pharmacol. 1978;14(4):633–43.
18. Hermida-Ameijeiras A, et al. Autoxidation and MAO-mediated metabolism of dopamine as a potential cause of oxidative stress: role of ferrous and ferric ions. Neurochem Int. 2004;45(1):103–16.
19. Nappi AJ, Vass E. Comparative studies of enhanced iron-mediated production of hydroxyl radical by glutathione, cysteine, ascorbic acid, and selected catechols. Biochim Biophys Acta. 1997;1336(2):295–302.
20. Klegeris A, Korkina LG, Greenfield SA. Autoxidation of dopamine: a comparison of luminescent and spectrophotometric detection in basic solutions. Free Radic Biol Med. 1995;18(2):215–22.
21. Levine RL, et al. Methionine residues as endogenous antioxidants in proteins. Proc Natl Acad Sci U S A. 1996;93(26):15036–40.
22. Requejo R, et al. Cysteine residues exposed on protein surfaces are the dominant intramitochondrial thiol and may protect against oxidative damage. FEBS J. 2010;277(6):1465–80.
23. Conway KA, et al. Kinetic stabilization of the alpha-synuclein protofibril by a dopamine-alpha-synuclein adduct. Science. 2001;294(5545):1346–9.
24. Girotto S, et al. Dopamine-derived quinones affect the structure of the redox sensor DJ-1 through modifications at Cys-106 and Cys-53. J Biol Chem. 2012;287(22):18738–49.
25. LaVoie MJ, et al. Dopamine covalently modifies and functionally inactivates parkin. Nat Med. 2005;11(11):1214–21.
26. Cerchiaro G, Bolin C, Cardozo-Pelaez F. Hydroxyl radical oxidation of guanosine 5'-triphosphate (GTP): requirement for a GTP-Cu(II) complex. Redox Rep. 2009;14(2):82–92.
27. Cooke MS, et al. Oxidative DNA damage: mechanisms, mutation, and disease. FASEB J. 2003;17(10):1195–214.
28. Isobe C, Abe T, Terayama Y. Levels of reduced and oxidized coenzyme Q-10 and 8-hydroxy-2'-deoxyguanosine in the cerebrospinal fluid of patients with living Parkinson's disease demonstrate that mitochondrial oxidative damage and/or oxidative DNA damage contributes to the neurodegenerative process. Neurosci Lett. 2010;469(1):159–63.
29. Qureshi GA, et al. Increased cerebrospinal fluid concentration of nitrite in Parkinson's disease. Neuroreport. 1995;6(12):1642–4.
30. Berg D, Youdim MB, Riederer P. Redox imbalance. Cell Tissue Res. 2004;318(1):201–13.
31. Gao HM, Liu B, Hong JS. Critical role for microglial NADPH oxidase in rotenone-induced degeneration of dopaminergic neurons. J Neurosci. 2003;23(15):6181–7.
32. Wu XF, et al. The role of microglia in paraquat-induced dopaminergic neurotoxicity. Antioxid Redox Signal. 2005;7(5–6):654–61.
33. Scatena R, Bottoni P, Giardina B. Advances in mitochondrial medicine. In: Advances in experimental medicine and biology, vol. xii. Dordrecht: Springer Verlag; 2012. p. 461.
34. Hayes WJ, Laws ER. Handbook of pesticide toxicology. San Diego: Academic; 1991.
35. Greenamyre JT, et al. Complex I and Parkinson's disease. IUBMB Life. 2001;52(3–5):135–41.

36. Mizuno Y, et al. Deficiencies in complex I subunits of the respiratory chain in Parkinson's disease. Biochem Biophys Res Commun. 1989;163(3):1450–5.
37. Parker Jr WD, Boyson SJ, Parks JK. Abnormalities of the electron transport chain in idiopathic Parkinson's disease. Ann Neurol. 1989;26(6):719–23.
38. Jin H, et al. Targeted toxicants to dopaminergic neuronal cell death. Methods Mol Biol. 2015;1254:239–52.
39. Chiba K, Trevor A, Castagnoli Jr N. Metabolism of the neurotoxic tertiary amine, MPTP, by brain monoamine oxidase. Biochem Biophys Res Commun. 1984;120(2):574–8.
40. Kitahama K, et al. Distribution of type B monoamine oxidase immunoreactivity in the cat brain with reference to enzyme histochemistry. Neuroscience. 1991;44(1):185–204.
41. Ghosh A, et al. The peptidyl-prolyl isomerase Pin1 up-regulation and proapoptotic function in dopaminergic neurons: relevance to the pathogenesis of Parkinson disease. J Biol Chem. 2013;288(30):21955–71.
42. Jin H, et al. Histone hyperacetylation up-regulates protein kinase Cdelta in dopaminergic neurons to induce cell death: relevance to epigenetic mechanisms of neurodegeneration in Parkinson disease. J Biol Chem. 2014;289(50):34743–67.
43. Latchoumycandane C, et al. Dopaminergic neurotoxicant 6-OHDA induces oxidative damage through proteolytic activation of PKCdelta in cell culture and animal models of Parkinson's disease. Toxicol Appl Pharmacol. 2011;256(3):314–23.
44. Bove J, et al. Toxin-induced models of Parkinson's disease. NeuroRx. 2005;2(3):484–94.
45. Blum D, et al. Molecular pathways involved in the neurotoxicity of 6-OHDA, dopamine and MPTP: contribution to the apoptotic theory in Parkinson's disease. Prog Neurobiol. 2001;65(2):135–72.
46. Catala A. Reactive oxygen species, lipid peroxidation, and protein oxidation. In: Chemistry research and applications. New York: Nova Science Publishers; 2014.
47. Halliwell B, Gutteridge JM. Oxygen toxicity, oxygen radicals, transition metals and disease. Biochem J. 1984;219(1):1–14.
48. Aluru N, et al. Targeted mutagenesis of aryl hydrocarbon receptor 2a and 2b genes in Atlantic killifish (Fundulus heteroclitus). Aquat Toxicol. 2015;158:192–201.
49. Esterbauer H, Schaur RJ, Zollner H. Chemistry and biochemistry of 4-hydroxynonenal, malonaldehyde and related aldehydes. Free Radic Biol Med. 1991;11(1):81–128.
50. Farooqui AA, Horrocks LA. Glycerophospholipids in brain: phospholipase A2 in neurological disorders, vol. xvi. New York: Springer; 2007. p. 394.
51. Farooqui T, Farooqui AA. Lipid-mediated oxidative stress and inflammation in the pathogenesis of Parkinson's disease. Parkinsons Dis. 2011;2011:247467.
52. Ullery JC, Marnett LJ. Protein modification by oxidized phospholipids and hydrolytically released lipid electrophiles: Investigating cellular responses. Biochim Biophys Acta. 2012;1818(10):2424–35.
53. Liu W, et al. 4-hydroxynonenal induces a cellular redox status-related activation of the caspase cascade for apoptotic cell death. J Cell Sci. 2000;113(Pt 4):635–41.
54. Ahmed I, et al. Differential modulation of growth and glutathione metabolism in cultured rat astrocytes by 4-hydroxynonenal and green tea polyphenol, epigallocatechin-3-gallate. Neurotoxicology. 2002;23(3):289–300.
55. Yin G, et al. Lipid peroxidation-mediated inflammation promotes cell apoptosis through activation of NF-kappaB pathway in rheumatoid arthritis synovial cells. Mediators Inflamm. 2015;2015:460310.
56. Cao ZG, et al. Comparison of 4-hydroxynonenal-induced p53-mediated apoptosis in prostate cancer cells LNCaP and DU145. Contemp Oncol (Pozn). 2014;18(1):22–8.
57. Raza H, John A. 4-hydroxynonenal induces mitochondrial oxidative stress, apoptosis and expression of glutathione S-transferase A4-4 and cytochrome P450 2E1 in PC12 cells. Toxicol Appl Pharmacol. 2006;216(2):309–18.
58. Yoritaka A, et al. Immunohistochemical detection of 4-hydroxynonenal protein adducts in Parkinson disease. Proc Natl Acad Sci U S A. 1996;93(7):2696–701.
59. Zarkovic K. 4-hydroxynonenal and neurodegenerative diseases. Mol Aspects Med. 2003;24(4–5):293–303.

60. Day BJ, et al. A mechanism of paraquat toxicity involving nitric oxide synthase. Proc Natl Acad Sci U S A. 1999;96(22):12760–5.
61. Duval DL, et al. Characterization of hepatic nitric oxide synthase: identification as the cytokine-inducible form primarily regulated by oxidants. Mol Pharmacol. 1996;50(2):277–84.
62. Hunot S, et al. Nitric oxide synthase and neuronal vulnerability in Parkinson's disease. Neuroscience. 1996;72(2):355–63.
63. Liberatore GT, et al. Inducible nitric oxide synthase stimulates dopaminergic neurodegeneration in the MPTP model of Parkinson disease. Nat Med. 1999;5(12):1403–9.
64. Itzhak Y, et al. Resistance of neuronal nitric oxide synthase-deficient mice to cocaine-induced locomotor sensitization. Psychopharmacology (Berl). 1998;140(3):378–86.
65. Dehmer T, et al. Deficiency of inducible nitric oxide synthase protects against MPTP toxicity in vivo. J Neurochem. 2000;74(5):2213–6.
66. Ara J, et al. Inactivation of tyrosine hydroxylase by nitration following exposure to peroxynitrite and 1-methyl-4-phenyl-1,2,3,6-tetrahydropyridine (MPTP). Proc Natl Acad Sci U S A. 1998;95(13):7659–63.
67. Przedborski S, Vila M. The last decade in Parkinson's disease research. Basic sciences. Adv Neurol. 2001;86:177–86.
68. Hirsch EC. Altered regulation of iron transport and storage in Parkinson's disease. J Neural Transm Suppl. 2006;71:201–4.
69. van der Vliet A, et al. Oxidative stress in cystic fibrosis: does it occur and does it matter? Adv Pharmacol. 1997;38:491–513.
70. Hampton MB, Kettle AJ, Winterbourn CC. Inside the neutrophil phagosome: oxidants, myeloperoxidase, and bacterial killing. Blood. 1998;92(9):3007–17.
71. Uttara B, et al. Oxidative stress and neurodegenerative diseases: a review of upstream and downstream antioxidant therapeutic options. Curr Neuropharmacol. 2009;7(1):65–74.
72. Lan J, Jiang DH. Desferrioxamine and vitamin E protect against iron and MPTP-induced neurodegeneration in mice. J Neural Transm. 1997;104(4–5):469–81.
73. Perumal AS, et al. Vitamin E attenuates the toxic effects of 6-hydroxydopamine on free radical scavenging systems in rat brain. Brain Res Bull. 1992;29(5):699–701.
74. Fahn S. A pilot trial of high-dose alpha-tocopherol and ascorbate in early Parkinson's disease. Ann Neurol. 1992;32(Suppl):S128–32.
75. Miyake Y, et al. Dietary intake of antioxidant vitamins and risk of Parkinson's disease: a case-control study in Japan. Eur J Neurol. 2011;18(1):106–13.
76. Parkinson Study G. Effects of tocopherol and deprenyl on the progression of disability in early Parkinson's disease. N Engl J Med. 1993;328(3):176–83.
77. Scheider WL, et al. Dietary antioxidants and other dietary factors in the etiology of Parkinson's disease. Mov Disord. 1997;12(2):190–6.
78. Anantharam V, et al. Pharmacological inhibition of neuronal NADPH oxidase protects against 1-methyl-4-phenylpyridinium (MPP+)-induced oxidative stress and apoptosis in mesencephalic dopaminergic neuronal cells. Neurotoxicology. 2007;28(5):988–97.
79. Philippens IH, et al. Oral treatment with the NADPH oxidase antagonist apocynin mitigates clinical and pathological features of parkinsonism in the MPTP marmoset model. J Neuroimmune Pharmacol. 2013;8(3):715–26.
80. Dranka BP, et al. Diapocynin prevents early Parkinson's disease symptoms in the leucine-rich repeat kinase 2 (LRRK2R(1)(4)(4)(1)G) transgenic mouse. Neurosci Lett. 2013;549:57–62.
81. Ghosh A, et al. Anti-inflammatory and neuroprotective effects of an orally active apocynin derivative in pre-clinical models of Parkinson's disease. J Neuroinflammation. 2012;9:241.
82. Ndiaye M, et al. Red wine polyphenol-induced, endothelium-dependent NO-mediated relaxation is due to the redox-sensitive PI3-kinase/Akt-dependent phosphorylation of endothelial NO-synthase in the isolated porcine coronary artery. FASEB J. 2005;19(3):455–7.
83. Sun AY, et al. Botanical phenolics and brain health. Neuromolecular Med. 2008;10(4):259–74.
84. Sun AY, et al. Resveratrol as a therapeutic agent for neurodegenerative diseases. Mol Neurobiol. 2010;41(2–3):375–83.

85. Ansari MA, et al. Protective effect of quercetin in primary neurons against Abeta(1–42): relevance to Alzheimer's disease. J Nutr Biochem. 2009;20(4):269–75.
86. Haleagrahara N, Siew CJ, Ponnusamy K. Effect of quercetin and desferrioxamine on 6-hydroxydopamine (6-OHDA) induced neurotoxicity in striatum of rats. J Toxicol Sci. 2013;38(1):25–33.
87. Karuppagounder SS, et al. Quercetin up-regulates mitochondrial complex-I activity to protect against programmed cell death in rotenone model of Parkinson's disease in rats. Neuroscience. 2013;236:136–48.
88. Sabogal-Guaqueta AM, et al. The flavonoid quercetin ameliorates Alzheimer's disease pathology and protects cognitive and emotional function in aged triple transgenic Alzheimer's disease model mice. Neuropharmacology. 2015;93C:134–45.
89. Zhu JT, et al. Flavonoids possess neuroprotective effects on cultured pheochromocytoma PC12 cells: a comparison of different flavonoids in activating estrogenic effect and in preventing beta-amyloid-induced cell death. J Agric Food Chem. 2007;55(6):2438–45.
90. Magalingam KB, Radhakrishnan A, Haleagrahara N. Protective effects of flavonol isoquercitrin, against 6-hydroxy dopamine (6-OHDA)-induced toxicity in PC12 cells. BMC Res Notes. 2014;7:49.
91. Haleagrahara N, et al. Neuroprotective effect of bioflavonoid quercetin in 6-hydroxydopamine-induced oxidative stress biomarkers in the rat striatum. Neurosci Lett. 2011;500(2):139–43.
92. Ross MF, et al. Lipophilic triphenylphosphonium cations as tools in mitochondrial bioenergetics and free radical biology. Biochemistry (Mosc). 2005;70(2):222–30.
93. Jin H, et al. Mitochondria-targeted antioxidants for treatment of Parkinson's disease: preclinical and clinical outcomes. Biochim Biophys Acta. 2014;1842(8):1282–94.
94. Solesio ME, et al. The mitochondria-targeted anti-oxidant MitōQ reduces aspects of mitochondrial fission in the 6-OHDA cell model of Parkinson's disease. Biochim Biophys Acta. 2013;1832(1):174–82.
95. Ghosh A, et al. Neuroprotection by a mitochondria-targeted drug in a Parkinson's disease model. Free Radic Biol Med. 2010;49(11):1674–84.
96. Smith RA, et al. Selective targeting of an antioxidant to mitochondria. Eur J Biochem. 1999;263(3):709–16.
97. Jauslin ML, et al. Mitochondria-targeted antioxidants protect Friedreich Ataxia fibroblasts from endogenous oxidative stress more effectively than untargeted antioxidants. FASEB J. 2003;17(13):1972–4.
98. Siler-Marsiglio KI, et al. Mitochondrially targeted vitamin E and vitamin E mitigate ethanol-mediated effects on cerebellar granule cell antioxidant defense systems. Brain Res. 2005;1052(2):202–11.
99. Dickey JS, et al. Mito-tempol and dexrazoxane exhibit cardioprotective and chemotherapeutic effects through specific protein oxidation and autophagy in a syngeneic breast tumor preclinical model. PLoS One. 2013;8(8), e70575.
100. Pung YF, et al. Resolution of mitochondrial oxidative stress rescues coronary collateral growth in Zucker obese fatty rats. Arterioscler Thromb Vasc Biol. 2012;32(2):325–34.
101. Dranka BP, et al. A novel mitochondrially-targeted apocynin derivative prevents hyposmia and loss of motor function in the leucine-rich repeat kinase 2 (LRRK2(R1441G)) transgenic mouse model of Parkinson's disease. Neurosci Lett. 2014;583:159–64.
102. Murphy MP, et al. Superoxide activates uncoupling proteins by generating carbon-centered radicals and initiating lipid peroxidation: studies using a mitochondria-targeted spin trap derived from alpha-phenyl-N-tert-butylnitrone. J Biol Chem. 2003;278(49):48534–45.
103. Zhao K, et al. Cell-permeable peptide antioxidants targeted to inner mitochondrial membrane inhibit mitochondrial swelling, oxidative cell death, and reperfusion injury. J Biol Chem. 2004;279(33):34682–90.
104. Zhao K, et al. Transcellular transport of a highly polar 3+ net charge opioid tetrapeptide. J Pharmacol Exp Ther. 2003;304(1):425–32.

105. Szeto HH. Cell-permeable, mitochondrial-targeted, peptide antioxidants. AAPS J. 2006;8(2):E277–83.
106. Szeto HH. Mitochondria-targeted cytoprotective peptides for ischemia-reperfusion injury. Antioxid Redox Signal. 2008;10(3):601–19.
107. Yang L, et al. Mitochondria targeted peptides protect against 1-methyl-4-phenyl-1,2,3,6-tetrahydropyridine neurotoxicity. Antioxid Redox Signal. 2009;11(9):2095–104.

Simultaneous Activation of Nrf2 and Elevation of Dietary and Endogenous Antioxidants for Prevention and Improved Management of Parkinson's Disease

Kedar N. Prasad

1 Introduction

Parkinson's disease (PD) is a slow progressive neurological disorder associated with defects in the function of extrapyramidal system which controls voluntary movements. This disease is also associated with non-motor deficits and neurological symptoms, including impaired olfaction, autonomic failure, cognitive impairment, and psychiatric symptoms. PD is the commonest form of neurodegenerative disease after Alzheimer's disease. It is estimated that in normal individuals about 3–5 % of DA neurons are lost every decade; however, in PD patients, the rate of loss is greater than that found in normal individuals [1]. The analysis of autopsied samples of PD brain revealed that about 70–75 of DA neurons are lost at the time the disease becomes detectable. This suggests that the brain possesses a high degree of plasticity with respect to DA neuron function.

About one million people suffer from Parkinson Disease and about 60,000 new cases are diagnosed annually in the USA (Parkinson Disease Foundation, 2013). About 3–4 million people remained undiagnosed. Average age of individuals for developing PD is around 60 years. Incidence rate of PD increases with age. Men are one and half time more likely to develop PD than women.

Despite extensive laboratory and clinical research, it has hitherto not been possible to significantly reduce the incidence or the rate of progression of PD. The current used drugs improve some symptoms for a limited period of time. At present, there are no effective strategies to reduce the incidence or the progression of this disease.

K.N. Prasad, Ph.D. (✉)
Antioxidant Research Institute, Premier Micronutrient Corporation,
245 El Faisan Drive, San Rafael, CA 94903, USA
e-mail: knprasad@comcast.net

S.C. Bondy, A. Campbell (eds.), *Inflammation, Aging, and Oxidative Stress*,
Oxidative Stress in Applied Basic Research and Clinical Practice,
DOI 10.1007/978-3-319-33486-8_15

The currently drug treatment strategies are based on the symptoms rather than the causes of the disease. These drugs improve some symptoms of PD for a certain period of time. Therefore, additional approaches are needed for reducing the incidence, progression and improved management of PD. In order to accomplish this goal, it is essential to identify external and internal agents that enhance the risk of PD and agents that reduce oxidative stress and chronic inflammation simultaneously.

External agents: Some external agents, such as exposure to manganese (Mn) [2], pesticide, herbicides and solvents, such as trichloroethylene (TCE), perchloroethylene (PERC) and carbon tetrachloride (CCl) [3, 4] or accumulation of free iron in the substantia nigra [5] have been associated with the increased risk of PD. In 1980, increased incidence of PD-like disease was seen among users of the designer drug, meperidene, which contains 1-methyl-4-phenyl 1,2,3,6 tetrahydropyridine (MPTP), a neurotoxic byproduct formed during the synthesis of this drug [6]. At least one of the mechanisms of action of damage produced by above neurotoxins on DA neurons is mediated by free radicals.

Internal factors: Several laboratory and human studies have identified major biochemical and genetic defects which play a central role in the initiation and progression of PD. These defects include increased chronic oxidative stress [7–13], chronic inflammation [14, 15] and glutamate [16–18]. Mutated alpha-synuclein, DJ-1, Parkin, Pten-induced kinase-1 (Pink-1) or leucine-rich repeat kinase-2 (LRRK2) associated with familial PD [19–22], increased the sensitivity of DA neurons to oxidative stress [23–26]. It appears that increased oxidative stress is one of the earliest biochemical defects which initiate chronic inflammation and together they participate in the initiation and progression of PD. The levels of glutamate increase late on, and it participates in the progression of symptoms of the disease. Therefore, reducing oxidative stress, chronic inflammation and glutamate levels appears to be one of the rational choices for reducing the incidence, progression, and in combination with drug therapy, improving management of this disease.

An activation of a nuclear transcriptional factor Nrf2 (nuclear factor-erythroid 2-related factor-2), which increases the expression of target genes coding for antioxidant enzymes and phase-2-detoxifying enzymes, is being used as a target for developing new therapeutic agents that reduce oxidative stress in chronic neurological diseases. However, this strategy alone may not be sufficient to reduce the levels of oxidative stress and chronic inflammation optimally, because the levels of dietary and endogenous antioxidants also decrease in a high oxidative environment of PD; therefore, their levels must also be increased simultaneously. Antioxidants are known to reduce oxidative stress by scavenging free radicals, but their role in reducing inflammation [27–34], and glutamate release [35–40] and its neurotoxicity [17, 18], are not widely appreciated. The role of B-vitamins in reducing the release of glutamate [41, 42] is also not widely known. The antioxidant enzymes reduce oxidative stress in part by a mechanism that is different from that of antioxidant chemicals; they destroy free radicals by catalysis, whereas antioxidant chemicals remove them by scavenging. On the other hand, phase-2-detoxifying enzymes remove damaged proteins that could interfere with neuronal function from the brain. Therefore, the levels of antioxidant enzymes and phase-2-detoxifying enzymes, and dietary and endogenous antioxidants must be elevated simultaneously in order to reduce oxidative stress and chronic inflammation optimally. The levels of dietary and endogenous antioxidants can easily be increased by supplementation; however, increasing the levels of antioxidant enzymes and detoxifying enzymes is complex requiring an activation of Nrf2 and it's binding with antioxidant response elements (AREs) in the nucleus.

Normally, in response to increased reactive oxygen species (ROS), Nrf2 is activated and translocated from the cytoplasm to the nucleus where it binds with the AREs which then increased the expression of target genes coding for antioxidant enzymes and phase-2-detoxifying enzymes [43, 44]. During chronic oxidative stress which is found in patients with PD, Nrf2 may become resistant to ROS. This is evidenced by the fact that increased oxidative continues to occur in the presence of Nrf2 in PD. The question arises as to how to activate ROS-resistant Nrf2.

Although most laboratory studies showed that treatment with individual antioxidants and certain phytochemicals decreased neurotoxin-induced elevated levels of oxidative damage and loss of DA neurons in animal models of PD. Laboratory studies also revealed that the neuroprotective effects of these agents require the presence of Nrf2. In contrast to the effects of individual antioxidants on animal models of PD, supplementation with vitamin E and coenzyme Q10 in patients with PD have produced no effect in case with vitamin E [45, 46] and limited beneficial effects on some symptoms in case with coenzyme Q10 [47]. Supplementation with individual antioxidants in human PD is unlikely to increase the levels of antioxidant enzymes and detoxifying enzymes as well as dietary and endogenous chemicals simultaneously. A review on the effect of individual agents on the progression and symptoms of PD topic was published earlier [48].

This chapter describes the evidence to show that increased oxidative stress and chronic inflammation play an important role in the initiation and progression of PD, and that an elevation of the levels of antioxidant enzymes and phase-2-detoxifying enzymes, and dietary and endogenous antioxidant chemicals simultaneously may be necessary for reducing these biochemical defects optimally. In addition, the regulation of activation of Nrf2 which is essential for increasing the levels of antioxidant enzymes and phase-2-detoxifying enzymes is briefly discussed, and agents that can activate Nrf2 are identified. Laboratory and clinical studies with the individual agents on the prevention and improved management of PD are reviewed in order to emphasize that supplementation with a single agent may not be sufficient to activate Nrf2 and enhance the levels of multiple dietary and endogenous antioxidant simultaneously. This review proposes a mixture of micronutrients that may accomplish the above goal.

2 Evidence for Increased Oxidative Stress in PD

Human autopsied brain tissue: Evidence for increased oxidative stress has been documented by increased levels of markers of oxidative damage in brain tissue [7, 8], decreased levels of antioxidants enzymes [9, 10], and antioxidants [11, 12].

In addition, the NADOH oxidases (NOXs) are the major source of reactive oxygen species (ROS). Elevated levels of NOX-1 were found in the dopaminergic neurons of the substantia nigra in the autopsied brain tissues [13].

Animal studies: 6-hydroxydopamine (6-OHDA) and MPTP are commonly used as chemical animal models for PD, because they selective damage DA neurons and induce behavioral changes similar to those observed in human PD. The effects of these neurotoxins are mediated via free radicals [49, 50]. In addition, the levels of NOX-1 in the rat dopaminergic cells (N27 neuron) were increased after treatment with 6-hydroxydopamine. Injection of 6-hydroxydopamine directly into the striatum increased the levels of NOX-1 in dopaminergic neurons of the rat substantia nigra.

Mitochondria are the major site for producing free radicals and at the same time they are most sensitive to free radical damage. Mitochondrial dysfunction induced by free radicals plays a central role in most neurodegenerative diseases including PD [51, 52]. Mitochondrial dysfunction can be induced by diverse groups of external and internal agents. External agents include MPTP, insecticides and pesticides, whereas internal agents include increased oxidative stress and chronic inflammation, and mutated or aggregated alpha-synuclein, mutated PINK1, DJ-1 and PARKIN genes [22, 53, 54]. Importance of mitochondrial dysfunction in the pathogenesis of PD is further suggested by the fact that rotenone, an inhibitor of mitochondrial complex-1, induces clinical and biochemical features of human PD in animal models.

3 Evidence for Increased Chronic Inflammation in PD

Studies on chronic inflammation in human and animal PD models: Microglia initiated chronic inflammation responses also play an important role in the mechanism of degeneration of DA neurons in PD [14, 15]. Aggregated or nitrated alpha-synuclein activates microglia which release pro-inflammatory cytokines and other neurotoxic factors that contribute to the degeneration of DA neurons [55, 56]. Activated microglia also release nitric oxide and superoxide that promote inflammation as well as formation of abnormal alpha-synuclein (excessive amount or mutated form) that cause degeneration of DA neurons in transgenic mice model of PD [57].

In the autopsied brain samples of PD brains, the number of activated microglia cells increased in the substantia nigra during the progression of PD. The levels of pro-inflammatory cytokines IL-6 and TNF-alpha increased in both PD and Lewy body disease [58]. The presence of extracellular neuromelanin serves as a source of chronic inflammation that aggravates the rate of degeneration of DA neurons. Activated microglia may produce excessive amounts of pro-inflammatory cytokines, complement proteins, prostaglandins, adhesion molecules and reactive oxygen species (ROS) all of which are neurotoxic.

Cyclooxygenase (COX) is the rate-limiting enzyme in the synthesis of prostaglandins that are neurotoxic in excessive amounts [59]. The inducible isoforms COX-2 is up-regulated in the DA neurons of the autopsied brain samples of PD patients. The levels of COX-2 are also increased in DA neurons of chemical-induced animal PD models. The studies presented in this section clearly show that chronic inflammation plays an important role in degeneration and apoptosis of DA neurons in PD.

Neuromelanin granules accumulate in the substantia nigra of PD patients. Neuromelanin can cause degeneration in DA neurons by generating H_2O_2 when it is intact, or by releasing redox active metals such as iron, if it is disintegrated. In addition, dying DA neurons can release melanin that can initiate chronic inflammatory responses by activating microglia cells.

4 Evidence for Increased Glutamate

Glutamate is a major excitatory transmitter in the mammalian central nervous system, and is neurotoxic when present in excess at the synapses. With the depletion of nigrostriatal DA neurons, the glutamatergic projections from subthalamic nucleus to the basal ganglia output nuclei become overactive [16]. The glutamatergic activity also increased in the striatal region of the PD brain. One of the neurotoxic effects of glutamate is mediated via free radicals [17, 18].

Increased glutamate signaling in the substantia nigra played an important role in the mechanisms of neuronal death induced by chronic treatment of mice with MPTP [60]. This is substantiated by the fact that inhibitors of glutamate receptors ameliorated abnormality in motor movements in the animal model of PD [61, 62]. Antioxidants also block the toxic effects of glutamate [17, 18] confirming that neurotoxic effect of glutamate is mediated via free radicals.

5 Mutations or Over-expression of Certain Genes in PD

Mutations in certain genes, such as DJ-1, alpha-synuclein, PINK-1and PARKIN, have been identified in familial PD [19–22], and account for about 2–3 % of all cases of PD. The transgenic animal models confirm the role of these mutations in the pathology of PD [63, 64].

6 DJ-1 Gene

The wild-type DJ-1 acts as a stress sensor and increases its level in response to elevated oxidative stress [65]. Over expression of the wild-type DJ-1gene protects DA neurons against oxidative stress induced by H_2O_2 [66], dopamine and

6-hydroxydopamine [67, 68]. DJ-1 is very sensitive to oxidative stress and oxidized form of DJ-1 is considered as a biomarker for neurodegenerative diseases including sporadic PD [69]. Indeed, increased levels of oxidized DJ-1 have been found in the autopsied brain tissues of sporadic as well as in familial PD [70]. Treatment of DA neurons with 6-hydroxydopamine (6-OHDA) caused a translocation of DJ-1 from the cytoplasm to the nucleus [71]. Treatment of DA neurons with n-acetylcysteine blocked the above effect of 6-OHDA. This suggests that increased oxidative stress may be involved in translocation of wild-type DJ-1 from the cytoplasm to the nucleus. In contrast to the effect of 6-OHDA treatment on the wild-type DJ-1; however, 6-OHDA treatment of DA neurons failed to translocate mutated DJ-1 to the nucleus. Thus, the wild-type DJ-1 response to oxidative stress is analogous to Nrf2 which in response to increased ROS translocates itself from the cytoplasm to the nucleus.

Mutated DJ-1 makes DA neurons more vulnerable to oxidative stress-induced apoptosis [72]. It also promotes aggregation of alpha-synuclein that impairs mitochondrial function causing DA neurons to degenerate slowly [23].

7 Alpha-Synuclein Gene

The over-expression of wild-type alpha-synuclein caused degenerative changes in human DA neurons [73], and in transgenic rat DA neurons [74, 75] by inducing mitochondrial dysfunction which increased oxidative stress [76, 77]. The over-expression of human wild-type alpha-synuclein in differentiated neuroblastoma cells decreased their viability and increased their sensitivity to oxidative stress and neurotoxins such as H_2O_2, nitric oxide and prostaglandin E2 [78]. Increased oxidative stress also causes aggregation of alpha-synuclein that is toxic to DA neurons. It has been shown that not only insoluble aggregated alpha-synuclein but also soluble oligomer aggregates of alpha-synuclein are toxic to DA neurons [79].

Over-expression of wild type alpha-synuclein or mutated alpha-synuclein (A53T) increased the sensitivity of DA neurons to MPP+ which induced mitochondrial dysfunction, and 6-hydroxydopamine which increased oxidative stress in human neuroblastoma cells in culture [24]. Antioxidants such as Edaravone protected only against MPP+-induced toxicity, and epigallocatechin-3-o-gallate protected only against 6-hydroxydopamine-induced neurodegeneration [24]. This study suggests that one type of antioxidant is not sufficient to reduce neurodegeneration induced by diverse groups of toxins.

8 PINK1 Gene

One of the functions of wild-type PINK1 is to protect mitochondria against a variety of stress signaling pathways. Genetic ablation of wild-type PINK1caused loss of mitochondrial membrane potential, decrease in ATP synthesis, complex IV activity and mitochondrial electron transport chain function [25]. Impairment of mitochondrial electron transport chain and increased frequency of deletions of mitochondrial DNA which codes some of the subunits of mitochondrial electron transport chain have been found in the autopsied samples of the substantia nigra of PD brains. Like mutated alpha-synuclein, mutated PINK1 also impairs mitochondrial function.

9 PARKIN Gene

The wild-type PARKIN improves mitochondrial function [80]. Mutations in this gene can impair mitochondrial function and dynamics [26]. Nitric oxide and oxidative stress inhibits PARKIN function which can lead to degeneration of DA neurons. Loss of PARKIN gene in human midbrain DA neurons increases the transcription of monoamine oxidases, and oxidative stress reduces DA uptake and enhances spontaneous release of DA. Insertion of a wild-type PARKIN gene into these DA neurons prevents the above changes [81].

10 Regulation of Activation of Nrf2

Nrf2: The nuclear transcriptional factor, Nrf2 (nuclear factor-erythroid-2-related factor 2) belongs to the Cap′N′Collar (CNC) family that contains a conserved basic leucine zipper (bZIP) transcriptional factor [82]. Normally, Nrf2 is associated with Kelch-like ECH associated protein 1 (Keap1) which acts as an inhibitor of Nrf2 [83]. Keap1 protein serves as an adaptor to link Nrf2 to the ubiquitin ligase Cul-Rbx1 complex for degradation by proteasomes and maintains the steady levels of Nrf2 in the cytoplasm. Nrf2-keap1 complex is primarily located in the cytoplasm, Keap1 acts as a sensor for ROS/electrophilic stress.

Activation of Nrf2 by a ROS-dependent mechanism: Normally, increased ROS activates Nrf2 which dissociates itself from Keap1-Cul-Rbx1 complex in the cytoplasm and translocates itself in the nucleus where it heterodimerizes with a small Maf protein, binds with antioxidant response elements (AREs) leading to increased expression of target genes coding for antioxidant enzymes and phase-2-detoxyfying enzymes [43, 44, 84]. Acute oxidative stress such as observed during exercise activates Nrf2 by a ROS-dependent mechanism [85]. Pretreatment with n-acetylcysteine (NAC) prevented the ROS-induced activation of Nrf2 [86]. Since NAC scavenged all ROS, ROS was not available for activating the Nrf2/ARE pathway. This observation should not be interpreted to mean that antioxidant impairs the normal response of activating Nrf2 by ROS. This result suggests that during acute oxidative stress, supplementation with a single antioxidant may not be needed for reducing oxidative damage. As a matter of fact, repeated administration of a single antioxidant before strenuous exercise could be harmful because a single antioxidant would be oxidized in a high oxidative environment and would act as a pro-oxidant rather than as an antioxidant.

Activation of Nrf2 by a ROS-independent mechanism: In contrast to acute oxidative stress, Nrf2 becomes resistant to ROS during chronic oxidative stress [87–89], suggesting that activation of Nrf2 by a ROS-independent mechanism exists. This is evidenced by the fact that increased chronic oxidative stress occurs in neurodegenerative diseases such as PD. Therefore, identification of agents which can activate Nrf2 by a ROS-independent mechanism would be useful in reducing the incidence and progression of PD.

11 Binding of Nrf2 with ARE

An activation of Nrf2 alone is not sufficient to increase the levels of antioxidant enzymes and phase-2-detoxifying enzymes. Activated Nrf2 must bind with ARE in the nucleus for increasing the expression of target genes for antioxidant enzymes and phase-2-detoxyfying enzymes. This binding ability of Nrf2 with ARE was impaired in aged rats and this defect was restored by supplementation with alpha-lipoic acid [90]. It is unknown whether the binding ability of Nrf2 with ARE is impaired in PD.

12 Regulation of the Levels and Activity of Nrf2

Nrf2 regulates Keap1 levels by controlling its transcription, whereas Keap1 regulates Nrf2 levels by controlling its degradation by proteasome [91]. Immediate early response-3 (IER-3) gene, a multifunctional stress response gene, also regulates Nrf2 activity. Deletion of IER-3 gene increases Nrf2 activity, whereas overexpression of IER-3 decreases it [92].

13 Epigenetically Regulation of the Levels of Nrf2

The levels of Nrf2 are regulated epigenetically by methylation of CpG (cytosine-phosphate-guanosine) and acetylation of histone3. Hypermethylation of CpG [93] and hyperacetylation of histone3 [94] increase the expression of Nrf2, whereas hypomethylation of CpG and hypoacetylation of histone3 decrease it. Therefore, any agent that can cause hypermethylation of CpG or hyperacetylation of histone3 could be useful in prevention and improved management of PD.

14 Antioxidants Regulating Nrf2 Activation and Free Radicals Levels

Antioxidants scavenge free radicals directly at varying levels; however, some can activate Nrf2 by a ROS-independent mechanism as well as scavenge free radicals, while others can activate Nrf2 by a ROS-dependent mechanism. They are described here.

Increased oxidative stress, chronic inflammation and glutamate play an important role in the initiation and progression of Parkinson disease (PD). Mutations in DJ-1, alpah-synuclein, PINK-1 or PARKIN gene associated with the familial PD impairs mitochondrial function that increases oxiadtive stress.Thus, attenuation of these biochemical defects may reduce the incidnce, progression, and in combination with standard therapy, improve the management of this disease. I propose that an elevation of the levels of antioxidant enzymes, phase-2-detoxifying enzymes, and dietary and antioxidant compounds simultaneously may be needed to optimally reduce oxidative stress, chronic inflammation and glutamate. A mixture of micronutrients is proposed to attain this goal.

1. **Antioxidants scavenge Free radicals:** All dietary and endogenous antioxidant chemicals reduce varying levels of oxidative stress by directly scavenging free radicals.
2. **Antioxidants which activate Nrf2 by a ROS-independent mechanism and scavenge free radicals**: Some examples are vitamin E and genistein [95], alpha-lipoic acid,[90], curcumin [96], resveratrol [97, 98], omega-3-fatty acids,[99, 100], glutathione [101], NAC [102], and coenzyme Q10 [103]. Several plant-derived phytochemicals, such as epigallocatechin-3-gallate, carestol, kahweol, cinnamonyl-based compounds, zerumbone, lycopene and carnosol [82, 104, 105], genistein [95], allicin, a major organosulfur compound found in garlic [106], sulforaphane, a organosulfur compound, found in cruciferous vegetables,[107], and kavalactones (methysticin, kavain and yangonin) [108] can activate Nrf2 by a ROS-independent mechanism.
3. **Antioxidant which activates Nrf2 by a ROS-dependent mechanism:** L-carnitine activates Nrf2 by a ROS-dependent mechanism [109] probably by generating transient ROS.

15 Reducing Oxidative Stress Optimally

From the above groups of agents, a mixture of selected micronutrients containing multiple dietary antioxidants (vitamins A, C and E, beta-carotene and selenium), polyphenolic compounds (curcumin and resveratrol), endogenous antioxidants (alpha-lipoic acid, L-carnitine, and coenzyme Q10), a synthetic antioxidant N-acetylcysteine (NAC), and omega-3-fatty acids is proposed for reducing oxidative stress optimally by enhancing the levels of antioxidant enzymes and phase-2-detoxifying enzymes through activating the Nrf2/ARE pathway, and multiple dietary and endogenous antioxidant chemicals simultaneously. Animal and limited human studies suggest that the proposed micronutrients in the mixture can cross blood-brain barrier at varying levels. This is supported by the observation in which supplementation with individual antioxidants improved neurological symptoms.

16 Reducing Chronic Inflammation Optimally

Activation of Nrf2 also suppresses inflammation [110, 111]. Some individual antioxidant chemicals from the above groups reduce chronic inflammation [27–34]. Therefore, the proposed mixture of micronutrients for reducing oxidative stress may also decrease chronic inflammation optimally.

17 Reducing Glutamate Release Optimally

Increased pro-inflammatory stimuli and oxidative stress following TBI cause microglia to release excessive amounts of glutamate, which contributes to loss of neurons [35]. Release of glutamate was blocked by antioxidants, such as vitamin E [35], tempol, a superoxide dismutase mimetic, and edaravone, a synthetic antioxidant [36], quercetin [37], glutathione and vitamin E [38], alpha-lipoic acid [39] and coenzyme Q10 [40]. In addition to antioxidants, Vitamin B-6 [41], vitamin B12 [42] and vitamin B2 (riboflavin) [112] also reduce release of glutamate. Antioxidants such as vitamin E [17] and coenzyme Q10 [18] also protected neurons against glutamate-induced degeneration and death. Therefore, the proposed mixture of micronutrients in combination with B-vitamins may decrease the levels of glutamate optimally.

18 Nrf2 in Parkinson Disease

6-hydroxydopamine (6-OHDA) and 1-methyl-4-phenyl-1, 2, 3, 6-tetrahydropyridine (MPT) are neurotoxins commonly used for inducing Parkinson disease-like symptoms in animals and degeneration and death of dopaminergic neurons in cell culture models. These neurotoxins produced above effects by increasing the levels of chronic oxidative stress and inflammation. . Pre-treatment of human neuroblastoma cells in culture and mice with naringenin, a natural flavonoid, protected neurons against 6-OHDA-induced toxicity. The protective effect of naringenin was dependent upon the presence of Nrf2, because deletion of Nrf2 blocked its protective effect [113]. Puerarin treatment reduced neurodegeneration in the substantia nigra of rats following injection with 6-OHDA by activating the Nrf2/ARE pathway and elevating the levels of brain-derived neurotrophic factor (BDNF) [114]. Licochalcone E (Lico-E), a Glycyrrhiza inflate-derived chaocone, reduced inflammation in microglia cells and protected human neuroblastoma cells against 6-OHDA- and MPTP-induced neurotoxicity by activating the Nrf2/ARE pathway [115]. Resveratrol reduced paraquat-induced ROS and inflammation by activating the Nrf2/ARE pathway [116]. Several other phytochemicals, such as gastrodin, a main constituent of Chinese herbal medicine *Gastrodia elata* [117], tetramethylpyrazine, an extract of Ligusticum wallichii Franchat (ChuanXiong) [118], Mangiferin, a polyphenolic compound [119] protected against MPTP-induced oxidative stress, neurodegenerative changes and motor deficits by activating the Nrf2/ARE pathway.

Importance of Nrf2 in neuroprotection is demonstrated by the fact that deficiency of Nrf2 enhanced alpha-synuclein aggregation and chronic inflammation that contribute to the degeneration of DA neurons in PD [89]. Nrf2 deficiency also increased the sensitivity of mice to MPTP-induced behavior and biochemical changes. Overexpression of Nrf2 in astrocyte was sufficient to provide neuroprotection in MPTP mouse model of PD [88].

The above studies suggest that activation Nrf2 is essential for neuroprotection in laboratory experiments; however, in human PD, activation of Nrf2 alone may not be sufficient for prevention or improved management, because the levels of dietary and endogenous antioxidants decrease in a high oxidative environment; and therefore, they must also be elevated simultaneously.

19 Studies with Antioxidants, Polyphenolic Compounds and Nicotinamide in PD

In vitro studies with antioxidants and nicotinamide: Certain antioxidants such as vitamin A, beta-carotene and coenzyme Q10 inhibited formation of alpha-synuclein fibrils and destabilized preformed alpha-synuclein fibrils in a dose-dependent manner, whereas vitamin B2, vitamin B6, vitamin C and vitamin E were ineffective in vitro [120]. Melatonin, deprenyl and vitamin E inhibited auto-oxidation of DA in a dose-dependent manner, whereas vitamin C was ineffective [121]. Glutamate-induced degeneration of DA neurons was blocked by an analog of NAC (N-acetylcysteine amide) [122], vitamin E [17] and coenzyme Q10 [18]. Treatment of human neuroblastoma cells in culture with a relatively high dose of nicotinamide (vitamin B3) protected from MPP+-induced toxicity by decreasing the activities of complex I and alpha-ketoglutarate dehydrogenase, and the levels of reactive oxygen specie (ROS), and oxidation products of DNA and protein [123].

Studies with vitamin E in animal models of PD: Pre-treatment of rats with d-alpha tocopherol or dl-alpha-tocopherol significantly reduced 6-OHDA-induced behavior and biochemical abnormalities [124, 125]. Intramuscular administration of d-alpha-tocopheryl succinate, the most effective form of vitamin E [126], protected 6-OHDA-induced death of locus coeruleus neurons as well as behavioral and biochemical defects in rats [127, 128].

Studies with L-carnitine in animal models of PD: Quinolinic acid, an excitotoxin and free radical generator, and 3-nitropropionic acid and rotenone (mitochondrial toxins) induced oxidative damage to DA neurons, and behavioral alterations in animals similar to those observed in PD patients. Administration of L-carnitine, coenzyme Q10, vitamin E, alpha-lipoic acid or resveratrol reduced oxidative damage and behavior abnormalities in animal model of PD induced by these diverse groups of neurotoxins [129, 130].

Studies with N-acetylcysteine (NAC) in animal models of PD: Supplementation with NAC decreased the levels of alpha-synuclein in mice overexpressing wild-type human alpha-synuclein in the brain, and partially prevented the loss of dopaminergic terminals in these mice [131].

Studies with nicotinamide in animal models of PD: Nicotinamide, an inhibitor of histone deacetylase, preserves the activity of silent information regulator-1 (SIRT-1), a regulator of mitochondrial biogenesis [132]. This vitamin inhibits oxidative damage and improves mitochondrial function and thus can protect neurodegeneration and improve motor functions. In addition, in *Drosophila melanogaster* model of PD (an alpha-synuclein transgenic fly), nicotinamide treatment significantly improved the motor function (climbing ability) [123].

Studies with fish oil, melatonin or vitamin E in animal models of PD: Pre-treatment treatment of mice with fish oil, melatonin or vitamin E decreased the levels of MPTP-induced elevation of the activity of COX-2 and lipid peroxides in the homogenates of midbrain. Treatment with fish oil was more effective in reducing MPTP-induced rise in COX-2 activity than vitamin E or melatonin, whereas melatonin was more effective in reducing MPTP-induced rise in lipid peroxides than fish oil or vitamin E [133]. These results suggest that different antioxidants affect markers of increased oxidative stress and chronic inflammation differently.

Studies with omega-3-fatty acids in animal models of PD: Omega-3-fatty acids restricted diet increased the levels of NO in the striatum of young and adult rats but not in the substantia nigra; however, increased lipoperoxidation and decreased catalase activity were found in both regions of the brain, while total SOD activity was lowered in the striatum. In addition, fewer tyrosine hydroxylase- and brain-derived nerve growth factor-positive cells present in the substantia nigra compared to the controls

Studies with resveratrol in animal models of PD: Resveratrol is also an activator of SIRT1, and thus, stimulated mitochondrial biogenesis in mice and reduces production of reactive oxygen species [134]. Pretreatment of fibroblasts obtained from patients with early-onset PD carrying mutated Park2 gene with resveratrol enhanced mitochondrial oxidative function by activating Peroxisome proliferator-activated receptor-gamma coactivator-1α (PGC-1α) [135].

Studies with curcumin and a mixture of antioxidants in animal models of PD: Both curcumin and a mixture of dietary and endogenous antioxidants (a gift from Premier Micronutrient Corporation, Nashville, TN) reduced the incidence of death and hypokinesia induced by MPTP treatment in mice. Although both curcumin and an antioxidant mixture markedly blocked MPTP-induced depletion of tyrosine hydroxylase (TH) activity, only the antioxidant mixture enhanced the TH activity [136]. This suggested that an antioxidant mixture treatment was more effective than the curcumin treatment in preventing MPTP-induced depletion of TH in mice.

Treatment of neuroblastoma cells (PC-12derived from the rat and SH-SY5Y derived from the human) with curcumin prevented mutated alpha-synuclein (A53T)-induced cell death, by inhibiting oxidative stress, mitochondrial depolarization, cytochrome c release, and caspase-9 and caspase-3 activation [137, 138]

Studies with quercetin and an iron chelator in animal models of PD: Pretreatment of rats with a flavonoid quercetin and a chelator of iron desferrioxamine reduced 6-OHDA-induced increase in the levels of protein carbonyl, glutathione, dopamine, and SOD in the striatum [139].

Studies with certain phytochemicals in animal models of PD: Treatment with some phytochemicals such as methanol extract of *Garcinia indica* fruits [140], silymarin, derived from the seeds of the plant *Silyburn marianum* [141], and silibinin, a major constituent of silymarin [142] reduced MPTP- and 6-OHDA-induced toxicity on dopaminergic neurons in the substantia nigra of rats. These agents exhibit antioxidant and anti-inflammation activities.

20 Antioxidant Studies on Human PD

Epidemiological studies with vitamin E: Several epidemiologic studies suggested that diet rich in vitamin E may reduce the risk of PD [143, 144].

Intervention studies with vitamin E or coenzyme Q10: The results of epidemiologic study on the benefit of vitamin E in reducing the risk of PD could not be confirmed by an intervention study. For example, Deprenyl and Tocopherol Antioxidative Therapy of Parkinsonism (DATATOP), a randomized, double-blind, placebo-controlled, multi-center clinical trial, was initiated in 1989 in order to evaluate the efficacy of deprenyl (10 mg per day) and dl-tocopherol (2000 IU per day) individually and in combination in patients with early stage of PD when no therapy was required. The primary outcome was prolongation of the time needed for levodopa therapy. After a follow-up period of 8.2 years, deprenyl significantly delay the time when levodopa therapy was needed, but alpha tocopherol was ineffective [45, 46].

In another clinical study, supplementation with coenzyme Q10 at the highest dose of 1200 mg per day that was safe and well-tolerated by patients revealed that the incidence of disability in patients receiving coenzyme Q10 was less than in placebo controls; the benefit was greater in patients receiving the highest dosage [47]. Reviews of several open and controlled clinical studies revealed that daily supplementation with coenzyme Q10 either had no effect or had minimal benefit in early stage PD patients [145, 146].

In an Open-labeled clinical trial, the efficacy of high doses of alpha-tocopherol and ascorbate was tested in early phase PD patients who were not taking levodopa. The primary outcome was delay of the time when levodopa therapy was needed. The results showed that these antioxidants delayed the time when levodopa therapy was needed by 2.5 years [147].

21 Problems Associated with a Single Antioxidant in Human PD Studies

(1) High risk populations of PD have high levels of oxidative stress and chronic inflammation. Individual antioxidants are easily oxidized; therefore, administered antioxidant would be oxidized under such an oxidative environment and then may act as a pro-oxidant rather than as an antioxidant; (2); individual antioxidants cannot elevate the levels of antioxidant enzymes and phase-2-detoxifying enzymes by activating Nrf2 and dietary and endogenous antioxidant chemicals simultaneously; (3) distribution of dietary and endogenous antioxidant chemicals differs from one organ to another; they differ from one cell to another within the same organ; they also differ from one compartment to another within the same cell. Therefore, supplement with one may not reduce oxidative stress and inflammation in the brain optimally; (4) a single antioxidant cannot reduce oxidative stress in both lipid and aqueous environments of the cells.

22 Proposed Criteria to be included in the Experimental Designs of Clinical Studies on Antioxidants in PD

(1) High risk populations, such as individuals 50 years or older or with a family history of PD can be used for the prevention study. Patients with an early phase of PD not requiring medications can be used for reducing the progression study, and those PD patients receiving medications can be utilized for the improved management study; (2) experimental designs should be randomized, double-blinded and placebo-controlled; (3) number of participants in the study should be high, generally in hundreds, for a meaningful statistical analysis and conclusions; (4) levels of markers of oxidative stress and chronic inflammation in subset population of smaller sample size before and after treatments with antioxidants should be measured; (5) daily oral supplementation with the proposed mixture of micronutrients that can enhance the levels of antioxidant enzymes and phase-2-detoxifying enzymes through activation of Nrf2, and the levels of dietary and endogenous antioxidant chemicals simultaneously should be included; (6) doses of antioxidants (high, but safe) and dose-schedules of twice-a-day should be adopted; (7) primary end-points, such as incidence of PD (for the prevention study) and time interval between micronutrient treatment and the need for levodopa therapy (for the progression study), and improvement in tremor and rigidity (for the management study) are appropriate; and (8) treatment and observation periods of generally 5 years (for the prevention study) and 2–3 years for the progression study and observation period of 3–5 years (for the improved management study,) are adequate.

The experimental designs of most previous clinical studies on antioxidants and PD have taken into consideration only criteria 1,2, 3, 7 and 8, but the criteria 4, 5, and 6 were not included which may account for the inconsistent results.

23 Proposed Prevention Strategies for PD

Primary prevention: The major objective of primary prevention is to protect healthy individuals from developing PD. Individuals 50 years or older, and individuals with a family history of PD who have not developed any clinical symptoms of PD are suitable for the primary prevention study. The proposed mixture of micronutrients is recommended for the study of primary prevention. This recommendation is based on the hypothesis that increased oxidative stress and chronic inflammation play a key role in the initiation of neurodegenerative changes in the brain that lead to PD, and that this mixture may enhance the levels of antioxidant enzymes and phase-2-detoxifying enzymes through the activation of the Nrf2/ARE pathway, and dietary and endogenous antioxidants simultaneously.

The gene HOP (TUM-1) is essential for the development of Drosophila melanogaster (fruit fly). A mutation in this gene markedly increases the risk of developing a leukemia-like tumor in female flies. In collaboration with Dr Bhattacharya of NASA Moffat Field, CA, we observed that whole-body irradiation of these flies with proton radiation dramatically increased the incidence of cancer compared to that in un-irradiated flies. Treatment with a mixture of antioxidants before and after irradiation blocked the proton radiation-induced cancer in fruit flies [148]. This finding is of particular interest, because it suggests that heritable genetic basis of the disease can be prevented by antioxidant treatment.

Secondary prevention: The purpose of secondary prevention is to stop or slow the progression of PD. Individuals who exhibit early sign of PD, but are not taking any medication, can be included in the secondary prevention study. The mixture of micronutrients recommended for the primary prevention study is also recommended for the secondary prevention study.

24 Proposed Strategies for Improving the Management of PD

Current drug and surgical treatments of PD: The major objective of treatment is to slow down the progression of disease and improve the symptoms of the disease. The treatment of PD includes medications that elevate dopamine in the surviving DA neurons. Levodopa therapy that increased the levels of DA in DA neurons is considered a gold standard for the treatment of PD, but its toxicity becomes a limiting factor, and the treatment is discontinued after about 5 years. The reasons for this effect of levodopa are unknown. In vitro studies have suggested that treatment of neuronal cells in culture with L-dopa is very toxic. This is due to the fact that L-dopa generates excessive amounts of free radicals during auto-oxidation as well as during oxidative metabolism of its product, DA; however, from animal studies it appears that there is no evidence of similar effects of L-dopa in vivo [149]. In a randomized, double-blind, placebo-controlled trial involving 361 patients with an early stage PD, the effects of various doses of levodopa for a period of 40 weeks were investigated [150]. The results showed that the patients receiving the highest dose of levodopa had significantly more dyskinesia, hypertonia, infection, headache, and nausea than those receiving placebo. The clinical data showed that levodopa treatment either slowed the progression of PD or has improved the symptoms of the disease. However, neuroimaging data suggested that levodopa treatment increased the rate of loss of nigrostriatal DA nerve terminals or it reduced the levels of DA transporter more than that produced by placebo treatment. A further investigation on this issue revealed that dose of levodopa is a factor in producing motor complications of dyskinesia and wearing-off, and these can develop as early as 5–6 months at high levodopa doses [151]. After about 5 years, L-dopa has a potential to cause increased oxidative damage peripherally and/or centrally; therefore, it appears rational to propose that reducing oxidative damage during levodopa therapy may improve the efficacy of this therapy by reducing the side-effects of levodopa. This would then allow levodopa treatment to be effective for a period longer than that currently expected. Furthermore, if the oxidation of L-dopa is reduced by decreasing oxidative stress; it would then be possible to reduce the dosage of levodopa without sacrificing its efficacy.

In some cases, surgery may be appropriate, if the disease doesn't respond to a standard drug therapy. A therapy called deep brain stimulation (DBS) has now been approved by the U.S. Food and Drug Administration. Current drugs used in the treatment of PD or surgical intervention do not reduce oxidative stress or chronic inflammation; and therefore DA neurons continue to die despite drug or surgical treatment.

In a rat PD model (induced by rotenone), the efficacy of oral L-dopa therapy with or without various doses of coenzyme Q10 was evaluated [152]. The results showed that L-dopa therapy improved the symptoms and restored striatal DA levels, but it did not show any significant effect on striatal mitochondrial complex I activity, ATP levels or the expression of Bcl2. Administration of coenzyme Q10 at a high dose with L-dopa increased striatal complex I activity, ATP levels, DA levels and Bcl2 expression compared to coenzyme Q10 treatment at low doses with L-dopa.

Proposed mixture of micronutrients in combination with drug therapy: The mixture of micronutrients recommended for primary prevention can be used in combination with standard drug therapy. This strategy may reduce the progression and prolong the beneficial effects of current drugs in patients with PD by protecting surviving DA neurons from damage produced by increased oxidative stress and chronic inflammation.

25 Safety of Ingredients in the Proposed Mixture

All ingredients in the proposed mixture of micronutrients are safe and come under category of "Food Supplement", and therefore, do not require FDA approval for their use. However, a few of them could produce harmful effects at higher doses in some individuals when consumed daily over an extended period. Vitamin A at doses of 10,000 IU or more per day can cause birth defects in pregnant women, and beta-carotene at doses 50 mg or more can produce bronzing of the skin that is reversible on discontinuation. Vitamin C as ascorbic acid at high doses (10 g or more per day) can cause diarrhea in some individuals. Vitamin E at high doses (2000 IU or more per day) can induce defects in blood clotting after long-term consumption. Vitamin B6 at high doses (50 mg or more per day) may produce peripheral neuropathy, and selenium at doses 400 μg or more per day can cause skin and liver toxicity after long-term consumption. Coenzyme Q10 at daily doses are 30–400 mg has no known toxicity. N-acetylcysteine doses of 250–1500 mg and alpha-lipoic acid doses of 600 mg are used in humans without reported toxicity.

In conclusion, increased oxidative stress and chronic inflammation play a central role in the initiation and progression of PD in humans. An accumulation of excess levels of glutamate participates in the progression of PD. A mixture of micronutrients containing dietary and endogenous antioxidant chemicals, curcumin, resveratrol, and nicotinamide and omega-3-fatty acids may increase the levels of antioxidant enzymes and phase-2-detoxifying enzymes by activating a nuclear transcriptional factor Nrf2 and dietary and endogenous antioxidants simultaneously. This mixture of micronutrients may reduce the incidence of PD and the progression in patients with early phase PD. The same mixture of micronutrients, in combination with the levodopa therapy, may prolong the efficacy of this drug for a longer period of time, and reduce long-term side effects which appear to be due to increased oxidative damage to neurons.

Financial Conflict The author is an independent consultant to nutrition industries.

References

1. Mandel S, Grunblatt E, Riederer P, Gerlach M, Levites Y, Youdim MB. Neuroprotective strategies in Parkinson's disease : an update on progress. CNS Drugs. 2003;17:729–62.
2. Racette BA, Aschner M, Guilarte TR, Dydak U, Criswell SR, Zheng W. Pathophysiology of manganese-associated neurotoxicity. Neurotoxicology. 2012;33:881–6.
3. Pezzoli G, Cereda E. Exposure to pesticides or solvents and risk of Parkinson disease. Neurology. 2013;80:2035–41.
4. Goldman SM, Quinlan PJ, Ross GW, Marras C, Meng C, Bhudhikanok GS, et al. Solvent exposures and Parkinson disease risk in twins. Ann Neurol. 2012;71:776–84.
5. Andersen JK. Iron dysregulation and Parkinson's disease. J Alzheimers Dis. 2004;6:S47–52.
6. Ballard PA, Tetrud JW, Langston JW. Permanent human parkinsonism due to 1-methyl-4-phenyl-1,2,3,6-tetrahydropyridine (MPTP): seven cases. Neurology. 1985;35:949–56.
7. Dexter DT, Holley AE, Flitter WD, Slater TF, Wells FR, Daniel SE, et al. Increased levels of lipid hydroperoxides in the parkinsonian substantia nigra: an HPLC and ESR study. Mov Disord. 1994;9:92–7.
8. Ebadi M, Srinivasan SK, Baxi MD. Oxidative stress and antioxidant therapy in Parkinson's disease. Prog Neurobiol. 1996;48:1–19.
9. Ambani LM, Van Woert MH, Murphy S. Brain peroxidase and catalase in Parkinson disease. Arch Neurol. 1975;32:114–8.

10. Kish SJ, Morito C, Hornykiewicz O. Glutathione peroxidase activity in Parkinson's disease brain. Neurosci Lett. 1985;58:343–6.
11. Fitzmaurice PS, Ang L, Guttman M, Rajput AH, Furukawa Y, Kish SJ. Nigral glutathione deficiency is not specific for idiopathic Parkinson's disease. Mov Disord. 2003;18:969–76.
12. Sofic E, Sapcanin A, Tahirovic I, Gavrankapetanovic I, Jellinger K, Reynolds GP, et al. Antioxidant capacity in postmortem brain tissues of Parkinson's and Alzheimer's diseases. J Neural Transm Suppl. 2006;39–43.
13. Choi DH, Cristovao AC, Guhathakurta S, Lee J, Joh TH, Beal MF, et al. NADPH oxidase 1-mediated oxidative stress leads to dopamine neuron death in Parkinson's disease. Antioxid Redox Signal. 2012;16:1033–45.
14. Whitton PS. Inflammation as a causative factor in the aetiology of Parkinson's disease. Br J Pharmacol. 2007;150:963–76.
15. McGeer PL, McGeer EG. Glial reactions in Parkinson's disease. Mov Disord. 2008;23:474–83.
16. Blandini F, Porter RH, Greenamyre JT. Glutamate and Parkinson's disease. Mol Neurobiol. 1996;12:73–94.
17. Schubert D, Kimura H, Maher P. Growth factors and vitamin E modify neuronal glutamate toxicity. Proc Natl Acad Sci U S A. 1992;89:8264–7.
18. Sandhu JK, Pandey S, Ribecco-Lutkiewicz M, Monette R, Borowy-Borowski H, Walker PR, et al. Molecular mechanisms of glutamate neurotoxicity in mixed cultures of NT2-derived neurons and astrocytes: protective effects of coenzyme Q10. J Neurosci Res. 2003;72:691–703.
19. Gandhi PN, Chen SG, Wilson-Delfosse AL. Leucine-rich repeat kinase 2 (LRRK2): a key player in the pathogenesis of Parkinson's disease. J Neurosci Res. 2009;87:1283–95.
20. Giaime E, Sunyach C, Druon C, Scarzello S, Robert G, Grosso S, et al. Loss of function of DJ-1 triggered by Parkinson's disease-associated mutation is due to proteolytic resistance to caspase-6. Cell Death Differ. 2010;17(1):158–69.
21. Fitzgerald JC, Plun-Favreau H. Emerging pathways in genetic Parkinson's disease: autosomal-recessive genes in Parkinson's disease--a common pathway? FEBS J. 2008;275:5758–66.
22. Dodson MW, Guo M. Pink1, Parkin, DJ-1 and mitochondrial dysfunction in Parkinson's disease. Curr Opin Neurobiol. 2007;17:331–7.
23. Batelli S, Albani D, Rametta R, Polito L, Prato F, Pesaresi M, et al. DJ-1 modulates alpha-synuclein aggregation state in a cellular model of oxidative stress: relevance for Parkinson's disease and involvement of HSP70. PLoS One. 2008;3, e1884.
24. Ma L, Cao TT, Kandpal G, Warren L, Fred Hess J, Seabrook GR, et al. Genome-wide microarray analysis of the differential neuroprotective effects of antioxidants in neuroblastoma cells overexpressing the familial Parkinson's disease alpha-synuclein A53T mutation. Neurochem Res. 2010;35(1):130–42.
25. Gegg ME, Cooper JM, Schapira AH, Taanman JW. Silencing of PINK1 expression affects mitochondrial DNA and oxidative phosphorylation in dopaminergic cells. PLoS One. 2009;4, e4756.
26. Bueler H. Impaired mitochondrial dynamics and function in the pathogenesis of Parkinson's disease. Exp Neurol. 2009;218:235–46.
27. Abate A, Yang G, Dennery PA, Oberle S, Schroder H. Synergistic inhibition of cyclooxygenase-2 expression by vitamin E and aspirin. Free Radic Biol Med. 2000;29:1135–42.
28. Devaraj S, Tang R, Adams-Huet B, Harris A, Seenivasan T, de Lemos JA, et al. Effect of high-dose alpha-tocopherol supplementation on biomarkers of oxidative stress and inflammation and carotid atherosclerosis in patients with coronary artery disease. Am J Clin Nutr. 2007;86:1392–8.
29. Fu Y, Zheng S, Lin J, Ryerse J, Chen A. Curcumin protects the rat liver from CCl4-caused injury and fibrogenesis by attenuating oxidative stress and suppressing inflammation. Mol Pharmacol. 2008;73:399–409.
30. Lee HS, Jung KK, Cho JY, Rhee MH, Hong S, Kwon M, et al. Neuroprotective effect of curcumin is mainly mediated by blockade of microglial cell activation. Pharmazie. 2007;62:937–42.

31. Peairs AT, Rankin JW. Inflammatory response to a high-fat, Low-carbohydrate weight loss diet: effect of antioxidants. Obesity (Silver Spring). 2008;16(7):1573–8.
32. Rahman S, Bhatia K, Khan AQ, Kaur M, Ahmad F, Rashid H, et al. Topically applied vitamin E prevents massive cutaneous inflammatory and oxidative stress responses induced by double application of 12-O-tetradecanoylphorbol-13-acetate (TPA) in mice. Chem Biol Interact. 2008;172:195–205.
33. Suzuki YJ, Aggarwal BB, Packer L. Alpha-lipoic acid is a potent inhibitor of NF-kappa B activation in human T cells. Biochem Biophys Res Commun. 1992;189:1709–15.
34. Zhu J, Yong W, Wu X, Yu Y, Lv J, Liu C, et al. Anti-inflammatory effect of resveratrol on TNF-alpha-induced MCP-1 expression in adipocytes. Biochem Biophys Res Commun. 2008;369:471–7.
35. Barger SW, Goodwin ME, Porter MM, Beggs ML. Glutamate release from activated microglia requires the oxidative burst and lipid peroxidation. J Neurochem. 2007;101:1205–13.
36. Dohare P, Hyzinski-Garcia MC, Vipani A, Bowens NH, Nalwalk JW, Feustel PJ, et al. The neuroprotective properties of the superoxide dismutase mimetic tempol correlate with its ability to reduce pathological glutamate release in a rodent model of stroke. Free Radic Biol Med. 2014;77:168–82.
37. Lu CW, Lin TY, Wang SJ. Quercetin inhibits depolarization-evoked glutamate release in nerve terminals from rat cerebral cortex. Neurotoxicology. 2013;39:1–9.
38. Hurtado O, De Cristobal J, Sanchez V, Lizasoain I, Cardenas A, Pereira MP, et al. Inhibition of glutamate release by delaying ATP fall accounts for neuroprotective effects of antioxidants in experimental stroke. FASEB J. 2003;17:2082–4.
39. Santos PS, Campelo LM, Freitas RL, Feitosa CM, Saldanha GB, Freitas RM. Lipoic acid effects on glutamate and taurine concentrations in rat hippocampus after pilocarpine-induced seizures. Arq Neuropsiquiatr. 2011;69:360–4.
40. Chang Y, Huang SK, Wang SJ. Coenzyme Q10 inhibits the release of glutamate in rat cerebrocortical nerve terminals by suppression of voltage-dependent calcium influx and mitogen-activated protein kinase signaling pathway. J Agric Food Chem. 2012;60:11909–18.
41. Yang TT, Wang SJ. Pyridoxine inhibits depolarization-evoked glutamate release in nerve terminals from rat cerebral cortex: a possible neuroprotective mechanism? J Pharmacol Exp Ther. 2009;331:244–54.
42. Hung KL, Wang CC, Huang CY, Wang SJ. Cyanocobalamin, vitamin B12, depresses glutamate release through inhibition of voltage-dependent Ca2+ influx in rat cerebrocortical nerve terminals (synaptosomes). Eur J Pharmacol. 2009;602:230–7.
43. Hayes JD, Chanas SA, Henderson CJ, McMahon M, Sun C, Moffat GJ, et al. The Nrf2 transcription factor contributes both to the basal expression of glutathione S-transferases in mouse liver and to their induction by the chemopreventive synthetic antioxidants, butylated hydroxyanisole and ethoxyquin. Biochem Soc Trans. 2000;28:33–41.
44. Chan K, Han XD, Kan YW. An important function of Nrf2 in combating oxidative stress: detoxification of acetaminophen. Proc Natl Acad Sci U S A. 2001;98:4611–6.
45. Shoulson I. DATATOP: a decade of neuroprotective inquiry. Parkinson study group. Deprenyl and tocopherol antioxidative therapy of parkinsonism. Ann Neurol. 1998;44:S160–6.
46. Group TPS. Effects of tocopherol and deprenyl on the progression of disability in early Parkinson's disease. N Engl J Med. 1993;176–83.
47. Shults CW, Oakes D, Kieburtz K, Beal MF, Haas R, Plumb S, et al. Effects of coenzyme Q10 in early Parkinson disease: evidence of slowing of the functional decline. Arch Neurol. 2002;59:1541–50.
48. Prasad KN, Cole WC, Kumar B. Multiple antioxidants in the prevention and treatment of Parkinson's disease. J Am Coll Nutr. 1999;18:413–23.
49. Salazar J, Mena N, Hunot S, Prigent A, Alvarez-Fischer D, Arredondo M, et al. Divalent metal transporter 1 (DMT1) contributes to neurodegeneration in animal models of Parkinson's disease. Proc Natl Acad Sci U S A. 2008;105:18578–83.
50. Ebadi M, Sharma SK. Peroxynitrite and mitochondrial dysfunction in the pathogenesis of Parkinson's disease. Antioxid Redox Signal. 2003;5:319–35.

51. Arduino DM, Esteves AR, Cardoso SM, Oliveira CR. Endoplasmic reticulum and mitochondria interplay mediates apoptotic cell death: relevance to Parkinson's disease. Neurochem Int. 2009;55:341–8.
52. Gubellini P, Picconi B, Di Filippo M, Calabresi P. Downstream mechanisms triggered by mitochondrial dysfunction in the basal ganglia: from experimental models to neurodegenerative diseases. Biochim Biophys Acta. 2010;1802(1):151–61.
53. Gautier CA, Kitada T, Shen J. Loss of PINK1 causes mitochondrial functional defects and increased sensitivity to oxidative stress. Proc Natl Acad Sci U S A. 2008;105:11364–9.
54. Lee SJ. Alpha-synuclein aggregation: a link between mitochondrial defects and Parkinson's disease? Antioxid Redox Signal. 2003;5:337–48.
55. Reynolds AD, Glanzer JG, Kadiu I, Ricardo-Dukelow M, Chaudhuri A, Ciborowski P, et al. Nitrated alpha-synuclein-activated microglial profiling for Parkinson's disease. J Neurochem. 2008;104:1504–25.
56. Roodveldt C, Christodoulou J, Dobson CM. Immunological features of alpha-synuclein in Parkinson's disease. J Cell Mol Med. 2008;12:1820–9.
57. Gao HM, Kotzbauer PT, Uryu K, Leight S, Trojanowski JQ, Lee VM. Neuroinflammation and oxidation/nitration of alpha-synuclein linked to dopaminergic neurodegeneration. J Neurosci. 2008;28:7687–98.
58. Sawada M, Imamura K, Nagatsu T. Role of cytokines in inflammatory process in Parkinson's disease. J Neural Transm Suppl. 2006;373–81.
59. Prasad KN, Hovland AR, La Rosa FG, Hovland PG. Prostaglandins as putative neurotoxins in Alzheimer's disease. Proc Soc Exp Biol Med. 1998;219:120–5.
60. Meredith GE, Totterdell S, Beales M, Meshul CK. Impaired glutamate homeostasis and programmed cell death in a chronic MPTP mouse model of Parkinson's disease. Exp Neurol. 2009;219:334–40.
61. Bonsi P, Cuomo D, Picconi B, Sciamanna G, Tscherter A, Tolu M, et al. Striatal metabotropic glutamate receptors as a target for pharmacotherapy in Parkinson's disease. Amino Acids. 2007;32:189–95.
62. Ossowska K, Konieczny J, Wardas J, Pietraszek M, Kuter K, Wolfarth S, et al. An influence of ligands of metabotropic glutamate receptor subtypes on parkinsonian-like symptoms and the striatopallidal pathway in rats. Amino Acids. 2007;32:179–88.
63. Li Y, Liu W, Oo TF, Wang L, Tang Y, Jackson-Lewis V, et al. Mutant LRRK2(R1441G) BAC transgenic mice recapitulate cardinal features of Parkinson's disease. Nat Neurosci. 2009;12:826–8.
64. Giasson BI, Van Deerlin VM. Mutations in LRRK2 as a cause of Parkinson's disease. Neurosignals. 2008;16:99–105.
65. Ariga H, Takahashi-Niki K, Kato I, Maita H, Niki T, Iguchi-Ariga SM. Neuroprotective function of DJ-1 in Parkinson's disease. Oxid Med Cell Longev. 2013;2013:683920.
66. Gu L, Cui T, Fan C, Zhao H, Zhao C, Lu L, et al. Involvement of ERK1/2 signaling pathway in DJ-1-induced neuroprotection against oxidative stress. Biochem Biophys Res Commun. 2009;383:469–74.
67. Lev N, Ickowicz D, Barhum Y, Lev S, Melamed E, Offen D. DJ-1 protects against dopamine toxicity. J Neural Transm. 2009;116:151–60.
68. Lev N, Ickowicz D, Melamed E, Offen D. Oxidative insults induce DJ-1 upregulation and redistribution: implications for neuroprotection. Neurotoxicology. 2008;29:397–405.
69. Bandopadhyay R, Kingsbury AE, Cookson MR, Reid AR, Evans IM, Hope AD, et al. The expression of DJ-1 (PARK7) in normal human CNS and idiopathic Parkinson's disease. Brain. 2004;127:420–30.
70. Kitamura Y, Watanabe S, Taguchi M, Takagi K, Kawata T, Takahashi-Niki K, et al. Neuroprotective effect of a new DJ-1-binding compound against neurodegeneration in Parkinson's disease and stroke model rats. Mol neurodegener. 2011;6:48.
71. Kim SJ, Park YJ, Hwang IY, Youdim MB, Park KS, Oh YJ. Nuclear translocation of DJ-1 during oxidative stress-induced neuronal cell death. Free Radic Biol Med. 2012;53:936–50.

72. Xu J, Zhong N, Wang H, Elias JE, Kim CY, Woldman I, et al. The Parkinson's disease-associated DJ-1 protein is a transcriptional co-activator that protects against neuronal apoptosis. Hum Mol Genet. 2005;14:1231–41.
73. Zhou W, Schaack J, Zawada WM, Freed CR. Overexpression of human alpha-synuclein causes dopamine neuron death in primary human mesencephalic culture. Brain Res. 2002;926:42–50.
74. Lo Bianco C, Ridet JL, Schneider BL, Deglon N, Aebischer P. alpha-Synucleinopathy and selective dopaminergic neuron loss in a rat lentiviral-based model of Parkinson's disease. Proc Natl Acad Sci U S A. 2002;99:10813–8.
75. Galvin JE. Interaction of alpha-synuclein and dopamine metabolites in the pathogenesis of Parkinson's disease: a case for the selective vulnerability of the substantia nigra. Acta Neuropathol. 2006;112:115–26.
76. Lucking CB, Brice A. Alpha-synuclein and Parkinson's disease. Cell Mol Life Sci. 2000;57:1894–908.
77. el-Agnaf OM, Irvine GB. Aggregation and neurotoxicity of alpha-synuclein and related peptides. Biochem Soc Trans. 2002;30:559–65.
78. Prasad JE, Kumar B, Andreatta C, Nahreini P, Hanson AJ, Yan XD, et al. Overexpression of alpha-synuclein decreased viability and enhanced sensitivity to prostaglandin E(2), hydrogen peroxide, and a nitric oxide donor in differentiated neuroblastoma cells. J Neurosci Res. 2004;76:415–22.
79. Kalia LV, Kalia SK, McLean PJ, Lozano AM, Lang AE. Alpha-Synuclein oligomers and clinical implications for Parkinson disease. Ann Neurol. 2013;73:155–69.
80. Mitsui T, Kuroda Y, Kaji R. Parkin and mitochondria. Brain Nerve. 2008;60:923–9.
81. Jiang H, Ren Y, Yuen EY, Zhong P, Ghaedi M, Hu Z, et al. Parkin controls dopamine utilization in human midbrain dopaminergic neurons derived from induced pluripotent stem cells. Nat Commun. 2012;3:668.
82. Jaramillo MC, Zhang DD. The emerging role of the Nrf2-Keap1 signaling pathway in cancer. Genes Dev. 2013;27:2179–91.
83. Williamson TP, Johnson DA, Johnson JA. Activation of the Nrf2-ARE pathway by siRNA knockdown of Keap1 reduces oxidative stress and provides partial protection from MPTP-mediated neurotoxicity. Neurotoxicology. 2012;33:272–9.
84. Itoh K, Chiba T, Takahashi S, Ishii T, Igarashi K, Katoh Y, et al. An Nrf2/small Maf heterodimer mediates the induction of phase II detoxifying enzyme genes through antioxidant response elements. Biochem Biophys Res Commun. 1997;236:313–22.
85. Muthusamy VR, Kannan S, Sadhaasivam K, Gounder SS, Davidson CJ, Boeheme C, et al. Acute exercise stress activates Nrf2/ARE signaling and promotes antioxidant mechanisms in the myocardium. Free Radic Biol Med. 2012;52:366–76.
86. Romanque P, Cornejo P, Valdes S, Videla LA. Thyroid hormone administration induces rat liver Nrf2 activation: suppression by N-acetylcysteine pretreatment. Thyroid. 2011;21:655–62.
87. Ramsey CP, Glass CA, Montgomery MB, Lindl KA, Ritson GP, Chia LA, et al. Expression of Nrf2 in neurodegenerative diseases. J Neuropathol Exp Neurol. 2007;66:75–85.
88. Chen PC, Vargas MR, Pani AK, Smeyne RJ, Johnson DA, Kan YW, et al. Nrf2-mediated neuroprotection in the MPTP mouse model of Parkinson's disease: Critical role for the astrocyte. Proc Natl Acad Sci U S A. 2009;106:2933–8.
89. Lastres-Becker I, Ulusoy A, Innamorato NG, Sahin G, Rabano A, Kirik D, et al. alpha-Synuclein expression and Nrf2 deficiency cooperate to aggravate protein aggregation, neuronal death and inflammation in early-stage Parkinson's disease. Hum Mol Genet. 2012;21:3173–92.
90. Suh JH, Shenvi SV, Dixon BM, Liu H, Jaiswal AK, Liu RM, et al. Decline in transcriptional activity of Nrf2 causes age-related loss of glutathione synthesis, which is reversible with lipoic acid. Proc Natl Acad Sci U S A. 2004;101:3381–6.
91. Niture SK, Kaspar JW, Shen J, Jaiswal AK. Nrf2 signaling and cell survival. Toxicol Appl Pharmacol. 2010;244:37–42.

92. Stachel I, Geismann C, Aden K, Deisinger F, Rosenstiel P, Schreiber S, et al. Modulation of nuclear factor E2-related factor-2 (Nrf2) activation by the stress response gene immediate early response-3 (IER3) in colonic epithelial cells: a novel mechanism of cellular adaption to inflammatory stress. J Biol Chem. 2014;289:1917–29.
93. Khor TO, Fuentes F, Shu L, Paredes-Gonzalez X, Yang AY, Liu Y, et al. Epigenetic DNA methylation of anti-oxidative stress regulator Nrf2 in human prostate cancer. Cancer Prev Res (Phila). 2014;7(12):1186–97.
94. Yu S, Khor TO, Cheung KL, Li W, Wu TY, Huang Y, et al. Nrf2 expression is regulated by epigenetic mechanisms in prostate cancer of TRAMP mice. PLoS One. 2010;5, e8579.
95. Xi YD, Yu HL, Ding J, Ma WW, Yuan LH, Feng JF, et al. Flavonoids protect cerebrovascular endothelial cells through Nrf2 and PI3K from beta-amyloid peptide-induced oxidative damage. Curr Neurovasc Res. 2012;9:32–41.
96. Trujillo J, Chirino YI, Molina-Jijon E, Anderica-Romero AC, Tapia E, Pedraza-Chaverri J. Renoprotective effect of the antioxidant curcumin: recent findings. Redox Biol. 2013;1:448–56.
97. Steele ML, Fuller S, Patel M, Kersaitis C, Ooi L, Munch G. Effect of Nrf2 activators on release of glutathione, cysteinylglycine and homocysteine by human U373 astroglial cells. Redox Biol. 2013;1:441–5.
98. Kode A, Rajendrasozhan S, Caito S, Yang SR, Megson IL, Rahman I. Resveratrol induces glutathione synthesis by activation of Nrf2 and protects against cigarette smoke-mediated oxidative stress in human lung epithelial cells. Am J Physiol Lung Cell Mol Physiol. 2008;294:L478–88.
99. Gao L, Wang J, Sekhar KR, Yin H, Yared NF, Schneider SN, et al. Novel n-3 fatty acid oxidation products activate Nrf2 by destabilizing the association between Keapl and Cullin3. J Biol Chem. 2007;282:2529–37.
100. Saw CL, Yang AY, Guo Y, Kong AN. Astaxanthin and omega-3 fatty acids individually and in combination protect against oxidative stress via the Nrf2-ARE pathway. Food Chem Toxicol. 2013;62:869–75.
101. Song J, Kang SM, Lee WT, Park KA, Lee KM, Lee JE. Glutathione protects brain endothelial cells from hydrogen peroxide-induced oxidative stress by increasing nrf2 expression. Exp neurobiol. 2014;23:93–103.
102. Ji L, Liu R, Zhang XD, Chen HL, Bai H, Wang X, et al. N-acetylcysteine attenuates phosgene-induced acute lung injury via up-regulation of Nrf2 expression. Inhal Toxicol. 2010;22:535–42.
103. Choi HK, Pokharel YR, Lim SC, Han HK, Ryu CS, Kim SK, et al. Inhibition of liver fibrosis by solubilized coenzyme Q10: role of Nrf2 activation in inhibiting transforming growth factor-beta1 expression. Toxicol Appl Pharmacol. 2009;240:377–84.
104. Chun KS, Kundu J, Kundu JK, Surh YJ. Targeting Nrf2-Keap1 signaling for chemoprevention of skin carcinogenesis with bioactive phytochemicals. Toxicol Lett. 2014;229:73–84.
105. Jeong WS, Jun M, Kong AN. Nrf2: a potential molecular target for cancer chemoprevention by natural compounds. Antioxid Redox Signal. 2006;8:99–106.
106. Li XH, Li CY, Lu JM, Tian RB, Wei J. Allicin ameliorates cognitive deficits ageing-induced learning and memory deficits through enhancing of Nrf2 antioxidant signaling pathways. Neurosci Lett. 2012;514:46–50.
107. Bergstrom P, Andersson HC, Gao Y, Karlsson JO, Nodin C, Anderson MF, et al. Repeated transient sulforaphane stimulation in astrocytes leads to prolonged Nrf2-mediated gene expression and protection from superoxide-induced damage. Neuropharmacology. 2011;60:343–53.
108. Wruck CJ, Gotz ME, Herdegen T, Varoga D, Brandenburg LO, Pufe T. Kavalactones protect neural cells against amyloid beta peptide-induced neurotoxicity via extracellular signal-regulated kinase 1/2-dependent nuclear factor erythroid 2-related factor 2 activation. Mol Pharmacol. 2008;73:1785–95.
109. Zambrano S, Blanca AJ, Ruiz-Armenta MV, Miguel-Carrasco JL, Revilla E, Santa-Maria C, et al. The renoprotective effect of L-carnitine in hypertensive rats is mediated by modulation of oxidative stress-related gene expression. Eur J Nutr. 2013;52:1649–59.

110. Li W, Khor TO, Xu C, Shen G, Jeong WS, Yu S, et al. Activation of Nrf2-antioxidant signaling attenuates NFkappaB-inflammatory response and elicits apoptosis. Biochem Pharmacol. 2008;76:1485–9.
111. Kim J, Cha YN, Surh YJ. A protective role of nuclear factor-erythroid 2-related factor-2 (Nrf2) in inflammatory disorders. Mutat Res. 2010;690:12–23.
112. Wang SJ, Wu WM, Yang FL, Hsu GS, Huang CY. Vitamin B2 inhibits glutamate release from rat cerebrocortical nerve terminals. Neuroreport. 2008;19:1335–8.
113. Lou H, Jing X, Wei X, Shi H, Ren D, Zhang X. Naringenin protects against 6-OHDA-induced neurotoxicity via activation of the Nrf2/ARE signaling pathway. Neuropharmacology. 2014;79:380–8.
114. Li R, Liang T, Xu L, Zheng N, Zhang K, Duan X. Puerarin attenuates neuronal degeneration in the substantia nigra of 6-OHDA-lesioned rats through regulating BDNF expression and activating the Nrf2/ARE signaling pathway. Brain Res. 2013;1523:1–9.
115. Kim SS, Lim J, Bang Y, Gal J, Lee SU, Cho YC, et al. Licochalcone E activates Nrf2/antioxidant response element signaling pathway in both neuronal and microglial cells: therapeutic relevance to neurodegenerative disease. J Nutr Biochem. 2012;23:1314–23.
116. He X, Wang L, Szklarz G, Bi Y, Ma Q. Resveratrol inhibits paraquat-induced oxidative stress and fibrogenic response by activating the nuclear factor erythroid 2-related factor 2 pathway. J Pharmacol Exp Ther. 2012;342:81–90.
117. Wang XL, Xing GH, Hong B, Li XM, Zou Y, Zhang XJ, et al. Gastrodin prevents motor deficits and oxidative stress in the MPTP mouse model of Parkinson's disease: involvement of ERK1/2-Nrf2 signaling pathway. Life Sci. 2014;114:77–85.
118. Lu C, Zhang J, Shi X, Miao S, Bi L, Zhang S, et al. Neuroprotective effects of tetramethylpyrazine against dopaminergic neuron injury in a rat model of Parkinson's disease induced by MPTP. Int J Biol Sci. 2014;10:350–7.
119. Kavitha M, Nataraj J, Essa MM, Memon MA, Manivasagam T. Mangiferin attenuates MPTP induced dopaminergic neurodegeneration and improves motor impairment, redox balance and Bcl-2/Bax expression in experimental Parkinson's disease mice. Chem Biol Interact. 2013;206:239–47.
120. Ono K, Yamada M. Vitamin A potently destabilizes preformed alpha-synuclein fibrils in vitro: implications for Lewy body diseases. Neurobiol Dis. 2007;25:446–54.
121. Khaldy H, Escames G, Leon J, Vives F, Luna JD, Acuna-Castroviejo D. Comparative effects of melatonin, L-deprenyl, Trolox and ascorbate in the suppression of hydroxyl radical formation during dopamine autoxidation in vitro. J Pineal Res. 2000;29:100–7.
122. Penugonda S, Mare S, Goldstein G, Banks WA, Ercal N. Effects of N-acetylcysteine amide (NACA), a novel thiol antioxidant against glutamate-induced cytotoxicity in neuronal cell line PC12. Brain Res. 2005;1056:132–8.
123. Jia H, Li X, Gao H, Feng Z, Zhao L, Jia X, et al. High doses of nicotinamide prevent oxidative mitochondrial dysfunction in a cellular model and improve motor deficit in a Drosophila model of Parkinson's disease. J Neurosci Res. 2008;86:2083–90.
124. Cadet JL, Katz M, Jackson-Lewis V, Fahn S. Vitamin E attenuates the toxic effects of intrastriatal injection of 6-hydroxydopamine (6-OHDA) in rats: behavioral and biochemical evidence. Brain Res. 1989;476:10–5.
125. Heim C, Kolasiewicz W, Kurz T, Sontag KH. Behavioral alterations after unilateral 6-hydroxydopamine lesions of the striatum. Effect of alpha-tocopherol. Pol J Pharmacol. 2001;53:435–48.
126. Prasad KN, Kumar B, Yan XD, Hanson AJ, Cole WC. Alpha-tocopheryl succinate, the most effective form of vitamin E for adjuvant cancer treatment: a review. J Am Coll Nutr. 2003;22:108–17.
127. Pasbakhsh P, Omidi N, Mehrannia K, Sobhani AG, Ragerdi Kashani I, Abbasi M, et al. The protective effect of vitamin E on locus coeruleus in early model of Parkinson's disease in rat: immunoreactivity evidence. Iran Biomed J. 2008;12:217–22.
128. Roghani M, Behzadi G. Neuroprotective effect of vitamin E on the early model of Parkinson's disease in rat: behavioral and histochemical evidence. Brain Res. 2001;892:211–7.

129. Silva-Adaya D, Perez-De La Cruz V, Herrera-Mundo MN, Mendoza-Macedo K, Villeda-Hernandez J, Binienda Z, et al. Excitotoxic damage, disrupted energy metabolism, and oxidative stress in the rat brain: antioxidant and neuroprotective effects of L-carnitine. J Neurochem. 2008;105:677–89.
130. Virmani A, Gaetani F, Binienda Z. Effects of metabolic modifiers such as carnitines, coenzyme Q10, and PUFAs against different forms of neurotoxic insults: metabolic inhibitors, MPTP, and methamphetamine. Ann N Y Acad Sci. 2005;1053:183–91.
131. Clark J, Clore EL, Zheng K, Adame A, Masliah E, Simon DK. Oral N-acetyl-cysteine attenuates loss of dopaminergic terminals in alpha-synuclein overexpressing mice. PLoS One. 2010;5, e12333.
132. Liu D, Pitta M, Mattson MP. Preventing NAD(+) depletion protects neurons against excitotoxicity: bioenergetic effects of mild mitochondrial uncoupling and caloric restriction. Ann N Y Acad Sci. 2008;1147:275–82.
133. Ortiz GG, Pacheco-Moises FP, Gomez-Rodriguez VM, Gonzalez-Renovato ED, Torres-Sanchez ED, Ramirez-Anguiano AC. Fish oil, Melatonin and vitamin E attenuates midbrain cyclooxygenase-2 activity and oxidative stress after administration of 1-methyl-4-phenyl-1,2,3,6-tetrahydropyridine. Matab Brain Dis. 2013;28(4):705–9.
134. Guarente L. Sirtuins in aging and disease. Cold Spring Harb Symp Quant Biol. 2007;72:483–8.
135. Ferretta A, Gaballo A, Tanzarella P, Piccoli C, Capitanio N, Nico B, et al. Effect of resveratrol on mitochondrial function: implications in parkin-associated familiar Parkinson's disease. Biochim Biophys Acta. 1842;2014:902–15.
136. Muthian G, King J, Mackey V, Prasad K, Charlton CG. Blockage of the proposed precipitating stage for Parkinson's disease by antioxidants. A potential preventive measure for PD. FASEB J. 2008;715:2.
137. Liu Z, Yu Y, Li X, Ross CA, Smith WW. Curcumin protects against A53T alpha-synuclein-induced toxicity in a PC12 inducible cell model for Parkinsonism. Pharmacol Res. 2011;63:439–44.
138. Jiang TF, Zhang YJ, Zhou HY, Wang HM, Tian LP, Liu J, et al. Curcumin ameliorates the neurodegenerative pathology in A53T alpha-synuclein cell model of Parkinson's disease through the downregulation of mTOR/p70S6K signaling and the recovery of macroautophagy. J Neuroimmune Pharmacol. 2013;8:356–69.
139. Haleagrahara N, Siew CJ, Ponnusamy K. Effect of quercetin and desferrioxamine on 6-hydroxydopamine (6-OHDA) induced neurotoxicity in striatum of rats. J Toxicol Sci. 2013;38:25–33.
140. Antala BV, Patel MS, Bhuva SV, Gupta S, Rabadiya S, Lahkar M. Protective effect of methanolic extract of Garcinia indica fruits in 6-OHDA rat model of Parkinson's disease. Indian J Pharmacol. 2012;44:683–7.
141. Perez HJ, Carrillo SC, Garcia E, Ruiz-Mar G, Perez-Tamayo R, Chavarria A. Neuroprotective effect of silymarin in a MPTP mouse model of Parkinson's disease. Toxicology. 2014;319:38–43.
142. Geed M, Garabadu D, Ahmad A, Krishnamurthy S. Silibinin pretreatment attenuates biochemical and behavioral changes induced by intrastriatal MPP+ injection in rats. Pharmacol Biochem Behav. 2014;117:92–103.
143. Zhang SM, Hernan MA, Chen H, Spiegelman D, Willett WC, Ascherio A. Intakes of vitamins E and C, carotenoids, vitamin supplements, and PD risk. Neurology. 2002;59:1161–9.
144. Etminan M, Gill SS, Samii A. Intake of vitamin E, vitamin C, and carotenoids and the risk of Parkinson's disease: a meta-analysis. Lancet Neurol. 2005;4:362–5.
145. Weber CA, Ernst ME. Antioxidants, supplements, and Parkinson's disease. Ann Pharmacother. 2006;40:935–8.
146. Storch A, Jost WH, Vieregge P, Spiegel J, Greulich W, Durner J, et al. Randomized, double-blind, placebo-controlled trial on symptomatic effects of coenzyme Q(10) in Parkinson disease. Arch Neurol. 2007;64:938–44.
147. Fahn S. A pilot trial of high-dose alpha-tocopherol and ascorbate in early Parkinson's disease. Ann Neurol. 1992;32(Suppl):S128–32.

148. Prasad KN. Micronutrients in the prevention and improvement of the standard therapy for Alzheimer's disease. In: Micronutrients in Health and Disease. Boca Raton, FL: CRE Press; 2011. p. 285–96.
149. Melamed E, Offen D, Shirvan A, Djaldetti R, Barzilai A, Ziv I. Levodopa toxicity and apoptosis. Ann Neurol. 1998;44:S149–54.
150. Fahn S, Oakes D, Shoulson I, Kieburtz K, Rudolph A, Lang A, et al. Levodopa and the progression of Parkinson's disease. N Engl J Med. 2004;351:2498–508.
151. Fahn S. Does levodopa slow or hasten the rate of progression of Parkinson's disease? J Neurol. 2005;252 Suppl 4:IV37–42.
152. Abdin AA, Hamouda HE. Mechanism of the neuroprotective role of coenzyme Q10 with or without L-dopa in rotenone-induced parkinsonism. Neuropharmacology. 2008;55:1340–6.

Modulation of Hallmarks of Brain Aging by Environmental Enrichment

Nancy Bartolotti and Orly Lazarov

1 Introduction

The dogma that the brain is immutable has been significantly challenged during the last 100 years. We now know that the brain has a remarkable level of plasticity, i.e., the ability to change in response to a stimulus. This change could be either short or long-term. The neuropsychologist Donald Hebb was the first to show that rats raised as pets were better able to perform complex tasks requiring learning and memory compared to standard housed lab rats [1, 2]. The difference in the pet rats' environment included a wider area of habituation and exploration, increased physical activity and novel experiences. The superior performance of the pet rats in the learning and memory task was the first evidence that the environment can modulate brain function. Later on, Hebb's successors and others discovered that experience in a complex environment enhanced synaptic formation, hippocampal neurogenesis, and number of dendrites and dendritic branching, as well as increased the weight of the cortex and induced neurotrophic factor expression and more [3–7]. These critical elements of the environment were then modeled in a lab setting by housing rodents in a large cage with opportunities for social interaction, toys and novel exploration, as well as access to running wheels for physical exercise. The set up was termed "environmental enrichment" or "complex environment". This raises the intriguing question of whether the aging process can be counteracted by environmental enrichment. We will review the evidence that suggests that the application of environmental enrichment protects the brain against aging related decline. While most of the available information comes from experiments in rodents, it is worth

N. Bartolotti • O. Lazarov, Ph.D. (✉)
Department of Anatomy and Cell Biology, College of Medicine, The University of Illinois at Chicago, Chicago, IL 60612, USA
e-mail: olazarov@uic.edu; http://lazarovlab.anat.uic.edu/home.html

S.C. Bondy, A. Campbell (eds.), *Inflammation, Aging, and Oxidative Stress*, Oxidative Stress in Applied Basic Research and Clinical Practice, DOI 10.1007/978-3-319-33486-8_16

noting the increasing number of human studies examining the therapeutic effect of exercise and cognitive stimulation for the elderly and those suffering from aging-related neurodegenerative disease and cognitive impairments like Alzheimer's disease. For example, some studies have shown that increased physical activity enhances cognitive function in the elderly [8–11]. Similarly, the role of cognitively complex experiences such as bilingualism, advanced education, and cognitively stimulating leisure activities like reading, puzzle solving, and playing a musical instrument has received substantial interest, and current evidence suggests that cognitively complex experiences may have profound effects on the human brain as well [12–15]. One challenge in human studies is separating out and identifying individual lifestyle factors that are responsible for preserving cognitive function. Human studies often rely on correlative experimental designs and it can be unclear whether greater participation in an active and complex lifestyle in old age is a cause or effect of greater cognitive functioning. However, in rodents the lifestyle factors can be more carefully controlled and the neuroanatomical effects more rigorously measured. Therefore, studies in rodents are useful for identifying the key mechanisms and pathways responsible for the effects of complex experience on cognition during aging. Once identified, therapies targeting these pathways can be developed. In the following sections, we discuss the research concerning the effects of environmental enrichment (EE) on brain plasticity during aging in rodents. EE may modulate cognitive function in aging through a variety of pathways. Here, we discuss the evidence for how EE modulates oxidative stress, inflammation, neurotrophic factor signaling, neurogenesis, and synaptic plasticity.

2 Oxidative Stress

Regulation of free radical species is a critical physiological process. The dysfunction of this regulation is thought to be an important characteristic of aging and may be one of the determining factors in the longevity of an organism [16]. Reactive oxygen species (ROS) are free radicals produced in the brain by microglia as a means of defense against pathogens [17], and by mitochondria as a result of energy generation [18]. Though ROS perform critical physiological functions, in excess they can also be highly toxic, and so endogenous mechanisms such as antioxidants exist in the body to regulate ROS [19]. Oxidative stress due to increased ROS may lead to several negative effects such as alterations in gene expression and cell dysfunction (for a more detailed discussion of ROS in cell dysfunction see [20]). The brain is thought to be particularly susceptible to change in oxidative state due the high concentration of lipids and high demand for oxygen [22]. During aging, production of ROS is increased, concomitantly with diminished production of antioxidants and mitochondrial homeostasis [24]. ROS can also directly disrupt mitochondrial function [26]. While ROS can be produced by the cell itself, they can also result from exposure to external factors like radiation or ultraviolet light adding to the accumulation of oxidative stress during aging [28, 29]. It is thought that because evolutionary pressure prioritizes the survival of an organism during its reproductive ages, the production of ROS is more tightly regulated compared to old age [30].

Another possible mechanism for mitochondrial dysfunction during aging is a reduction in neurotrophic factor production, for example, BDNF, GDNF, and IGF-1 [21, 31, 32]. Neurotrophic factors are implicated in the maintenance of brain function on several levels. First, they regulate neuronal survival and neurotransmission. Second, they play a role in regulation of ROS. While ROS are able to disrupt calcium homeostasis in neurons and interrupt critical calcium signaling, exposure to growth factors like BDNF, bFGF and NGF prevents this interruption, evidently through activation of the antioxidants superoxide dismutase (SOD) and glutathione peroxidase [33, 34]. In addition, BDNF is thought to enhance mitochondrial function in the brain by regulating oxygen usage efficiency [35].

How, though, does an increase in oxidative stress during aging lead to cognitive decline? One mechanism may be by interfering with cellular function. Some studies suggest that at certain level, ROS are toxic to cells and may lead to increased cell dysfunction, leading to cognitive decline [36]. Other studies suggest that the ROS-induced pro-survival signaling ERK, PI3K, Akt pathways are less effective during aging [29, 37, 38].

Another way oxidative stress may be contributing to cognitive decline may be by altering brain plasticity. "Plasticity" typically refers to structural changes in the brain such as dendritic branching and arborization, synaptic formation and strengthening, long-term potentiation (LTP) or formation of new connective pathways (for a more in-depth discussion of plasticity see [39]). It is thought that decreased plasticity may bear more responsibility for cognitive decline during aging than outright loss of neurons [40]. Mitochondrial function may be a critical component of plasticity by promoting release of neurotransmitters, dendritic branching, and LTP [41–46]. Indeed, mitochondria have been shown to respond to synaptic signaling in the brain and increased energy production from mitochondria may be a critical component of LTP maintenance [48]. Therefore, inappropriate mitochondrial function due to damage from ROS could have profound effects on brain plasticity and thus cognitive function.

Reduced CREB signaling with age is another proposed mechanism for the cognitive decline during aging [49, 50]. The CREB signaling complex, consisting of pCREB and transcription cofactors such as CBP and p300, initiate the transcription of genes thought to be important for synaptic plasticity and learning and memory such as c-fos, BDNF, and Egr-1 [51]. CREB activation is known to be important for learning and memory, and this includes activation in mitochondrial associated CREB [52]. ROS may also directly reduce expression of the CREB protein via protein kinase D, and reduce the availability of this critical component for learning processes [53, 54]. In support of the role of ROS and CREB signaling, antioxidants have been shown to increase CREB phosphorylation via mitochondrial PKA [55].

A large body of studies suggests that EE reduces oxidative stress and enhances the production of endogenous antioxidants such as SOD in a mouse model of the aging-related disorder Alzheimer's disease [56, 57]. Experience in an EE has been shown to act upon several of the pro-oxidative stress pathways mentioned above.

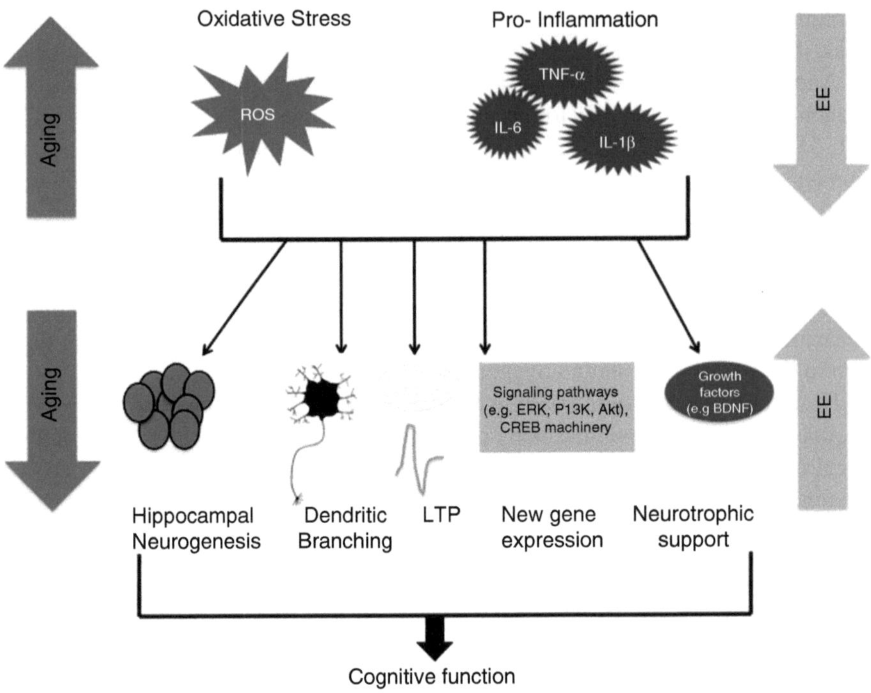

Fig. 1 Aging and environmental enrichment have inverse effects on many processes important for cognitive function. During aging, ROS and inflammation are increased, impacting critical aspects of brain plasticity, such as neurogenesis, dendritic branching, LTP, and neurotrophic support, leading to decreased cognitive function. EE, on the other hand, leads to decreased levels of ROS and inflammation, and also increases neurogenesis, dendritic branching, LTP, and neurotrophic support, leading to increased cognitive function

Specifically, EE protects mitochondria from aging related defects and, in turn, improvement in mitochondrial activity in aged rodents has been correlated with enhanced memory performance [58, 59]. In addition, ERK and PI3K pathways are more active following EE [60, 61]. EE can enhance BDNF signaling, which may in turn promote proper mitochondrial function and may allow for retention of the machinery necessary for plasticity [61–63]. In aged rodents, access to a running wheel improved hippocampal function and learning and memory [64]. In rats, running on a wheel enhanced expression of BDNF mRNA [65]. Experience of rodents in EE increases the phosphorylation and activation of CREB (pCREB) [61], a critical signal for the formation of memories. Therefore, EE acts on many of the pathways that are impaired by aging, offering the intriguing possibility of a non-invasive therapy to counteract aging-related cognitive decline (Fig. 1).

3 Inflammation

The term "inflammaging" is meant to describe the fundamental role of inflammation in the aging process [66]. This hypothesis proposes that a robust inflammatory response is useful when fighting off infections when the organism is younger, but becomes pathological in old age when both the pro-inflammatory response is more active and the anti-inflammatory response is attenuated, leading to an overall higher levels of inflammation [66]. In the aging brain, ROS can disrupt microglia function and lead to their activation. In turn, activated microglia will produce higher levels of proinflammatory cytokines [67–70]. Evidence for increased inflammation in the aged brain shows increased levels of pro-inflammatory chemokines and cytokines like tumor necrosis factor α (TNFα), interleukin-1β (IL-1β) and interleukin-6 (IL-6), as well as a decrease in anti-inflammatory cytokines like interleukin-2 (IL-2, for review see [71]). The increased presence of ROS during aging is thought to contribute to cell damage and lead to chronic inflammation via NF-kappa-B [72]. Recent studies show that adults with higher levels of circulating pro-inflammatory cytokines such as IL-6 were more likely to exhibit impairments in memory and reduced volume of white and gray matter in the cortex and hippocampus, a brain structure critical for learning and memory [74]. Therefore, increased neuroinflammation may in part underlie the cognitive decline associated with aging [75] (Fig. 2).

Inflammation may underlie attenuation of neurogenesis in aging. Neurogenesis in the adult brain occurs in the subgranular zone of the dentate gyrus in the hippocampus and in the subventricular zone [76, 77]. The generation of new neurons is thought to be an important part of cognitive function. A large body of studies suggests that neurogenesis decreases during adulthood and aging [76]. Microglia play a role in the regulation of neurogenesis. Thus, their activation during aging may compromise neurogenesis. In that regard, TNF-α, IL-6, and IL-1β have all been shown to decrease neurogenesis in the hippocampus, although the mechanism is yet to be elucidated [78–80]. Interestingly, anti-inflammatory cytokines like interferon γ (IFN-γ) and interleukin-4 (IL-4) may promote neurogenesis in the hippocampus by stimulating neurotrophin release from microglia [82]. One recent study showed that exposing old mice to young blood enhanced neurogenesis in the old mice [83], suggesting an important role for systemic blood factors in the regulation of neurogenesis and brain plasticity. In a follow up study, exposure of older mice to young blood increased pCREB expression in the hippocampus of these mice [84]. These observations raise the intriguing possibility that modulation of systemic inflammation during aging may enhance cognitive function.

EE and exercise have a profound effect on neuroinflammation [85]. First, numerous studies suggest that EE or physical activity enhance neurogenesis in older mice and improve memory performance [86, 87]. Cognitively complex tasks that enhance neurogenesis and neuronal density may help to mitigate the effects of aging [88], possibly by preventing age-related alterations in the immune system via oxidative stress regulation [89]. Cytokine and chemokine expression is altered by EE [90].

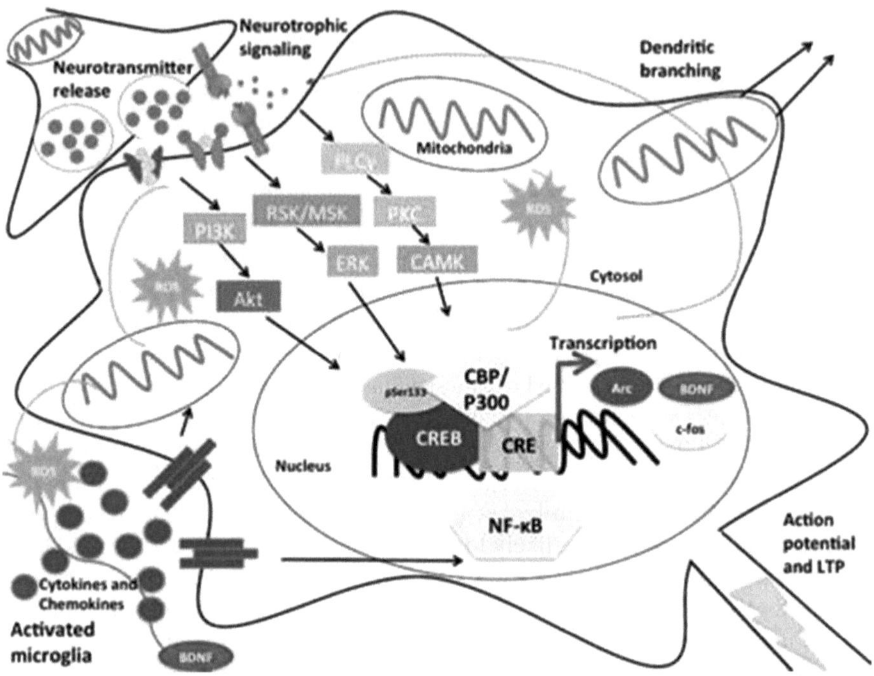

Fig. 2 Reactive oxygen species and inflammation are important regulators of plasticity. Mitochondria may aid in many aspects of plasticity including neurotransmitter release. Activation of signaling pathways such as PI3K, ERK and Akt can result from neurotransmission, or via Reactive oxygen species (ROS) release from mitochondria. Activation of these pathways can lead to transcription of immediate early genes and neurotrophins, such as BDNF, that are important for learning and memory. These processes can then lead to LTP and dendritic branching. Similarly, inflammatory cytokines and ROS released by microglia can also impact neuronal function, such as via NF-kappa-B

Emerging evidence suggests that a powerful environmental modulator of inflammation may be physical activity. Physical activity may directly modulate the immune system through regulation of NF-kappa-B and release of anti-inflammatory cytokines following muscle contractions, as well as lead to a decrease in pro-inflammatory factors [92–94]. Participation of older adults in aerobic exercise reduced circulating levels of pro-inflammatory cytokines such as IL-6 and IL-18 [95]. In aged rodents, physical activity led to an increase in anti-inflammatory factors in the hippocampus [91]. Exercise also rescues the ERK and Akt pathways, which are impaired during aging [93], and in humans exercise is associated with greater oxygen consumption in addition to greater hippocampal volume [96]. One possible mechanism underlying the effect of physical activity on inflammation is increased oxygen consumption, which may facilitate angiogenesis (the generation of blood vessels). In turn, angiogenesis may enhance plasticity by supporting neurogenesis [97]. Another interesting possibility is the role of systemic factors. Recent evidence suggests that exercise may mitigate the effects of aging by altering systemic inflammation of the gut.

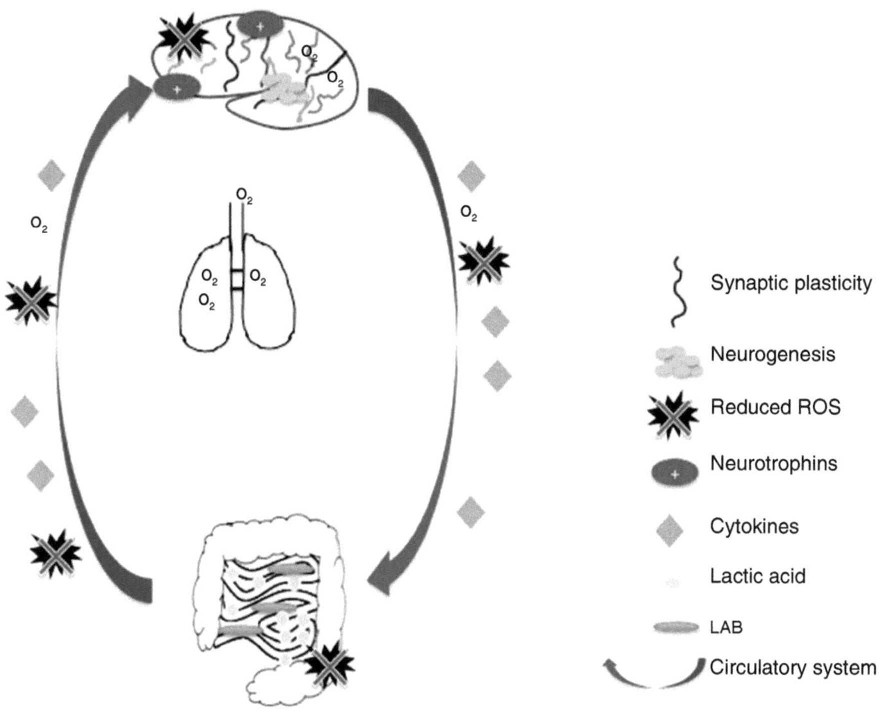

Fig. 3 Speculative mechanism for the systemic effects of exercise on cognitive function. Exercise enhances circulation, which can increase oxygen consumption, leading to an increase in angiogenesis, neurogenesis and neuronal function. In addition, enhanced circulation may result in better clearance of ROS, pro-inflammatory cytokines and compost. Physical activity can also lead to increased lactic acid, which promotes cultivation of lactic acid bacteria (LAB) in the gut. A healthy microbiome can have many positive effects on the brain such as reduced ROS via increased antioxidant production, and increased levels of anti-inflammatory cytokines. This may allow for greater brain plasticity and production of neurotrophic factors, which in may then propagate reduced inflammation back to the gut, allowing continued microbiome health

Inflammation of the intestines correlates with increased circulating proinflammatory cytokines such as IL-6 and TNF-α and reduces hippocampal neurogenesis in adult rodents [98]. Early studies have also demonstrated that the microbiome is altered with inflammation and these changes are correlated with changes in cognition, suggesting a destructive cycle may be present during aging [99, 100] (Fig. 3). Oxidative stress, which increases during aging, may also alter the microbiome [101–103], and even laboratory mice raised in controlled environments exhibit changes in microbiota composition with age [104]. Taken together, recent evidence suggests that the microbiome is altered with aging and inflammatory state, and this may further increase inflammation and contribute to cognitive decline during aging. The mechanism underlying the communication between the gut microbiome and the brain is yet to be elucidated, but proposed mechanisms include communication via the enteric nervous system and vagus nerve, cells of the immune system, the endocrine system, and bacterial metabolites [105, 106]. Further support for the role of the microbiome in aging related inflammation and cognitive decline comes from preliminary studies in rodents which suggest that treatment with probiotics may reverse some of the inflammatory effects from the aging microbiome and perhaps enhance diminished antioxidant production [107–110]. Treatment with probiotics was also shown to rescue age-related impairments in LTP [111]. In the SAMP8 mouse model of accelerated aging, ingestion of saccharides intended to promote microflora health improved performance on memory tasks and reduced levels of proinflammatory cytokines such as IL-6 [112]. Additional evidence from mice suggests that treatment with probiotics may be particularly beneficial for enhancing cognitive functioning when the microbiome is disrupted [113]. Interestingly, recent studies indicate that exercise may be a modifier of microbiota [114–116]. One proposed mechanism is that exercise increases lactic acid, which may promote the cultivation of lactic acid bacteria (LAB) [117]. In one recent study, aged rodents treated with Lactobacillus pentosus, a LAB, showed improvements in memory as well as rescue of age-related decline in BDNF, neurogenesis, and activation of the CREB and Akt signaling pathways via NF-kappa-B and reduction of pro-inflammatory factors like TNF-α and IL-6 and ROS in the hippocampus [107, 118–121]. In addition, LAB may promote activity of antioxidants like SOD, thereby reducing oxidative stress [109]. Therefore, the microbiome may change with age, possibly due to changes in the inflammation and oxidative stress. EE, and particularly physical exercise, may reverse this process. However, these observations require further mechanistic support.

4 Implications for Human Aging

Translating the effects of environmental enrichment into human activity and lifestyle should require the consideration of the evolutionary difference between rodents and humans. Some of the environmental factors, such as odor, may have a different level of effect in humans compared to rodents. However, it would be reasonable to assume that factors that support brain plasticity and counteract the aging processes, such as exercise, would have beneficial effects on cognitive function in humans as well. Studies investigating the role of environmental factors and cognitive functioning during aging are more complex compared to the experimental rodent. However, there is some promising evidence from studies focusing on humans. For example, greater levels of physical activity have been repeatedly observed to enhance hippocampal volume and function in older individuals [8–10] and longitudinal studies have shown that greater participation in physical activity correlates with greater cognitive function in the elderly [11]. While brain volume deteriorates with age, individuals with greater aerobic fitness exhibit greater brain volume, suggesting that physical activity may exert a protective effect against aging-related brain degeneration [122]. One recent study also showed that physical activity is correlated with better cognitive performance in the elderly when compared to individuals with primarily sedentary activities [123]. Furthermore, greater physical activity may also result in a reduced risk for Alzheimer's disease (AD), brain disorder characterized by progressive memory loss and cognitive decline and for which age is the greatest risk factor [124, 125].

Some studies have also been done on the impact of socialization on aging. Socialization is thought to be a key aspect of the EE paradigm [126]. A healthy social life may also be a factor in slowing cognitive decline during aging in humans [11] and these benefits may include protection against dementia, either directly or by promoting overall health [127]. Similarly, infrequent participation in social activities later in life, as well as living alone may increase dementia risk [128, 129]. Interestingly, feelings of loneliness due to reduced social connections may be the driving force behind this phenomenon [130]. Individuals who report feelings of loneliness had double the risk for AD in a longitudinal study, and also experienced a faster rate of cognitive decline [131]. Social interactions that accompany participation in religious activities may be one way in which elderly people can increase socialization and help to slow cognitive decline [132–137]. The availability of community areas and activities for the elderly may improve frequency of socialization and improve cognitive function [138].

While decreasing inflammation and other facets of aging may be one way of promoting healthy aging and reducing risk of dementia, individuals may also be able to maintain brain function even in the presence of brain pathology. Individuals who have experienced cognitively complex environments during their life time may be able to cultivate a "reserve" of additional brain matter or processing efficiency allowing their brains to function longer even in the presence of inflammation or neuropathology [139]. Thus, it is hypothesized that cognitively complex experiences may protect against aging by strengthening and/or building up additional brain matter, such as synapses and dendritic spines [140]. Higher levels of education have previously been associated with greater grey matter density, particularly in areas that are usually most vulnerable to aging related degeneration such as the hippocampus and entorhinal cortex [141]. In a study of religious order members, a population with relatively fewer variables in lifestyle, individuals with lower levels of education experienced cognitive impairments at an earlier age [142]. Likewise, education can offer a protective effect in diverse populations [143, 144]. While the hippocampus degenerates with age, individuals who report a life time of greater involvement in cognitively complex activities exhibit reduced rates of hippocampal degeneration during aging [145]. It has been hypothesized that proper hippocampal functioning may be the difference between elderly individuals who remain cognitively intact and those who succumb to dementia [146].

Several cognitively complex tasks have been shown to increase gray matter density such as learning a second language or a musical instrument [147, 148]. Speaking two or more languages may be another form of cognitive complexity that may offer protection against degeneration and lifetime bilinguals appear to retain white matter volume and integrity and gray matter during aging [149–153]. Interestingly, bilingual individuals seem to exhibit signs of dementia approximately 4 years later than monolinguals [154]. Even when white matter begins to degenerate bilinguals are able to preserve cognitive functioning longer than monolinguals [155]. Learning a second language in adulthood may also provide a protective effect against dementia [156]. However, the extent of the protection from bilingualism remains controversial and requires further study and careful consideration of study design [157, 158]. Cognitive training has been shown to increase myelination in older adults, which results in enhanced connectivity [159–161]. These observations are all in line with observations from animal studies of EE, in which greater brain mass and density are observed [5–7].

Taken together, these studies suggest that the brain retains plasticity even into old age, and cognitively stimulating tasks may enhance cognitive function in the elderly [162]. More studies are warranted for the understanding of the mechanisms underlying the benefits of cognitively complex experiences and physical activity for the diminution of aging processes, and cognitive decline in particular. In addition, it will be important to determine the effective "dose" and duration of different environmental factors for an optimal effect on brain function.

5 Conclusion

Environmental factors have a clear and meaningful effect on many aspects of aging and on cognitive function specifically. Data from mouse models indicate that reduced oxidative stress and preserved homeostasis of oxygen species, proper regulation of inflammation, increased neurogenesis, increased expression of neurotrophic factors, and enhanced brain size, density and connectivity, are part of the processes that are modulated by the environment. Future challenges include identifying targets in signaling pathways that play a role in these processes for developing therapeutics to preserve cognitive function in the elderly.

Bibliography

1. Hebb DO. The effects of early experience on problem-solving at maturity. Am Psychol. 1947;2:306–7.
2. Hebb DO. The organization of behavior: a neuropsychological theory. New York: Wiley; 1949.
3. Bennett EL, et al. Chemical and anatomical plasticity brain. Science. 1964;146(3644):610–9.
4. Hamilton LD, Bennett JL, Silver J. Nandrolone phenpropionate in the treatment of geriatric patients with chronic brain damage. J Am Geriatr Soc. 1964;12:373–8. (see http://www.ncbi.nlm.nih.gov/pubmed/14136815).
5. Volkmar FR, Greenough WT. Rearing complexity affects branching of dendrites in the visual cortex of the rat. Science. 1972;176(4042):1445–7.
6. Globus A, et al. Effects of differential experience on dendritic spine counts in rat cerebral cortex. J Comp Physiol Psychol. 1973;82(2):175–81.

7. Turner AM, Greenough WT. Differential rearing effects on rat visual cortex synapses. I. Synaptic and neuronal density and synapses per neuron. Brain Res. 1985;329(1–2):195–203.
8. Niemann C, Godde B, Voelcker-Rehage C. Not only cardiovascular, but also coordinative exercise increases hippocampal volume in older adults. Front Aging Neurosci. 2014;6:170.
9. Shah T, et al. A combination of physical activity and computerized brain training improves verbal memory and increases cerebral glucose metabolism in the elderly. Transl Psychiatry. 2014;4:e487.
10. Varma VR, et al. Low-intensity daily walking activity is associated with hippocampal volume in older adults. Hippocampus. 2015;25(5):605–15.
11. Fratiglioni L, Paillard-Borg S, Winblad B. An active and socially integrated lifestyle in late life might protect against dementia. Lancet Neurol. 2004;3(6):343–53.
12. Iacono D, et al. The Nun Study: clinically silent AD, neuronal hypertrophy, and linguistic skills in early life. Neurology. 2009;73(9):665–73.
13. Steffener J, et al. The role of education and verbal abilities in altering the effect of age-related gray matter differences on cognition. PLoS One. 2014;9(3):e91196.
14. Landau SM, et al. Association of lifetime cognitive engagement and low β-amyloid deposition. Arch Neurol. 2012;69(5):623–9.
15. Wilson RS, et al. Life-span cognitive activity, neuropathologic burden, and cognitive aging. Neurology. 2013;81(4):314–21.
16. Harman D. Aging: a theory based on free radical and radiation chemistry. J Gerontol. 1956;11(3):298–300.
17. Colton CA, Gilbert DL. Production of superoxide anions by a CNS macrophage, the microglia. FEBS Lett. 1987;223(2):284–8.
18. Chance B, Sies H, Boveris A. Hydroperoxide metabolism in mammalian organs. Physiol Rev. 1979;59(3):527–605.
19. Fridovich I. Superoxide anion radical ($O_2^{-\bullet}$), superoxide dismutases, and related matters. J Biol Chem. 1997;272(30):18515–7.
20. Fleury C, Mignotte B, Vayssiere JL. Mitochondrial reactive oxygen species in cell death signaling. Biochimie. 2002;84(2–3):131–41.
21. Bishop NA, Lu T, Yankner BA. Neural mechanisms of ageing and cognitive decline. Nature. 2010;464(7288):529–35.
22. Floyd RA, Hensley K. Oxidative stress in brain aging. Implications for therapeutics of neurodegenerative diseases. Neurobiol Aging. 2002;23(5):795–807.
23. Gamba P, et al. Oxidized cholesterol as the driving force behind the development of Alzheimer's disease. Front Aging Neurosci. 2015;7:119.
24. Smith CD, et al. Excess brain protein oxidation and enzyme dysfunction in normal aging and in Alzheimer disease. Proc Natl Acad Sci U S A. 1991;88(23):10540–3.
25. Gemma C, et al. Oxidative stress and the aging brain: from theory to prevention. In: Riddle DR, editor. Brain aging: models, methods, and mechanisms. Boca Raton: CRC; 2007.
26. Bandy B, Davison AJ. Mitochondrial mutations may increase oxidative stress—implications for carcinogenesis and aging. Free Radic Biol Med. 1990;8(6):523–39.
27. Beckman KB, Ames BN. The free radical theory of aging matures. Physiol Rev. 1998;78(2):547–81.
28. Heck DE, et al. UVB light stimulates production of reactive oxygen species—unexpected role for catalase. J Biol Chem. 2003;278(25):22432–6.
29. Finkel T, Holbrook NJ. Oxidants, oxidative stress and the biology of ageing. Nature. 2000;408(6809):239–47.
30. Gavrilov LA, Gavrilova NS. Evolutionary theories of aging and longevity. Scientific WorldJournal. 2002;2:339–56.
31. Silhol M, et al. Age-related changes in brain-derived neurotrophic factor and tyrosine kinase receptor isoforms in the hippocampus and hypothalamus in male rats. Neuroscience. 2005;132(3):613–24.

32. Matsunaga W, Isobe K, Shirokawa T. Involvement of neurotrophic factors in aging of noradrenergic innervations in hippocampus and frontal cortex. Neurosci Res. 2006;54(4):313–8.
33. Mattson MP, et al. Neurotrophic factors attenuate glutamate-induced accumulation of peroxides, elevation of intracellular Ca^{2+} concentration, and neurotoxicity and increase antioxidant enzyme-activities in hippocampal-neurons. J Neurochem. 1995;65(4):1740–51.
34. McCord JM, Fridovich I. Superoxide dismutase. An enzymic function for erythrocuprein (hemocuprein). J Biol Chem. 1969;244(22):6049–55.
35. Markham A, et al. BDNF increases rat brain mitochondrial respiratory coupling at complex I, but not complex II. Eur J Neurosci. 2004;20(5):1189–96.
36. Cortopassi G, Wang E. Modelling the effects of age-related mtDNA mutation accumulation; complex I deficiency, superoxide and cell death. Biochim Biophys Acta. 1995;1271(1): 171–6.
37. Ikeyama S, et al. Loss in oxidative stress tolerance with aging linked to reduced extracellular signal-regulated kinase and Akt kinase activities. FASEB J. 2002;16(1):114–6.
38. Guyton KZ, et al. Age-related changes in activation of mitogen-activated protein kinase cascades by oxidative stress. J Investig Dermatol Symp Proc. 1998;3(1):23–7.
39. Markham JA, Greenough WT. Experience-driven brain plasticity: beyond the synapse. Neuron Glia Biol. 2004;1:351–63.
40. Burke SN, Barnes CA. Neural plasticity in the ageing brain. Nat Rev Neurosci. 2006;7(1): 30–40.
41. Levy M, et al. Mitochondrial regulation of synaptic plasticity in the hippocampus. J Biol Chem. 2003;278(20):17727–34.
42. Wang P, et al. A Drosophila temperature-sensitive seizure mutant in phosphoglycerate kinase disrupts ATP generation and alters synaptic function. J Neurosci. 2004;24(19): 4518–29.
43. Kim HY, et al. Mitochondrial Ca^{2+} uptake is essential for synaptic plasticity in pain. J Neurosci. 2011;31(36):12982–91.
44. Kimura R, et al. Acute exposure to the mitochondrial complex I toxin rotenone impairs synaptic long-term potentiation in rat hippocampal slices. CNS Neurosci Ther. 2012;18(8): 641–6.
45. David G, Barrett EF. Stimulation-evoked increases in cytosolic [Ca^{2+}] in mouse motor nerve terminals are limited by mitochondrial uptake and are temperature-dependent. J Neurosci. 2000;20(19):7290–6.
46. Verstreken P, et al. Synaptic mitochondria are critical for mobilization of reserve pool vesicles at Drosophila neuromuscular junctions. Neuron. 2005;47(3):365–78.
47. Mattson MP. Mitochondrial regulation of neuronal plasticity. Neurochem Res. 2007;32(4–5): 707–15.
48. Williams JM, et al. Synaptic activity-dependent modulation of mitochondrial gene expression in the rat hippocampus. Mol Brain Res. 1998;60(1):50–6.
49. Kudo K, et al. Age-related disturbance of memory and CREB phosphorylation in CA1 area of hippocampus of rats. Brain Res. 2005;1054(1):30–7.
50. Porte Y, Buhot MC, Mons N. Alteration of CREB phosphorylation and spatial memory deficits in aged 129T2/Sv mice. Neurobiol Aging. 2008;29(10):1533–46.
51. Zhang X, et al. Genome-wide analysis of cAMP-response element binding protein occupancy, phosphorylation, and target gene activation in human tissues. Proc Natl Acad Sci U S A. 2005;102(12):4459–64.
52. Bevilaqua LRM, et al. Experience-dependent increase in cAMP-responsive element binding protein in synaptic and nonsynaptic mitochondria of the rat hippocampus. Eur J Neurosci. 1999;11(10):3753–6.
53. Waldron RT, Rozengurt E. Oxidative stress induces protein kinase D activation in intact cells. Involvement of Src and dependence on protein kinase C. J Biol Chem. 2000;275(22): 17114–21.
54. Ozgen N, et al. Reactive oxygen species decrease cAMP response element binding protein expression in cardiomyocytes via a protein kinase D1-dependent mechanism that does not require Ser133 phosphorylation. Mol Pharmacol. 2009;76(4):896–902.

55. Ryu H, et al. Antioxidants modulate mitochondrial PKA and increase CREB binding to D-loop DNA of the mitochondrial genome in neurons. Proc Natl Acad Sci U S A. 2005;102(39):13915–20.
56. Herring A, et al. Environmental enrichment enhances cellular plasticity in transgenic mice with Alzheimer-like pathology. Exp Neurol. 2009;216(1):184–92.
57. Cechetti F, et al. Environmental enrichment prevents behavioral deficits and oxidative stress caused by chronic cerebral hypoperfusion in the rat. Life Sci. 2012;91(1–2):29–36.
58. Arnaiz SL, et al. Enriched environment, nitric oxide production and synaptic plasticity prevent the aging-dependent impairment of spatial cognition. Mol Aspects Med. 2004;25(1–2): 91–101.
59. Lores-Arnaiz S, et al. Extensive enriched environments protect old rats from the aging dependent impairment of spatial cognition, synaptic plasticity and nitric oxide production. Behav Brain Res. 2006;169(2):294–302.
60. Jain V, et al. Enriched environment prevents hypobaric hypoxia induced memory impairment and neurodegeneration: role of BDNF/PI3K/GSK3beta pathway coupled with CREB activation. PLoS One. 2013;8(5):e62235.
61. Hu YS, et al. Molecular mechanisms of environmental enrichment: impairments in Akt/GSK3beta, neurotrophin-3 and CREB signaling. PLoS One. 2013;8(5):e64460.
62. Falkenberg T, et al. Increased expression of brain-derived neurotrophic factor messenger-RNA in rat hippocampus is associated with improved spatial memory and enriched environment. Neurosci Lett. 1992;138(1):153–6.
63. Ickes BR, et al. Long-term environmental enrichment leads to regional increases in neurotrophin levels in rat brain. Exp Neurol. 2000;164(1):45–52.
64. Gibbons TE, et al. Voluntary wheel running, but not a diet containing (-)-epigallocatechin-3-gallate and beta-alanine, improves learning, memory and hippocampal neurogenesis in aged mice. Behav Brain Res. 2014;272:131–40.
65. Cotman CW, Berchtold NC. Exercise: a behavioral intervention to enhance brain health and plasticity. Trends Neurosci. 2002;25(6):295–301.
66. Franceschi C, et al. Inflammaging and anti-inflammaging: a systemic perspective on aging and longevity emerged from studies in humans. Mech Ageing Dev. 2007;128(1):92–105.
67. Streit WJ, et al. Dystrophic microglia in the aging human brain. Glia. 2004;45(2):208–12.
68. Luo XG, Ding JQ, Chen SD. Microglia in the aging brain: relevance to neurodegeneration. Mol Neurodegener. 2010;5:12.
69. Lucin KM, Wyss-Coray T. Immune activation in brain aging and neurodegeneration: too much or too little? Neuron. 2009;64(1):110–22.
70. Chen J, et al. Neuroinflammation and disruption in working memory in aged mice after acute stimulation of the peripheral innate immune system. Brain Behav Immun. 2008;22(3): 301–11.
71. Chung HY, et al. Molecular inflammation: underpinnings of aging and age-related diseases. Ageing Res Rev. 2009;8(1):18–30.
72. Li NX, Karin M. Is NF-kappa B the sensor of oxidative stress? FASEB J. 1999;13(10): 1137–43.
73. Makarov SS. NF-kappaB as a therapeutic target in chronic inflammation: recent advances. Mol Med Today. 2000;6(11):441–8.
74. Marsland AL, et al. Brain morphology links systemic inflammation to cognitive function in midlife adults. Brain Behav Immun. 2015;48:195–204.
75. Wyss-Coray T. Inflammation in Alzheimer disease: driving force, bystander or beneficial response? Nat Med. 2006;12(9):1005–15.
76. Kuhn HG, Dickinson-Anson H, Gage FH. Neurogenesis in the dentate gyrus of the adult rat: age-related decrease of neuronal progenitor proliferation. J Neurosci. 1996;16(6):2027–33.
77. Alvarez-Buylla A, Garcia-Verdugo JM. Neurogenesis in adult subventricular zone. J Neurosci. 2002;22(3):629–34.
78. Ben-Hur T, et al. Effects of proinflammatory cytokines on the growth, fate, and motility of multipotential neural precursor cells. Mol Cell Neurosci. 2003;24(3):623–31.

79. Vallieres L, et al. Reduced hippocampal neurogenesis in adult transgenic mice with chronic astrocytic production of interleukin-6. J Neurosci. 2002;22(2):486–92.
80. Koo JW, Duman RS. IL-1 beta is an essential mediator of the antineurogenic and anhedonic effects of stress. Proc Natl Acad Sci U S A. 2008;105(2):751–6.
81. Carpentier PA, Palmer TD. Immune influence on adult neural stem cell regulation and function. Neuron. 2009;64(1):79–92.
82. Butovsky O, et al. Microglia activated by IL-4 or IFN-gamma differentially induce neurogenesis and oligodendrogenesis from adult stem/progenitor cells. Mol Cell Neurosci. 2006; 31(1):149–60.
83. Villeda SA, et al. The ageing systemic milieu negatively regulates neurogenesis and cognitive function. Nature. 2011;477(7362):90–4.
84. Villeda SA, et al. Young blood reverses age-related impairments in cognitive function and synaptic plasticity in mice. Nat Med. 2014;20(6):659–63.
85. Deleidi M, Jaggle M, Rubino G. Immune aging, dysmetabolism, and inflammation in neurological diseases. Front Neurosci. 2015;9:172.
86. Kempermann G, Gast D, Gage FH. Neuroplasticity in old age: sustained fivefold induction of hippocampal neurogenesis by long-term environmental enrichment. Ann Neurol. 2002;52(2): 135–43.
87. Olson AK, et al. Environmental enrichment and voluntary exercise massively increase neurogenesis in the adult hippocampus via dissociable pathways. Hippocampus. 2006;16(3): 250–60.
88. Wilson RS, et al. Neural reserve, neuronal density in the locus ceruleus, and cognitive decline. Neurology. 2013;80(13):1202–8.
89. Arranz L, et al. Environmental enrichment improves age-related immune system impairment: long-term exposure since adulthood increases life span in mice (vol 13, pg 415, 2010). Rejuvenation Res. 2010;13(5):627.
90. Williamson LL, Chao A, Bilbo SD. Environmental enrichment alters glial antigen expression and neuroimmune function in the adult rat hippocampus. Brain Behav Immun. 2012;26(3): 500–10.
91. Gomes da Silva S, et al. Exercise-induced hippocampal anti-inflammatory response in aged rats. J Neuroinflammation. 2013;10:61.
92. Handschin C, Spiegelman BM. The role of exercise and PGC1alpha in inflammation and chronic disease. Nature. 2008;454(7203):463–9.
93. Febbraio MA. Exercise and inflammation. J Appl Physiol. 2007;103(1):376–7.
94. Pinto A, et al. Effects of physical exercise on inflammatory markers of atherosclerosis. Curr Pharm Des. 2012;18(28):4326–49.
95. Kohut ML, et al. Aerobic exercise, but not flexibility/resistance exercise, reduces serum IL-18 CRP, and IL-6 independent of beta-blockers, BMI, and psychosocial factors in older adults. Brain Behav Immun. 2006;20(3):201–9.
96. Erickson KI, et al. Exercise training increases size of hippocampus and improves memory. Proc Natl Acad Sci U S A. 2011;108(7):3017–22.
97. Van der Borght K, et al. Physical exercise leads to rapid adaptations in hippocampal vasculature: temporal dynamics and relationship to cell proliferation and neurogenesis. Hippocampus. 2009;19(10):928–36.
98. Zonis S, et al. Chronic intestinal inflammation alters hippocampal neurogenesis. J Neuroinflammation. 2015;12:65.
99. Bajaj JS, et al. Linkage of gut microbiome with cognition in hepatic encephalopathy. Am J Physiol Gastrointest Liver Physiol. 2012;302(1):G168–75.
100. Ohland CL, et al. Effects of Lactobacillus helveticus on murine behavior are dependent on diet and genotype and correlate with alterations in the gut microbiome. Psychoneuroendocrinology. 2013;38(9):1738–47.
101. Duncan SH, Flint HJ. Probiotics and prebiotics and health in ageing populations. Maturitas. 2013;75(1):44–50.

102. Lynch DB, et al. Diet-microbiota-health interactions in older subjects: implications for healthy aging. Interdiscip Top Gerontol. 2015;40:141–54.
103. Patrignani P, Tacconelli S, Bruno A. Gut microbiota, host gene expression, and aging. J Clin Gastroenterol. 2014;48 Suppl 1:S28–31.
104. Langille MG, et al. Microbial shifts in the aging mouse gut. Microbiome. 2014;2(1):50.
105. Caracciolo B, et al. Cognitive decline, dietary factors and gut-brain interactions. Mech Ageing Dev. 2014;136–137:59–69.
106. Collins SM, Surette M, Bercik P. The interplay between the intestinal microbiota and the brain. Nat Rev Microbiol. 2012;10(11):735–42.
107. Jeong JJ, et al. Orally administrated Lactobacillus pentosus var. plantarum C29 ameliorates age-dependent colitis by inhibiting the nuclear factor-kappa B signaling pathway via the regulation of lipopolysaccharide production by gut microbiota. PLoS One. 2015;10(2):e0116533.
108. Jeong JJ, et al. The probiotic mixture IRT5 ameliorates age-dependent colitis in rats. Int Immunopharmacol. 2015;26(2):416–22.
109. Sharma R, et al. Dietary supplementation of milk fermented with probiotic Lactobacillus fermentum enhances systemic immune response and antioxidant capacity in aging mice. Nutr Res. 2014;34(11):968–81.
110. Xiao J, et al. Lactobacillus casei-01 facilitates the ameliorative effects of proanthocyanidins extracted from lotus seedpod on learning and memory impairment in scopolamine-induced amnesia mice. PLoS One. 2014;9(11):e112773.
111. Distrutti E, et al. Modulation of intestinal microbiota by the probiotic VSL#3 resets brain gene expression and ameliorates the age-related deficit in LTP. PLoS One. 2014;9(9):e106503.
112. Nakamura S, et al. Daily feeding of fructooligosaccharide or glucomannan delays onset of senescence in SAMP8 mice. Gastroenterol Res Pract. 2014;2014:303184.
113. Smith CJ, et al. Probiotics normalize the gut-brain-microbiota axis in immunodeficient mice. Am J Physiol Gastrointest Liver Physiol. 2014;307(8):G793–802.
114. Choi JJ, et al. Exercise attenuates PCB-induced changes in the mouse gut microbiome. Environ Health Perspect. 2013;121(6):725–30.
115. Lambert JE, et al. Exercise training modifies gut microbiota in normal and diabetic mice. Appl Physiol Nutr Metab. 2015;40(7):749–52.
116. Petriz BA, et al. Exercise induction of gut microbiota modifications in obese, non-obese and hypertensive rats. BMC Genomics. 2014;15:511.
117. Klaenhammer T, et al. Discovering lactic acid bacteria by genomics. Antonie Van Leeuwenhoek. 2002;82(1–4):29–58.
118. Jeong JJ, et al. Lactobacillus pentosus var. plantarum C29 ameliorates age-dependent memory impairment in Fischer 344 rats. Lett Appl Microbiol. 2015;60(4):307–14.
119. Jeong JJ, et al. Probiotic mixture KF attenuates age-dependent memory deficit and lipidemia in Fischer 344 rats. J Microbiol Biotechnol. 2015;25(9):1532–6.
120. Woo JY, et al. Lactobacillus pentosus var. plantarum C29 ameliorates memory impairment and inflammaging in a D-galactose-induced accelerated aging mouse model. Anaerobe. 2014;27:22–6.
121. Mika A, et al. Exercise is more effective at altering gut microbial composition and producing stable changes in lean mass in juvenile versus adult male F344 rats. PLoS One. 2015; 10(5):e0125889.
122. Colcombe SJ, et al. Aerobic fitness reduces brain tissue loss in aging humans. J Gerontol A Biol Sci Med Sci. 2003;58(2):176–80.
123. Steinberg SI, et al. Exercise, sedentary pastimes, and cognitive performance in healthy older adults. Am J Alzheimers Dis Other Demen. 2015;30(3):290–8.
124. Okonkwo OC, et al. Physical activity attenuates age-related biomarker alterations in preclinical AD. Neurology. 2014;83(19):1753–60.
125. Scarmeas N, et al. Physical activity, diet, and risk of Alzheimer disease. JAMA. 2009; 302(6):627–37.

126. Johansson BB, Ohlsson AL. Environment, social interaction, and physical activity as determinants of functional outcome after cerebral infarction in the rat. Exp Neurol. 1996; 139(2):322–7.
127. Lautenschlager NT, Almeida OP. Physical activity and cognition in old age. Curr Opin Psychiatry. 2006;19(2):190–3.
128. Saczynski JS, et al. The effect of social engagement on incident dementia: the Honolulu-Asia Aging Study. Am J Epidemiol. 2006;163(5):433–40.
129. Fratiglioni L, et al. Influence of social network on occurrence of dementia: a community-based longitudinal study. Lancet. 2000;355(9212):1315–9.
130. Holwerda TJ, et al. Feelings of loneliness, but not social isolation, predict dementia onset: results from the Amsterdam Study of the Elderly (AMSTEL). J Neurol Neurosurg Psychiatry. 2014;85(2):135–42.
131. Wilson RS, et al. Loneliness and risk of Alzheimer disease. Arch Gen Psychiatry. 2007;64(2):234–40.
132. Coin A, et al. Does religiosity protect against cognitive and behavioral decline in Alzheimer's dementia? Curr Alzheimer Res. 2010;7(5):445–52.
133. Corsentino EA, et al. Religious attendance reduces cognitive decline among older women with high levels of depressive symptoms. J Gerontol A Biol Sci Med Sci. 2009;64(12): 1283–9.
134. Hill TD, et al. Religious attendance and cognitive functioning among older Mexican Americans. J Gerontol B Psychol Sci Soc Sci. 2006;61:3–9.
135. Kaufman Y, et al. Cognitive decline in Alzheimer disease: impact of spirituality, religiosity, and QOL. Neurology. 2007;68(18):1509–14.
136. Reyes-Ortiz CA, et al. Church attendance mediates the association between depressive symptoms and cognitive functioning among older Mexican Americans. J Gerontol A Biol Sci Med Sci. 2008;63(5):480–6.
137. Paganini-Hill A, Kawas CH, Corrada MM. Lifestyle factors and dementia in the oldest-old: the 90+ study. Alzheimer Dis Assoc Disord. 2016;30(1):21–6.
138. Clarke P, et al. Does the neighborhood environment modify genetic risk for cognitive decline? Gerontologist. 2014;54:64.
139. Stern Y. What is cognitive reserve? Theory and research application of the reserve concept. J Int Neuropsychol Soc. 2002;8(3):448–60.
140. Steffener J, Stern Y. Exploring the neural basis of cognitive reserve in aging. Biochim Biophys Acta. 2012;1822(3):467–73.
141. Serra L, et al. Neuroanatomical correlates of cognitive reserve in Alzheimer disease. Rejuvenation Res. 2011;14(2):143–51.
142. Tyas SL, et al. Healthy ageing in the Nun Study: definition and neuropathologic correlates. Age Ageing. 2007;36(6):650–5.
143. Farfel JM, et al. Very low levels of education and cognitive reserve: a clinicopathologic study. Neurology. 2013;81(7):650–7.
144. Meng X, D'Arcy C. Education and dementia in the context of the cognitive reserve hypothesis: a systematic review with meta-analyses and qualitative analyses. PLoS One. 2012; 7(6):e38268.
145. Valenzuela MJ, et al. Lifespan mental activity predicts diminished rate of hippocampal atrophy. PLoS One. 2008;3(7):e2598.
146. Persson J, et al. Preserved hippocampus activation in normal aging as revealed by fMRI. Hippocampus. 2011;21(7):753–66.
147. James CE, et al. Musical training intensity yields opposite effects on grey matter density in cognitive versus sensorimotor networks. Brain Struct Funct. 2014;219(1):353–66.
148. Mechelli A, et al. Neurolinguistics: structural plasticity in the bilingual brain. Nature. 2004;431(7010):757.
149. Olsen RK, et al. The effect of lifelong bilingualism on regional grey and white matter volume. Brain Res. 2015;1612:128–39.

150. Luk G, et al. Lifelong bilingualism maintains white matter integrity in older adults. J Neurosci. 2011;31(46):16808–13.
151. Gold BT, et al. Lifelong bilingualism maintains neural efficiency for cognitive control in aging. J Neurosci. 2013;33(2):387–96.
152. Ljungberg JK, et al. A longitudinal study of memory advantages in bilinguals. PLoS One. 2013;8(9):e73029.
153. Grady CL, et al. Brain network activity in monolingual and bilingual older adults. Neuropsychologia. 2015;66:170–81.
154. Bialystok E, Craik FIM, Freedman M. Bilingualism as a protection against the onset of symptoms of dementia. Neuropsychologia. 2007;45(2):459–64.
155. Gold BT, Johnson NF, Powell DK. Lifelong bilingualism contributes to cognitive reserve against white matter integrity declines in aging. Neuropsychologia. 2013;51(13):2841–6.
156. Bak TH, et al. Does bilingualism influence cognitive aging? Ann Neurol. 2014;75(6): 959–63.
157. Yeung CM, et al. Is bilingualism associated with a lower risk of dementia in community-living older adults? Cross-sectional and prospective analyses. Alzheimer Dis Assoc Disord. 2014;28(4):326–32.
158. Lawton DM, Gasquoine PG, Weimer AA. Age of dementia diagnosis in community dwelling bilingual and monolingual Hispanic Americans. Cortex. 2015;66:141–5.
159. Engvig A, et al. Memory training impacts short-term changes in aging white matter: a longitudinal diffusion tensor imaging study. Hum Brain Mapp. 2012;33(10):2390–406.
160. Lövdén M, et al. Experience-dependent plasticity of white-matter microstructure extends into old age. Neuropsychologia. 2010;48(13):3878–83.
161. Takeuchi H, et al. Training of working memory impacts structural connectivity. J Neurosci. 2010;30(9):3297–303.
162. Erickson KI, et al. Training-induced plasticity in older adults: effects of training on hemispheric asymmetry. Neurobiol Aging. 2007;28(2):272–83.

Protective Effect of Exercise on Age-Related Oxidant and Inflammatory Events

Nada Sallam and Ismail Laher

1 Exercise and Ageing

There is much evidence indicating that regular exercise counteracts the negative effects of ageing. For example, regular exercise is associated with reduced risks of all-cause and cardiovascular mortality [1–3], and also with increased longevity [4–7]. Furthermore, exercise reduces the risk of cardiovascular diseases [8], type 2 diabetes [9], metabolic syndrome [10], colon cancer [11], obesity [12], osteoporosis (Kelley 1998; Marques et al. 2012), sarcopenia [13], anxiety (Wipfli et al. 2008), and cognitive impairment [14]. Most importantly, exercise improves the quality of life of elderly people [15].

2 Ageing and Oxidative Stress

Ageing is associated with oxidative stress that is mainly due to defective (leaky) mitochondria [16], probably resulting from lower cytochrome C oxidase (complex IV) activity (Navarro et al. 2003) and peroxidative damage of mitochondrial membrane lipids [17]. Hence more electrons escape from the mitochondria, generating a

N. Sallam, Ph.D.
Department of Pharmacology and Toxicology, Faculty of Pharmacy, Cairo University, Cairo, Egypt

I. Laher, Ph.D. (✉)
Department of Pharmacology and Therapeutics, Faculty of Medicine, The University of British Columbia, 2176 Health Sciences Mall, Vancouver, BC, Canada V6T 1Z3
e-mail: ilaher@mail.ubc.ca

S.C. Bondy, A. Campbell (eds.), *Inflammation, Aging, and Oxidative Stress*, Oxidative Stress in Applied Basic Research and Clinical Practice,
DOI 10.1007/978-3-319-33486-8_17

long trail of reactive oxygen species (ROS) [18, 19] leading to progressive mitochondrial dysfunction and further exacerbating ROS generation, thus creating a vicious cycle of oxidative damage. Age-associated increases in ROS production occur in skeletal muscles [20] and other organs such as the heart, liver, brain, and kidney [5, 18, 21].

Reduced protein synthesis limits antioxidant defense mechanisms and repair capacity in aged individuals, which further contributes to the state of oxidative stress. The free radicals theory of aging hypothesizes that oxidative stress causes damage to macromolecules, including lipids, proteins and nucleic acids that overwhelms cellular antioxidant defense and repair mechanisms, leading to progressive deleterious changes over time [22, 23].

3 Aging and Inflammation

Age-associated progressive loss of muscle mass and strength, known as sarcopenia [24, 25], increases the incidence of muscle injury [6] and subsequently increases the infiltration of immune cells into the injured muscles. Activated immune cells release ROS, reactive nitrogen species (RNS) and proinflammatory mediators during the respiratory burst [26]. Similarly, the injured muscles generate and release proinflammatory mediators, which bind with membrane receptors and activate specific ROS-generating enzymes such as lipooxygenase, NADPH oxidase, and xanthine oxidase [27–32].

Sarcopenia can also lead to reduced physical activity and increased adiposity. Accumulation of excess adipose tissues induces a state of low-grade but chronic inflammation through the release of a multitude of pro-inflammatory cytokines including tumor necrosis factor-alpha (TNF-α), interleukin-6 (IL-6) and interleukin-1beta (IL-1β) [33–35]. Indeed, ageing is associated with increased levels of circulating pro-inflammatory cytokines such as TNF-α, IL-6, and interleukin-1 receptor agonist (IL-1ra) and systemic inflammatory biomarkers such as C-reactive protein (CRP) as well as higher count of inflammatory cells (neutrophil and monocytes) [36–38]. Hence, aging is associated with a state of oxidative stress and chronic inflammation. The major events related to age-associated oxidative stress and inflammation is shown in Fig. 1.

4 Oxidative Stress and Inflammation Overlapping Signaling Pathways

Oxidative stress and inflammation share common and overlapping signaling pathways. ROS initiate and augment inflammation, and are also products of inflammation. During the inflammatory response, particularly during the respiratory burst, immune cells generate ROS and RNS via NADPH oxidase and nitric oxide synthase

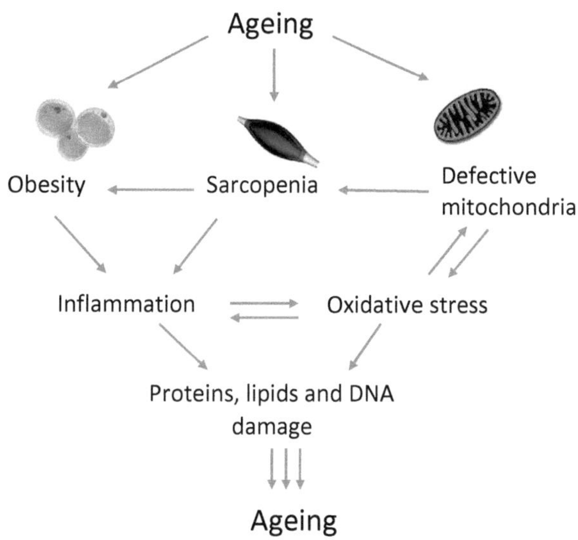

Fig. 1 Some important events related to oxidative stress and inflammation during the ageing process

(NOS), and also release proinflammatory cytokines such as TNF-α, IL-1β and IL-6 [26, 39, 40]. Similarly, the injured tissues release proinflammatory cytokines. These cytokines/proinflammatory mediators bind to membrane receptors and activate specific ROS-generating enzymes, such as lipooxygenase, NADPH oxidase, myeloperoxidase and xanthine oxidase [27–32] and specific RNS generating pathways such as NOS, protein kinase B (Akt) and Sph1P (sphingosine-1-phosphate) [41–43].

ROS overproduction activates redox-sensitive transcription factors, namely nuclear factor kappa B (NF-kB) and activator protein-1 (AP-1) via stress kinases such as extracellular signal regulated kinases (ERKs), c-jun N-terminal kinases (JNKs), mitogen activated protein kinase p38 (MAPK p38), protein kinase C (PKC), phosphatidylinositol-4,5-bisphosphate 3-kinase (PI3K)/Akt, Src family kinases (SFKs) leading to increased expression of inflammatory target proteins genes such as matrix metalloproteinase-9 (MMP-9), intercellular adhesion molecule-1 (ICAM-1), vascular cell adhesion molecule-1 (VCAM-1), inducible nitric oxide synthase (iNOS), cyclooxygenase-2 (COX-2), prostaglandin E_2 (PGE_2) and cytosolic phospholipase A2 (cPLA2) (Kim et al. 2008, 2014; Lee et al. 2012) [44–50].

Interestingly, NF-kB has been shown to regulate the transcription of TNF-α gene [51] as well as other proinflammatory mediators such as IL-1, and interleukin 8 (IL-8) [50]. Furthermore, many of NF-kB-induced proteins such as NOS, COX and PGE_2 are prominent sources of ROS and RNS [52] forming an auto-activating loop which feeds the vicious cycle of inflammation and oxidative stress. In short, proinflammatory mediators such as TNF-α, IL-1, IL-6 activate redox-sensitive transcription factors such as NF-kB and AP-1 through redox signaling, resulting in the generation of large amounts of these proinflammatory mediators and ROS (Fig. 2). Indeed, aging is associated with adverse health conditions characterized by elevated levels of both oxidative stress and inflammatory markers such as atherosclerosis,

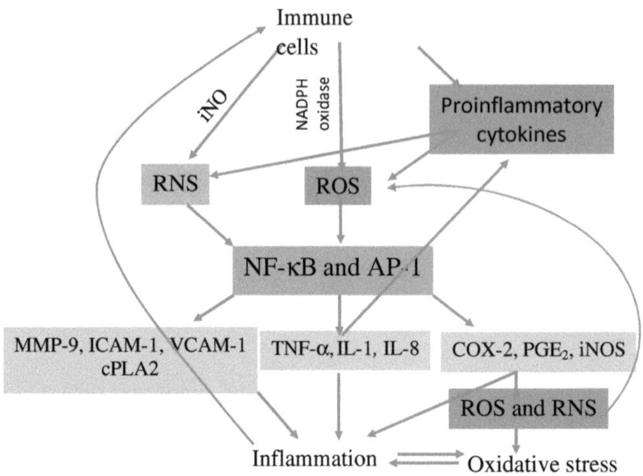

Fig. 2 Overlapping signaling pathways of oxidative stress and inflammation in aging. *AP-1* activator protein-1, *COX-2* cyclooxygenase-2, *cPLA2* cytosolic phospholipase A2, *ICAM-1* intercellular adhesion molecule-1, *IL-1* interleukin-1, *IL-8* interleukin-8, *iNOS* inducible nitric oxide synthase, *MMP-9* matrix metalloproteinase-9, *NF-kB* nuclear factor kappa B, *PGE*$_2$ prostaglandin E$_2$, *RNS* reactive nitrogen species, *ROS* reactive oxygen species, *TNF-α* tumor necrosis factor-alpha, *VCAM-1* vascular cell adhesion molecule-1

metabolic syndrome, sarcopenia, arthritis, and chronic obstructive pulmonary disease [53].

Not surprisingly, ROS can also induce proteins such as heat-shock proteins (HSPs), HSP70 in particular [54] and heme oxygenate 1 oxygenase (HO-1) (Lee et al. 2012) that can protect cells and tissues from the deleterious effects of inflammation. However, in aging the balance of antioxidant/anti-inflammatory to oxidant/inflammatory proteins is tilted towards the latter

5 Exercise: Anti-inflammatory and Antioxidant Effects

Regular exercise reduces the risk of a wide range of oxidative stress and inflammation associated diseases including cardiovascular diseases [8], type 2 diabetes [9], metabolic syndrome [10], cancer [11], obesity [12], and sarcopenia [13].

5.1 Anti-inflammatory Effects of Exercise

Acute bouts of exercise cause transient damage to the contracting skeletal muscles, so triggering an inflammatory response that increases the levels of pro-inflammatory cytokines and acute-phase reactants in the blood [55, 56]. However, regular exercise is associated with reduced levels of systemic inflammatory markers such as CRP,

IL-6 and TNF-α that occur independently of weight loss in young and middle aged adults [57–62], as well as in the elderly [57, 63–70]. Also, many interventional studies report that exercise reduces inflammatory markers, particularly CRP, TNF-α, monocyte chemoattractant protein-1 (MCP-1) and (IL-8), soluble TNF-α receptor 2 (sTNFR2) and soluble IL-6 receptor (sIL-6R), and increases the anti-inflammatory mediators interleukin-10 (IL-10), interleukin-4 (IL-4), and transforming growth factor beta 1 (TGFβ1) [71–80]. However, only a few randomized controlled trials were conducted to confirm that [81–84]. These benefits of exercise were also evident in the elderly [81, 85–89]. It is likely that exercise causes the most significant anti-inflammatory effects in patients with high baseline inflammatory biomarkers, particularly when exercise is associated with weight loss.

However, it is worth noting that some interventional and randomized controlled trials studies did not detect a significant effect of regular exercise on systemic inflammatory biomarkers in adults [59, 90–92], or in aged adults [93–96]. A meta-analysis conducted in 2006 found only five randomized controlled trials that examined the effects of regular aerobic exercise (at least 4 weeks duration) in adults and concluded that aerobic exercise did not reduce CRP levels [97]. It is likely that these discrepancies may be attributed to the smaller sample size used in the clinical trials examined.

On the other hand, the effects of resistance exercise on inflammatory mediators are mostly negative [98–100], although Brooks et al. [101] reported that 16-week resistance training reduced CRP and increased adiponectin levels in older diabetic patients. Clearly, the effects of exercise depends on the type (aerobic/resistance), intensity (mild/moderate/intense/exhaustive), and frequency (sessions per day/week/month) of exercise, and also on the subject's basic condition and endurance capacity.

5.1.1 Anti-inflammatory Signaling Pathways of Exercise

The signaling pathways underlying the anti-inflammatory effects of exercise are complex and not completely understood but for the sake of convenience, can be divided into three main pathways according to the site of action. The main sites of action are the adipose tissue, the immune system and skeletal muscles.

- ***Anti-inflammatory effects of exercise on adipose tissue***

Obesity is associated with chronic inflammation. Adipose tissue, particularly visceral fat depots, and macrophages trapped in the adipose tissue release pro-inflammatory cytokines such as IL-6 and TNF-α [33–35, 102]. Exercise increases energy expenditure and burns off some of the body fat, which can result in weight loss, particularly visceral fat loss [12, 103, 104]. Subsequently, the production and release of pro-inflammatory adipokines such as IL-6 and TNF-α are reduced [79, 105–107]. Exercise also induces the release of adiponectin from adipose tissues [108, 109]. Adiponectin exerts anti-apoptotic, anti-inflammatory and anti-oxidative activities [110, 111]. Exercise inhibits the infiltration of M1-type macrophages into adipose tissue and also induces the switch of macrophages from the more inflammatory phenotype M1-type to the less inflammatory phenotype M2-type in obese mice [112].

- ***Anti-inflammatory effects of exercise on the immune system***

Aerobic exercise downregulates the innate immune response and activates the adaptive immune system with consequent suppression of inflammation. Exercise modulates the immune system by reducing the number of inflammatory CD14+CD16+ monocytes [113], increasing the number of CD4CD25 regulatory T cells [114, 115], increasing the dominance of Type 2 helper T cell over Type 1 helper T cell [116–118], and reducing the expression of toll like receptor-4 (TLR4) on monocyte surfaces [119, 120]. TLR4 signaling participates in several innate immunity and inflammatory processes [121]. Exercise also reduces the production of proinflammatory cytokines such as interferon gamma (INFγ), TNFα, IL-1α, IL-8, MCP-1 and the receptors for TNF-α (sTNFR2) and IL-6 (sIL-6R). In addition, exercise releases anti-inflammatory cytokines such as IL-10, IL-4, TGFβ1 and adiponectin [71–76, 78–81, 85–89, 108, 122].

Exercise is a positive stressor to the body; it stimulates the sympathetic nervous system and the hypothalamic–pituitary–adrenal axis. Therefore, exercise increases serum glucocorticoid levels [123] to cause a subsequent inhibition of the immune system [124].

- ***Anti-inflammatory effects of exercise on skeletal muscles***

By improving muscle mass and strength, exercise renders skeletal muscles less vulnerable to acute injury and the associated inflammatory responses [125–129]. Also, by stimulating mitochondrial biogenesis [130] and enhancing mitochondrial oxidative capacity [131], exercise mitigates mitochondrial aging and interrupts the vicious cycle of oxidative damage.

Exercise induces the release of several cytokines (myokines) from skeletal muscle, most notably IL-6 [132, 133]. IL-6 triggers the release of several anti-inflammatory cytokines such as IL-1 receptor antagonist (IL-1ra) and IL-10, in addition to cortisol [134, 135]. IL-10 inhibits the synthesis of several pro-inflammatory cytokines such as TNF-α and IL-1β [136]. Exercise also reduces TNF-α and IL-1β production in skeletal muscles [137–141], and upregulates the expression of the anabolic myokine IL-15 [136, 142] and HSPs in skeletal muscles [143–145]. **The mechanisms underlying the anti-inflammatory actions of exercise are summarized in Fig. 3**.

5.2 *Anti-oxidant Effects of Exercise*

There is little doubt that generation of ROS is increased acutely during exercise. However, the incidence of diseases associated with oxidative stress is reduced by regular exercise. Exercise training attenuates oxidative damage in the brain [5,

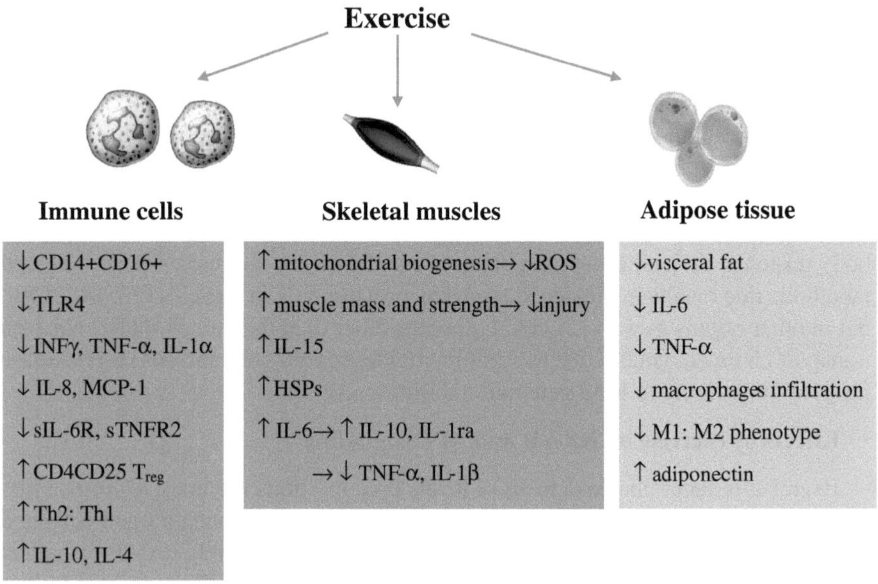

Fig. 3 Signaling pathways underlying the anti-inflammatory actions of exercise. *HSPs* heat shock proteins, *IL-1α* interleukin-1 alpha, *IL-1ra* interleukin-1 receptor antagonist, *IL-1β* interleukin-1 beta, *IL-6* interleukin-6, *IL-8* interleukin-8, *IL-10* interleukin-10, *IL-15* interleukin-15, *INFγ* interferon gamma, *M1* macrophage phenotype 1, *M2* macrophage phenotype 2, *ROS* reactive oxygen species, *sTNFR2* soluble TNF-α receptor 2, *sIL-6R* soluble IL-6 receptor, *TLR4* toll like receptor-4, *TGFβ1* transforming growth factor beta 1, *TNF-α* tumor necrosis factor-alpha, *Th1* Type 1 helper T cell, *Th2* Type 2 helper T cell

146–149], liver [5, 150–152] (Radak et al. 2004), kidney [5], skeletal muscles [153] and heart [5, 154].

Importantly, regular exercise ameliorates age-associated oxidative stress in the heart [154, 155], liver (Radak et al. 2004), and skeletal muscle [156] (Radak et al. 2002). In the study of Navarro et al. [5], exercise reduced age-associated mitochondrial oxidative damage and upregulated mitochondrial NADH-cytochrome-*c* reductase and cytochrome oxidase activities in brain, heart, liver, and kidney of 52 week old rats. However, exercise caused an increase in oxidative damage in skeletal muscles [157] and hearts [158] of aged rats.

In elderly people, exercise reduced serum levels of myeloperoxidase, a marker of inflammation and oxidative stress [125] and thiobarbituric-reactive acid substances, a marker of lipid peroxidation [159]. However, de Gonzalo-Calvo et al. [160] reported that although regular exercise increased protein carbonyl content and lipid peroxidation levels in the plasma and erythrocytes of long-term trained elderly men, their overall health condition was markedly improved. Another clinical study showed that 8 weeks of walking exercise did not significantly change low density lipoprotein (LDL) oxidation or nitration in the elderly [161].

5.2.1 Anti-oxidant Signaling Pathways of Exercise

By suppressing inflammatory pathways, exercise inhibits prominent sources of ROS and RNS generation and thus exerts beneficial antioxidant effects. Exercise also upregulates the antioxidant defense mechanisms and repair proteins in the body via redox-sensitive transcription factors, mainly NF-kB, AP-1 and peroxisome proliferator-activated receptor gamma coactivator 1-alpha (PGC-1α).

The energy demand of contracting muscles increases during exercise and the body responds by increasing oxygen uptake and delivery to muscles. The increased metabolic rate results in greater ROS production not only in muscles [20, 148, 149], but in other organs as well [21, 162]. Sources other than the mitochondrial electron transport chain enzymes, such as xanthine oxidase [163–165] and NADPH oxidase [20, 166] contribute to ROS generation during exercise.

- **Effects of exercise on NF-kB and AP-1 signaling**

Exercise-induced increase in ROS levels preconditions the body against oxidative damage by evoking an adaptive process that is mediated via mitogen-activated protein kinases (MAPK p38, ERK 1 and ERK 2) [163, 167–169], cAMP-response-element binding (CREB) [170, 171], and synapsin [170, 171]. These effects lead to activation of redox-sensitive transcription factors such as NF-kB [163, 172, 173] and AP-1 [169, 173], resulting in increased expression of antioxidant enzymes [174] such as superoxide dismutase [163, 173] and catalase [173], repair proteins such as heat shock proteins HSP25, HSP60, HSP72, HSP70, heat shock cognate 70 HSC70 [157, 173–176], proteasomes complex and NOS [150, 163]. These signaling cascades were demonstrated in skeletal muscles [163, 172], brain [170, 171], leukocytes [177] and hearts [169] of experimental animals as well as in humans [167, 177] and in aged animals [173, 176, 178] and humans [177]. However, other studies reported that exercise-induced activation of NF-kB and AP-1 [173] and upregulation of HSP70 were attenuated in fast skeletal muscles of old rats [179]. Interestingly, ageing also increased ROS production and NF-KB activity in the livers of aged rats; these effects were attenuated by exercise (Radak et al. 2004) [150].

- **Effects of exercise on PGC-1α signaling**

Exercise stimulates mitochondrial biogenesis [130] and ameliorates the age-associated decline in mitochondrial oxidative capacity in skeletal muscles [131], and other organs [180] (Navarro et al. 2003) via PGC-1α signaling [181, 182]. PGC-1α is a redox-sensitive transcription factor that is activated by 5′-AMP-activated protein kinase (AMPK), [168, 183–185] to trigger the transcription of nuclear respiratory factor 1 (NRF-1) and expression of mitochondrial transcription factor A (mtTFA), a key regulator of mitochondrial DNA replication [186]. PGC-1α also increases the expression of antioxidant proteins such as glutathione peroxidase (GPX) and SOD-2 [187]. Safdar et al. [188] reported that exercise reversed most of the multisystem pathology and premature mortality in mice which were genetically modified to accumulate mitochondrial mutations. The effects of exercise on AMPK, and PGC-1α were preserved in the hippocampus of aging rats. However, results from Derbré et al. [181] suggest a blunted effect of exercise response in PGC-1α, NRF-1 in skeletal muscles of aged rats.

- **Effects of exercise on antioxidant enzymes expression and activity**

The NF-kB and PGC-1α signaling cascades converge to upregulate antioxidant defense mechanisms in cells to counteract and interrupt the vicious cycle of inflammation and oxidative stress associated with ageing. The most studied antioxidant enzymes systems in laboratory animals and in humans are SOD, catalase, GPX and glutathione reductase.

Regular exercise increases the activities of SOD in the brain [189, 190], erythrocytes [191–193], heart [154, 158, 194–196], tissues from the lung [197], and liver [197]. Exercise increases the protein expression of SOD in blood vessels [195, 198, 199, 200] (Lee et al. 2001), liver [201] and blood [202].

The activity of GPX was increased by exercise in the brain [189, 190], erythrocytes [191, 192, 203], blood (Elosua et al. 2008), liver, heart, lung [18, 197] and skeletal muscles containing a high percentage of type I or type IIa fibers of old rats [156]. Similarly, exercise increases the activities of plasma glutathione reductase (Elosua et al. 2008) and catalase in erythrocytes [193], heart [154, 158, 204] and liver [18]. Some studies reported no changes in the activities of SOD, CAT, or GPX in the brain [205] or skeletal muscle [153].

Several studies investigated the effects of exercise on antioxidant enzymes at old age. For example, exercise enhances the activities of SOD in the heart [154, 158, 195, 197], brain [189], and lung [197] and increases GPX activities in the brain, liver, heart, lung and skeletal muscles of old rats [156, 189, 197, 206] and in erythrocytes of elderly people [203]. Catalase activity is upregulated by exercise in the liver [197] and heart [154, 158] of aged rats. In the study of Navarro et al. [5], exercise reduced the extent of age-associated decline in SOD and catalase activities in brain, heart, liver, and kidney of 52 week old but not older mice (72 week old). Exercise also up-regulated the protein expression of SOD-1 and GPX in the hippocampus of aged rats [147].

Exercise-induced adaptation of antioxidant enzymes is highly isoform, tissue and time course specific. Exercise modulates the three SOD isoforms differently [195, 199, 201, 202, 207, 208] as the promoter region of SOD-2 contains more ROS-sensitive binding sites [209]. Exercise-induced protein expression of SOD is time dependent (Navarro et al. 2003); SOD-1 protein expression was increased in rat skeletal muscles 48 h post exercise, whereas SOD-2 protein content was increased after 10 and 24 h, but not 48 h [208].

- **Effects of exercise on repair mechanisms**

Exercise can also stimulate the proteasome complex, which is responsible for the degradation of oxidatively damaged proteins [150, 210, 211] (Radak et al. 2000), and therefore enhances the cellular repair processes. Exercise also modulates the activity of DNA repair enzymes, particularly oxoguanine DNA glycosylase (OGG1) and uracil DNA glycosylase (UDG), and thus reduces the accumulation of nuclear 8-hydroxydeoxyguanosine (8-OHdG) and mutations in skeletal muscles (Radak et al. 2002) [212, 213], but not brains of aged rats [214]. Exercise increased thioredoxin reductase 1 (TrxR1), one of the thioredoxin system enzymes with direct and indirect antioxidant effects, in peripheral blood mononuclear cells in humans [125] (Wadley et al. 2015).

Exercise increases the content of the brain-derived neurotrophic factor (BDNF), a critically important neurotrophic factor that is involved in higher cognitive function [170, 171].

Telomeres are considered '*the guardians of the genome*'. Telomere dysfunction activates p53, leading to suppression of PGC-1α and PGC-1β promoters with consequent metabolic and organ failure [215]. The leukocyte telomere was 200 nucleotides longer in people who exercise regularly, which roughly corresponds to a 10-year increase in longevity [216]. Exercise increases the activity of telomerase, and induces the expression of telomere repeat-binding factor 2 and Ku70 in the thoracic aorta and in leukocytes from mice and humans [217]. **The signaling pathways underlying the antioxidant actions of exercise is summarized in Fig. 4**.

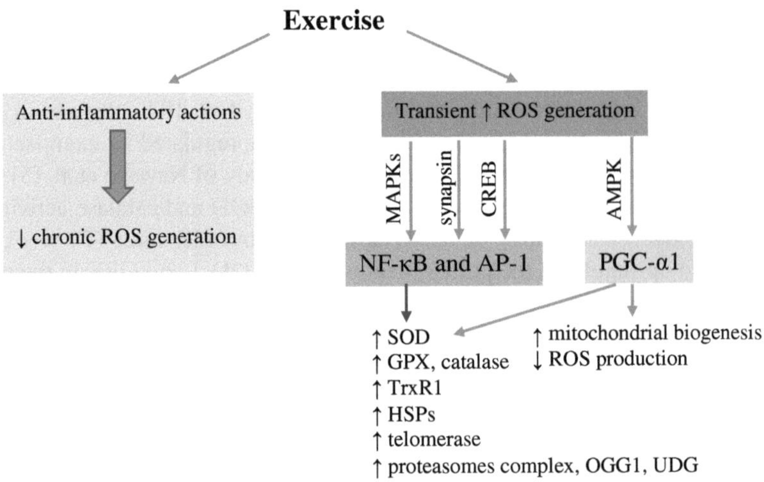

Fig. 4 Signaling pathways underlying the antioxidant actions of exercise. *AMPK* AMP-activated protein kinase, *AP-1* activator protein-1, *CREB* cAMP-response-element binding, *HSPs* heat shock proteins, *GPX* glutathione peroxidase, *MAPKs* mitogen activated protein kinases, *NF-kB* nuclear factor kappa B, *OGG1* oxoguanine DNA glycosylase, *PGC-1α* peroxisome proliferator-activated receptor gamma, coactivator 1-alpha, *SOD* superoxide dismutase, *ROS* reactive oxygen species, *TrxR1* Thioredoxin reductase 1, *UDG* uracil DNA glycosylase

References

1. Nocon M, Hiemann T, Müller-Riemenschneider F, Thalau F, Roll S, Willich SN. Association of physical activity with all-cause and cardiovascular mortality: a systematic review and meta-analysis. Eur J Cardiovasc Prev Rehabil. 2008;15(3):239–46.
2. Samitz G, Egger M, Zwahlen M. Domains of physical activity and all-cause mortality: systematic review and dose-response meta-analysis of cohort studies. Int J Epidemiol. 2011;40(5):1382–400.
3. Taylor RS, Brown A, Ebrahim S, Jolliffe J, Noorani H, Rees K, Skidmore B, Stone JA, Thompson DR, Oldridge N. Exercise-based rehabilitation for patients with coronary heart disease: systematic review and meta-analysis of randomized controlled trials. Am J Med. 2004;116(10):682–92.
4. Franco OH, de Laet C, Peeters A, Jonker J, Mackenbach J, Nusselder W. Effects of physical activity on life expectancy with cardiovascular disease. Arch Intern Med. 2005;165(20):2355–60.
5. Navarro A, Gomez C, López-Cepero JM, Boveris A. Beneficial effects of moderate exercise on mice aging: survival, behavior, oxidative stress, and mitochondrial electron transfer. Am J Physiol Regul Integr Comp Physiol. 2004;286(3):R505–11.
6. Paffenbarger Jr RS, Hyde RT, Wing AL, Lee IM, Jung DL, Kampert JB. The association of changes in physical-activity level and other lifestyle characteristics with mortality among men. N Engl J Med. 1993;328(8):538–45.
7. Sandvik L, Erikssen J, Thaulow E, Erikssen G, Mundal R, Rodahl K. Physical fitness as a predictor of mortality among healthy, middle-aged Norwegian men. N Engl J Med. 1993;328(8):533–7.
8. Wilund KR. Is the anti-inflammatory effect of regular exercise responsible for reduced cardiovascular disease? Clin Sci (Lond). 2007;112(11):543–55.
9. LaMonte MJ, Blair SN, Church TS. Physical activity and diabetes prevention. J Appl Physiol. 2005;99(3):1205–13.
10. Gaesser GA. Exercise for prevention and treatment of cardiovascular disease, type 2 diabetes, and metabolic syndrome. Curr Diab Rep. 2007;7:14–9.
11. Thune I, Furberg AS. Physical activity and cancer risk: dose-response and cancer, all sites and site-specific. Med Sci Sports Exerc. 2001;33(6 Suppl):S530–50. discussion S609–10.
12. Ross R, Dagnone D, Jones PJ, Smith H, Paddags A, Hudson R, Janssen I. Reduction in obesity and related comorbid conditions after diet-induced weight loss or exercise-induced weight loss in men. A randomized, controlled trial. Ann Intern Med. 2000;133(2):92–103.
13. Glass D, Roubenoff R. Recent advances in the biology and therapy of muscle wasting. Ann N Y Acad Sci. 2010;1211:25–36.
14. Tseng CN, Gau BS, Lou MF. The effectiveness of exercise on improving cognitive function in older people: a systematic review. J Nurs Res. 2011;19(2):119–31.
15. Ozaki A, Uchiyama M, Tagaya H, Ohida T, Ogihara R. The Japanese Centenarian Study: autonomy was associated with health practices as well as physical status. J Am Geriatr Soc. 2007;55(1):95–101.
16. Navarro A, López-Cepero JM, Sánchez del Pino MJ. Skeletal muscle and aging. Front Biosci. 2001;6:D26–44.
17. Chen JJ, Yu BP. Alterations in mitochondrial membrane fluidity by lipid peroxidation products. Free Radic Biol Med. 1994;17(5):411–8.
18. Kim JD, McCarter RJ, Yu BP. Influence of age, exercise, and dietary restriction on oxidative stress in rats. Aging (Milano). 1996;8(2):123–9.
19. Miquel J, Fleming J. Theoretical and experimental support for an oxygen radical mitochondrial injury hypothesis of cell aging. In: Johnson J, Walford R, Harman D, Miquel J, editors. Biology of aging. New York: Alan R. Liss Inc.; 1986. p. 51–76.
20. Bejma J, Ji LL. Aging and acute exercise enhance free radical generation in rat skeletal muscle. J Appl Physiol. 1999;87(1):465–70.

21. Bejma J, Ramires P, Ji LL. Free radical generation and oxidative stress with ageing and exercise: differential effects in the myocardium and liver. Acta Physiol Scand. 2000;169(4):343–51.
22. Harman D. Aging: a theory based on free radical and radiation chemistry. J Gerontol. 1956;11:298–300.
23. Kregel KC, Zhang HJ. An integrated view of oxidative stress in aging: basic mechanisms, functional effects, and pathological considerations. Am J Physiol Regul Integr Comp Physiol. 2007;292(1):R18–36.
24. Rogers MA, Evans WJ. Changes in skeletal muscle with aging: effects of exercise training. In: Holloszy JO, editor. Exercise and sport sciences reviews, vol. 21. Baltimore: Williams & Wilkins; 1993. p. 65–102.
25. Roubenoff R. Physical activity, inflammation, and muscle loss. Nutr Rev. 2007;65:S208–12.
26. Cannon JG, Blumberg JB. Acute phase immune responses in exercise. In: Sen CK, Packer L, Hanninen O, editors. Exercise and oxygen toxicity. New York: Elsevier Science; 1994. p. 447–79.
27. Flohé L, Brigelius-Flohé R, Saliou C, Traber MG, Packer L. Redox regulation of NF-kappa B activation. Free Radic Biol Med. 1997;22(6):1115–26.
28. Gao X, Belmadani S, Picchi A, Xu X, Potter BJ, Tewari-Singh N, Capobianco S, Chilian WM, Zhang C. Tumor necrosis factor-alpha induces endothelial dysfunction in Lepr(db) mice. Circulation. 2007;115(2):245–54.
29. Li JM, Fan LM, Christie MR, Shah AM. Acute tumor necrosis factor alpha signaling via NADPH oxidase in microvascular endothelial cells: role of $p47^{phox}$ phosphorylation and binding to TRAF4. Mol Cell Biol. 2005;25:2320–30.
30. Picchi A, Gao X, Belmadani S, Potter BJ, Focardi M, Chilian WM, Zhang C. Tumor necrosis factor-alpha induces endothelial dysfunction in the prediabetic metabolic syndrome. Circ Res. 2006;99(1):69–77.
31. Reeves EP, Nagl M, Godovac-Zimmermann J, Segal AW. Reassessment of the microbicidal activity of reactive oxygen species and hypochlorous acid with reference to the phagocytic vacuole of the neutrophil granulocyte. J Med Microbiol. 2003;52:643–51.
32. Thabut G, El-Benna J, Samb A, Corda S, Megret J, Leseche G, Vicaut E, Aubier M, Boczkowski J. Tumor necrosis factor-α increases airway smooth muscle oxidants production through a NADPH oxidase-like system to enhance myosin light chain phosphorylation and contractility. J Biol Chem. 2002;277:22814–21.
33. Fried SK, Bunkin DA, Greenberg AS. Omental and subcutaneous adipose tissues of obese subjects release interleukin-6: depot difference and regulation by glucocorticoid. J Clin Endocrinol Metab. 1998;83:847–50.
34. Mohamed-Ali V, Pinkney JH, Coppack SW. Adipose tissue as an endocrine and paracrine organ. Int J Obes Relat Metab Disord. 1998;22:1145–58.
35. Trayhurn P, Wood IS. Adipokines: inflammation and the pleiotropic role of white adipose tissue. Br J Nutr. 2004;92(3):347–55.
36. Bruunsgaard H, Andersen-Ranberg K, Jeune B, Pedersen AN, Skinhøj P, Pedersen BK. A high plasma concentration of TNF-alpha is associated with dementia in centenarians. J Gerontol A Biol Sci Med Sci. 1999;54:M357–64.
37. Bruunsgaard H. The clinical impact of systemic low-level inflammation in elderly populations. With special reference to cardiovascular disease, dementia and mortality. Dan Med Bull. 2006;53:285–309. PubMed: 17092450.
38. Vasto S, Candore G, Balistreri CR, Caruso M, Colonna-Romano G, Grimaldi MP, Listi F, Nuzzo D, Lio D, Caruso C. Inflammatory networks in ageing, age-related diseases and longevity. Mech Ageing Dev. 2007;128:83–91.
39. Pizza FX, Hernandez IJ, Tidball JG. Nitric oxide synthase inhibition reduces muscle inflammation and necrosis in modified muscle use. J Leukoc Biol. 1998;64(4):427–33.
40. Pyne DB. Regulation of neutrophil function during exercise. Sports Med. 1994;17:245–58.
41. De Palma C, Meacci E, Perrotta C, Bruni P, Clementi E. Endothelial nitric oxide synthase activation by tumor necrosis factor alpha through neutral sphingomyelinase 2, sphingosine kinase 1, and sphingosine 1 phosphate receptors: a novel pathway relevant to the pathophysiology of endothelium. Arterioscler Thromb Vasc Biol. 2006;26(1):99–105.

42. Zemse SM, Chiao CW, Hilgers RH, Webb RC. Interleukin-10 inhibits the in vivo and in vitro adverse effects of TNF-alpha on the endothelium of murine aorta. Am J Physiol Heart Circ Physiol. 2010;299(4):H1160–7.
43. Zhong L, You J, Sun Q. The role of NF-*k*B in the TNF-α-induced endothelial cell apoptosis. Zhonghua Yixue Zazhi. 1999;79:863–6.
44. Lee IT, Yang CM. Role of NADPH oxidase/ROS in pro-inflammatory mediators-induced airway and pulmonary diseases. Biochem Pharmacol. 2012;84(5):581–90.
45. Lee CW, Lin CC, Lee IT, Lee HC, Yang CM. Activation and induction of cytosolic phospholipase A2 by TNF-α mediated through Nox2, MAPKs, NF-kB, and p300 in human tracheal smooth muscle cells. J Cell Physiol. 2011;226:2103–14.
46. Lee S, Park Y, Dellsperger KC, Zhang C. Exercise training improves endothelial function via adiponectin-dependent and independent pathways in type 2 diabetic mice. Am J Physiol Heart Circ Physiol. 2011;301(2):H306–14.
47. Lin CP, Huang PH, Tsai HS, Wu TC, Leu HB, Liu PL, Chen YH. Monascus purpureus-fermented rice inhibits tumor necrosis factor-α-induced upregulation of matrix metalloproteinase 2 and 9 in human aortic smooth muscle cells. J Pharm Pharmacol. 2011;63(12):1587–94.
48. Luo SF, Chang CC, Lee IT, Lee CW, Lin WN, Lin CC, Yang CM. Activation of ROS/NF-kappaB and Ca^{2+}/CaM kinase II are necessary for VCAM-1 induction in IL-1beta-treated human tracheal smooth muscle cells. Toxicol Appl Pharmacol. 2009;237(1):8–21.
49. Phalitakul S, Okada M, Hara Y, Yamawaki H. Vaspin prevents TNF-a-induced intracellular adhesion molecule-1 via inhibiting reactive oxygen species-dependent NF-kB and PKCu activation in cultured rat vascular smooth muscle cells. Pharmacol Res. 2011;64:493–500.
50. Rahman I, MacNee W. Oxidative stress and regulation of glutathione in lung inflammation. Eur Respir J. 2000;16:534–54.
51. Liu H, Sidiropoulos P, Song G, Pagliari LJ, Birrer MJ, Stein B, Anrather J, Pope RM. TNF-alpha gene expression in macrophages: regulation by NF-kappa B is independent of c-Jun or C/EBP beta. J Immunol. 2000;164(8):4277–85.
52. Wolin MS. Reactive oxygen species and vascular signal transduction mechanisms. Microcirculation. 1996;3(1):1–17.
53. Chung HY, Cesari M, Anton S, Marzetti E, Giovannini S, Seo AY, Carter C, Yu BP, Leeuwenburgh C. Molecular inflammation: underpinnings of aging and age-related diseases. Ageing Res Rev. 2009;8(1):18–30.
54. Jacquier-Sarlin MR, Fuller K, Dinh-Xuan AT, Richard MJ, Polla BS. Protective effects of hsp70 in inflammation. Experientia. 1994;50(11–12):1031–8.
55. Moldoveanu AI, Shephard RJ, Shek PN. The cytokine response to physical activity and training. Sports Med. 2001;31:115–44.
56. Zaldivar F, Wang-Rodriguez J, Nemet D, Schwindt C, Galassetti P, Mills PJ, Wilson LD, Cooper DM. Constitutive pro- and anti-inflammatory cytokine and growth factor response to exercise in leukocytes. J Appl Physiol. 2006;100(4):1124–33.
57. Abramson JL, Vaccarino V. Relationship between physical activity and inflammation among apparently healthy middle-aged and older US adults. Arch Intern Med. 2002;162:1286–92.
58. Albert MA, Glynn RJ, Ridker PM. Effect of physical activity on serum C reactive protein. Am J Cardiol. 2004;93:221–5.
59. Beavers KM, Brinkley TE, Nicklas BJ. Effect of exercise training on chronic inflammation. Clin Chim Acta. 2010;411(11–12):785–93.
60. Church TS, Barlow CE, Earnest CP, Kampert JB, Priest EL, Blair SN. Associations between cardiorespiratory fitness and C-reactive protein in men. Arterioscler Thromb Vasc Biol. 2002;22:1869–76.
61. King DE, Carek P, Mainous III AG, Pearson WS. Inflammatory markers and exercise: differences related to exercise type. Med Sci Sports Exerc. 2003;35:575–81.
62. Pischon T, Hankinson SE, Hotamisligil GS, Rifai N, Rimm EB. Leisure-time physical activity and reduced plasma levels of obesity-related inflammatory markers. Obes Res. 2003;11:1055–64.
63. Colbert LH, Visser M, Simonsick EM, Tracy RP, Newman AB, Kritchevsky SB, Pahor M, Taaffe DR, Brach J, Rubin S, Harris TB. Physical activity, exercise, and inflammatory mark-

ers in older adults: findings from the Health, Aging and Body Composition Study. J Am Geriatr Soc. 2004;52(7):1098–104.
64. Elosua R, Bartali B, Ordovas JM, Corsi AM, Lauretani F, Ferrucci L. Association between physical activity, physical performance, and inflammatory biomarkers in an elderly population: the InCHIANTI study. J Gerontol A Biol Sci Med Sci. 2005;60:760–7.
65. Geffken D, Cushman M, Burke G, Polak J, Sakkinen P, Tracy R. Association between physical activity and markers of inflammation in a healthy elderly population. Am J Epidemiol. 2001;153:242–50.
66. Jankord R, Jemiolo B. Influence of physical activity on serum IL-6 and IL-10 levels in healthy older men. Med Sci Sports Exerc. 2004;36:960–4.
67. Reuben DB, Judd-Hamilton L, Harris TB, Seeman TE. The associations between physical activity and inflammatory markers in high-functioning older persons: MacArthur Studies of Successful Aging. J Am Geriatr Soc. 2003;51:1125–30.
68. Taaffe DR, Harris TB, Ferrucci L, Rowe J, Seeman TE. Cross-sectional and prospective relationships of interleukin-6 and C-reactive protein with physical performance in elderly persons: MacArthur studies of successful aging. J Gerontol A Biol Sci Med Sci. 2000;55(12):M709–15. PubMed PMID: 11129392.
69. Wannamethee SG, Lowe GD, Whincup PH, Rumley A, Walker M, Lennon L. Physical activity and hemostatic and inflammatory variables in elderly men. Circulation. 2002;105:1785–90.
70. Yu Z, Ye X, Wang J, Qi Q, Franco OH, Rennie KL, Pan A, Li H, Liu Y, Hu FB, Lin X. Associations of physical activity with inflammatory factors, adipocytokines, and metabolic syndrome in middle-aged and older Chinese people. Circulation. 2009;119(23):2969–77.
71. Conraads VM, Beckers P, Bosmans J, De Clerck LS, Stevens WJ, Vrints CJ, et al. Combined endurance/resistance training reduces plasma TNF-alpha receptor levels in patients with chronic heart failure and coronary artery disease. Eur Heart J. 2002;23:1854–60.
72. Kohut ML, McCann DA, Russell DW, Konopka DN, Cunnick JE, Franke WD, Castillo MC, Reighard AE, Vanderah E. Aerobic exercise, but not flexibility/resistance exercise, reduces serum IL-18, CRP, and IL-6 independent of beta-blockers, BMI, and psychosocial factors in older adults. Brain Behav Immun. 2006;20(3):201–9. Epub 2006 Feb 28.
73. Larsen AI, Aukrust P, Aarsland T, Dickstein K. Effect of aerobic exercise training on plasma levels of tumor necrosis factor alpha in patients with heart failure. Am J Cardiol. 2001;88:805–8.
74. LeMaitre JP, Harris S, Fox KA, Denvir M. Change in circulating cytokines after 2 forms of exercise training in chronic stable heart failure. Am Heart J. 2004;147:100–5.
75. Mattusch F, Dufaux B, Heine O, Mertens I, Rost R. Reduction of the plasma concentration of C-reactive protein following nine months of endurance training. Int J Sports Med. 2000;21:21–4.
76. Smith JK, Dykes R, Douglas JE, Krishnaswamy G, Berk S. Long-term exercise and atherogenic activity of blood mononuclear cells in persons at risk of developing ischemic heart disease. JAMA. 1999;281:1722–7.
77. Stefanov T, Vekova A, Bonova I, Tzvetkov S, Kurktschiev D, Blüher M, Temelkova-Kurktschiev T. Effects of supervised vs non-supervised combined aerobic and resistance exercise programme on cardiometabolic risk factors. Cent Eur J Public Health. 2013;21(1):8–16.
78. Straczkowski M, Kowalska I, Dzienis-Straczkowska S, Stepién A, Skibińska E, Szelachowska M, Kinalska I. Changes in tumor necrosis factor-alpha system and insulin sensitivity during an exercise training program in obese women with normal and impaired glucose tolerance. Eur J Endocrinol. 2001;145(3):273–80.
79. Troseid M, Lappegard KT, Claudi T, Damas JK, Morkrid L, Brendberg R, Mollnes TE. Exercise reduces plasma levels of the chemokines MCP-1 and IL-8 in subjects with the metabolic syndrome. Eur Heart J. 2004;25:349–55.
80. Tsukui S, Kanda T, Nara M, Nishino M, Kondo T, Kobayashi I. Moderate intensity regular exercise decreases serum tumor necrosis factor-alpha and HbA1c levels in healthy women. Int J Obes Relat Metab Disord. 2000;24:1207–11.

81. Campbell PT, Campbell KL, Wener MH, et al. A yearlong exercise intervention decreases CRP among obese postmenopausal women. Med Sci Sports Exerc. 2009;41:1533–9.
82. Kadoglou NP, Iliadis F, Angelopoulou N, Perrea D, Ampatzidis G, Liapis CD, Alevizos M. The anti-inflammatory effects of exercise training in patients with type 2 diabetes mellitus. Eur J Cardiovasc Prev Rehabil. 2007;14:837–43.
83. Tisi PV, Hulse M, Chulakadabba A, Gosling P, Shearman CP. Exercise training for intermittent claudication: does it adversely affect biochemical markers of the exercise-induced inflammatory response? Eur J Vasc Endovasc Surg. 1997;14:344–50.
84. Walther C, Möbius-Winkler S, Linke A, Bruegel M, Thiery J, Schuler G, Halbrecht R. Regular exercise training compared with percutaneous intervention leads to a reduction of inflammatory markers and cardiovascular events in patients with coronary artery disease. Eur J Cardiovasc Prev Rehabil. 2008;15(1):107–12.
85. Greiwe JS, Cheng B, Rubin DC, Yarasheski KE, Semenkovich CF. Resistance exercise decreases skeletal muscle tumor necrosis factor alpha in frail elderly humans. FASEB J. 2001;15:475–82.
86. Milani RV, Lavie CJ, Mehra MR. Reduction in C-reactive protein through cardiac rehabilitation and exercise training. J Am Coll Cardiol. 2004;43:1056–61.
87. Nicklas BJ, Hsu FC, Brinkley TJ, Church T, Goodpaster BH, Kritchevsky SB, Pahor M. Exercise training and plasma C-reactive protein and interleukin-6 in elderly people. J Am Geriatr Soc. 2008;56(11):2045–52.
88. Wegge JK, Roberts CK, Ngo TH, Barnard RJ. Effect of diet and exercise intervention on inflammatory and adhesion molecules in postmenopausal women on hormone replacement therapy and at risk for coronary artery disease. Metabolism. 2004;53:377–81.
89. You T, Berman DM, Ryan AS, Nicklas BJ. Effects of hypocaloric diet and exercise training on inflammation and adipocyte lipolysis in obese postmenopausal women. J Clin Endocrinol Metab. 2004;89:1739–46.
90. Gielen S, Adams V, Möbius-Winkler S, Linke A, Erbs S, Yu J, Kempf W, Schubert A, Schuler G, Hambrecht R. Anti-inflammatory effects of exercise training in the skeletal muscle of patients with chronic heart failure. J Am Coll Cardiol. 2003;42(5):861–8.
91. Marcell TJ, McAuley KA, Traustadottir T, Reaven PD. Exercise training is not associated with improved levels of C-reactive protein or adiponectin. Metabolism. 2005;54:533–41.
92. Rauramaa R, Halonen P, Väisänen SB, Lakka TA, Schmidt-Trucksäss A, Berg A, Penttilä IM, Rankinen T, Bouchard C. Effects of aerobic physical exercise on inflammation and atherosclerosis in men: the DNASCO Study: a six-year randomized, controlled trial. Ann Intern Med. 2004;140(12):1007–14.
93. Fairey AS, Courneya KS, Field CJ, Bell GJ, Jones LW, Martin BS, Mackey JR. Effect of exercise training on C-reactive protein in postmenopausal breast cancer survivors: a randomized controlled trial. Brain Behav Immun. 2005;19(5):381–8.
94. Hammett CJ, Oxenham HC, Baldi JC, Doughty RN, Ameratunga R, French JK, White HD, Stewart RA. Effect of six months' exercise training on C-reactive protein levels in healthy elderly subjects. J Am Coll Cardiol. 2004;44(12):2411–3.
95. Kemmler W, Von SS, Engelke K, Kalender WA. Exercise decreases the risk of metabolic syndrome in elderly females. Med Sci Sports Exerc. 2009;41:297–305.
96. Nicklas BJ, Ambrosius W, Messier SP, Miller GD, Penninx BW, Loeser RF, Palla S, Bleecker E, Pahor M. Diet-induced weight loss, exercise, and chronic inflammation in older, obese adults: a randomized controlled clinical trial. Am J Clin Nutr. 2004;79(4):544–51.
97. Kelley GA, Kelley KS. Effects of aerobic exercise on C-reactive protein, body composition, and maximum oxygen consumption in adults: a meta-analysis of randomized controlled trials. Metabolism. 2006;55:1500–7.
98. Brochu M, Malita MF, Messier V, Doucet E, Strychar I, Lavoie JM, Prud'homme D, Rabasa-Lhoret R. Resistance training does not contribute to improving the metabolic profile after a 6-month weight loss program in overweight and obese postmenopausal women. J Clin Endocrinol Metab. 2009;94(9):3226–33.

99. Levinger I, Goodman C, Peake J, Garnham A, Hare DL, Jerums G, Selig S. Inflammation, hepatic enzymes and resistance training in individuals with metabolic risk factors. Diabet Med. 2009;26(3):220–7.
100. Rall LC, Roubenoff R, Cannon JG, Abad LW, Dinarello CA, Meydani SN. Effects of progressive resistance training on immune response in aging and chronic inflammation. Med Sci Sports Exerc. 1996;28:1356–65.
101. Brooks N, Layne JE, Gordon PL, Roubenoff R, Nelson ME, Castaneda-Sceppa C. Strength training improves muscle quality and insulin sensitivity in Hispanic older adults with type 2 diabetes. Int J Med Sci. 2007;4:19–27. PubMed: 17211497.
102. Cancello R, Henegar C, Viguerie N, Taleb S, Poitou C, Rouault C, Coupaye M, Pelloux V, Hugol D, Bouillot JL, Bouloumié A, Barbatelli G, Cinti S, Svensson PA, Barsh GS, Zucker JD, Basdevant A, Langin D, Clément K. Reduction of macrophage infiltration and chemoattractant gene expression changes in white adipose tissue of morbidly obese subjects after surgery-induced weight loss. Diabetes. 2005;54(8):2277–86.
103. Giannopoulou I, Ploutz-Snyder LL, Carhart R, Weinstock RS, Fernhall B, Goulopoulou S, Kanaley JA. Exercise is required for visceral fat loss in postmenopausal women with type 2 diabetes. J Clin Endocrinol Metab. 2005;90(3):1511–8.
104. Ross R, Bradshaw AJ. The future of obesity reduction: beyond weight loss. Nat Rev Endocrinol. 2009;5:319–25.
105. Esposito EK, Giugliano G, Scuderi N, Giugliano D. Role of adipokines in the obesity inflammation relationship: the effect of fat removal. Plast Reconstr Surg. 2006;118:1048–57.
106. Forsythe LK, Wallace JM, Livingstone MB. Obesity and inflammation: the effects of weight loss. Nutr Res Rev. 2008;21:117–33.
107. You T, Nicklas BJ. Chronic inflammation: role of adipose tissue and modulation by weight loss. Curr Diabetes Rev. 2006;2:29–37.
108. de Lemos ET, Reis F, Baptista S, Pinto R, Sepodes B, Vala H, Rocha-Pereira P, Silva AS, Teixeira F. Exercise training is associated with improved levels of C-reactive protein and adiponectin in ZDF (type 2) diabetic rats. Med Sci Monit. 2007;13(8):BR168–74.
109. Mujumdar PP, Duerksen PJ, Firek AF, Hessingere DA. Long-term, progressive, aerobic training increases adiponectin in middle-aged, overweight, untrained males and females. Scand J Clin Lab Invest. 2011;71:101–7.
110. Andrade-Oliveira V, Câmara NO, Moraes-Vieira PM. Adipokines as drug targets in diabetes and underlying disturbances. J Diabetes Res. 2015;2015:681612. Epub 2015 Apr 8. Review.
111. Hui X, Lam KS, Vanhoutte PM, Xu A. Adiponectin and cardiovascular health: an update. Br J Pharmacol. 2012;165(3):574–90.
112. Kawanishi N, Yano H, Yokogawa Y, Suzuki K. Exercise training inhibits inflammation in adipose tissue via both suppression of macrophage infiltration and acceleration of phenotypic switching from M1 to M2 macrophages in high-fat-diet-induced obese mice. Exerc Immunol Rev. 2010;16:105–18.
113. Timmerman KL, Flynn MG, Coen PM, Markofski MM, Pence BD. Exercise training-induced lowering of inflammatory (CD14+CD16+) monocytes: a role in the anti-inflammatory influence of exercise? J Leukoc Biol. 2008;84:1271–8.
114. Wang J, Song H, Tang X, Yang Y, Vieira VJ, Niu Y, Ma Y. Effect of exercise training intensity on murine T-regulatory cells and vaccination response. Scand J Med Sci Sports. 2012;22(5):643–52.
115. Yeh SH, Chuang H, Lin LW, Hsiao CY, Eng HL. Regular tai chi chuan exercise enhances functional mobility and CD4CD25 regulatory T cells. Br J Sports Med. 2006;40(3):239–43.
116. Lancaster GI, Halson SL, Khan Q, Drysdale P, Wallace F, Jeukendrup AE, Drayson MT, Gleeson M. Effects of acute exhaustive exercise and chronic exercise training on type 1 and type 2 T lymphocytes. Exerc Immunol Rev. 2004;10:91–106.
117. Steensberg A, Toft AD, Bruunsgaard H, Sandmand M, Halkjaer-Kristensen J, Pedersen BK. Strenuous exercise decreases the percentage of type 1 T cells in the circulation. J Appl Physiol. 2001;91:1708–12.

118. Yeh SH, Chuang H, Lin LW, Hsiao CY, Wang PW, Liu RT, Yang KD. Regular Tai Chi Chuan exercise improves T cell helper function of patients with type 2 diabetes mellitus with an increase in T-bet transcription factor and IL-12 production. Br J Sports Med. 2009;43(11):845–50.
119. McFarlin BK, Flynn MG, Campbell WW, Craig BA, Robinson JP, Stewart LK, Timmerman KL, Coen PM. Physical activity status, but not age, influences inflammatory biomarkers and toll-like receptor 4. J Gerontol A Biol Sci Med Sci. 2006;61(4):388–93.
120. Stewart LK, Flynn MG, Campbell WW, Craig BA, Robinson JP, McFarlin BK, Timmerman KL, Coen PM, Felker J, Talbert E. Influence of exercise training and age on CD14+ cell-surface expression of toll-like receptor 2 and 4. Brain Behav Immun. 2005;19(5):389–97.
121. Takeda K, Akira S. Toll-like receptors in innate immunity. Int Immunol. 2005;17:1–14.
122. Oberbach A, Tönjes A, Klöting N, Fasshauer M, Kratzsch J, Busse MW, Paschke R, Stumvoll M, Blüher M. Effect of a 4 week physical training program on plasma concentrations of inflammatory markers in patients with abnormal glucose tolerance. Eur J Endocrinol. 2006;154(4):577–85.
123. Goto S, Radák Z, Nyakas C, Chung HY, Naito H, Takahashi R, Nakamoto H, Abea R. Regular exercise: an effective means to reduce oxidative stress in old rats. Ann N Y Acad Sci. 2004;1019:471–4.
124. Cupps TR, Fauci AS. Corticosteroid-mediated immunoregulation in man. Immunol Rev. 1982;65:133–55.
125. Beltran Valls MR, Dimauro I, Brunelli A, Tranchita E, Ciminelli E, Caserotti P, Duranti G, Sabatini S, Parisi P, Parisi A, Caporossi D. Explosive type of moderate-resistance training induces functional, cardiovascular, and molecular adaptations in the elderly. Age (Dordr). 2014;36(2):759–72.
126. Fitts RH, Troup JP, Witzmann FA, Holloszy JO. The effect of ageing and exercise on skeletal muscle function. Mech Ageing Dev. 1984;27(2):161–72.
127. Narici MV, Reeves ND, Morse CI, Maganaris CN. Muscular adaptations to resistance exercise in the elderly. J Musculoskelet Neuronal Interact. 2004;4(2):161–4.
128. Peterson MD, Rhea MR, Sen A, Gordon PM. Resistance exercise for muscular strength in older adults: a meta-analysis. Ageing Res Rev. 2010;9(3):226–37.
129. Peterson MD, Sen A, Gordon PM. Influence of resistance exercise on lean body mass in aging adults: a meta-analysis. Med Sci Sports Exerc. 2011;43:249–58.
130. Holloszy JO. Biochemical adaptations in muscle: effects of exercise on mitochondrial oxygen uptake and respiratory enzyme activity in skeletal muscle. J Biol Chem. 1967;242:2278–82.
131. Short KR, Vittone JL, Bigelow ML, Proctor DN, Rizza RA, Coenen-Schimke JM, Nair KS. Impact of aerobic exercise training on age-related changes in insulin sensitivity and muscle oxidative capacity. Diabetes. 2003;52(8):1888–96.
132. Febbraio MA, Pedersen BK. Muscle-derived interleukin-6: mechanisms for activation and possible biological roles. FASEB J. 2002;16:1335–47.
133. Steensberg A, van Hall G, Osada T, Sacchetti M, Saltin B, Klarlund Pedersen B. Production of interleukin-6 in contracting human skeletal muscles can account for the exercise-induced increase in plasma interleukin-6. J Physiol. 2000;529(Pt 1):237–42.
134. Bruunsgaard H. Physical activity and modulation of systemic low-level inflammation. J Leukoc Biol. 2005;78:819–35.
135. Steensberg A, Fischer CP, Keller C, Møller K, Pedersen BK. IL-6 enhances plasma IL-1ra, IL-10, and cortisol in humans. Am J Physiol Endocrinol Metab. 2003;285(2):E433–7.
136. Nielsen AR, Mounier R, Plomgaard P, Mortensen OH, Penkowa M, Speerschneider T, Pilegaard H, Pedersen BK. Expression of interleukin-15 in human skeletal muscle effect of exercise and muscle fibre type composition. J Physiol. 2007;584:305–12.
137. Pedersen BK, Fischer CP. Physiological roles of muscle-derived interleukin-6 in response to exercise. Curr Opin Clin Nutr Metab Care. 2007;10(3):265–71.
138. Pedersen BK, Steensberg A, Fischer C, Keller C, Keller P, Plomgaard P, Febbraio M, Saltin B. Searching for the exercise factor: is IL-6 a candidate? J Muscle Res Cell Motil. 2003;24(2–3):113–9.

139. Pedersen M, Bruunsgaard H, Weis N, Hendel HW, Andreassen BU, Eldrup E, Dela F, Pedersen BK. Circulating levels of TNF-alpha and IL-6-relation to truncal fat mass and muscle mass in healthy elderly individuals and in patients with type-2 diabetes. Mech Ageing Dev. 2003;124(4):495–502.
140. Schindler R, Mancilla J, Endres S, Ghorbani R, Clark SC, Dinarello CA. Correlations and interactions in the production of interleukin-6 (IL-6), IL-1, and tumor necrosis factor (TNF) in human blood mononuclear cells: IL-6 suppresses IL-1 and TNF. Blood. 1990;75(1):40–7.
141. Starkie R, Ostrowski SR, Jauffred S, Febbraio M, Pedersen BK. Exercise and IL-6 infusion inhibit endotoxin-induced TNF-alpha production in humans. FASEB J. 2003;17:884–6.
142. Quinn LS, Anderson BG, Drivdahl RH, Alvarez B, Argilés JM. Overexpression of interleukin-15 induces skeletal muscle hypertrophy in vitro: implications for treatment of muscle wasting disorders. Exp Cell Res. 2002;280:55–63.
143. Febbraio MA, Ott P, Nielsen HB, et al. Exercise induces hepatosplanchnic release of heat shock protein 72 in humans. J Physiol. 2002;544(pt 3):957–62.
144. Thompson HS, Clarkson PM, Scordilis SP. The repeated bout effect and heat shock proteins: intramuscular HSP27 and HSP70 expression following two bouts of eccentric exercise in humans. Acta Physiol Scand. 2002;174:47–56.
145. Walsh RC, Koukoulas I, Garnham A, Moseley PL, Hargreaves M, Febbraio MA. Exercise increases serum Hsp72 in humans. Cell Stress Chaperones. 2001;6:386–93.
146. Engesser-Cesar C, Anderson AJ, Basso DM, Edgerton VR, Cotman CW. Voluntary wheel running improves recovery from a moderate spinal cord injury. J Neurotrauma. 2005;22(1):157–71.
147. Marosi K, Bori Z, Hart N, Sárga L, Koltai E, Radák Z, Nyakas C. Long-term exercise treatment reduces oxidative stress in the hippocampus of aging rats. Neuroscience. 2012;226:21–8.
148. Radak Z, Sasvari M, Nyakas C, Kaneko T, Tahara S, Ohno H, Goto S. Single bout of exercise eliminates the immobilization-induced oxidative stress in rat brain. Neurochem Int. 2001;39(1):33–8.
149. Radak Z, Taylor AW, Ohno H, Goto S. Adaptation to exercise-induced oxidative stress: from muscle to brain. Exerc Immunol Rev. 2001;7:90–107.
150. Goto S, Naito H, Kaneko T, Chung HY, Radák Z. Hormetic effects of regular exercise in aging: correlation with oxidative stress. Appl Physiol Nutr Metab. 2007;32(5):948–53.
151. Nakamoto H, Kaneko T, Tahara S, Hayashi E, Naito H, Radak Z, Goto S. Regular exercise reduces 8-oxodG in the nuclear and mitochondrial DNA and modulates the DNA repair activity in the liver of old rats. Exp Gerontol. 2007;42(4):287–95.
152. Viboolvorakul S, Niimi H, Wongeak-in N, Eksakulkla S, Patumraj S. Increased capillary vascularity in the femur of aged rats by exercise training. Microvasc Res. 2009;78(3):459–63.
153. Leeuwenburgh C, Fiebig R, Chandwaney R, Ji LL. Aging and exercise training in skeletal muscle: responses of glutathione and antioxidant enzyme systems. Am J Physiol. 1994;267(2 Pt 2):R439–45.
154. Starnes JW, Taylor RP, Park Y. Exercise improves postischemic function in aging hearts. Am J Physiol Heart Circ Physiol. 2003;285(1):H347–51.
155. Quindry J, French J, Hamilton K, Lee Y, Mehta JL, Powers S. Exercise training provides cardioprotection against ischemia-reperfusion induced apoptosis in young and old animals. Exp Gerontol. 2005;40(5):416–25.
156. Hammeren J, Powers S, Lawler J, Criswell D, Martin D, Lowenthal D, Pollock M. Exercise training-induced alterations in skeletal muscle oxidative and antioxidant enzyme activity in senescent rats. Int J Sports Med. 1992;13(5):412–6.
157. Thomas MM, Vigna C, Betik AC, Tupling AR, Hepple RT. Initiating treadmill training in late middle age offers modest adaptations in Ca^{2+} handling but enhances oxidative damage in senescent rat skeletal muscle. Am J Physiol Regul Integr Comp Physiol. 2010;298(5):R1269–78.
158. Pushpalatha K, Nishanth K, Sathyavelu RK. Myocardial antioxidant status and oxidative stress after combined action of exercise training and ethanol in two different age groups of male albino rats. Acta Biol Hung. 2007;58(2):173–85.

159. Vincent HK, Bourguignon C, Vincent KR. Resistance training lowers exercise-induced oxidative stress and homocysteine levels in overweight and obese older adults. Obesity. 2006;14(11):1921–30.
160. de Gonzalo-Calvo D, Fernández-García B, de Luxán-Delgado B, Rodríguez-González S, García-Macia M, Suárez FM, Solano JJ, Rodríguez-Colunga MJ, Coto-Montes A. Chronic training increases blood oxidative damage but promotes health in elderly men. Age (Dordr). 2013;35(2):407–17.
161. Aldred S, Rohalu M. A moderate intensity exercise program did not increase the oxidative stress in older adults. Arch Gerontol Geriatr. 2011;53(3):350–3.
162. Davies KJ, Quintanilha AT, Brooks GA, Packer L. Free radicals and tissue damage produced by exercise. Biochem Biophys Res Commun. 1982;107:1198–205.
163. Gomez-Cabrera MC, Borras C, Pallardo FV, Sastre J, Ji LL, Vina J. Decreasing xanthine oxidase-mediated oxidative stress prevents useful cellular adaptations to exercise in rats. J Physiol. 2005;567:113–20.
164. Hellsten Y, Frandsen U, Orthenblad N, Sjødin B, Richter EA. Xanthine oxidase in human skeletal muscle following eccentric exercise: a role in inflammation. J Physiol. 1997;498(Pt 1):239–48.
165. Radak Z, Asano K, Inoue M, Kizaki T, Oh-Ishi S, Suzuki K, Taniguchi N, Ohno H. Superoxide dismutase derivative reduces oxidative damage in skeletal muscle of rats during exhaustive exercise. J Appl Physiol. 1995;79(1):129–35.
166. Ji LL, Leeuwenburgh C, Leichtweis S, Gore M, Fiebig R, Hollander J, Bejma J. Oxidative stress and aging. Role of exercise and its influences on antioxidant systems. Ann N Y Acad Sci. 1998;854:102–17.
167. Aronson D, Violan MA, Dufresne SD, Zangen D, Fielding RA, Goodyear LJ. Exercise stimulates the mitogen-activated protein kinase pathway in human skeletal muscle. J Clin Invest. 1997;99(6):1251–7.
168. Gibala MJ, McGee SL, Garnham AP, Howlett KF, Snow RJ, Hargreaves M. Brief intense interval exercise activates AMPK and p38 MAPK signaling and increases the expression of PGC-1alpha in human skeletal muscle. J Appl Physiol. 2009;106(3):929–34.
169. Iemitsu M, Maeda S, Jesmin S, Otsuki T, Kasuya Y, Miyauchi T. Activation pattern of MAPK signaling in the hearts of trained and untrained rats following a single bout of exercise. J Appl Physiol. 2006;101:151–63.
170. Vaynman S, Ying Z, Gomez-Pinilla F. Hippocampal BDNF mediates the efficacy of exercise on synaptic plasticity and cognition. Eur J Neurosci. 2004;20(10):2580–90.
171. Vaynman S, Ying Z, Gomez-Pinilla F. Interplay between brain-derived neurotrophic factor and signal transduction modulators in the regulation of the effects of exercise on synaptic-plasticity. Neuroscience. 2003;122(3):647–57.
172. Ji LL, Gomez-Cabrera MC, Steinhafel N, Vina J. Acute exercise activates nuclear factor (NF)-kappaB signaling pathway in rat skeletal muscle. FASEB J. 2004;18(13):1499–506.
173. Vasilaki A, McArdle F, Iwanejko LM, McArdle A. Adaptive responses of mouse skeletal muscle to contractile activity: the effect of age. Mech Ageing Dev. 2006;127(11):830–9. Epub 2006 Sep 22.
174. McArdle A, Jackson MJ. Exercise, oxidative stress and ageing. J Anat. 2000;197(Pt 4):539–41.
175. Huffman DM, Moellering DR, Grizzle WE, Stockard CR, Johnson MS, Nagy TR. Effect of exercise and calorie restriction on biomarkers of aging in mice. Am J Physiol Regul Integr Comp Physiol. 2008;294:R1618–27.
176. Naito H, Powers SK, Demirel HA, Aoki J. Exercise training increases heat shock protein in skeletal muscles of old rats. Med Sci Sports Exerc. 2001;33(5):729–34.
177. Simar D, Malatesta D, Koechlin C, Cristol JP, Vendrell JP, Caillaud C. Effect of age on Hsp72 expression in leukocytes of healthy active people. Exp Gerontol. 2004;39(10):1467–74.
178. McArdle A, Vasilaki A, Jackson M. Exercise and skeletal muscle ageing: cellular and molecular mechanisms. Ageing Res Rev. 2002;1(1):79–93.

179. Vasilaki A, Jackson MJ, McArdle A. Attenuated HSP70 response in skeletal muscle of aged rats following contractile activity. Muscle Nerve. 2002;25(6):902–5.
180. Boveris A, Navarro A. Systemic and mitochondrial adaptive responses to moderate exercise in rodents. Free Radic Biol Med. 2008;44:224–9.
181. Derbré F, Gomez-Cabrera MC, Nascimento AL, Sanchis-Gomar F, Martinez-Bello VE, Tresguerres JA, Fuentes T, Gratas-Delamarche A, Monsalve M, Viña J. Age associated low mitochondrial biogenesis may be explained by lack of response of PGC-1α to exercise training. Age (Dordr). 2012;34(3):669–79.
182. Jäger S, Handschin C, St-Pierre J, Spiegelman BM. AMP-activated protein kinase (AMPK) action in skeletal muscle via direct phosphorylation of PGC-1alpha. Proc Natl Acad Sci U S A. 2007;104(29):12017–22.
183. Lee-Young RS, Griffee SR, Lynes SE, Bracy DP, Ayala JE, McGuinness OP, Wasserman DH. Skeletal muscle AMP-activated protein kinase is essential for the metabolic response to exercise in vivo. J Biol Chem. 2009;284(36):23925–34.
184. O'Neill HM, Maarbjerg SJ, Crane JD, Jeppesen J, Jørgensen SB, Schertzer JD, Shyroka O, Kiens B, van Denderen BJ, Tarnopolsky MA, Kemp BE, Richter EA, Steinberg GR. AMP-activated protein kinase (AMPK) beta1beta2 muscle null mice reveal an essential role for AMPK in maintaining mitochondrial content and glucose uptake during exercise. Proc Natl Acad Sci U S A. 2011;108(38):16092–7.
185. Palacios OM, Carmona JJ, Michan S, Chen KY, Manabe Y, Ward III JL, Goodyear LJ, Tong Q. Diet and exercise signals regulate SIRT3 and activate AMPK and PGC-1alpha in skeletal muscle. Aging. 2009;1(9):771–83.
186. Ventura-Clapier R, Garnier A, Veksler V. Transcriptional control of mitochondrial biogenesis: the central role of PGC-1alpha. Cardiovasc Res. 2008;79(2):208–17.
187. St-Pierre J, Drori S, Uldry M, Silvaggi JM, Rhee J, Jager S, Handschin C, Zheng K, Lin J, Yang W, Simon DK, Bachoo R, Spiegelman BM. Suppression of reactive oxygen species and neurodegeneration by the PGC-1 transcriptional coactivators. Cell. 2006;127:397–408.
188. Safdar A, Bourgeois JM, Ogborn DI, Little JP, Hettinga BP, Akhtar M, Thompson JE, Melov S, Mocellin NJ, Kujoth GC, Prolla TA, Tarnopolsky MA. Endurance exercise rescues progeroid aging and induces systemic mitochondrial rejuvenation in mtDNA mutator mice. Proc Natl Acad Sci U S A. 2011;108:4135–40.
189. Devi SA, Kiran TR. Regional responses in antioxidant system to exercise training and dietary vitamin E in aging rat brain. Neurobiol Aging. 2004;25:501–8.
190. Somani SM, Husain K. Exercise training alters kinetics of antioxidant enzymes in rat tissues. Biochem Mol Biol Int. 1996;38:587–95.
191. Marzatico F, Pansarasa O, Bertorelli L, Somenzini L, Della Valle G. Blood free radical antioxidant enzymes and lipid peroxides following long-distance and lactacidemic performances in highly trained aerobic and sprint athletes. J Sports Med Phys Fitness. 1997;37(4):235–9.
192. Miyazaki H, Oh-ishi S, Ookawara T, Kizaki T, Toshinai K, Ha S, Haga S, Ji LL, Ohno H. Strenuous endurance training in humans reduces oxidative stress following exhausting exercise. Eur J Appl Physiol. 2001;84(1–2):1–6.
193. Siems WG, Brenke R, Sommerburg O, Grune T. Improved antioxidative protection in winter swimmers. QJM. 1999;92(4):193–8.
194. Hamilton KL, Powers SK, Sugiura T, Kim S, Lennon S, Tumer N, Mehta JL. Short-term exercise training can improve myocardial tolerance to I/R without elevation in heat shock proteins. Am J Physiol Heart Circ Physiol. 2001;281(3):H1346–52.
195. Lawler JM, Kwak HB, Kim JH, Suk MH. Exercise training inducibility of MnSOD protein expression and activity is retained while reducing prooxidant signaling in the heart of senescent rats. Am J Physiol Regul Integr Comp Physiol. 2009;296(5):R1496–502.
196. Yamashita N, Hoshida S, Otsu K, Asahi M, Kuzuya T, Hori M. Exercise provides direct biphasic cardioprotection via manganese superoxide dismutase activation. J Exp Med. 1999;189(11):1699–706.
197. Gündüz F, Sentürk UK, Kuru O, Aktekin B, Aktekin MR. The effect of one year's swimming exercise on oxidant stress and antioxidant capacity in aged rats. Physiol Res. 2004;53(2):171–6.

198. de Moraes C, Davel AP, Rossoni LV, Antunes E, Zanesco A. Exercise training improves relaxation response and SOD-1 expression in aortic and mesenteric rings from high caloric diet-fed rats. BMC Physiol. 2008;8:12.
199. Rush JW, Turk JR, Laughlin MH. Exercise training regulates SOD-1 and oxidative stress in porcine aortic endothelium. Am J Physiol Heart Circ Physiol. 2003;284:H1378–87.
200. Moien-Afshari F, Ghosh S, Elmi S, Khazaei M, Rahman MM, Sallam N, Laher I. Exercise restores coronary vascular function independent of myogenic tone or hyperglycemic status in db/db mice. Am J Physiol Heart Circ Physiol. 2008;295(4):H1470–80.
201. Chang SP, Chen YH, Chang WC, Liu IM, Cheng JT. Increase of anti-oxidation by exercise in the liver of obese Zucker rats. Clin Exp Pharmacol Physiol. 2004;31:506–11.
202. Ookawara T, Haga S, Ha S, Oh-Ishi S, Toshinai K, Kizaki T, Ji LL, Suzuki K, Ohno H. Effects of endurance training on three superoxide dismutase isoenzymes in human plasma. Free Radic Res. 2003;37(7):713–9.
203. Rousseau AS, Margaritis I, Arnaud J, Faure H, Roussel AM. Physical activity alters antioxidant status in exercising elderly subjects. J Nutr Biochem. 2006;17(7):463–70.
204. Lennon SL, Quindry JC, Hamilton KL, French JP, Hughes J, Mehta JL, Powers SK. Elevated MnSOD is not required for exercise-induced cardioprotection against myocardial stunning. Am J Physiol Heart Circ Physiol. 2004;287(2):H975–80.
205. Radak Z, Asano K, Inoue M, Kizaki T, Ohishi S, Suzuki K, Taniguchi N, Ohno H. Acute bout of exercise does not alter the antioxidant enzyme status and lipid peroxidation in rat hippocampus and cerebellum. Pathophysiology. 1995;2:243–5.
206. Ji LL, Wu E, Thomas DP. Effect of exercise training on antioxidant and metabolic functions in senescent rat skeletal muscle. Gerontology. 1991;37(6):317–25.
207. Higuchi M, Cartier LJ, Chen M, Holloszy JO. Superoxide dismutase and catalase in skeletal muscle: adaptive response to exercise. J Gerontol. 1985;40:281–6.
208. Hollander J, Fiebig R, Gore M, Ookawara T, Ohno H, Ji LL. Superoxide dismutase gene expression is activated by a single bout of exercise in rat skeletal muscle. Pflugers Arch. 2001;442:426–34.
209. Ho YS, Howard AJ, Crapo JD. Molecular structure of a functional rat gene for manganese-containing superoxide dismutase. Am J Respir Cell Mol Biol. 1991;4:278–86.
210. Grune T, Merker K, Jubg T, Sitte N, Davies KJ. Protein oxidation and degradation during postmitotic senescence. Free Radic Biol Med. 2005;39:1208–15.
211. Sitte N, Merker K, Grune T. Proteasome-dependent degradation of oxidized proteins in MRC-5 fibroblasts. FEBS Lett. 1998;440:399–402.
212. Radak Z, Kumagai S, Nakamoto H, Goto S. 8-Oxoguanosine and uracil repair of nuclear and mitochondrial DNA in red and white skeletal muscle of exercise-trained old rats. J Appl Physiol. 2007;102(4):1696–701.
213. Radak Z, Apor P, Pucsok J, Berkes I, Ogonovszky H, Pavlik G, Nakamoto H, Goto S. Marathon running alters the DNA base excision repair in human skeletal muscle. Life Sci. 2003;72:1627–33.
214. Radak Z, Toldy A, Szabo Z, Siamilis S, Nyakas C, Silye G, Jakus J, Goto S. The effects of training and detraining on memory, neurotrophins and oxidative stress markers in rat brain. Neurochem Int. 2006;49(4):387–92.
215. Sahin E, Colla S, Liesa M, Moslehi J, Muller FL, Guo M, Cooper M, Kotton D, Fabian AJ, Walkey C, Maser RS, Tonon G, Foerster F, Xiong R, Wang YA, Shukla SA, Jaskelioff M, Martin ES, Heffernan TP, Protopopov A, Ivanova E, Mahoney JE, Kost-Alimova M, Perry SR, Bronson R, Liao R, Mulligan R, Shirihai OS, Chin L, DePinho RA. Telomere dysfunction induces metabolic and mitochondrial compromise. Nature. 2011;470:359–65.
216. Cherkas LF, Hunkin JL, Kato BS, Richards JB, Gardner JP, Surdulescu GL, Kimura M, Lu X, Spector TD, Aviv A. The association between physical activity in leisure time and leukocyte telomere length. Arch Intern Med. 2008;168(2):154–8.
217. Werner C, Furster T, Widmann T, Poss J, Roggia C, Hanhoun M, Scharhag J, Buchner N, Meyer T, Kindermann W, Haendeler J, Bohm M, Laufs U. Physical exercise prevents cellular senescence in circulating leukocytes and in the vessel wall. Circulation. 2009;120:2438–47.

218. Bruunsgaard H, Pedersen M, Pedersen BK. Aging and proinflammatory cytokines. Curr Opin Hematol. 2001;8:131–6. PubMed: 11303144.
219. Cannon JG, St Pierre BA. Cytokines in exertion-induced skeletal muscle injury. Mol Cell Biochem. 1998;179(1–2):159–67.
220. Das UN. Anti-inflammatory nature of exercise. Nutrition. 2004;20:323–6.
221. Elosua R, Molina L, Fito M, Arquer A, Sanchez-Quesada JL, Covas MI, Ordoñez-Llanos J, Marrugat J. Response of oxidative stress biomarkers to a 16-week aerobic physical activity program, and to acute physical activity, in healthy young men and women. Atherosclerosis. 2003;167(2):327–34. Erratum in: Atherosclerosis. 2008 Apr;197(2):967.
222. Fleg JL, Schulman SP, O'Connor FC, Gerstenblith G, Becker LC, Fortney S, Goldberg AP, Lakatta EG. Cardiovascular responses to exhaustive upright cycle exercise in highly trained older men. J Appl Physiol. 1994;77(3):1500–6.
223. Flynn M, McFarlin BK, Markofski MA. The anti-inflammatory actions of exercise training. Am J Lifestyle Med. 2007;1:220–35.
224. Gohil K, Rothfuss L, Lang J, Packer L. Effect of exercise training on tissue vitamin E and ubiquinone content. J Appl Physiol. 1987;63(4):1638–41.
225. Hack V, Strobel G, Rau JP, Weicker H. The effect of maximal exercise on the activity of neutrophil granulocytes in highly trained athletes in a moderate training period. Eur J Appl Physiol Occup Physiol. 1992;65(6):520–4.
226. Handel ML, McMorrow LB, Gravallese EM. Nuclear factor-kappa B in rheumatoid synovium. Localization of p50 and p65. Arthritis Rheum. 1995;38:1762–70.
227. Huang G, Gibson CA, Tran ZV, Osness WH. Controlled endurance exercise training and VO_{2max} changes in older adults: a meta-analysis. Prev Cardiol. 2005;8:217–25.
228. Ji LL. Exercise at old age: does it increase or alleviate oxidative stress? Ann N Y Acad Sci. 2001;928:236–47.
229. Keller C, Keller P, Giralt M, Hidalgo J, Pedersen BK. Exercise normalises overexpression of TNF-α in knockout mice. Biochem Biophys Res Commun. 2004;321:179–82.
230. Kim SR, Bae YH, Bae SK, Choi KS, Yoon KH, Koo TH, Jang HO, Yun I, Kim KW, Kwon YG, Yoo MA, Bae MK. Visfatin enhances ICAM-1 and VCAM-1 expression through ROS-dependent NF-kB activation in endothelial cells. Biochim Biophys Acta. 1783;2008:886–95.
231. Kim SY, Moon KA, Jo HY, Jeong S, Seon SH, Jung E, Cho YS, Chun E, Lee KY. Anti-inflammatory effects of apocynin, an inhibitor of NADPH oxidase, in airway inflammation. Immunol Cell Biol. 2012;90(4):441–8.
232. Kohrt WM, Malley MT, Coggan AR, Spina RJ, Ogawa T, Ehsani AA, Bourey RE, Martin 3rd WH, Holloszy JO. Effects of gender, age, and fitness level on response of VO_{2max} to training in 60-71 yr olds. J Appl Physiol. 1991;71(5):2004–11.
233. Konopka AR, Sreekumaran Nair K. Mitochondrial and skeletal muscle health with advancing age. Mol Cell Endocrinol. 2013;379(1–2):19–29.
234. Lambert CP, Wright NR, Finck BN, Villareal DT. Exercise but not diet-induced weight loss decreases skeletal muscle inflammatory gene expression in frail obese elderly persons. J Appl Physiol. 2008;105(2):473–8.
235. Lee IT, Luo SF, Lee CW, Wang SW, Lin CC, Chang CC, Chen YL, Chau LY, Yang CM. Overexpression of HO-1 protects against TNF-alpha-mediated airway inflammation by down-regulation of TNFR1-dependent oxidative stress. Am J Pathol. 2009;175(2):519–32.
236. Lee EG, Boone DL, Chai S, Libby SL, Chien M, Lodolce JP, Ma A. Failure to regulate TNF-induced NF-kappaB and cell death responses in A20-deficient mice. Science. 2000;289(5488):2350–4.
237. Pedersen BK, Ostrowski K, Rohde T, Bruunsgaard H. The cytokine response to strenuous exercise. Can J Physiol Pharmacol. 1998;76(5):505–11.
238. Petrone WF, English DK, Wong K, McCord JM. Free radicals and inflammation: superoxide-dependent activation of a neutrophil chemotactic factor in plasma. Proc Natl Acad Sci U S A. 1980;77(2):1159–63.

239. Radák Z, Chung HY, Naito H, Takahashi R, Jung KJ, Kim HJ, Goto S. Age-associated increase in oxidative stress and nuclear factor kappaB activation are attenuated in rat liver by regular exercise. FASEB J. 2004;18(6):749–50.
240. Radák Z, Naito H, Kaneko T, Tahara S, Nakamoto H, Takahashi R, Cardozo-Pelaez F, Goto S. Exercise training decreases DNA damage and increases DNA repair and resistance against oxidative stress of proteins in aged rat skeletal muscle. Pflugers Arch. 2002;445(2):273–8.
241. Radák Z, Sasvári M, Nyakas C, Pucsok J, Nakamoto H, Goto S. Exercise preconditioning against hydrogen peroxide-induced oxidative damage in proteins of rat myocardium. Arch Biochem Biophys. 2000;376(2):248–51.
242. Starnes JW, Cantu G, Farrar RP, Kehrer JP. Skeletal muscle lipid peroxidation in exercised and food-restricted rats during aging. J Appl Physiol. 1989;67(1):69–75.
243. Yu BP. Cellular defenses against damage from reactive oxygen species. Physiol Rev. 1994;74:139–62.

Deacceleration of Brain Aging by Melatonin

Rüdiger Hardeland

1 Introduction

Aging is a multifaceted process that involves numerous mechanisms. It seems important to distinguish between basal, lingering, steadily ongoing changes and other events preferably occurring at advanced age that cause new pathophysiologically relevant alterations and often strongly or even dramatically accelerate aging. The first category of changes is, surprisingly, poorly understood, although some pertinent theories consider these in terms of accumulating damage, which is not always easily demonstrable in early stages. Notably, the processes with an early onset are, phenomenologically but not necessarily mechanistically, already evident in relatively young individuals. For instance, a decline in physical fitness is obvious in many individuals over 30 years, also in the absence of unfavorable lifestyle habits. This will certainly not be explained by a number of mechanisms discussed in gerontology, such as thymic involution, immune remodeling, inflammaging, or telomere attrition. Whether or not these changes that steadily continue as a basal aging process can be attenuated and, thus, aging be decelerated, remains largely unknown. This statement is not ruled out by studies on mutations leading to senescence acceleration, because it is usually uncertain or even unlikely that the accelerated process is identical with basal aging in wild-type animals or normal human individuals. For instance, a laminopathy such as progeria, which causes a dramatic acceleration of aging in children, cannot be taken as a model of normal aging, in which there is no evidence for a progressing instability of the nuclear envelope. Much more

R. Hardeland (✉)
Johann Friedrich Blumenbach Institute of Zoology and Anthropology,
University of Göttingen, Berliner Str. 28, D-37073 Göttingen, Germany
e-mail: rhardel@gwdg.de

S.C. Bondy, A. Campbell (eds.), *Inflammation, Aging, and Oxidative Stress*,
Oxidative Stress in Applied Basic Research and Clinical Practice,
DOI 10.1007/978-3-319-33486-8_18

information is available about the stronger changes typically occurring later in life, including the various smaller or larger individual catastrophes that especially affect the brain and, indeed, accelerate aging. Beyond the relatively uncertain deceleration of normal aging, the question arising is, therefore, whether and to what extent the accelerating processes can be attenuated, in other words, whether deaccelerating treatments are possible.

Numerous ideas have been forwarded and numerous studies conducted concerning the possibility of life extension by pharmacological or nutraceutical means. One of the compounds that have been frequently discussed in this context is melatonin (*N*-acetyl-5-methoxytryptamine). This indoleamine has become known as a hormone of the pineal gland, but it is meanwhile known to be produced by many extrapineal tissues, in which it sometimes attains considerably higher concentrations than in the pineal or in the circulation [1, 2]. With regard to this distribution and the presence of its receptors in many organs and cell types, melatonin is relevant to almost the entire body. Moreover, intracellular actions are facilitated by its capability of crossing membranes. The systemic role of melatonin also extends to the control of other hormones, modulation of the immune system, multiple antioxidant effects, and a substantial role in the circadian multioscillator system, which in turn regulates a plethora of functions [2, 3] (Fig. 1).

The influence of melatonin is particularly apparent in the brain, in which melatonin participates, by virtue of multiple mechanisms, in the regulation of excitation levels, secretion of neurotrophic factors, and cerebrovascular blood flow [2, 4]. The central nervous system (CNS) is not only supplied with melatonin via the circulation, but also receives this hormone directly via the pineal recess, from where it enters the third ventricle in much higher concentrations than present in the blood [5–7]. Moreover, melatonin seems to be synthesized in several CNS areas other than pineal gland and retina [2, 4, 8]. These conclusions have not only been based on elevated levels and upregulation of the usually rate-limiting enzyme of melatonin biosynthesis, aralkylamine *N*-acetyltransferase (alias arylalkylamine *N*-acetyltransferase, AANAT). Recently, melatonin formation has been reported to be directly stimulated in the cerebellum [9]. Cultured cortical astrocytes from neonatal rats have also been shown to synthesize melatonin, at rates attaining about one third of those of pinealocytes under same conditions [10]. Moreover, melatonin biosynthesis has been demonstrated in the prenatal, developing brain, including stages at which the pineal gland is not yet functional [11]. However, it is still uncertain whether these findings are partially applicable to the adult CNS. Unfortunately, the reported levels of brain melatonin are highly divergent and urgently require clarification [4]. Nevertheless, there is good reason to assume that melatonin is an important regulator molecule in the brain, as a hormone and also as an intrinsic compound.

The relevance of melatonin to the CNS is also supported by numerous publications on neuroprotection, which have been multiply reviewed [12–19]. Although many of the reported findings have been obtained using supraphysiological concentrations, various findings summarized in those papers indicate that natural levels of melatonin are, in fact, involved in natural neuroprotective mechanisms. As a consequence, reductions of circulating melatonin levels that have been repeatedly

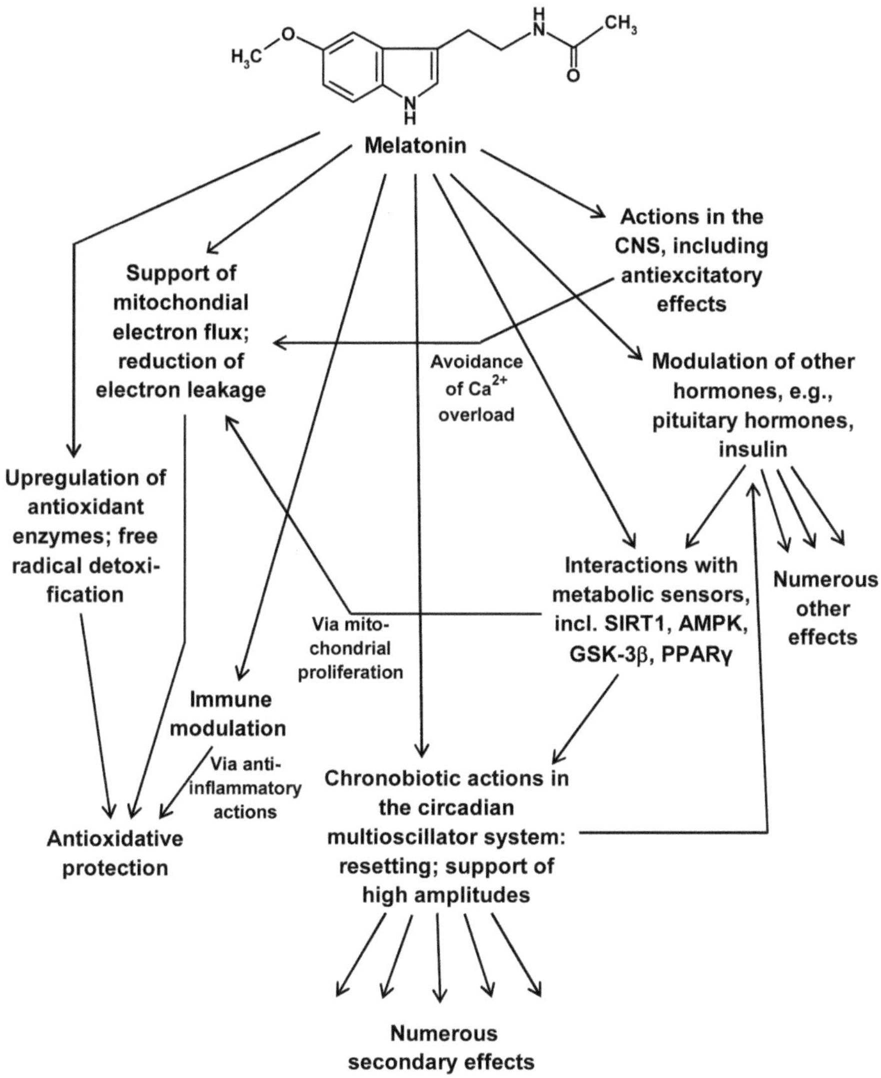

Fig. 1 The pleiotropy of melatonin within a functional network involved in the maintenance of the health state (overview based on selected aspects)

described in the context of aging and various diseases [20–22] would imply a gradual loss of neuroprotection, with consequences to enhanced vulnerability and accelerated deterioration of the CNS.

Attempts have been made to counteract these processes related to aging and age-associated diseases by melatonin supplementation, also in the hope of thereby extending the lifespan. However, as summarized elsewhere [23], convincing life-prolonging effects were mainly obtained in non-vertebrate organisms, whereas most findings in mammals either remained marginal or were questionable because of methodological reasons. Moreover, an extension of lifespan by slowing aging processes has to be distinguished from chemopreventive actions, which especially prolong life in strains of rodents that predominantly die from cancer [1, 20, 23]. A reliable extension of lifespan by melatonin was demonstrated in the senescence-accelerated SAMP8 mouse strain [24]. These mice are known to be more vulnerable to oxidative stress. Although the idea of life extension by reducing oxidative damage is highly suggestive, it remains to be clarified, also in this context, whether the processes driving accelerated senescence are identical with the normal causes of aging.

However, focusing exclusively on life extension may be strongly misleading with regard to the real value of a geroprotector. A typical observation made in rodents treated during aging with melatonin is that of maintenance of a better health state, referred to as the "Methuselah syndrome" [23]. Compared to untreated age-matched controls, these old animals display a higher mobility, a glossy fur, absence of skin inflammation and low osteoporosis [23]. These findings are strongly in favor of a particular value of melatonin in healthy aging. Taking into account that the progression of aging differs between tissues, deacceleration of brain aging may be specifically supported by melatonin and contribute to the maintenance of cognitive, behavioral, sensory and motor functions, even in the absence of major lifespan-extending effects. This assumption is supported by a number of findings which will be subsequently discussed in this chapter.

2 Reductions of Melatonin Secretion in Aging and Aging-Associated Diseases

In the course of aging, the nocturnal maximum of melatonin has been reported to substantially decrease [25–28]. This reduction is usually observed in the blood plasma, but can be also detected in the pineal gland and in other body fluids, such as saliva [29] and cerebrospinal fluid [30, 31], and also in the urinary metabolite, 6-sulfatoxymelatonin [32, 33]. However, these changes are interindividually highly variable. At advanced age, several subjects exhibit low nighttime values close to those found during daytime, whereas others show only moderate reductions and have maintained a fairly well pronounced rhythm of plasma melatonin. These differences have to be mainly seen under two aspects. First, a breakdown of the melatonin biosynthetic capacity in the pineal gland can reflect a progressive dysfunctionality of the circadian master clock, the suprachiasmatic nucleus (SCN), because of ongoing degenerative processes in the SCN or reduced retinal input, e.g., because of pupillary miosis or reduced crystalline lens transmission of blue light, which is absorbed by melanopsin of retinal ganglion cells that are mainly responsible for the photic input to the SCN [3, 21]. Disturbances in the neural connection between SCN and pineal gland may lead to a similar result. The importance of age-dependent deviations in the central circadian system is further supported by the observation that the nocturnal maximum of plasma melatonin is frequently phase-advanced relative to young subjects [28]. A second cause of reduced melatonin during senescence and even earlier in life may be sought in changes caused by various age-associated diseases [20–22]. Among these, diabetes type 2 and related metabolic disorders with insulin resistance, heart diseases, some types of cancer, painful conditions, macular degeneration and, most importantly, neurodegeneration because of Alzheimer's disease (AD) or frontotemporal dementia (Pick's disease) have to be mentioned. From this point of view, it is necessary to clarify in the individual case whether reduced melatonin levels are really due to normal aging or rather caused by some of these aging-related disorders or diseases. As a third possible cause of impaired melatonin secretion, pineal calcification has been identified [34, 35], but this does not seem to represent a general phenomenon.

Although neurodegenerative diseases differ in many pathological and pathophysiological details from normal aging, they nevertheless do exhibit traits of a brain-specific advanced and accelerated aging. This becomes especially obvious under the viewpoint of brain inflammaging, in which inflammation is initiated and fueled by intertwined processes of microglia activation, Ca^{2+} overload, nitroxidative stress, mitochondrial malfunction, fibrillogenesis, cytoskeletal disorganization, enhanced release of proinflammatory mediators because of both an age-related immune remodeling and the senescence-associated secretory phenotype (SASP), as exhibited by DNA-damaged astrocytes and endothelial cells [36]. These deteriorating and multiply interacting processes tend to form vicious cycles that can aggravate the condition of the CNS and accelerate brain aging. Notably, several of these pathological deviations can be counteracted or gradually normalized by melatonin [36, 37]. Even though it would be speculative to assume a causality between the age-related reduction of melatonin release and the onset of a brain inflammatory pathology, a decline in melatonin availability seems to represent an aggravating factor. This may be especially the case in AD and other senile dementias, in which the levels of melatonin are frequently more strongly decreased than in age-matched controls [21, 26, 28, 31, 38–44].

3 Brain Inflammaging

Inflammatory processes can be initiated in the aging brain by a number of pathophysiologically relevant changes or events, which are assumed the drive CNS senescence. In a nondiseased subject, the progression of inflammaging may remain moderate for quite a number of years. However, as soon as additional proinflammatory alterations take place, both brain inflammation and resulting neurodegeneration can be substantially accelerated and aggravated. This concerns especially the formation of clinically relevant amounts of Aβ peptides and tau hyperphosphorylation with its numerous consequences for peripheral mitochondrial function, energy supply and neuronal connectivity.

Under subclinical conditions of solely age-related low-grade brain inflammation, the following processes deserve particular attention. The inevitable age-associated remodeling of the immune system, which is a consequence of progressive thymic involution, life-long repeated exposure to foreign antigens and exhaustion of several subtypes of leukocytes [36, 37, 45–51] can lead to a proinflammatory phenotype that makes the brain more susceptible to inflammation initiation. In the case of such an immune risk profile (IRP), elevated levels of proinflammatory cytokines and other inflammatory mediators are typically observed.

Low-grade brain inflammation can be enhanced by various mechanisms. The inflammatory state may still remain in a subclinical range, but contribute to the progression of aging. Moreover, the same processes can be involved in the development of pathological changes of clinical relevance. One of the inflammation-promoting mechanisms is SASP, which has turned out to be a sustained source of inflammatory signals and elevated formation of free radicals [52–55]. SASP represents the potentially problematic side of an otherwise favorable mechanism that serves the mitotic arrest of DNA-damaged cells, a way of keeping these cells alive and metabolically active but preventing them from entering a neoplastic development. However, these arrested cells which display the so-called DNA damage response (DDR) steadily release proinflammatory cytokines. Importantly, SASP is a feature of many nonimmune cells, which, however, stimulate and attract immune cells, thereby contributing to an increased formation of inflammation-induced reactive oxygen and nitrogen species. With regard to the CNS, SASP has been shown to occur in aging astrocytes [56]. Moreover, SASP was demonstrated in endothelial cells from vessels outside the brain [57, 58], but the existence of this mechanism in CNS circulation system appears highly likely.

A process central to brain inflammation is microglia activation. Microglia-associated inflammation can occur at different degrees of severity and is relevant already at low-grade. This variability should be seen on the background of basal microglia activities. Contrary to earlier belief, these cells are not generally inactive nor do they behave as a uniformly responding entity. Even ramified microglia is known to be continuously active in terms of movement and safeguarding the CNS microenvironment [59]. Moreover, microglia activation can lead to different phenotypes, which may be either neurodestructive and phagocytically active or, alternately, primarily neuroprotective and also growth promoting [59]. Various mechanisms and signals can lead to the stimulation of microglia and are based on a complex network of, sometimes mutual, interactions with astrocytes and neurons. For instance, glutamate excitotoxicity can cause microglia activation [60–63], whereas, on the other hand, primary immune responses that activate microglia may initiate excitotoxicity [64–66]. A further complexity results from the frequently occurring coactivation of microglial cells and astrocytes. The involvement of astrocytes may not only further stimulate the microglia, e.g., by nitroxidative stress or inflammatory signals such as NO or SASP-associated cytokines, but also promote neuronal excitation, by impaired glutamate uptake and enhanced NO release with consequences to Ca^{2+} uptake and mitochondrial function [56, 65, 67, 68]. From a certain level on, proinflammatory processes in different cell types can lead to vicious cycles based on positive feedback loops between neurons, astrocytes and microglia. This may expand the grade and area of inflammation and become further aggravated by recruitment of other immune cells. A particular role can be attributed to the assembly of inflammasomes formed as different subtypes in neurons (NLRP1 and AIM2), astrocytes (NLRP2) and microglia (NLRP3), which are known to cause the release of proinflammatory cytokines such as IL-1β and IL-18 and to induce apoptotic or pyroptotic cell death [69]. As soon as cells are dying, they liberate histone H1, which acts as an additional pro-inflammatory signal to microglial cells and also as a chemoattractant [70].

Brain inflammation and neuronal overexcitation are connected in several ways, among which the enhanced formation of free radicals seems to be of particular importance. The role of oxidative stress has been demonstrated in this context in neurological diseases [71], but may be likewise applicable to respective subclinical changes. There are mainly three sources for enhanced production of reactive oxygen species, (i) leukocytes including microglial cells that are activated in the course of inflammation and primarily form superoxide by NADPH oxidase (Nox) and hypochlorite by myeloperoxidase, (ii) other Nox subforms expressed in astrocytes, neurons and endothelial cells, and (iii) mitochondria. In addition to leukocytes, neurons have also been shown to express myeloperoxidase and, although their contribution to oxidant formation under basal and subclinical inflammatory conditions is uncertain, neurons were reported to upregulate this enzyme in AD [72]. Mitochondria seem to play an additional crucial part in brain inflammation. Notably, several theories of aging have focused on these organelles, with regard to oxidant formation, but also under various additional aspects such as apoptotic cell death, mitophagy with its consequences to peripheral mitochondrial depletion, interconnections with metabolic sensing and the role of aging suppressors, as summarized elsewhere [37, 73, 74]. These considerations have been also specifically discussed in the context of the aging brain [36, 75]. Instead of repeating all these details in full length, only findings that are critical to brain inflammaging and neurodegeneration shall be addressed. Free radicals are generated in mitochondria by electron dissipation from the electron transport chain (ETC), a process that is reinforced by damage to the respirasomes as it particularly occurs under conditions of enhanced NO formation. Details on actions of the NO radical, $^{\bullet}NO$, on peroxynitrite-derived free radicals ($^{\bullet}OH$, $CO_3^{\bullet-}$, $^{\bullet}NO_2$) and nitrosating metabolites such as N_2O_3, other NO congeners or nitrosothiols have been summarized elsewhere [37, 73, 76–78]. These changes can lead to a vicious cycle of free radical formation and may end up in apoptosis or mitophagy. Notably, some frequently discussed consequences may not be as relevant to damage and aging as formerly believed. First, a breakdown of the mitochondrial membrane potential ($\Delta\Psi_{mt}$), as it occurs during a superoxide flash, does not necessarily result in an immediate initiation of apoptosis or mitophagy, but only does this after prolonged duration [36, 79]. Second, the crucial step of cardiolipin peroxidation is not mainly caused by free radicals directly, but is typically catalyzed by the peroxidase activity of the cytochrome c/cardiolipin complex [80–83]. However, decreases in mitochondrial levels of reduced glutathione (GSH), which can be caused by excess of free radicals, favor this process and have been shown to be counteracted by overexpression of the mitochondrial subform of glutathione peroxidase, GPx4 [84]. Therefore, free radicals may act upstream of cardiolipin peroxidation. Third, the damage of mitochondrial DNA (mtDNA) by free radicals has been overrated. Apart from the fact that mitochondrial chromosomes are not naked, but rather densely associated with proteins different from histones (cf. ref. [37]), mtDNA mutator mice showed an age-related accumulation of mitochondrial mutations, but no substantial increase in free radical formation [85]. Subforms of NADPH oxidase as another relevant source of free radicals are associated with processes of normal and pathological types of aging, with neuroinflammation and are subject to activation by various proinflammatory signals and factors known to induce neurodegeneration [86–92]. Accumulating evidence speaks for a crucial role of Nox isoenzymes in oxidative damage as a consequence of microglia activation und also directs attention to the multiple links between microglia, astrocytes and neurons in aging and neurodegeneration.

4 Prevention of Inflammation Initiation by Melatonin

Melatonin has been shown to possess several properties that antagonize stimuli which may lead to the initiation of brain inflammation. These effects of the pineal hormone are exerted in microglial cells, astrocytes and neurons as well (Fig. 2). They also concern the major sources of free radicals as outlined in the preceding section. Moreover, they extend to the counteraction of proinflammatory pathological changes observed in neurodegenerative diseases.

In neurons, melatonin displays several antiexcitatory and antiexcitotoxic actions that are based on multiple mechanisms. These effects have been mainly studied in the context of sedation and are, at least partially, also related to melatonin's anticonvulsive [93–98], anxiolytic, antihyperalgesic and antinociceptive properties [99–106]. However, they have to be clearly distinguished from the sleep-inducing effects [36] mediated by melatonin receptors in the SCN and in the thalamus (cf. refs. [107, 108]). The antiexcitatory actions of melatonin, which are also observed in nocturnally active animals, in which the pineal hormone is not sleep-related but associated with elevated physical and neuronal activity, seem to serve the prevention of overexcitation, with all its negative consequences of nitroxidative stress, mitochondrial impairments and microglia activation. In mechanistic terms, the following antiexcitatory effects in the CNS have been described, which are, however, differently relevant to its various parts and areas: (i) modulation of GABA and glutamate receptors [97, 109], including secondary decreases in cytosolic Ca^{2+} via $GABA_c$ [110] or metabotropic glutamate $mGlu_3$ receptors [111]; (ii) inhibition of high voltage-gated Ca^{2+} channels [112]; (iii) upregulation of $GABA_a$ receptors [113]; (iv) reduction of free cytosolic Ca^{2+} by upregulation of Ca^{2+}-buffering proteins [114, 115]; (v) increases in outward K^+ currents [116–118]; (vi) potentiation of glycine receptor-mediated inhibitory post-synaptic currents, to date observed in retinal ganglion cells [119–121]; (vii) modulation of the opioid system [99, 104–106]; (viii) inhibition of neuronal nitric oxide synthase (nNOS), with a

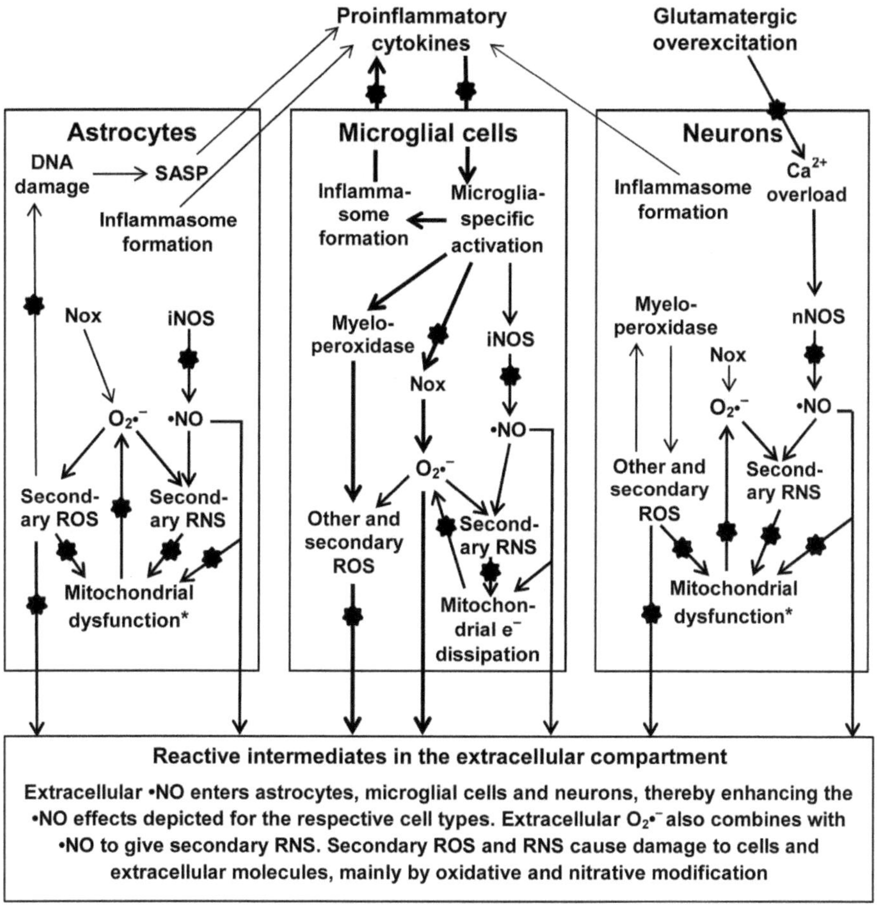

Fig. 2 Overview of the proinflammatory network in the CNS and protective actions by melatonin with relevance to the avoidance of accelerated brain aging. The relative contribution of processes are, according to current knowledge, indicated by variations in thickness of *arrows*. *With further consequences to apoptosome formation/apoptosis or mitophagy/mitochondrial depletion/peripheral energy depletion and, in neurons, losses of connectivity. ***Bold asterisk*** This symbol indicates processes known to be blocked or attenuated by melatonin. *RNS* reactive nitrogen species, *ROS* reactive oxygen species

possible contribution by its metabolite, N^1-acetyl-5-methoxykynuramine (AMK) [122–126]; (ix) with regard to the excitatory action of the diffusible NO released from other cells, downregulation of astrocytic and microglial inducible NOS (iNOS), again with a possible contribution by AMK [125–127]. Some of the summarized actions are presumably region-specific, whereas other seem to be relevant to most parts of the CNS. Nevertheless, all these effects collectively indicate a primarily antiexcitatory role of melatonin, which may be important for the avoidance of proinflammatory signaling toward the microglia.

A prevention of inflammation initiation by melatonin may be deduced from various effects on the inhibition or reduction of microglia activation. However, a discussion on this issue should always take into account the Janus-headed properties of melatonin in the immune system, in which it may behave in either a pro- or an antiinflammatory way [36, 37, 128]. These differences depend on dosage, time course, proinflammatory interventions and experimental models. In the context of aging, including the aging brain, the majority of findings indicate a rather antiinflammatory action, although this may be surprising with regard to usually lower doses applied, which lead under other conditions preferably to proinflammatory responses [36, 37]. Direct effects on microglia have, however, been usually studied using rather strong insults. Oxidative stress induced by fluoride in the BV-2 microglial cell line caused increases in the activities of iNOS and NADPH oxidase, release of TNFα and IL-1β and downstream effects on c-Jun N-terminal kinase (JNK) phosphorylation, effects that were all antagonized by melatonin [129]. In another microglial cell line, HAPI, melatonin inhibited amphetamine- or methamphetamine-induced upregulations of iNOS [130], TNFα, IL-1β and IL-6 [131], findings that are in line with similar results obtained in nonmicroglial cells [36]. Various other data exist for high-grade inflammation induced by infection, transient focal ischemia, traumatic brain injury, or endotoxemia, i.e., in models that may be fundamentally different from processes of aging, as summarized elsewhere [36]. In BV-2 cells, expression of the chemokines CCL2, CCL5, and CCL9 in response to bacterial lipopolysaccharide (LPS) was reduced by melatonin [132]. In traumatic brain injury, melatonin inhibited microglia activation, upregulation of TNFα and IL-1β and favored dephosphorylation in the mTOR pathway [133]. In *Klebsiella pneumoniae* meningitis, melatonin strongly inhibited hippocampal microglia activation and a rise in cytosolic Ca^{2+} [134]. Various other findings on the prevention of proinflammatory responses, by inhibiting upregulation of iNOS, TNFα, IL-1β and IL-6, have been obtained in CNS tissues, but may mainly reflect the suppression of local microglia activation (further details in refs. [36, 135–137]). Collectively, the majority of findings seem to indicate an inhibition of microglia activation by melatonin. However, the counteractions observed upon treatments leading to high-grade inflammation usually require supraphysiological doses of melatonin. In a recent study on aging mice, dietary melatonin failed to reduce LPS-induced rises of TNFα in the brain [137]. In conclusion, more studies on the role of melatonin in microglia are required in the future and have to be carried out under conditions of low-grade inflammation.

The prevention of inflammation initiation comprises the maintenance of low rates of free radical formation, an aspect that is also valid for the limitation of sustained inflammation. The information on melatonin effects on Nox enzymes is still relatively limited, especially with regard to the CNS. In a study on bile duct ligation in rats, Nox expression and superoxide formation were shown to be upregulated in prefrontal cortex and hippocampus, effects that were reduced by melatonin [138]. Treatment with $A\beta_{1\text{-}42}$ peptide causes an activation of microglial Nox, which was, again, inhibited by melatonin. The mechanism was based on a reduced phosphorylation of the Nox subunit $p47^{phox}$ via the phosphatidylinositol 3-kinase (PI3K)/Akt cascade, which prevents its translocation to the plasma membrane and, thus, its association with the subunits gp91^{phox} and p67^{phox} [139]. Although these brain-related effects have not been studied under basal conditions in normal aging, they demonstrate, at least, the capacity of melatonin to limit Nox activation and, thereby, to avoid enhanced superoxide formation by this enzyme. Therefore, this type of action by melatonin should considered in the future as possible contribution to the prevention of low-grade brain inflammation, too. A contribution of Nox activation to brain inflammaging may be even important under basal conditions, since mild, diet-related changes have been demonstrated [87].

Much more is known about the protective and free radical-reducing effects of melatonin in mitochondria. Because these actions have been multiply reviewed (e.g., in refs. [1, 2, 36, 37, 75–77, 140–144]), the pertinent findings shall not be repeated here in all details, but rather summarized with regard to the key features. As outlined in these reviews, melatonin improves the electron flux through the ETC and, thus, decreases superoxide ($O_2^{\bullet -}$) formed by electron dissipation at Complexes I and III. A high-affinity melatonin binding site detected in the amphipathic ramp of Complex I may participate in the regulation of electron flow. Enhanced activities of Complexes I, III and IV have been repeatedly demonstrated upon melatonin treatment, in some studies at physiological or near-physiological concentrations. The improvements seem to also comprise protection from $^{\bullet}NO$ binding to iron/sulfur clusters and heme irons, from nitrosation, nitration and oxidation of respirasomal subunits. Melatonin also enhanced the expression of some respirasomal proteins, such as subunits ND1 and ND4 of Complex I and subunits 1–3 of Complex IV. The reduction of electron overflow and backflow does not only cause lower formation rates of superoxide and, also, peroxynitrite ($ONOO^-$) that derives from the combination of $O_2^{\bullet -}$ and $^{\bullet}NO$. Thereby, the vicious cycle of self-stimulating free radical generation is interrupted, but, additionally, respiratory efficiency and ATP synthesis are improved. The protection from oxidative damage is also supported by enhanced synthesis of reduced glutathione (GSH). This leads to an increased ratio of reduced/oxidized glutathione (GSH/GSSG) and is further associated with an upregulation of the mitochondrial subform of glutathione peroxidase, GPx4, as well as protection of glutathione peroxidase and glutathione reductase from oxidative inactivation. Sometimes, upregulations of other mitochondrial antioxidant enzymes have been also described. At least under conditions of high-grade inflammation, the down-regulation of iNOS as a source of ETC-blocking and -damaging intermediates contributes significantly to the ameliorating properties of melatonin. Reduction of NO formation, whether intra- or extramitochondrial, by melatonin also avoids dysfunction and damage of this organelle by Ca^{2+} overload. Another potentially crucial effect of melatonin concerns the inhibition of cardiolipin peroxidation. Contrary to earlier belief, this does not seem to be based on a reduction of free radical attack to cardiolipin, but may be related to redox processes upstream of the cytochrome c/cardiolipin interaction that leads to the cardiolipin peroxidase properties (cf. preceding section). This may be concluded from the prevention of cardiolipin peroxidation by upregulation of GPx4 and its exacerbation by GPx4 knockout. Whether melatonin additionally inhibits the cardiolipin peroxidase directly, as shown for other low molecular weight antioxidants, or indirectly via modulation of the cytochrome c-deacetylating sirtuin isoform SIRT5 is not yet known. With regard to the initiation of apoptosis by cytochrome c release in the course of cardiolipin peroxidation, its inhibition represents an early antiapoptotic action of melatonin. Another early step concerns the $\Delta\Psi_{mt}$ breakdown. Melatonin was shown to directly inhibit the mitochondrial permeability transition pore (mtPTP), with an IC_{50} of 0.8 μM [145], a concentration that would require mitochondrial accumulation of melatonin, which has, in fact, been observed in other, differently designed studies. More recently, melatonin was reported to reduce the duration of the $\Delta\Psi_{mt}$ breakdown, in astrocytic mitochondria, down to times not yet inducing apoptosis [79]. The antiapoptotic actions of melatonin further extend to the balance between anti- and proapoptotic intramitochondrial factors, such as Bcl-2 and Bax, as demonstrated in numerous studies. Collectively, the numerous mitochondrial effects observed are, at least, in nontumor cells, protective to these organelles and prevent proinflammatory processes by enhanced formation of free oxygen and nitrogen radicals.

5 Counteraction of Progression of Brain Aging

In principle, all antiexcitatory, antiinflammatory and mitochondria protecting actions that have been discussed in the preceding section are not only relevant to the initiation, but also to the progression of brain inflammaging. However, additional aspects are arising concerning the sustained inflammatory condition, which leads to a steadily aggravating progression of aging and to more severe pathological changes in neurodegenerative diseases. The attenuation of neurodegenerative processes by melatonin has been studied in numerous model systems and was multiply reviewed [12–19, 42, 146–148], whereas in humans only a few symptomatic improvements have been described. A weakness of several models has to be seen in necessity of using extremely high doses of melatonin to efficiently antagonize the severe insults caused by neurotrauma, ischemia/reperfusion or use of powerful neurotoxins including Aβ peptides and prion proteins. Therefore, the applicability of melatonin at reasonable, pharmacologically meaningful concentrations sometimes remains uncertain under the real pathological conditions in humans. Nevertheless, the entire body of evidence is strongly indicative of a protective potential of melatonin by reducing directly or indirectly brain inflammation and cellular or mitochondrial damage.

Improvements by melatonin have been studied in the brain of normally aging animals and in senescence-accelerated SAMP8 mice. Several data on normal aging in otherwise noncompromized animals refer to changes in oxidative damage, mitochondrial function and the antioxidant status. In the frontal and occipital rat cortex, an age-related increase in lipid peroxidation was reduced by melatonin, while corresponding losses in GSH levels were reversed [149]. Likewise, oxidative and nitrosylative protein damage in the aged mouse brain was also decreased by the pineal hormone [137]. In brain mitochondria of aging mice, an increase of Complex IV activity was described that was restored to normal by melatonin [150]. In aged rat mitochondria, melatonin counteracted Complex I dysfunction, reduced H_2O_2 formation from superoxide produced by electron leakage and prevented cardiolipin peroxidation [151]. In a comparable setting, an upregulation of superoxide dismutase was described [152]. However, a corresponding result was not observed by comparing neuronal cultures from young and aged rats, but, in this case, an age-related decrease of catalase was reversed by melatonin [153]. In the aging murine cortex, increases in nNOS expression and nitrotyrosine content of proteins were reduced by the pineal hormone [154]. More direct evidence for changes in inflammation-related parameters has also been reported, apart from additional studies in which animals were challenged by LPS, which would, however, cause high-grade inflammation not necessarily typical for normal aging. Importantly, age-dependent rises in the murine cortical expression of TNFα and IL-6 were strongly suppressed by dietary melatonin [154]. Under same conditions, a considerable increase in the expression of glial fibrillary acidic protein (GFAP) was widely reversed, indicating that a presumably inflammation-induced activation of astrocytes was efficiently blocked. In addition, various other immunity-related proteins of different function that showed age-dependent up- or downregulations were partially normalized by melatonin in their levels of expression [154]. Secretion of Aβ peptides also increases, more moderately, in the aging non-AD brain, a process that is otherwise stronger and crucial in AD in which it induces, besides other effects, a more pronounced stimulation of inflammation. In the cortex of non-AD mice, dietary melatonin reduced the levels of both $A\beta_{x\text{-}40}$ and $A\beta_{x\text{-}42}$ ($x = 1$ or higher) peptides [155]. Therefore, this potentially aggravating cause of brain inflammaging is efficiently counteracted in the normal brain, but may be also effective in early stages of AD, whereas, in advanced AD, neither melatonin nor any other approved treatment can halt the disease progression, as will be discussed later.

In the senescence-accelerated mouse strain SAMP8, age-related mitochondrial malfunction is a typical feature that comprises downregulation of several respirasomal subunits, decreased Complex I and IV activities, increased electron dissipation, reduced efficiency of ATP synthesis, and changes in other respiratory parameters. These observations have been predominantly made in peripheral organs, but corresponding data also exist for the CNS (summarized in ref. [37]). In the brains of 10 months-old SAMP8 mice, melatonin, when administered via the drinking water from weaning on, strongly enhanced Complex I activity, which was substantially reduced in the controls [156]. Similar but less pronounced effects were found in Complex II and III activities, whereas a rise in Complex IV activity was only observed in females, in the absence of a reduction in controls. These investigators also demonstrated improvements in ATP synthesis as well as GPx and glutathione reductase activities [156]. Moreover, another study revealed a decreased mitochondrial membrane fluidity in the brains of old SAMP8 mice, which was reversed by melatonin [157]. Further results that involve changes in sirtuin expression will be discussed in the next section. Another important aspect of the mitochondrial role in accelerated aging concerns the number and, even more, the intraneuronal distribution of these organelles. In the CA1 hippocampal layer of

SAMP8 mice, mitochondria were found to be reduced with regard to their amount and presence in the periphery. These impairments were, again, corrected by long-term treatment with melatonin [158]. The reductions observed without melatonin treatment are indicative of dysfunction-related mitophagy, perhaps associated with changes in the fusion/fission balance, and the loss of mitochondrial surface density can be expected to result in peripheral ATP deficiency with the further consequence of reduced neuronal connectivity [2].

Mitochondrial changes as observed in senescence-accelerated animals are even more pronounced in neurodegenerative diseases, mostly studied in AD and respective model systems. Apart from the classic hallmarks of AD, such as enhanced secretion of toxic Aβ peptides, formation of amyloid deposits, tau hyperphosphorylation and tangle formation, mitochondrial malfunction is another characteristic of this disease and, again, connected to inflammatory processes (Fig. 3). Increased oxidative damage to lipids, proteins and DNA, elevated levels of 3,3′-dityrosine and 3-nitrotyrosine residues in proteins can be interpreted as the result of interrelated neuroinflammation and mitochondrial malfunction (summarized in ref. [14]). At the mitochondrial level, reductions in $\Delta\Psi_m$ and ATP production along with increased electron dissipation are associated with downregulation of fusion-promoting proteins and increases in the fission-promoting Fis1, changes that lead to mitochondrial shortening, mitophagy, progressive peripheral mitochondrial depletion with local ATP deficiency, losses of spines at neurites and, thus, to reduced neuronal connectivity [2]. Aβ peptides and its oligomers are a main source of toxicity, whereas amyloid plaques may especially cause damage to cells by direct contact and by binding of copper and zinc, which become intracellularly depleted and thereby reduce the formation of functionally active Cu,Zn-superoxide dismutase [159]. The disease-promoting effects of Aβ peptides are multiple and also comprise (i) excitotoxicity [159], (ii) copper binding that leads to hydroxyl radical formation via redox cycling in a Fenton-like reaction, (iii) activation of microglia with upregulation of iNOS, TNFα and IL-6, of IL-18, a key regulator of neuroinflammation and important player in neurodegeneration, which in turn also promotes Aβ formation, and of IL-15, which stimulates the activities and proliferation of cytotoxic T-cells, NK-cells and B-cells, which further enhance microglia activation (summarized in ref. [37]). Various additional changes in AD pathology have been described that extend, e.g., to impaired actions of neurotrophins and alterations in tyrosine kinase receptors [14]. Mitochondria from AD patients seem to be more vulnerable to Aβ, as shown in cybrid cells (cytoplasmic hybrid cells: mitochondrial DNA-depleted recipients of mitochondria from other sources) containing mitochondria from individuals suffering from AD [160]. Moreover, cybrids from a sporadic AD with reduced cytochrome c oxidase activity showed various signs of mitochondrial dysfunction (summarized in ref. [14]). In human SK-N-SH neuroblastoma cells, oxidative damage of mitochondrial DNA by exposure to $A\beta_{1-42}$ peptide was reportedly prevented by melatonin in a study based on radioautography of gels, but without complete documentation of quantitative data, whereas the restoration of cell viability was convincingly demonstrated [161]. After intrahippocampal injection of $A\beta_{1-42}$, dietary melatonin was shown to reverse the decrease in membrane fluidity and to restore the membrane lipid composition in mitochondria isolated from this region [162].

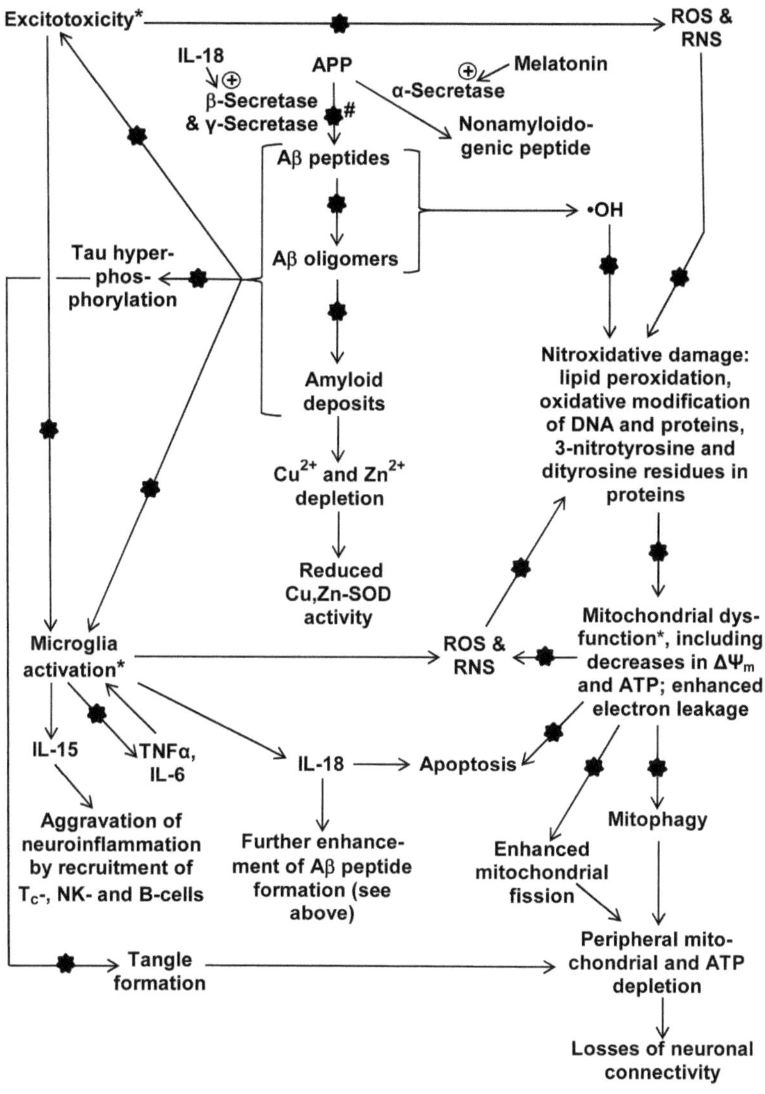

Fig. 3 Overview of protective actions by melatonin that are assumed to attenuate the progression of Alzheimer's disease, at least, under conditions of early onset of treatment. *Encircled plus sign*: upregulation. ***Bold asterisk*** Processes known to be blocked or attenuated by melatonin. # Reduction of Aβ peptide formation by competition with melatonin-induced α-secretase. *IL* interleukin, *˙OH* hydroxyl radical, *SOD* superoxide dismutase, *TNF* tumor necrosis factor, other abbreviations as in Fig. 2

With regard to the multiple processes affected in AD, the combination of numerous properties exhibited by the highly pleiotropic regulator molecule, melatonin, are of particular interest. These actions exceed by far the multiply reviewed antioxidant actions which comprise, apart from free radical scavenging, modulation of various redox-relevant enzymes and various effects that reduce free-radical formation (e.g., ref. [2, 12–14, 18, 163]). In addition to the mitochondrial effects discussed above and their consequences for decreasing apoptotic cell death [15, 16, 18, 145, 146, 164], several reports have demonstrated additional AD-specific ameliorations. In vitro, melatonin behaved as an antifibrillogenic agent [147, 165], and also counteracted the profibrillogenic action of apolipoprotein E4 [166]. These effects were frequently related to findings on protection of cultured neurons and animals against Aβ peptide toxicity (summarized in ref. [14]). However, this interpretation does not consider the higher toxicity of Aβ peptides and oligomers compared to plaque fibrils. Nevertheless, reduced fibrillogenesis was also observed in transgenic AD mice treated with dietary melatonin [167–169]. In the animals, these changes may not only reflect the inhibition of fibril formation from Aβ peptides already present, but also reductions in the secretion of Aβ peptides and improved elimination of amyloid deposits. This conclusion is supported by substantially decreased Aβ peptide levels in the brains of melatonin-treated animals [155, 170]. Effects of melatonin on β- and γ-secretase had been hypothesized in the context of the Ca^{2+} hypothesis of AD, in conjunction with the known prevention of Ca^{2+} overload by melatonin [148]. More recently, a direct effect of melatonin on α-secretase activity has been described, which leads to the formation of the nonamyloidogenic and neuroprotective fragment sAPPα [171]. This action, which was studied in human βAPP overexpressing HEK293 or N2a cells, was transmitted by melatonin membrane receptors via ERK1/2 activation and induction of the sheddases, ADAM10 or ADAM17.

Additional data concern another hallmark of AD, tau hyperphosphorylation, which was also reduced by melatonin [14, 170, 172]. In cultured hippocampal slices, this effect was associated with the prevention of the tau-phosphorylating protein kinase GSK-3β (glycogen synthase kinase 3β) [172]. These effects are well compatible with similar findings on the prevention of hyperphosphorylation induced by other agents such as wortmannin [14, 173] or haloperidol [174]. The latter compound, otherwise known as a neuroleptic, suppresses melatonin biosynthesis and, thereby, favors cdk-5 activation and neurofilament hyperphosphorylation, which is reversed by exogenous melatonin [174]. Other AD-related effects of melatonin observed after intrahippocampal injection of $A\beta_{1-42}$ consisted in the reversal of the upregulation of the calcium-binding protein S-100β and the oxidative stress-related transcription factor NF-kB as well as an increase in synaptophysin expression [175]. Moreover, with regard to approved AD therapies with cholinesterase inhibitors [159], a further observation may be of interest, which describes the supportive action of melatonin for the expression of nicotinic α-bungarotoxin-sensitive ACh receptors [176], a finding that should be associated with numerous downstream effects. Finally, melatonin was also shown to antagonize elevations of proinflammatory cytokines such as TNFα and IL-6, as induced by Aβ administration [172] or resulting from APP transgene expression [169], results that support the mainly anti-inflammatory action of the pineal hormone under otherwise proinflammatory conditions in the brain, as discussed above.

A recently emerged relationship connects AD and the associated neuroinflammation with insulin resistance [177, 178] (further literature in ref. [36]). This new information may, again, indicate a role of melatonin, which has been repeatedly shown to suppress insulin resistance and other symptoms of metabolic syndrome under various experimental conditions [21, 37] including pancreatic senescence acceleration in SAMP8 mice [179]. Notably, insulin resistance was also induced by pinealectomy and by impaired melatonergic signaling. Therefore, an age-related decrease in nocturnal melatonin levels may favor neuroinflammation and the development of AD. It would be of utmost importance to analyze whether melatonin's counteraction of insulin resistance is relevant to an attenuation of symptoms and, perhaps, reduced disease progression in AD. Even if this is not successful in advanced AD, it may be worth to be studied in earlier stages and, perhaps, mild subclinical alterations that occur in normal aging, which can be associated with moderate, but clearly demonstrable Aβ secretion, amyloid deposits and low-grade inflammatory processes [155].

6 The Multifaceted Relationship of Melatonin and Aging Suppressor Genes

The idea that melatonin may exert antiaging effects via upregulation of aging suppressor genes has been discussed in particular with regard to sirtuins, klothos and FoxOs, especially FoxO3a [73]. The sirtuin subform SIRT3 and klotho (α-klotho) are known to modulate the FoxO pathway and to cooperate in the regulation of mitochondrial mass, which can be of greatest importance to both metabolic signaling and avoidance of age- or inflammation-associated mitochondrial depletion [37]. With regard to klothos and FoxOs, most of the evidence is rather indirect or not brain-related. Concerning klotho, brain-specific effects of melatonin have been only studied in klotho-deficient mutant mice, in which the pineal hormone attenuated, via the MT_2 receptor, prooxidant effects and changes in redox-related signaling in the hippocampus as well as memory deficits [180].

With regard to sirtuins, melatonin-related effects were focused on the subform SIRT1. Findings were highly divergent and appeared, at first glance, contradictory. However, the decisive difference seems to be that between tumor and nontumor cells. In various tumor cell lines studied, melatonin strongly decreased SIRT1 expression and also promoted apoptosis, contrary to the otherwise antiapoptotic actions in other cells including neurons. The discrepancy can be solved on the basis of the chronobiological role of SIRT1, which acts as an accessory component of the cellular circadian core oscillator (notably including peripheral oscillators), some of whose genes display the property of tumor suppressors and are epigenetically downregulated in transformed cells to allow tumor cell proliferation. This leads to substantially dysregulated oscillators that keep SIRT1 strongly and steadily upregulated, an effect reversed by melatonin. The fundamental difference in SIRT1 expression between tumor and oscillating nontumor cells has been recently described in a model [181].

Contrary to the findings in tumor cells, melatonin upregulated SIRT1 expression in a number of studies related to the aging CNS. This was first observed in the brain of senescence-accelerated SAMP8 mice [182] and later also in the hippocampus of sleep-deprived rats [183]. More detailed information was obtained in a comparison of young and aged neurons [153], in which SIRT1 upregulation by melatonin was associated with enhanced deacetylation of various SIRT1 substrates, such as PGC-1α, FoxO1, NFkB, and p53, changes that were largely reversed by the SIRT1 inhibitor sirtinol. The melatonin-induced deacetylation of PGC-1α may also be indicative of a stimulation of mitochondrial proliferation. Corresponding effects by melatonin and counteraction by sirtinol were obtained in cultured neurons from SAMP8 mice, in which the pineal hormone also attenuated actions of prooxidant agents and mitochondrial toxins [184]. Upregulations of SIRT1 were also observed in the dentate gyrus of male [185] and old, ovariectomized female rats, along with antiinflammatory and antiapoptotic actions [186]. Recently, increased SIRT1 expression by melatonin was described in a murine ischemia-reperfusion model [187]. Another observation concerning vascular protection by melatonin may be, again, relevant to the brain. In this study, the pineal hormone also prevented an age-related decrease in SIRT1 [188]. Several additional findings on melatonin-induced upregulation of SIRT1 have been obtained in peripheral organs such as heart, liver and pancreas, but shall not be discussed here in detail. Collectively, all the aging-associated findings in nontumor cells are in favor of a positive relationship between melatonin and SIRT1 expression.

7 Conditions and Limits of Efficacy

An aging-decelerating agent should be expected to extend the lifespan. In fact, substantial increases by melatonin were observed in several non-vertebrate organisms [23], up to 50 % in the rotifer *Philodina acuticornis* [1], a gerontological model organism. However, the outcome in normally aging mammals has been rather moderate and was not always convincingly demonstrable [23]. In the normally aging mouse strain SAMR1, melatonin extended the mean lifespan from about 20 to 23 months and the maximal lifespan from about 25 to 26 months [24]. For fundamental reasons, such findings have to be distinguished from comparable data obtained in mouse strains that predominantly die from cancer [23]. In these animals, a longer survival seems to mainly reflect, according to more recent data, the oncostatic and proapoptotic actions of the pineal hormone in cancer cells [37]. Notably, such observations strongly contrast with findings on anti-cancer effects in cancer-prone C3H/He mice [189]. More profound life-extending effects of melatonin were described in the senescence-accelerated mouse strain SAMP8, in which the mean lifespan was increased from about 16 to 22 months and the maximal lifespan from about 23 to 27 months [24]. These findings indicate that melatonin interferes with the processes causing an acceleration of aging and, thus, displays deaccelerating properties, whereas the effects in SAMR1 were much less expressed and do not support a profound aging-decelerating capacity. However, in SAMR1 mice, the extension of the mean, but not so much the maximal lifespan may be interpreted in terms of healthy aging. This would be well in accordance with the observations of a healthier appearance, lower tendency to inflammation and low osteoporosis in melatonin-treated animals [23, 154], as already mentioned in the introductory section.

The numerous beneficial effects reported for melatonin in neurodegenerative diseases, especially in experimental models, indicate profound neuroprotective properties. This was confirmed in transgenic AD mice treated with melatonin, starting at the age of 4 months, i.e., 2 months before the rise in Aβ levels and 5–6 months before appearance of amyloid deposits [167]. Marked reductions of protein nitration, $A\beta_{x\text{-}40}$ and $A\beta_{x\text{-}42}$ peptide concentrations were accompanied by a significantly improved survival (around 16 months about 93 % survivors in the treated vs. about 70 % in the untreated group). However, in a comparable study in which melatonin treatment was started at 14 months, no reductions of Aβ peptides were observed [190]. Several insights may be deduced from these results. First, melatonin is capable of counteracting the development of AD symptoms in vivo, which is in good agreement with other studies [168–170, 191, 192]. Second, by interference with the pathological mechanisms of AD, melatonin attenuates the acceleration of aging and, thus, acts in a deaccelerating way. Again, this property should not be confused with a deceleration of basal aging processes, even though there may be a certain overlap between normal and disease-related aging mechanisms. Third, there seems to be a point of no return, from which on a treatment with melatonin is no longer promising. In mechanistic terms, this may be a stage in which damage to neurons and astrocytes including mitochondrial malfunction with all its further consequences, and also activated microglia and high levels of Aβ peptides will continue to promote inflammatory signaling and further Aβ formation in a vicious cycle. With regard to the treatment of AD patients, this conclusion has a pessimistic aspect insofar as treatments usually start after symptoms have already become severe. Earlier and, therefore, more promising treatments may, however, be possible on the basis of screening for combinations of genetic and other risk factors.

8 A Brief Outlook at Aging-Relevant Melatonin Effects Beyond Brain Inflammaging

Losses of neurons that contribute to aging may be corrected to a limited and site-specific degree by neurotrophic factors. The respective actions of melatonin on the expression of neurotrophins have been recently summarized [4]. At first glance, some of these findings appear contradictory. In experiments using neurotoxins such as 3-nitropropionic acid, MPTP or Aβ, neurotrophins like GDNF (glial cell line-derived neurotrophic factor) or BDNF (brain-derived neurotrophic factor) were upregulated and these effects counteracted by melatonin. However, other studies showed upregulations of these factors and also of CDNF (conserved dopamine neurotrophic factor), persephin and MANF (mesencephalic astrocyte-derived neurotrophic factor) by melatonin. Decreases in NGF (nerve growth factor) secretion were induced by Aβ or H_2O_2, but reversed by melatonin. The apparent inconsistency of these findings may, however, be resolved. Increased neurotrophin secretion in response to neurotoxins may reflect attempts of repair and reductions by melatonin might be interpreted as a return to normal. Otherwise, melatonin seems to be rather a neurotrophin-promoting agent. In this way, it may contribute to the maintenance of neuronal functionality and favor interneuronal connectivity. Whether or not additional effects on stem cells may be of importance, as in various peripheral organs [2], is still uncertain and, if existing, seems to be widely limited to certain regions with low, but not negligible neurogenesis, such as the hippocampus.

Another potentially important aspect of melatonin's actions concerns its chronobiological role. For quite some time, this has been rarely seen in its connection to health. However, there is increasing evidence for its relevance. Mutations of core oscillator proteins are associated with various diseases and these changes are also affecting the numerous peripheral circadian oscillators [3]. Melatonin seems to coordinate the phasing of central and various peripheral oscillators, as far as they are not already directly steered by the SCN [3]. Moreover, factors such as the aging suppressor SIRT1, the aging-, mitochondria- and inflammation-related metabolic sensors AMPK, GSK-3β (also known as a hyperphosphorylating kinase) and peroxisome proliferator-activated receptor-γ (PPARγ) are meanwhile classified as accessory components of core oscillators and are all modulated by melatonin [36, 37]. A further relationship to chronobiology became apparent in comparisons of SAMR1 and SAMP8 mice, which exhibit different free-running circadian periods and respond differently to melatonin in the adaptation to the light/dark cycle [193]. Another strong hint for the relevance of the circadian system concerns the deterioration and functional decomposition of overt circadian rhythms at advanced age [44, 194]. These deteriorations are even more pronounced in AD [39, 44]. The role of the circadian system in aging became also apparent by comparing lifespans of wild-type, homo- and heterozygous Syrian hamsters carrying the short-period mutation *tau*. Homozygous *tau/tau* hamsters had either a moderately reduced [195] or slightly extended [196] mean lifespan. However, an obviously disturbed circadian machinery in heterozygous *tau*/+ hamsters shorted the mean survival down to only 10.9 months, compared to 17.5 months in homozygous wild-type animals [195]. The progressive breakdown of the circadian system, with its countless consequences for the coordination of physiological functions may be regarded as another aging-accelerating process. At least in the earlier phases of rhythmic malfunction, when in humans the signs of frequent sleep disturbances and nocturia become apparent, well-timed melatonin may have a beneficial effect by enhancing circadian amplitudes and harmonizing the phase relationships between, at least, the melatonin-sensitive oscillators. The importance of the circadian system in antagonizing aging is underlined by one of the most convincing findings on demonstrable rejuvenation by transplantation of juvenile SCN tissue to old Syrian hamsters [195], which does not only restore various overt circadian rhythms, as also shown in other studies (cf. ref. [3]), but leads to a more juvenile physical appearance and to an expanded lifespan in the recipient old hamsters [195]. However, an eventual contribution of melatonin to the restoration of rhythms and life extension after SCN transplantation remains to be elucidated.

9 Conclusion

Melatonin counteracts various processes that can promote aging, in particular, supracritical neuronal, astrocytic and microglial activations that have the potential of causing overexcitation and brain inflammaging. It displays beneficial properties concerning the improvement of mitochondrial function, reduction of free radical formation and limitation of aging-related increases of proinflammatory cytokines. In addition, Aβ formation, fibrillogenesis and tau hyperphosphorylation are inhibited by melatonin. On the background of age- and disease-associated decreases of the pineal secretory capacity, the more or less pronounced partial deficiency of melatonin may substantially contribute to an acceleration of normal and, even more, disease-promoted aging. Whether or not melatonin already decelerates basal aging that occurs already earlier in life remains uncertain in the present state of our knowledge. With regard to the processes of brain inflammaging, the potential of melatonin is amply documented. However, the efficacy of the pineal hormone in terms of life extension seems to be limited, since increases in lifespan of laboratory mammals have frequently remained at borderline and are only in a few studies convincingly demonstrated. Whether such findings can be translated to humans, remains uncertain on chronobiological grounds. Successful investigations have been conducted using nocturnally active rodents, in which melatonin can be easily administered via the drinking water. However, this procedure is not practicable in the diurnally active human. The use of controlled-release preparations of melatonin or of synthetic, longer acting melatonergic agonists appears to be insufficient with regard persistent effects overnight, as judged from sleep data, in which the improvements by melatonin remained marginal, even when they were statistically demonstrable [107]. Moreover, a fundamental uncertainty results from the differences in the relationship between melatonin secretion and phases of activity. While in nocturnal rodents high melatonin levels are associated with enhanced locomotor and behavioral activities, their brains may respond differently to the pineal hormone compared to the diurnally active human, in whom melatonin is associated with rest. Therefore, in many functions controlled by circadian oscillators, melatonin should exert opposite effects in laboratory rodents and humans. On the other hand, similarities in melatonin's action have to be expected as far as the stimulation of the circadian system is concerned, such as favoring of high rhythm amplitudes and coordination between rhythms in the multioscillator machinery. With regard to neurodegenerative diseases, in particular, AD, another limitation of melatonin's efficacy has become apparent, concerning the onset of treatment. At least in the transgenic AD mice, a successful treatment was possible when the administration of melatonin started before AD-typic alterations such as increased Aβ secretion [167], but a later onset, after the development of symptoms, remained unsuccessful [190]. However, the results of early-beginning treatment indicate that melatonin may be able to counteract the initiation of brain inflammaging and also to gradually inhibit the progression of AD. It would be of importance to elucidate whether this can be interpreted in terms of a more general capacity of deaccelerating aging, which usually proceeds at increasing velocity in the course of senescence.

At the present state of our knowledge, life extension by melatonin may be possible, but only in a marginal range. It may be more important to support healthy aging instead of speculating on a prolonged lifespan. In humans, more pronounced extensions of lifespan have anyway been achieved during the last decades by improvements in nutrition, lifestyle and medical care. The numerous physiological actions of melatonin including those in the CNS, as well as the associations of melatonergic dysfunction with numerous diseases [20–22] seem to be favorable for maintaining good health conditions during aging. This is clearly supported, at least in rodents, by the physical appearance of melatonin-treated animals [23, 154]. Whether or not these findings can be translated to humans, in which melatonin is administered shortly before sleep, would also depend on the contribution of the melatonin-modulated circadian system to health. The role of circadian oscillators in health maintenance and their disturbance in diseases is meanwhile becoming more and more apparent [3]. If melatonin—and other treatments supporting the circadian system, such as appropriately timed light exposure—will be demonstrated to favor healthy aging, a moderately longer average lifespan may result as kind of a side effect that reduces early deaths by age-associated diseases.

References

1. Hardeland R, Poeggeler B. Melatonin beyond its classical functions. Open Physiol J. 2008;1:1–23.
2. Hardeland R, Cardinali DP, Srinivasan V, Spence DW, Brown GM, Pandi-Perumal SR. Melatonin—a pleiotropic, orchestrating regulator molecule. Prog Neurobiol. 2011;93:350–84.
3. Hardeland R, Madrid JA, Tan D-X, Reiter RJ. Melatonin, the circadian multioscillator system and health: the need for detailed analyses of peripheral melatonin signaling. J Pineal Res. 2012;52:139–66.
4. Hardeland R, Poeggeler B. Melatonin and synthetic melatonergic agonists: actions and metabolism in the central nervous system. Cent Nerv Syst Agents Med Chem. 2012;12:189–216.
5. Tricoire H, Locatelli A, Chemineau P, Malpaux B. Melatonin enters the cerebrospinal fluid through the pineal recess. Endocrinology. 2002;143:84–90.
6. Tricoire H, Malpaux B, Møller M. Cellular lining of the sheep pineal recess studied by light-, transmission-, and scanning electron microscopy: morphologic indications for a direct secretion of melatonin from the pineal gland to the cerebrospinal fluid. J Comp Neurol. 2003;456:39–47.
7. Tricoire H, Møller M, Chemineau P, Malpaux B. Origin of cerebrospinal fluid melatonin and possible function in the integration of photoperiod. Reprod Suppl. 2003;61:311–21.
8. Acuña-Castroviejo D, Escames G, Venegas C, Díaz-Casado ME, Lima-Cabello E, López LC, Rosales-Corral S, Tan D-X, Reiter RJ. Extrapineal melatonin: sources, regulation and potential functions. Cell Mol Life Sci. 2014;71:2997–3025.

9. Pinato L, da Silveira Cruz-Machado S, Franco DG, Campos LM, Cecon E, Fernandes PA, Bittencourt JC, Markus RP. Selective protection of the cerebellum against intracerebroventricular LPS is mediated by local melatonin synthesis. Brain Struct Funct. 2015;220(2):827–40. doi:10.1007/s00429-013-0686-4 [Epub ahead of print, Dec. 22].
10. Liu Y-J, Zhuang J, Zhu H-Y, Shen Y-X, Tan Z-L, Zhou J-N. Cultured rat cortical astrocytes synthesize melatonin: absence of a diurnal rhythm. J Pineal Res. 2007;43:232–8.
11. Jimenez-Jorge S, Guerrero JM, Jimenez-Caliani AJ, Naranjo MC, Lardone PJ, Carrillo-Vico A, Osuna C, Molinero P. Evidence for melatonin synthesis in the rat brain during development. J Pineal Res. 2007;42:240–6.
12. Reiter RJ. Oxidative damage in the central nervous system: protection by melatonin. Prog Neurobiol. 1998;56:359–84.
13. Reiter RJ, Acuña-Castroviejo D, Tan D-X, Burkhardt S. Free radical-mediated molecular damage. Mechanisms for the protective actions of melatonin in the central nervous system. Ann N Y Acad Sci. 2001;939:200–15.
14. Srinivasan V, Pandi-Perumal SR, Cardinali DP, Poeggeler B, Hardeland R. Melatonin in Alzheimer's disease and other neurodegenerative disorders. Behav Brain Funct. 2006;2:15.
15. Wang X. The antiapoptotic activity of melatonin in neurodegenerative diseases. CNS Neurosci Ther. 2009;15:345–57.
16. He H, Dong W, Huang F. Anti-amyloidogenic and anti-apoptotic role of melatonin in Alzheimer disease. Curr Neuropharmacol. 2010;8:211–7.
17. Carpentieri A, Díaz de Barbosa G, Areco V, Peralta López N, Tolosa de Talamoni N. New perspectives in melatonin research. Pharmacol Res. 2012;65:437–44.
18. Pandi-Perumal SR, BaHammam AS, Brown GM, Spence DW, Bharti VK, Kaur C, Hardeland R, Cardinali DP. Melatonin oxidative defense: therapeutical implications for aging and neurodegenerative processes. Neurotox Res. 2013;23:267–300.
19. Shinozuka K, Staples M, Borlongan CV. Melatonin-based neuroprotection in stroke. Int J Mol Sci. 2013;14:8924–47.
20. Hardeland R. Neurobiology, pathophysiology, and treatment of melatonin deficiency and dysfunction. ScientificWorldJournal. 2012;2012:640389.
21. Hardeland R. Melatonin in aging and disease—multiple consequences of reduced secretion, options and limits of treatment. Aging Dis. 2012;3:194–225.
22. Hardeland R. Deviations of melatonin levels and signaling in aging and diseases—options and limits of treatment. In: Srinivasan V, Gobbi G, Shillcutt SD, Suzen S, editors. Melatonin: therapeutic value and neuroprotection. Boca Raton: CRC; 2014. p. 11–8.
23. Poeggeler B. Melatonin, aging, and age-related diseases: perspectives for prevention, intervention, and therapy. Endocrine. 2005;27:201–12.
24. Rodríguez MI, Escames G, López LC, López A, García JA, Ortiz F, Sánchez V, Romeu M, Acuña-Castroviejo D. Improved mitochondrial function and increased life span after chronic melatonin treatment in senescent prone mice. Exp Gerontol. 2008;43:749–56.
25. Sack RL, Lewy AJ, Erb DL, Vollmer WM, Singer CM. Human melatonin production decreases with age. J Pineal Res. 1986;3:379–88.
26. Skene DJ, Vivien-Roels B, Sparks DL, Hunsaker JC, Pévet P, Ravid D, Swaab DF. Daily variation in the concentration of melatonin and 5-methoxytryptophol in the human pineal gland: effect of age and Alzheimer's disease. Brain Res. 1990;528:170–4.
27. Karasek M, Reiter RJ. Melatonin and aging. Neuroendocrinol Lett. 2002;23 Suppl 1:14–6.
28. Skene DJ, Swaab DF. Melatonin rhythmicity: effect of age and Alzheimer's disease. Exp Gerontol. 2003;38:199–206.
29. Kripke DF, Youngstedt SD, Elliott JA, Tuunainen A, Rex KM, Hauger RL, Marler MR. Circadian phase in adults of contrasting ages. Chronobiol Int. 2005;22:695–709.
30. Brown GM, Young SN, Gauthier S, Tsui H, Grota LJ. Melatonin in human cerebrospinal fluid in daytime; its origin and variation with age. Life Sci. 1979;25:929–36.
31. Liu RY, Zhou JN, van Heerikhuize J, Hofman MA, Swaab DF. Decreased melatonin levels in postmortem cerebrospinal fluid in relation to aging, Alzheimer's disease, and apolipoprotein E-ε4/4 genotype. J Clin Endocrinol Metab. 1999;84:323–7.

32. Youngstedt SD, Kripke DF, Elliott JA, Klauber MR. Circadian abnormalities in older adults. J Pineal Res. 2001;31:264–72.
33. Mahlberg R, Tilmann A, Salewski L, Kunz D. Normative data on the daily profile of urinary 6-sulfatoxymelatonin in healthy subjects between the ages of 20 and 84. Psychoneuroendocrinology. 2006;31:634–41.
34. Schmid HA. Decreased melatonin biosynthesis, calcium flux, pineal gland calcification and aging: a hypothetical framework. Gerontology. 1993;39:189–99.
35. Kunz D, Schmitz S, Mahlberg R, Mohr A, Stoter C, Wolf KJ, Herrmann WM. A new concept for melatonin deficit: on pineal calcification and melatonin excretion. Neuropsychopharmacology. 1999;21:765–72.
36. Hardeland R, Cardinali DP, Brown GM, Pandi-Perumal SR. Melatonin and brain inflammaging. Prog Neurobiol. 2015;127–128:46–63.
37. Hardeland R. Melatonin and the theories of aging: a critical appraisal of melatonin's role in antiaging mechanisms. J Pineal Res. 2013;55:325–56.
38. Uchida K, Okamoto N, Ohara K, Morita Y. Daily rhythm of serum melatonin in patients with dementia of the degenerate type. Brain Res. 1996;717:154–9.
39. Mishima K, Tozawa T, Satoh K, Matsumoto Y, Hishikawa Y, Okawa M. Melatonin secretion rhythm disorders in patients with senile dementia of Alzheimer's type with disturbed sleep-waking. Biol Psychiatry. 1999;45:417–21.
40. Ohashi Y, Okamoto N, Uchida K, Iyo M, Mori N, Morita Y. Daily rhythm of serum melatonin levels and effect of light exposure in patients with dementia of the Alzheimer's type. Biol Psychiatry. 1999;45:1646–52.
41. Ferrari E, Fioravanti M, Magri F, Solerte SB. Variability of interactions between neuroendocrine and immunological functions in physiological aging and dementia of the Alzheimer's type. Ann N Y Acad Sci. 2000;917:582–96.
42. Srinivasan V, Pandi-Perumal SR, Maestroni GJM, Esquifino AI, Hardeland R, Cardinali DP. Role of melatonin in neurodegenerative diseases. Neurotox Res. 2005;7:293–318.
43. Wu YH, Fischer DF, Kalsbeek A, Garidou-Boof ML, van der Vliet J, van Heijningen C, Liu RY, Zhou JN, Swaab DF. Pineal clock gene oscillation is disturbed in Alzheimer's disease, due to functional disconnection from the "master clock". FASEB J. 2006;20:1874–6.
44. Wu YH, Swaab DF. Disturbance and strategies for reactivation of the circadian rhythm system in aging and Alzheimer's disease. Sleep Med. 2007;8:623–36.
45. Ponnappan S, Ponnappan U. Aging and immune function: molecular mechanisms to interventions. Antioxid Redox Signal. 2011;14:1551–85.
46. Franceschi C, Monti D, Barbieri D, Salvioli S, Grassilli E, Capri M, Troiano L, Guido M, Bonafè M, Tropea F, Salomoni P, Benatti F, Bellesia E, Macchioni S, Anderlini R, Sansoni P, Mariotti S, Wratten ML, Tetta C, Cossarizza A. Successful immunosenescence and the remodelling of immune responses with ageing. Nephrol Dial Transplant. 1996;11 Suppl 9:18–25.
47. Ginaldi L, De Martinis M, D'Ostillo A, Marini L, Loreto MF, Quaglino D. Immunological changes in the elderly. Aging (Milano). 1999;11:281–6.
48. Franceschi C, Bonafè M, Valensin S, Olivieri F, De Luca M, Ottaviani E, De Benedictis G. Inflamm-aging. An evolutionary perspective in immunosenescence. Ann N Y Acad Sci. 2000;908:244–54.
49. Sansoni P, Vescovini R, Fagnoni F, Biasini C, Zanni L, Telera A, Lucchini G, Passeri G, Monti D, Franceschi C, Passeri M. The immune system in extreme longevity. Exp Gerontol. 2008;43:61–5.
50. Candore G, Caruso C, Colonna-Romano G. Inflammation, genetic background and longevity. Biogerontology. 2010;11:565–73.
51. Franceschi C, Campisi J. Chronic inflammation (inflammaging) and its potential contribution to age-associated diseases. J Gerontol A Biol Sci Med Sci. 2014;69:S4–9.
52. Coppé JP, Patil CK, Rodier F, Sun Y, Muñoz DP, Goldstein J, Nelson PS, Desprez PY, Campisi J. Senescence-associated secretory phenotypes reveal cell-nonautonomous functions of oncogenic RAS and p53 tumor suppressor. PLoS Biol. 2008;6:2853–68.

53. Fumagalli M, d'Adda di Fagagna F. SASPense and DDRama in cancer and ageing. Nat Cell Biol. 2009;11:921–3.
54. Young ARJ, Narita M. SASP reflects senescence. EMBO Rep. 2009;10:228–30.
55. Coppé JP, Desprez PY, Krtolica A, Campisi J. The senescence-associated secretory phenotype: the dark side of tumor suppression. Annu Rev Pathol. 2010;5:99–118.
56. Salminen A, Ojala J, Kaamiranta K, Haapasalo A, Hiltunen M, Soininen H. Astrocytes in the aging brain express characteristics of senescence-associated secretory phenotype. Eur J Neurosci. 2011;34:3–11.
57. Olivieri F, Lazzarini R, Babini L, Prattichizzo F, Rippo MR, Tiano L, Di Nuzzo S, Graciotti L, Festa R, Brugè F, Orlando P, Silvestri S, Capri M, Palma L, Magnani M, Franceschi C, Littarru GP, Procopio AD. Anti-inflammatory effect of ubiquinol-10 on young and senescent endothelial cells via miR-146a modulation. Free Radic Biol Med. 2013;63:410–20.
58. Coleman PR, Chang G, Hutas G, Grimshaw M, Vadas MA, Gamble JR. Age-associated stresses induce an anti-inflammatory senescent phenotype in endothelial cells. Aging. 2013;5:913–24.
59. Blaylock RL. Immunology primer for neurosurgeons and neurologists part 2: innate brain immunity. Surg Neurol Int. 2013;4:118.
60. Campuzano O, Castillo-Ruiz MM, Acarin L, Castellano B, Gonzalez B. Distinct pattern of microglial response, cyclooxygenase-2, and inducible nitric oxide synthase expression in the aged rat brain after excitotoxic damage. J Neurosci Res. 2008;86:3170–83.
61. Arlicot N, Tronel C, Bodard S, Garreau L, de la Crompe B, Vandevelde I, Guilloteau D, Antier D, Chalon S. Translocator protein (18 kDa) mapping with [^{125}I]-CLINDE in the quinolinic acid rat model of excitotoxicity: a longitudinal comparison with microglial activation, astrogliosis, and neuronal death. Mol Imaging. 2014;13:4–11.
62. Lin TY, Huang WJ, Wu CC, Lu CW, Wang SJ. Acacetin inhibits glutamate release and prevents kainic-acid-induced neurotoxicity in rats. PLoS One. 2014;9:e88644.
63. Nomaru H, Sakumi K, Katogi A, Ohnishi YN, Kajitani K, Tsuchimoto D, Nestler EJ, Nakabeppu Y. Fosb gene products contribute to excitotoxic microglial activation by regulating the expression of complement C5a receptors in microglia. Glia. 2014;62:1284–98.
64. Takeuchi H, Mizuno T, Zhang G, Wang J, Kawanokuchi J, Kuno R, Suzumura A. Neuritic beading induced by activated microglia is an early feature of neuronal dysfunction toward neuronal death by inhibition of mitochondrial respiration and axonal transport. J Biol Chem. 2005;280:10444–54.
65. Brown GC. Mechanisms of inflammatory neurodegeneration: iNOS and NADPH oxidase. Biochem Soc Trans. 2007;35(Pt 5):1119–21.
66. Brown GC, Neher JJ. Inflammatory neurodegeneration and mechanisms of microglial killing of neurons. Mol Neurobiol. 2010;41:242–7.
67. Tilleux S, Hermans E. Neuroinflammation and regulation of glial glutamate uptake in neurological disorders. J Neurosci Res. 2007;85:2059–70.
68. Morales I, Guzmán-Martínez L, Cerda-Troncoso C, Farías GA, Maccioni RB. Neuroinflammation in the pathogenesis of Alzheimer's disease. A rational framework for the search of novel therapeutic approaches. Front Cell Neurosci. 2014;8:112.
69. de Rivero Vaccari JP, Dietrich WD, Keane RW. Activation and regulation of cellular inflammasomes: gaps in our knowledge for central nervous system injury. J Cereb Blood Flow Metab. 2014;34:369–75.
70. Gilthorpe JD, Oozeer F, Nash J, Calvo M, Bennett DL, Lumsden A, Pini A. Extracellular histone H1 is neurotoxic and drives a pro-inflammatory response in microglia. F1000Res. 2013;2:148.
71. Bondy SC. The relation of oxidative stress and hyperexcitation to neurological disease. Proc Soc Exp Biol Med. 1995;208:337–45.
72. Green PS, Mendez AJ, Jacob JS, Crowley JR, Growdon W, Hyman BT, Heinecke JW. Neuronal expression of myeloperoxidase is increased in Alzheimer's disease. J Neurochem. 2004;90:724–33.
73. Hardeland R, Coto-Montes A. New vistas on oxidative damage and aging. Open Biol J. 2010;3:39–52.

74. Kong Y, Trabucco SE, Zhang H. Oxidative stress, mitochondrial dysfunction and the mitochondria theory of aging. Interdiscip Top Gerontol. 2014;39:86–107.
75. Bondy SC, Sharman EH. Melatonin, oxidative stress, and the aging brain. In: Bondy SC, Maiese K, editors. Aging and age-related disorders. New York: Springer; 2010. p. 339–57.
76. Hardeland R. Neuroprotection by radical avoidance: search for suitable agents. Molecules. 2009;14:5054–102.
77. Hardeland R. Melatonin, mitochondrial electron flux and leakage: recent findings and resolution of contradictory results. Adv Stud Biol. 2009;1:207–30.
78. Hardeland R. Melatonin and its metabolites as anti-nitrosating and anti-nitrating agents. J Exp Integr Med. 2011;1:67–81.
79. Jou MJ. Melatonin preserves the transient mitochondrial permeability transition for protection during mitochondrial Ca^{2+} stress in astrocyte. J Pineal Res. 2011;50:427–35.
80. Basova LV, Kurnikov IV, Wang L, Ritov VB, Belikova NA, Vlasova II, Pacheco AA, Winnica DE, Peterson J, Bayir H, Waldeck DH, Kagan VE. Cardiolipin switch in mitochondria: shutting off the reduction of cytochrome c and turning on the peroxidase activity. Biochemistry. 2007;46:3423–34.
81. Kagan VE, Bayir A, Bayir H, Stoyanovsky D, Borisenko GG, Tyurina YY, Wipf P, Atkinson J, Greenberger JS, Chapkin RS, Belikova NA. Mitochondria-targeted disruptors and inhibitors of cytochrome c/cardiolipin peroxidase complexes: a new strategy in anti-apoptotic drug discovery. Mol Nutr Food Res. 2009;53:104–14.
82. Kagan VE, Bayir HA, Belikova NA, Kapralov O, Tyurina YY, Tyurin VA, Jiang J, Stoyanovski DA, Wipf P, Kochanek PM, Greenberger JS, Pitt B, Shvedova AA, Borisenko G. Cytochrome c/cardiolipin relations in mitochondria: a kiss of death. Free Radic Biol Med. 2009;46:1439–53.
83. Ji J, Kline AE, Amoscato A, Samhan-Arias AK, Sparvero LJ, Tyurin VA, Tyurina YY, Fink B, Manole MD, Puccio AM, Okonkwo DO, Cheng JP, Alexander H, Clark RS, Kochanek PM, Wipf P, Kagan VE, Bayır H. Lipidomics identifies cardiolipin oxidation as a mitochondrial target for redox therapy of brain injury. Nat Neurosci. 2012;15:1407–13.
84. Liang H, Ran Q, Jang YC, Holstein D, Lechleiter J, McDonald-Marsh T, Musatov A, Song W, Van Remmen H, Richardson A. Glutathione peroxidase 4 differentially regulates the release of apoptogenic proteins from mitochondria. Free Radic Biol Med. 2009;47:312–20.
85. Thompson LV. Oxidative stress, mitochondria and mtDNA-mutator mice. Exp Gerontol. 2006;41:1220–2.
86. Park KW, Baik HH, Jin BK. Interleukin-4-induced oxidative stress via microglial NADPH oxidase contributes to the death of hippocampal neurons in vivo. Curr Aging Sci. 2008;1:192–201.
87. Bruce-Keller AJ, White CL, Gupta S, Knight AG, Pistell PJ, Ingram DK, Morrison CD, Keller JN. NOX activity in brain aging: exacerbation by high fat diet. Free Radic Biol Med. 2010;49:22–30.
88. Zhang J, Malik A, Choi HB, Ko RW, Dissing-Olesen L, MacVicar BA. Microglial CR3 activation triggers long-term synaptic depression in the hippocampus via NADPH oxidase. Neuron. 2014;82:195–207.
89. Bordt EA, Polster BM. NADPH oxidase- and mitochondria-derived reactive oxygen species in proinflammatory microglial activation: a bipartisan affair? Free Radic Biol Med. 2014;76:34–46.
90. Sorce S, Nuvolone M, Keller A, Falsig J, Varol A, Schwarz P, Bieri M, Budka H, Aguzzi A. The role of the NADPH oxidase NOX2 in prion pathogenesis. PLoS Pathog. 2014;10:e1004531.
91. Rodriguez-Perez AI, Borrajo A, Rodriguez-Pallares J, Guerra MJ, Labandeira-Garcia JL. Interaction between NADPH-oxidase and Rho-kinase in angiotensin II-induced microglial activation. Glia. 2015;63:466–82.
92. Han BH, Zhou ML, Johnson AW, Singh I, Liao F, Vellimana AK, Nelson JW, Milner E, Cirrito JR, Basak J, Yoo M, Dietrich HH, Holtzman DM, Zipfel GJ. Contribution of reactive oxygen species to cerebral amyloid angiopathy, vasomotor dysfunction, and microhemorrhage in aged Tg2576 mice. Proc Natl Acad Sci U S A. 2015;112:E881–90.

93. Fariello RG, Bubenik GA, Brown GM, Grota LJ. Epileptogenic action of intraventricularly injected antimelatonin antibody. Neurology. 1977;27:567–70.
94. Golombek DA, Escolar E, Burin LJ, De Brito Sánchez MG, Fernández Duque D, Cardinali DP. Chronopharmacology of melatonin: inhibition by benzodiazepine antagonism. Chronobiol Int. 1992;9:124–31.
95. Golombek DA, Fernández Duque D, De Brito Sánchez MG, Burin L, Cardinali DP. Time-dependent anticonvulsant activity of melatonin in hamsters. Eur J Pharmacol. 1992;210:253–8.
96. Muñoz-Hoyos A, Sánchez-Forte M, Molina-Carballo A, Escames G, Martín-Medina E, Reiter RJ, Molina-Font JA, Acuña-Castroviejo D. Melatonin's role as an anticonvulsant and neuronal protector: experimental and clinical evidence. J Child Neurol. 1998;13:501–9.
97. Molina-Carballo A, Muñoz-Hoyos A, Sánchez-Forte M, Uberos-Fernández J, Moreno-Madrid F, Acuña-Castroviejo D. Melatonin increases following convulsive seizures may be related to its anticonvulsant properties at physiological concentrations. Neuropediatrics. 2007;38:122–5.
98. Solmaz I, Gürkanlar D, Gökçil Z, Göksoy C, Ozkan M, Erdoğan E. Antiepileptic activity of melatonin in guinea pigs with pentylenetetrazol-induced seizures. Neurol Res. 2009;31:989–95.
99. Golombek DA, Escolar E, Burin LJ, De Brito Sánchez MG, Cardinali DP. Time-dependent melatonin analgesia in mice: inhibition by opiate or benzodiazepine antagonism. Eur J Pharmacol. 1991;194:25–30.
100. Golombek DA, Escolar E, Cardinali DP. Melatonin-induced depression of locomotor activity in hamsters: time dependency and inhibition by the central type benzodiazepine antagonist Ro 15-1788. Physiol Behav. 1991;49:1091–8.
101. Golombek DA, Martíni M, Cardinali DP. Melatonin as an anxiolytic in rats: time dependence and interaction with the central GABAergic system. Eur J Pharmacol. 1993;237:231–6.
102. Pang CS, Tsang SF, Yang JC. Effects of melatonin, morphine and diazepam on formalin-induced nociception in mice. Life Sci. 2001;68:943–51.
103. Papp M, Litwa E, Gruca P, Mocaër E. Anxiolytic-like activity of agomelatine and melatonin in three animal models of anxiety. Behav Pharmacol. 2006;17:9–18.
104. Ulugol A, Dokmeci D, Guray G, Sapolyo N, Ozyigit F, Tamer M. Antihyperalgesic, but not antiallodynic, effect of melatonin in nerve-injured neuropathic mice: possible involvements of the L-arginine-NO pathway and opioid system. Life Sci. 2006;78:1592–7.
105. Srinivasan V, Pandi-Perumal SR, Spence DW, Moscovitch A, Trakht I, Brown GM, Cardinali DP. Potential use of melatonergic drugs in analgesia: mechanisms of action. Brain Res Bull. 2010;81:362–71.
106. Srinivasan V, Zakaria R, Jeet Singh H, Acuña-Castroviejo D. Melatonin and its agonists in pain modulation and its clinical application. Arch Ital Biol. 2012;150:274–89.
107. Hardeland R. New approaches in the management of insomnia: weighing the advantages of prolonged release melatonin and synthetic melatoninergic agonists. Neuropsychiatr Dis Treat. 2009;5:341–54.
108. Hardeland R. Investigational melatonin receptor agonists. Expert Opin Investig Drugs. 2010;19:747–64.
109. Rosenstein RE, Cardinali DP. Central gabaergic mechanisms as target for melatonin activity. Neurochem Int. 1990;17:373–9.
110. Prada C, Udin SB, Wiechmann AF, Zhdanova IV. Stimulation of melatonin receptors decreases calcium levels in Xenopus tectal cells by activating $GABA_C$ receptors. J Neurophysiol. 2005;94:968–78.
111. Prada C, Udin SB. Melatonin decreases calcium levels in retinotectal axons of *Xenopus laevis* by indirect activation of group III metabotropic glutamate receptors. Brain Res. 2005;1053:67–76.
112. Ayar A, Martin DJ, Ozcan M, Kelestimur H. Melatonin inhibits high voltage activated calcium currents in cultured rat dorsal root ganglion neurones. Neurosci Lett. 2001;313:73–7.
113. Naguib M, Gottomukkala V, Goldstein PA. Melatonin and anesthesia: a clinical perspective. J Pineal Res. 2007;42:12–21.

114. Koh PO. Melatonin regulates the calcium-buffering proteins, parvalbumin and hippocalcin, in ischemic brain injury. J Pineal Res. 2012;53:358–65.
115. Das A, Wallace IV G, Reiter RJ, Varma AK, Ray SK, Banik NL. Overexpression of melatonin membrane receptors increases calcium-binding proteins and protects VSC4.1 motoneurons from glutamate toxicity through multiple mechanisms. J Pineal Res. 2013;54:58–68.
116. Huan C, Zhou M, Zhang Z, Mei Y. Activation of melatonin receptor increases a delayed rectifier K^+ current in rate cerebellar granule cells. Brain Res. 2001;917:182–90.
117. Liu LY, Hoffman GE, Fei XW, Li Z, Zhang ZH, Mei YA. Delayed rectifier outward K^+ current mediates the migration of rat cerebellar granule cells stimulated by melatonin. J Neurochem. 2007;102:333–44.
118. Corthell JT, Olcese J, Trombley PQ. Melatonin in the mammalian olfactory bulb. Neuroscience. 2014;261:74–84.
119. Zhang M, Cao LH, Yang XL. Melatonin modulates glycine currents of retinal ganglion cells in rat. Neuroreport. 2007;18:1675–8.
120. Zhao WJ, Zhang M, Miao Y, Yang XL, Wang Z. Melatonin potentiates glycine currents through a PLC/PKC signalling pathway in rat retinal ganglion cells. J Physiol. 2010;588:2605–19.
121. Ji M, Zhao WJ, Dong LD, Miao Y, Yang XL, Sun YH, Wang Z. RGS2 and RGS4 modulate melatonin-induced potentiation of glycine currents in rat retinal ganglion cells. Brain Res. 2011;1411:1–8.
122. León J, Macías M, Escames G, Camacho E, Khaldy H, Martín M, Espinosa A, Gallo MA, Acuña-Castroviejo D. Structure-related inhibition of calmodulin-dependent neuronal nitric-oxide synthase activity by melatonin and synthetic kynurenines. Mol Pharmacol. 2000;58:967–75.
123. Acuña-Castroviejo D, Escames G, López LC, Hitos AB, León J. Melatonin and nitric oxide: two required antagonists for mitochondrial homeostasis. Endocrine. 2005;27:159–68.
124. Entrena A, Camacho ME, Carrión MD, López-Cara LC, Velasco G, León J, Escames G, Acuña-Castroviejo D, Tapias V, Gallo MA, Vivo A, Espinosa A. Kynurenamines as neural nitric oxide synthase inhibitors. J Med Chem. 2005;48:8174–81.
125. Jiménez-Ortega V, Cano P, Cardinali DP, Esquifino AI. 24-Hour variation in gene expression of redox pathway enzymes in rat hypothalamus: effect of melatonin treatment. Redox Rep. 2009;14:132–8.
126. Hardeland R, Tan D-X, Reiter RJ. Kynuramines, metabolites of melatonin and other indoles: the resurrection of an almost forgotten class of biogenic amines. J Pineal Res. 2009;47:109–26.
127. Tapias V, Escames G, López LC, López A, Camacho E, Carrión MD, Entrena A, Gallo MA, Espinosa A, Acuña-Castroviejo D. Melatonin and its brain metabolite N^1-acetyl-5-methoxykynuramine prevent mitochondrial nitric oxide synthase induction in Parkinsonian mice. J Neurosci Res. 2009;87:3002–10.
128. Carrillo-Vico A, Lardone PJ, Álvarez-Sánchez N, Rodríguez-Rodríguez A, Guerrero JM. Melatonin: buffering the immune system. Int J Mol Sci. 2013;14:8638–83.
129. Yan L, Liu S, Wang C, Wang F, Song Y, Yan N, Xi S, Liu Z, Sun G. JNK and NADPH oxidase involved in fluoride-induced oxidative stress in BV-2 microglia cells. Mediators Inflamm. 2013;2013:895975.
130. Tocharus J, Chongthammakun S, Govitrapong P. Melatonin inhibits amphetamine-induced nitric oxide synthase mRNA overexpression in microglial cell lines. Neurosci Lett. 2008;439:134–7.
131. Tocharus J, Khonthun C, Chongthammakun S, Govitrapong P. Melatonin attenuates methamphetamine-induced overexpression of pro-inflammatory cytokines in microglial cell lines. J Pineal Res. 2010;48:347–52.
132. Min KJ, Jang JH, Kwon TK. Inhibitory effects of melatonin on the lipopolysaccharide-induced CC chemokine expression in BV2 murine microglial cells are mediated by suppression of Akt-induced NF-kB and STAT/GAS activity. J Pineal Res. 2012;52:296–304.
133. Ding K, Wang H, Xu J, Lu X, Zhang L, Zhu L. Melatonin reduced microglial activation and alleviated neuroinflammation induced neuron degeneration in experimental traumatic brain injury: possible involvement of mTOR pathway. Neurochem Int. 2014;76:23–31.

134. Wu U-I, Mai F-D, Sheu J-N, Chen L-Y, Liu Y-T, Huang H-C, Chang H-M. Melatonin inhibits microglial activation, reduces pro-inflammatory cytokine levels, and rescues hippocampal neurons of adults rats with acute *Klebsiella pneumoniae* meningitis. J Pineal Res. 2011;50:159–70.
135. Wang Z, Wu L, You W, Ji C, Chen G. Melatonin alleviates secondary brain damage and neurobehavioral dysfunction after experimental subarachnoid hemorrhage: possible involvement of TLR4-mediated inflammatory pathway. J Pineal Res. 2013;55:399–408.
136. Chen J, Chen G, Li J, Qian C, Mo H, Gu C, Yan F, Yan W, Wang L. Melatonin attenuates inflammatory response-induced brain edema in early brain injury following a subarachnoid hemorrhage: a possible role for the regulation of pro-inflammatory cytokines. J Pineal Res. 2014;57:340–7.
137. Campbell A, Sharman E, Bondy SC. Age-related differences in the response of the brain to dietary melatonin. Age (Dordr). 2014;36:49–55.
138. Chen YC, Sheen JM, Tain YL, Chen CC, Tiao MM, Huang YH, Hsieh CS, Huang LT. Alterations in NADPH oxidase expression and blood-brain barrier in bile-duct ligation-treated young rats: effects of melatonin. Neurochem Int. 2012;60:751–8.
139. Zhou J, Zhang S, Zhao X, Wei T. Melatonin impairs NADPH oxidase assembly and decreases superoxide anion production in microglia exposed to amyloid-β_{1-42}. J Pineal Res. 2008;45:157–65.
140. Acuña-Castroviejo D, Escames G, Rodriguez MI, Lopez LC. Melatonin role in the mitochondrial function. Front Biosci. 2007;12:947–63.
141. Hardeland R, Poeggeler B, Pappolla MA. Mitochondrial actions of melatonin—an endeavor to identify their adaptive and cytoprotective mechanisms. Abh Sächs Akad Wiss Math-Nat Kl. 2009;65 Pt 3:14–31.
142. Paradies G, Petrosillo G, Paradies V, Reiter RJ, Ruggiero FM. Melatonin, cardiolipin and mitochondrial bioenergetics in health and disease. J Pineal Res. 2010;48:297–310.
143. Acuña Castroviejo D, López LC, Escames G, López A, García JA, Reiter RJ. Melatonin-mitochondria interplay in health and disease. Curr Top Med Chem. 2011;11:221–40.
144. Cardinali DP, Pagano ES, Scacchi Bernasconi PA, Reynoso R, Scacchi P. Melatonin and mitochondrial dysfunction in the central nervous system. Horm Behav. 2013;63:322–30.
145. Andrabi SA, Sayeed I, Siemen D, Wolf G, Horn TF. Direct inhibition of the mitochondrial permeability transition pore: a possible mechanism responsible for anti-apoptotic effects of melatonin. FASEB J. 2004;18:869–71.
146. Reiter RJ, Cabrera J, Sainz RM, Mayo JC, Manchester LC, Tan D-X. Melatonin as a pharmacological agent against neuronal loss in experimental models of Huntington's disease, Alzheimer's disease and Parkinsonism. Ann N Y Acad Sci. 1999;890:471–85.
147. Pappolla MA, Chyan YJ, Poeggeler B, Frangione B, Wilson G, Ghiso J, Reiter RJ. An assessment of the antioxidant and the antiamyloidogenic properties of melatonin: implications for Alzheimer's disease. J Neural Transm. 2000;107:203–31.
148. Rosales-Corral SA, Acuña-Castroviejo D, Coto-Montes A, Boga JA, Manchester LC, Fuentes-Broto L, Korkmaz A, Ma S, Tan D-X, Reiter RJ. Alzheimer's disease: pathological mechanisms and the beneficial role of melatonin. J Pineal Res. 2012;52:167–202.
149. Akbulut KG, Gonül B, Akbulut H. Exogenous melatonin decreases age-induced lipid peroxidation in the brain. Brain Res. 2008;1238:31–5.
150. Sharman EH, Bondy SC. Effects of age and dietary antioxidants on cerebral electron transport chain activity. Neurobiol Aging. 2001;22:629–34.
151. Petrosillo G, Fattoretti P, Matera M, Ruggiero FM, Bertoni-Freddari C, Paradies G. Melatonin prevents age-related mitochondrial dysfunction in rat brain via cardiolipin protection. Rejuvenation Res. 2008;11:935–43.
152. Öztürk G, Akbulut KG, Güney S, Acuña-Castroviejo D. Age-related changes in the rat brain mitochondrial antioxidative enzyme ratios: modulation by melatonin. Exp Gerontol. 2012;47:706–11.
153. Tajes M, Gutierrez-Cuesta J, Ortuño-Sahagun D, Camins A, Pallàs M. Anti-aging properties of melatonin in an in vitro murine senescence model: involvement of the sirtuin 1 pathway. J Pineal Res. 2009;47:228–37.

154. Bondy SC, Lahiri DK, Perreau VM, Sharman KZ, Campbell A, Zhou J, Sharman EH. Retardation of brain aging by chronic treatment with melatonin. Ann N Y Acad Sci. 2004;1035:197–215.
155. Lahiri DK, Chen D, Ge Y-W, Bondy SC, Sharman EH. Dietary supplementation with melatonin reduces levels of amyloid beta-peptides in the murine cerebral cortex. J Pineal Res. 2004;36:224–31.
156. Carretero M, Escames G, López LC, Venegas C, Dayoub JC, García L, Acuña-Castroviejo D. Long-term melatonin administration protects brain mitochondria from aging. J Pineal Res. 2009;47:192–200.
157. García JJ, Piñol-Ripoll G, Martínez-Ballarín E, Fuentes-Broto L, Miana-Mena FJ, Venegas C, Caballero B, Escames G, Coto-Montes A, Acuña-Castroviejo D. Melatonin reduces membrane rigidity and oxidative damage in the brain of SAMP8 mice. Neurobiol Aging. 2011;32:2045–54.
158. Cheng S, Ma C, Qu H, Fan W, Pang J, He H. Differential effects of melatonin on hippocampal neurodegeneration in different aged accelerated senescence prone mouse-8. Neuroendocrinol Lett. 2008;29:91–9.
159. Hardeland R. Cognitive enhancers in moderate to severe Alzheimer's disease. Clin Med Insights Ther. 2011;3:459–76.
160. Cardoso SM, Santana I, Swerdlow RH, Oliveira CR. Mitochondria dysfunction of Alzheimer's disease cybrids enhances Aβ toxicity. J Neurochem. 2004;89:1417–26.
161. Pappolla MA, Chyan YJ, Poeggeler B, Bozner P, Ghiso J, LeDoux SP, Wilson GL. Alzheimer β protein mediated oxidative damage of mitochondrial DNA: prevention by melatonin. J Pineal Res. 1999;27:226–9.
162. Rosales-Corral SA, Lopez-Armas G, Cruz-Ramos J, Melnikov VG, Tan D-X, Manchester LC, Munoz R, Reiter RJ. Alterations in lipid levels of mitochondrial membranes induced by amyloid-β: a protective role of melatonin. Int J Alzheimers Dis. 2012;2012:459806.
163. Hardeland R. Antioxidative protection by melatonin—multiplicity of mechanisms from radical detoxification to radical avoidance. Endocrine. 2005;27:119–30.
164. Sainz RM, Mayo JC, Rodriguez C, Tan D-X, Lopez-Burillo S, Reiter RJ. Melatonin and cell death: differential actions on apoptosis in normal and cancer cells. Cell Mol Life Sci. 2003;60:1407–26.
165. Cheng X, van Breemen RB. Mass spectrometry-based screening for inhibitors of β-amyloid protein aggregation. Anal Chem. 2005;77:7012–5.
166. Poeggeler B, Miravalle L, Zagorski MG, Wisniewski T, Chyan YJ, Zhang Y, Shao H, Bryant-Thomas T, Vidal R, Frangione B, Ghiso J, Pappolla MA. Melatonin reverses the profibrillogenic activity of apolipoprotein E4 on the Alzheimer amyloid Aβ peptide. Biochemistry. 2001;40:14995–5001.
167. Matsubara E, Bryant-Thomas T, Pacheco Quinto J, Henry TL, Poeggeler B, Herbert D, Cruz-Sanchez F, Chyan YJ, Smith MA, Perry G, Shoji M, Abe K, Leone A, Grundke-Ikbal I, Wilson GL, Ghiso J, Williams C, Refolo LM, Pappolla MA, Chain DG, Neria E. Melatonin increases survival and inhibits oxidative and amyloid pathology in a transgenic model of Alzheimer's disease. J Neurochem. 2003;85:1101–8.
168. Feng Z, Chang Y, Cheng Y, Zhang BL, Qu ZW, Qin C, Zhang JT. Melatonin alleviates behavioral deficits associated with apoptosis and cholinergic system dysfunction in the APP 695 transgenic mouse model of Alzheimer's disease. J Pineal Res. 2004;37:129–36.
169. Olcese JM, Cao C, Mori T, Mamcarz MB, Maxwell A, Runfeldt MJ, Wang L, Zhang C, Lin X, Zhang G, Arendash GW. Protection against cognitive deficits and markers of neurodegeneration by long-term administration of melatonin in a transgenic model of Alzheimer disease. J Pineal Res. 2009;47:82–96.
170. García-Mesa Y, Giménez-Llort L, López LC, Venegas C, Cristòfol R, Escames G, Acuña-Castroviejo D. Melatonin plus physical exercise are highly neuroprotective in the 3×Tg-AD mouse. Neurobiol Aging. 2012;33:1124.e13–29.
171. Shukla M, Htoo HH, Wintachai P, Hernandez JF, Dubois C, Postina R, Xu H, Checler F, Smith DR, Govitrapong P, Vincent B. Melatonin stimulates the nonamyloidogenic processing

of βAPP through the positive transcriptional regulation of ADAM10 and ADAM17. J Pineal Res. 2015;58:151–65.
172. Bender Hoppe J, Frozza RL, Horn AP, Argenta Comiran R, Bernardi A, Campos MM, Oliveira Battastini AM, Salbego C. Amyloid-β neurotoxicity in organotypic culture is attenuated by melatonin: involvement of GSK-3β, *tau* and neuroinflammation. J Pineal Res. 2010;48:230–8.
173. Deng YQ, Xu GG, Duan P, Zhang Q, Wang JZ. Effects of melatonin on wortmannin-induced tau hyperphosphorylation. Acta Pharmacol Sin. 2005;26:519–26.
174. Wang S, Zhu L, Shi H, Zheng H, Tian Q, Wang Q, Liu R, Wang JZ. Inhibition of melatonin biosynthesis induced neurofilament hyperphosphorylation with activation of cyclin-dependent kinase 5. Neurochem Res. 2007;32:1329–35.
175. Jun Z, Li Z, Fang W, Fengzhen Y, Puyuan W, Wenwen L, Zhi S, Bondy SC. Melatonin decreases levels of S100β and NFKB, increases levels of synaptophysin in a rat model of Alzheimer's disease. Curr Aging Sci. 2013;6:142–9.
176. Markus RP, Silva CL, Franco DG, Barbosa Jr EM, Ferreira ZS. Is modulation of nicotinic acetylcholine receptors by melatonin relevant for therapy with cholinergic drugs? Pharmacol Ther. 2010;126:251–62.
177. de la Monte SM, Tong M. Brain metabolic dysfunction at the core of Alzheimer's disease. Biochem Pharmacol. 2014;88:548–59.
178. Ferreira ST, Clarke JR, Bomfim TR, De Felice FG. Inflammation, defective insulin signaling, and neurological dysfunction in Alzheimer's disease. Alzheimers Dement. 2014;10 Suppl 1:S76–83.
179. Cuesta S, Kireev R, García C, Rancan L, Tresguerres JAF. Melatonin can improve insulin resistance and aging-induced pancreas alterations in senescence-accelerated prone male mice (SAMP8). Age (Dordr). 2013;35:659–71.
180. Shin EJ, Chung YH, Le HL, Jeong JH, Dang DK, Nam Y, Wie MB, Nah SY, Nabeshima YI, Nabeshima T, Kim HC. Melatonin attenuates memory impairment induced by klotho gene deficiency via interactive signaling between MT2 receptor, ERK, and Nrf2-related antioxidant potential. Int J Neuropsychopharmacol. 2015;18: pyu105. doi: 10.1093/ijnp/pyu105.
181. Hardeland R. Melatonin, noncoding RNAs, messenger RNA stability and epigenetics—evidence, hints, gaps and perspectives. Int J Mol Sci. 2014;15:18221–52.
182. Gutierrez-Cuesta J, Sureda FX, Romeu M, Canudas AM, Caballero B, Coto-Montes A, Camins A, Pallàs M. Chronic administration of melatonin reduced cerebral injury biomarkers in SAMP8. J Pineal Res. 2007;42:394–402.
183. Chang HM, Wu UI, Lan CT. Melatonin preserves longevity protein (sirtuin 1) expression in the hippocampus of total sleep-deprived rats. J Pineal Res. 2009;47:211–20.
184. Cristòfol R, Porquet D, Corpas R, Coto-Montes A, Serret J, Camins A, Pallàs M, Sanfeliu C. Neurons from senescence-accelerated SAMP8 mice are protected against frailty by the sirtuin 1 promoting agents melatonin and resveratrol. J Pineal Res. 2012;52:271–81.
185. Kireev RA, Vara E, Tresguerres JAF. Growth hormone and melatonin prevent age-related alteration in apoptosis processes in the dentate gyrus of male rats. Biogerontology. 2013;14:431–42.
186. Kireev RA, Vara E, Viña J, Tresguerres JAF. Melatonin and oestrogen treatments were able to improve neuroinflammation and apoptotic processes in dentate gyrus of old ovariectomized female rats. Age (Dordr). 2014;36:9707.
187. Yang Y, Jiang S, Dong Y, Fan C, Zhao L, Yang X, Li J, Di S, Yue L, Liang G, Reiter RJ, Qu Y. Melatonin prevents cell death and mitochondrial dysfunction via a SIRT1-dependent mechanism during ischemic-stroke in mice. J Pineal Res. 2015;58:61–70.
188. Rodella LF, Favero G, Rossini C, Foglio E, Bonomini F, Reiter RJ, Rezzani R. Aging and vascular dysfunction: beneficial melatonin effects. Age (Dordr). 2013;35:103–15.
189. Bulian D, Pierpaoli W. The pineal gland and cancer. I. Pinealectomy corrects congenital hormonal dysfunctions and prolongs life of cancer-prone C3H/He mice. J Neuroimmunol. 2000;108:131–5.

190. Quinn J, Kulhanek D, Nowlin J, Jones R, Praticò D, Rokach J, Stackman R. Chronic melatonin therapy fails to alter amyloid burden or oxidative damage in old Tg2576 mice: implications for clinical trials. Brain Res. 2005;1037:209–13.
191. Dragicevic N, Copes N, O'Neill-Moffitt G, Jin J, Buzzeo R, Mamcarz M, Tan J, Cao C, Olcese JM, Arendash GW, Bradshaw PC. Melatonin treatment restores mitochondrial function in Alzheimer's mice: a mitochondrial protective role of melatonin membrane receptor signaling. J Pineal Res. 2011;51:75–86.
192. Peng CX, Hu J, Hong XP, Wu YY, Zhu LQ, Wang JZ. Disease-modified glycogen synthase kinase-3β intervention by melatonin arrests the pathology and memory deficits in an Alzheimer's animal model. Neurobiol Aging. 2013;34:1555–63.
193. Asai M, Ikeda M, Akiyama M, Oshima I, Shibata S. Administration of melatonin in drinking water promotes the phase advance of light-dark cycle in senescence-accelerated mice, SAMR1 but not SAMP8. Brain Res. 2000;876:220–4.
194. Weinert D. The temporal order of mammals. Evidence for multiple central and peripheral control mechanisms and for endogenous and exogenous components: some implications for research on aging. Biol Rhythm Res. 2005;36:293–308.
195. Hurd MW, Ralph MR. The significance of circadian organization for longevity in the golden hamster. J Biol Rhythms. 1998;13:430–6.
196. Oklejewicz M, Daan S. Enhanced longevity in *tau* mutant Syrian hamsters, *Mesocricetus auratus*. J Biol Rhythms. 2002;17:210–6.

Redox Based-Peripheral Biomarkers in Alzheimer's Disease: Challenges, Limits and Prospects

Morgane Perrotte, Joel Pincemail, Mohammed Haddad, and Charles Ramassamy

1 Introduction

Alzheimer's disease (AD) is the most common form of dementia in the elderly. The World Health Organization currently estimates that more than 35 million people are afflicted by AD worldwide and this number is expected to dramatically triple by 2050 [1]. In the United States, approximately 7 million people older than 65 years are known to suffer from AD. According to the 2013 Alzheimer's Association (AA) report [1], the number of deaths related to AD has increased by 68 % during the 2000–2010 period [2]. The social and psychological burden associated with caring AD patients remains difficult to quantify but in the United States, but in the United States it was estimated to surpass $170 billion in 2010 and projected to exceed $1 trillion by 2050 [3]. A hypothetical intervention that could delay the onset of AD dementia by 5 years would result in a 57 % reduction in the number of patients with AD dementia, and reduce consequently the projected Medicare costs of AD from $627 to $344 billion dollars [4].

AD is characterized by a progressive neurodegeneration and loss of cognitive functions. AD is often difficult to differentiate from other forms of dementia, especially in the early clinical stages. Although the pathogenesis of AD is not yet fully

M. Perrotte • M. Haddad • C. Ramassamy (✉)
National Institute of Scientific Research- Armand Frappier Institute, Laval, QC, Canada

Institute of Nutrition and Functional Foods, Laval University, Laval, QC, Canada
e-mail: charles.ramassamy@iaf.inrs.ca

J. Pincemail
Liège CHU Department of Cardiovascular Surgery and Antioxidant Nutrition and Health Platform, Sart Tilman, Liège 4000, Belgium

S.C. Bondy, A. Campbell (eds.), *Inflammation, Aging, and Oxidative Stress*, Oxidative Stress in Applied Basic Research and Clinical Practice,
DOI 10.1007/978-3-319-33486-8_19

known, it is clear that the disease is caused by a combination of genetic and risk factors with complex etiology. The pathophysiological hallmarks of AD are extracellular fibrillary amyloid-β deposition and intraneuronal hyperphosphorylated tau (p- tau) protein aggregation particularly in the hippocampus, amygdala and frontal cortex. In AD, cognitive decline closely correlates to neurofibrillary tangles (NFTs), synapse loss, and neurodegeneration.

Clinically, AD can be divided into different phases: (1) a preclinical phase in which subjects are cognitively normal but have mild AD pathology, (2) a prodromal phase known as mild cognitive impairment (MCI), and (3) a phase when patients display dementia with impairments in multiple domains and loss of function in activities of daily living [5].

Thus, MCI can be considered the intermediate phase between normal aging and the early form of AD [6]. Persons with amnestic MCI show objective cognitive problems, involving memory, with no or low functional impairment in activities of daily living and very often evolves to AD [7]. Consequently MCI may be a useful AD's or prodromal phase of AD in which to analyze biomarkers for early and accurate disease detection.

In 2011, the National Institute on Aging (NIA) and the AA recognized the new guideline on the preclinical stages of AD [4] i.e. the pathophysiology of AD begins years to decades before the apparent clinical picture. The new guideline recognizes that the asymptomatic (preclinical) state consists of three distinct stages: stage 1 is characterized by asymptomatic amyloidosis in the brain evidenced by increased positron emission tomography (PET) Aß ligand binding that may begin as early as young adulthood and evolves slowly through midlife into old age. This amyloidosis is associated to low CSF Aβ42 levels. In addition to amyloidosis, the stage 2 includes early neurodegeneration as evidenced by (1) neuronal dysfunction based on neuroimaging with PET or MRI analysis, (2) cellular indicators of neurodegeneration, including high CSF tau or p-tau concentrations, or (3) structural brain changes based on MRI with cortical thinning and hippocampal atrophy. Finally, the stage 3 corresponds to amyloidosis, neurodegeneration, and subtle cognitive decline. The cognitive deficits in stage 3 do not significantly affect functional capacities and activities of daily living. Thus, the three preclinical stages are without the cognitive impairment that defines MCI and the cognitive and functional impairments that determines AD. This recognition of a preclinical state represents a major advancement because this stage represents a window of opportunity for the determination of valid and reliable biomarkers to define the disease stage and to monitor the progression of the underlying pathophysiology. Moreover, this or this slot is a timely disease-modifying or preventive therapies.

The aim of this chapter is to review the current state of knowledge on peripheral oxidative biomarkers for AD or MCI patients, their limits and prospects as well as the challenges for their standardisation and validation. This review will determine whether biomarkers related to oxidative stress can play an important role for early diagnosis and prognosis.

2 Current Clinical Biomarkers for AD and Their Limits

Currently, a definite diagnosis of AD can be made only by *post-mortem* neuropathological examination. Neuroimaging, genetic testing and chemical analysis from fluids, CSF, represent the current diagnostic markers used.

Valid biomarkers should be paralleled with the hypothetical pathophysiological sequence of AD and should be relevant to track the preclinical stages of AD. The biomarkers model proposed by Jack et al. [8], expands the preclinical phase with the following features: (1) Aβ accumulation become abnormal first and a substantial Aβ load accumulates before the appearance of clinical symptoms. The lag phase between Aβ accumulation and clinical symptoms may be for more than a decade. Brain Aβ accumulation is necessary but not sufficient to produce the clinical symptoms of MCI and dementia; (2) biomarkers of synaptic dysfunction, including functional MRI (fMRI), may demonstrate abnormalities very early, particularly in APOE gene ε4 allele carriers, who may manifest functional abnormalities before detectable Aβ deposition. The severity and change over time in these synaptic markers correlate with clinical symptoms during MCI and AD dementia, (3) structural MRI is thought to become abnormal a bit later, as a marker of neuronal loss, and MRI retains a close relationship with cognitive performance through the clinical phases of MCI and dementia, (4) none of the biomarkers are static and rates of change over time follow a nonlinear time course, which is hypothesized to be in sigmoid shape, and (5) anatomic information from imaging biomarkers provides useful disease staging information with disease progression. Several multicenter biomarker initiatives, including the Alzheimer's Disease Neuroimaging Initiative (ADNI); the Australian Imaging, Biomarkers and Lifestyle Flagship Study of Aging; as well as major biomarker studies in preclinical populations at several academic centers, are ongoing. These studies have already provided preliminary evidence that biomarker variations are consistent with AD pathophysiological process and are detectable before the emergence of clinical symptomatology and are predictive of subsequent cognitive decline.

The diagnostic markers currently used include increase in total tau and p-tau, and decrease in Aβ42 level and Aβ42/Aβ40 ratio in CSF from AD patients and from MCI subjects. The variations of these four biomarkers have been confirmed with large prospective studies, including the ADNI project [9] and are considered as reliable biomarkers of AD [10]. Therefore, a decline in Aß42 levels in combination with levels of the p-tau in the CSF has been advocated for use in the diagnosis of AD [11].

Neuroimaging is currently gaining high interest for the possibility to test several promising markers, as suggested by the ADNI project through the use of MRI or PET to analyze hippocampal volumetry, brain flow, decrease [18F]fluoro-2-deoxy-D-glucose uptake (FDG) or increased binding of amyloid-targeting PET trace in the living human brain [9, 12–14]. However, these diagnostic methods are limited by the cost and availability of the equipments, limiting the routinely use of these technique for the diagnosis of the asymptomatic early stages of AD. In addition, amyloid imaging and Aβ measurements in CSF are associated with a significant number of false-positive and false negative findings, with as many as 30 % of cognitively normal elderly showing signs of Aβ accumulation, and a substantial number of AD patients who present no signs of Aβ accumulation in CSF [15]. Moreover, there are some reports with inconsistent normal CSF biomarkers pattern but pathological amyloid-targeting PET or vice versa [16, 17]. Recently, in a study with 40 AD cases, the number of biomarkers was consistent in only 32.5 % [18]. Thus the accumulation of Aß is necessary but not sufficient to produce the clinical symptoms of MCI and dementia.

Despite major advances in clinical chemistry, brain imaging, and genomic analysis, the lack of highly specific diagnostic tools for early diagnosis in AD research represented a challenge. There is an urgent need for the identification of accurate biosignatures for preclinical AD, which will differentiate subjects at risk for developing prodromal or AD from those without risk of progression to dementia.

3 Peripheral Antioxidant Levels-Based Biomarkers

Although CSF represents the most suitable biological fluid to study neurodegenerative diseases since it can reflect the biochemical changes occurring in brain, its analysis is not always easily feasible for a large scale screening, because the procedures are invasive, uncomfortable and that can be followed by several side effects including headache, back pain, nausea and vomiting and still carry a small risk of infection and damage to the spinal cord. Compared with CSF, blood or serum–based biomarkers are minimally invasive, could increase diagnostic accuracy, and could be useful for prognosis and in monitoring therapeutic interventions, especially for large scale studies and for repeated measures. One of the earliest blood-based biomarker studies were conducted by Ray et al. [19] with the identification of 18 plasma proteins among the 120 known signaling proteins. Overall, these studies suggest consistent alteration of the blood proteome in AD patients. There is a growing body of evidence, however, that preclinical AD may have a biosignature that can be deduced from peripheral blood [19–26].

Although the amyloid cascade hypothesis represents the underlying pathogenesis for the familial form of AD, increasing evidence indicates that oxidative stress plays a key role in the pathophysiology of AD [27, 28]. Strong evidence supports the role of oxidative damage in MCI and AD [29] and recent studies extensively demonstrated the specific oxidative modification of selected macromolecules in the brain of AD patients, in the CSF and blood stream [30, 31] and how their dysfunction possibly correlates with the pathology. These observations strengthen the notion that oxidative stress can precede the development of the neuropathological hallmarks of AD [32]. Therefore, oxidative stress markers have been studied in several matrices, in brain tissue, CSF and in blood compartment (for a review see [33–35]).

However, the *in vivo* evaluation of oxidative stress always remains a great challenge. Indeed, it is very difficult to detect reactive oxygen species (ROS) since their short-living time renders them not amenable for direct assay. Therefore, most assays to identify *in vivo* oxidative stress are based on indirect methods. They can be divided in four categories: analysis of small molecular weight non enzymatic and enzymatic antioxidants, determination of trace elements (selenium, copper, zinc) and evidence of oxidative damage to lipids, DNA and proteins and identification of sources producing ROS.

Although the determination of antioxidants levels and oxidative markers in the brain reflects the direct redox status and the oxidative metabolism in different cell types within the brain, they cannot be used as biomarkers. As described above, blood analysis is more accurate than CSF for population-based disease screening.

The major contributors to antioxidant capacity of serum are uric acid, plasma proteins (particularly albumin, transferrin and ceruloplasmin), and vitamins C, E, vitamin A/carotenoids and polyphenols but their peripheral levels are controversial.

Several studies have found a reduction (25 %) in uric acid and (53 %) in ascorbic acid in serum/plasma from AD [36–42] while others do not observe this difference [43, 44]. High plasma urate was also related to a slower rate of cognitive decline [45]. Vitamin E represents an important lipophilic antioxidant and has been investigated in relation to cognitive impairment. Most of studies have reported a decrease in plasma and serum vitamin E in AD [30, 39, 46–51]. More recently, plasma levels of tocopherols and tocotrienols together with automated MRI measures can help to differentiate AD and MCI patients from control subjects, and to prospectively predict MCI conversion into AD [52]. The AddNeuroMed-Project with in 168 AD cases, 166 MCI, and 187 cognitively normal people shows that plasma tocopherols, tocotrienols, α-tocopherylquinone, and 5-nitro-γ-tocopherol in AD and MCI had lower levels of total tocopherols, total tocotrienols, and total vitamin E [53]. The measurement of plasma levels of vitamin E (alpha-, beta-, gamma, and delta-tocopherol; alpha-, beta-, gamma-, and delta-tocotrienol) from 232 subjects aged 80 years and over, from the Kungsholmen Project, with a follow-up of 6 years indicates that high plasma levels of vitamin E are associated with a reduced risk of AD in advanced age [54]. However, vitamin E supplementation alone does not seem to improve the course of AD [55, 56] likely because the contribution of vitamin E to serum antioxidant capacity is quite small, accounting for less than 10 % [57, 58]. However, dietary intakes of the three antioxidants (vitamin E, vitamin C, and β-carotene) can lower the risk of AD, with vitamin E exhibiting the most pronounced protective effects [59, 60]. The CAIDE (Cardiovascular Risk Factors, Aging, and Dementia) study showed lower risk of cognitive impairment in subjects with higher levels of γ-tocopherol, β-tocotrienol, and total tocotrienols [61].

Vitamin C concentration in plasma or serum is also decreased in plasma or serum from AD [36, 37, 39, 49, 62] with lower level being observed especially in moderate and severe AD [62]. Plasma vitamin C was positively associated with MMSE score [63]. In the critical review of the role of vitamin C on the prevention of cognitive decline and AD, the authors conclude that maintaining healthy vitamin C levels can have a protective function against age-related cognitive decline and AD and avoiding vitamin C deficiency is likely to be more beneficial than taking supplements [64]. The analysis of the level of vitamin A on serum and plasma indicate a controversial effect with significant reduction in AD [37, 39, 47, 51] while five studies reported no difference [49, 65]. For reduced GSH, some studies shown no difference between AD and controls [44, 66, 67] while others found lower levels of reduced GSH with a balance towards GSSG in AD [30, 68–70].

Plasma thiols represent a measure of the exposure of the plasma to oxidative stress causing oxidation of the –SH groups to –S–S– disulphide linkages or to species such as sulphuric acids (–SO3–). Only proteins containing reactive –SH groups will be exposed to this particular process protein sulphydryl groups (–SH) also are indicative of altered redox status as well as altered capacity to maintain correct structures of the proteins during stress condition. Plasma thiol levels were not different between the AD group and the normal control group and no correlation was found between plasma thiol level and overall cognitive function. However, a very significant negative correlation was found between age and plasma thiol concentrations [71].

With respect to non enzymatic antioxidants, their concentration can, however, be largely modulated by the diet. Therefore, the AD patients have to be imperatively fasted during at least 6–12 h before the blood collection. Moreover, it is very important to use a food frequency questionnaire (FFQ) to check if the antioxidant decrease cannot be attributed to possible malnutrition in AD patients. As an example, we have demonstrated that a daily intake of less than two fruits was associated with low plasma levels in vitamin C and ß-carotene [72]. Unfortunately in most of clinical studies poor information is given about their nutritional status.

Large discrepancies have been reported for antioxidant enzymes involving superoxide dismutase enzyme (SOD) and glutathione peroxidase. SOD activity has been described in both plasma/serum and erythrocytes. In these cells, SOD is primarily SOD1 while in plasma/serum the main isoform is SOD3 or eSOD which uses Cu/Zn as a cofactor. In plasma, SOD3 level in AD is controversial with no difference between controls and MCI [73] or AD groups [40, 48, 74–76]. Lustig et al. were the first to demonstrate elevated levels of Cu, Zn SOD in erythrocytes of patients with AD [77, 78] but when considering overall studies there is no significant difference between AD and controls [30, 39, 66, 79–84]. In plasma/serum, most of studies found no significant difference [28, 39, 66, 85, 86]. These controversial findings may be due to the difference in sample size or different inclusion criteria. The activity of eSOD was measured in plasma and was found to be correlated with glucose metabolism in a large area of the left temporal lobe including the superior, middle and inferior temporal gyrus and the fusiform gyrus [87]. Overall, in plasma/serum, there was no significant association of SOD activity with AD across different studies.

The activity of glutathione peroxidase was also determined in plasma, serum and erythrocytes from control and AD patients and data remain controversial [30, 39, 44, 48, 66, 70, 79, 82, 88–92]. Glutathione peroxidase activity was higher in AD compared to MCI and healthy control patients while no difference was observed between controls and MCI [89]. Recently, low baseline glutathione peroxidase activity was associated with an average loss of 1.19 MMSE points per 6 months [93].

Oxidized GSH (GSSG) is reduced back to GSH by glutathione reductase (GR). In erythrocytes, its activity was lower in AD and in MCI as compared to control patients but it is not different between MCI and AD [89]. In blood compartment, GR activity was decreased in AD as compared to control healthy patients [91].

Regarding catalase enzyme, its activity was conducted in erythrocytes in seven studies and overall, no significant variation was observed [30, 79, 89, 94, 95].

Some studies have also investigated the total antioxidant capacity (TAC) of blood from AD. The most popular of them being the Oxygen Radical Antioxidant Capacity (ORAC assay) and the Total Antioxidant Status (TAS assay) as the reflect of the balance between antioxidants and oxidants or the redox status [96, 97]. The principle is to generate *in vitro* an artificial source of oxidants and to evaluate the ability of a plasma sample to quench oxidative reactions using suitable colorimetric or fluorimetric probes. Decrease of TAS has been regularly described in peripheral fluids from AD patients [33, 34]. According to Wayner et al. [57], the contribution of plasma antioxidants to total ORAC value is, in decreasing order, protein, uric acid, vitamin C, vitamin E, bilirubin, and carotenoids. Moreover, the proteins, and more particularly albumin, account for 80–95 % of the antioxidant capacity of plasma. Because of this high contribution of proteins, results of the whole plasma are hard to interpret due to the many variables such as nutrition status (very variable in older persons), osmotic pressure, metabolism, catabolism and extracellular body fluid status. Moreover, it is well known that the ORAC value is strongly correlated with the plasma concentration of uric acid [57, 88], the value of which can be modulated by some diseases such as gout or renal dysfunction. At least, the ORAC value can also be influenced by some drugs known that have antioxidant capacity. For all these reasons, the determination of plasma ORAC value or TAS assay should also be considered with great caution in order to evidence the presence of an *in vivo* oxidative stress in pathologies such as AD [98].

4 Peripheral Oxidized Macromolecules-Based Biomarkers

As lipids (polyunsatured fatty acids or PUFA's), proteins and DNA are the primary targets for ROS, the detection of oxidized biological material in plasma or urine is another tool for evidencing *in vivo* oxidative stress. Common markers of protein oxidation are indexed by protein carbonyls, 3-nitrotyrosine and protein glutathionylation [99], lipid peroxidation indexed by thiobarbituric acid-reactive substances (TBARS), free fatty acid release, iso- and neuro-prostane formation, 2-propen-1-al (acrolein), and 4-hydroxy-2-trans-nonenal (HNE) [100] and DNA oxidation by 8-hydroxy-2-deoxyguanosine (8-OHdG) [37].

For nucleic acids, 8-hydroxy-2-deoxyguanosine (8-OHdG) is one of the most commonly used markers of oxidative nucleic acid damage and can be measured in lymphocytes, leukocytes, and the brain.

Most studies found higher levels of 8-OHdG in AD [37, 101]. Protein carbonyl groups are generated by direct oxidation of several amino acid side chains (i.e., Lys, Arg, Pro, Thr, His and others). His, Lys, and Cys residues could also react by

Michael addition reactions by hydrogen atom abstraction at alpha carbons with products of lipid peroxidation [102]. Protein carbonyls are also produced by glycation/glycoxidation of Lys or arginine amino groups, forming advanced glycation end products (AGEs) [103].

Protein carbonyls are chemically stable compared with the other products of oxidative stress, they are generally used as markers to determine the extent of oxidative modification both *in vivo* and *in vitro* conditions. The most common way to measure carbonylated proteins is the DNPH-based detection method, where samples are derivatized with 2,4-dinitrophenylhydrazine (DNPH), and are immunochemically detected with an antibody against the resulting protein hydrazone adduct; however spectrophotometric and HPLC quantitation of the DNPH adduct also could be used. Cysteine, methionine, phenylalanine, and tyrosine residues of proteins are particularly susceptible to reactive nitrogen species (RNS). Thus, protein nitration is another widely recognized marker of protein oxidation and numerous studies support the idea that nitrosative stress contributes to neurodegeneration in AD [104]. Several methods of analysis are available to measure levels of protein nitration including chemical analysis using HPLC and GC coupled to a mass spectrometer, the immunochemical detection of 3-nitrotyrosine. From a technical point of view, the procedure is time analysis with a great risk of auto-oxidation of proteins in the biological sample.

Protein carbonyls and nitrated protein can be measured in plasma, serum, CSF, and brain tissue. Conrad et al. were the first to measure the levels of oxidized proteins in plasma from AD patients compared to controls. They found that a significant increase of total carbonyl groups in plasma proteins with specific oxidation of a 78 kDa protein in subject with AD [105]. Several studies have described higher serum/plasma protein carbonyls level in AD despite varying patient selection criteria [28, 30, 31, 69, 105–107] while a few showed no significant difference between AD and controls [106, 108]. However, oxidation of low density lipoproteins is considerably increased (increased by 55 % compared to controls across five studies), reflecting the pro-oxidative environment in the lipid fraction rather than a pattern of protein-targeted oxidation [108, 109].

Protein glutathionylation is the formation of a mixed disulfide between protein cysteinyl residues and a small-molecular-weight thiol [110]. Glutathionylation can protect cysteine thiols against irreversible oxidation but can also modify the activity of proteins. Glutathionylation could represent an important redox signaling mechanism, allowing cells to sense and signal harmful stress conditions and activate appropriate responses [111]. Peripheral levels of S-glutathionylated protein in MCI and AD remain to be determined.

Non-enzymatic glycosylation is a common post translational modification of proteins *in vivo*, resulting from reactions between reducing sugar and amino groups on proteins; this process is called the "Maillard reaction" and results in the formation of AGEs. Measurement of AGEs can be performed by ELISA or by quantitative fluorescence spectroscopy [112]. In AD brain, AGEs have been shown to be colocalized with neurofibrillary tangles, senile plaques, microglia, and astrocytes but have been also measured in plasma.

In plasma, the Aβ peptide can reenter the CNS *via* the receptor for advanced glycation end products (RAGE) [113]. The soluble isoform of RAGE, sRAGE, lacks the transmembrane domain, and is present in plasma, binds Aβ in plasma and prevent the neurotoxic or proinflammatory responses of RAGE–Aβ interaction [114] or intracellular oxidative stress [115]. Levels of sRAGE were not different between AD and non-AD neurodegenerative dementia [116].

Advanced oxidation protein products are defined as dityrosine-containing cross-linked protein products, and are also considered as reliable markers to estimate the degree of oxidant-mediated protein damage [117]. They are formed during oxidative stress by the action of chlorinated oxidants, mainly hypochlorous acid and chloramines (produced by myeloperoxidase in activated neutrophils). They are supposed to be structurally similar to AGE-proteins and to exert similar biological activities as AGEs, The levels of AOPP (advanced oxidation protein products) were found to be significantly higher in plasma from AD patients than in the control group [118]. High AOPP levels are indicative of oxidative stress.

The reaction of PUFA's with ROS leads to the formation of lipoperoxides (lipid peroxidation) which can break down in a variety of secondary products that are aldehydes (malondialdehydes, MDA) usually expressed as TBARS, conjugated diene hydroperoxides, F2, F4-isoprostanes, 4-hydroxynonenal and acrolein [100]. It causes changes in the fluidity and permeability of cell membranes, impairs the activity of membrane-bound enzymes and modified the activity of neuronal survival and redox pathways [119, 120]. Results of TBARS on serum, plasma or blood cells from MCI [89, 90] and AD [28, 30, 38, 48, 66–68, 75, 76, 79, 86, 89, 90, 108, 121] are controversial and the link between the TBARS levels and the MMSE score is lacking.

The detection of MDA with the thiobarbituric test (TBAR's assay) was incontestably the favorite assay for most lipid peroxidation studies in AD patients [35]. However, the TBAR test is a great matter of controversy due to its low specificity and sensibility. Indeed, a large number of substances not related to lipid peroxidation such as glucose or hemoglobin strongly interfere with TBAR leading to the overestimation in addition to an extreme wide variation of the MDA levels [122]. Actually, most investigators recognized that it is scientifically unsound to equate increased plasma or serum TBARS levels (as often observed in AD patients) with the occurrence of a free radical disease. Instead of the TBAR's assay, it has been proposed to evaluate in plasma the "free" MDA by using high pressure liquid chromatography (HPLC). Even if it is more specific, we have to keep in mind that MDA is just a by—product of lipid peroxides and that it only reflects less than 1 % of the lipid peroxidation phenomenon. For all these reasons, the determination of MDA should be discouraged as a marker of oxidative stress in AD patients.

Since a few years, the measurement of isoprostanes has clearly emerged as "*gold standard*" reliable marker of *in vivo* lipid peroxidation [123]. Numerous studies demonstrated that isoprostane levels are elevated in plasma or urine AD patients and in animal models of the disease [124]. In a very nice work, Pratico et al. [125] found higher isoprostane levels in cerebrospinal fluid (CSF), plasma and urine of subjects with mild cognitive impairment (MCI) compared with cognitively normal elderly subjects.

Interestingly, isoprostane levels were more elevated in AD patients than in subjects having MCI. As AD is preceded by a prodromal phase characterized by MCI, these authors assume that the measurement of isoprostanes could be, therefore, a possible predictor for AD. Of great interest is also this recent observation showing that mice receiving isoprostane showed significant memory deficits, increase in tau phosphorylation, activation of the cyclin kinase 5 pathway and neuroinflammation [126]. Actually, the method of choice for measuring isoprostanes in biological fluid is the gas or liquid chromatography coupled with mass spectrometry (GC-MS and LC-MS).

4-hydroxynonenal (4-HNE), another marker of lipid peroxidation in serum was found to be markedly elevated in AD compared to controls [127, 128]. However, its application in a routine way is not easy because that requires a long purification procedure of the sample (chromatographic steps, extraction) that is critical and time-consuming with substantial loss of target compounds. To limit these inconveniences, several authors have proposed to evaluate isoprostanes in exhaled breath condensate (EBC) samples but such an interesting technology is far to be validated [129]. Some companies have also developed immunoassay methods for routinely use but they are still lacking in specificity [130]. 4-HNE can react with lysine, histidine, and cysteine residues in proteins to form Michael adducts and Schiff base products.

Acrolein is an α,β-unsaturated aldehyde with two functional groups that can participate in chemical reactions: the aldehyde group and the carbon-carbon double bond. Acrolein is the strongest electrophile among the unsaturated aldehydes and therefore displays strong reactivity with nucleophilic compounds and can form adducts with amino acids residues such as lysine, histidine and cysteine (see review by [131]). Several studies have demonstrated that acrolein levels are elevated in brain tissue such as in hippocampus from AD or MCI patients (see review by [100]). We have demonstrated that acrolein can also induced the depletion of glutathion, the generation of 4-HNE and protein carbonylation [119, 120]. Recently, we have demonstrated that the administration of acrolein could induce AD-like pathologies *in vitro* and *in vivo* [132].

5 Peripheral Biomarkers: Limits and Prospects

Currently, there are no fully validated blood-based biomarkers of AD. The development of a blood- based biomarker for AD represents a great challenge because blood profile represents an accumulation of the alterations in all tissues, and for non brain-specific markers, there is a substantial contribution that is not related to brain pathology. Also, blood comprises a liquid component (serum or plasma) and different types of cells, including the mononuclear leukocytes, which contain the deoxyribonucleic acid (DNA) and ribonucleic acid (RNA), erythrocytes, and platelets. Each blood component has its advantages for delineation of different biomarker species. Therefore, blood-based biomarkers are less specific to the brain activity compared to CSF biosignatures.

For instance, increase in tau and Aβ ratio in CSF from AD patients and from MCI subjects are considered as reliable biomarkers of AD [10]. However, the determination of peripheral Aβ and tau levels should be interpreted with caution when used for diagnosis because peripheral Aß is from nonneuronal tissues and, more importantly, Aβ is bound to a variety of proteins in blood [133]. In serum/plasma, using classic systems, tau is virtually undetectable in MCI and/or AD but plasma tau levels are elevated in a series of pathologies, such as ischemic stroke, Creutzfeldt-Jacobs disease [134, 135], and traumatic brain injury [136]. Consequently, to have strong impact, the identification of blood-based biomarker profiles should be combined with clinical informations.

Another challenge is the extent of the loss of BBB integrity remains unknown because in addition to reducing entry into the brain, the BBB also reduces exit of molecules from the brain [137]. However, in AD, there is some degree of loss of integrity of the BBB, potentially allowing the crossing of additional molecules into the blood [138].

The progression of the disease with different preclinical and clinical stages is another challenge to identify biomarker. This means that an optimal biosignature for one stage of preclinical disease may not be best for other stages. A single biomarker may not be sensitive to the underlying pathology at multiple stages of the preclinical disease.

Another challenging factor for studies biomarkers in AD as for other diseases is the validation of healthy control subjects for comparison with AD patients who, by their advancing age, are also at risk for AD and may have underlying pathology without clinical signs. The variation of antioxidants or oxidative markers in reference group consisted of *healthy* populations could be an issue [139] due to non-standardized criteria (medication, diet, physical activity...).

6 Importance of the Pre-analytical Procedure and Standardisation of Methods

The most important issues are the adoption of standardized protocols during sample collection and other pre-analytical processes, method validation including establishing prospective quality control protocols. Up to now, the determination for reliable biomarkers for AD in peripheral blood is very challenging due to difficulties of the standardization of the methods of analysis. To validate plasma or serum biomarkers, assays should be run using validated and standardized protocols because the determination of all oxidative stress biomarkers requires a rigorous pre-analytical procedure in order to avoid artefacts. The blood sample must be drawn very carefully in order to avoid haemolysis which can interfere in some assay (e.g. determination of oxidized LDL). Once drawn, the blood has to be immediately centrifuged at 3000 rpm during 15 min if possible at 4 °C. Indeed, the concentration of some antioxidants such as vitamin C, β-carotene and reduced glutathione may rapidly decrease within 30 min if the whole blood is kept at room temperature before centrifugation. For both vitamin C, isoprostanes and oxidized glutathione, it is also necessary to mix the plasma respectively with metaphosphoric acid, butylhydroxy-toluene (BHT) and N-ethylmaleimide in order to prevent spontaneous oxidation of the markers in the tube. At least, plasma or serum must be immediately frozen on dry ice and then kept at −80 °C until analysis which must be performed as soon as possible. Lack of standardization of procedures for blood collection, processing, and storage (time in storage, number of subsequent freeze-thaw cycles) leads to preanalytical variability (incubation times before separating plasma (or serum) from cells, type of blood collection tubes, temperature for collection and storage) causing inconsistencies in downstream analysis and results [140, 141].

In 2014, the international working group of the Standards for Alzheimer's Research in Blood biomarkers (STAR-B) and Blood-Based Biomarker Interest Group (BBBIG) have published the first set of guidelines in order to standardize preanalytical variables for blood-based biomarker studies [142, 143]. The principle of the guidelines for preanalytical methods for blood-based AD biomarker by BBBIG/STAR-B working group follows the regulatory good laboratory practice as defined by Clinical Laboratory Improvement Amendment in United States or international standards of Clinical Laboratory Standards Initiative. For the progress of blood-based biomarker studies in AD, the adoption of guidelines to standardize preanalytical methods across cohorts and laboratories is required. Therefore, the BBBIG/STAR-B guidelines are a good starting point toward standardized methods that will be essential to move putative blood-based biomarkers forward in future studies.

The downstream analytical methods should be validated on the same matrix (e.g., serum, plasma, urine…). Analytical validation should include calibration curves, intra- and interprecision and accuracy. Serum and plasma contain high amounts of albumin and immunoglobulins, successful protein biomarker discovery requires enrichment techniques for low abundance proteins and sensitive technological platforms to facilitate detection and quantification of protein biomarkers. Sample size for controls and AD groups should be large enough. The identification of new biomarkers should be validated, replicated and compared with existing gold standards and validated in larger studies and validated in different groups. Demographic information (age, gender, education…), APOE status should be included. Further complications in the interpretation of plasma profiles then arise from the use of different types of medication for patients with AD. For instance, a study with limited number of patients (10 controls and 21 AD) showed that the medication including Rivastigmine and Donepezil-Memantin could modify the pattern of oxidative markers [144]. Patients with AD have impaired systemic availability of several nutrients which could impact on peripheral antioxidant levels. Finally, the potential issue is the variability in results depending on the statistical or computational methods used to identify these biomarkers.

7 Conclusion

AD pathology could begin approximately 10–15 years before the appearance of cognitive impairment, thus, the identification of biomarkers that can detect AD pathology in its early stages would be valuable for monitoring the efficacy of disease interventions during clinical trials. In this regards, there is an intensive search for novel biomarker candidates using a plethora of approaches. Current CSF biomarkers and neuroimaging data are accurate for disease detection. However, for large scale studies and for repeated measures, peripheral biomarkers present more advantages. Thus, there is a concerted research effort toward the development of biomarker panels using blood as matrix.

Alterations of peripheral antioxidants balance and elevation of by-product of oxidation were evidenced in numerous studies with controversial results in AD and in mild cognitive patients (MCI) in both the brain and the blood stream.

More longitudinal and cross-validation studies with larger and more diverse ethnic cohorts are required to validate biomarker panels that could be used clinically.

Finally, the criteria used for the selection of controls, the standardisation of the pre-analytical procedure, and of the downstream of the analytical methods are necessary.

Support: Chaire Louise & André Charron Fondation universitaire sur la maladie d'Alzheimer,

INRS-Armand-Frappier, INAF, Univ.Laval

References

1. Thies W, Bleiler L. 2013 Alzheimer's disease facts and figures. Alzheimers Dement. 2013; 9:208–45.
2. Fiandaca MS, Mapstone ME, Cheema AK, Federoff HJ. The critical need for defining preclinical biomarkers in Alzheimer's disease. Alzheimers Dement. 2014;10:S196–212.
3. Stefanacci RG. The costs of Alzheimer's disease and the value of effective therapies. Am J Manag Care. 2011;17 Suppl 13:S356–62.
4. Sperling RA, Aisen PS, Beckett LA, Bennett DA, Craft S, Fagan AM, Iwatsubo T, Jack Jr CR, Kaye J, Montine TJ, et al. Toward defining the preclinical stages of Alzheimer's disease: recommendations from the National Institute on Aging-Alzheimer's Association workgroups on diagnostic guidelines for Alzheimer's disease. Alzheimers Dement. 2011;7:280–92.
5. Savva GM, Wharton SB, Ince PG, Forster G, Matthews FE, Brayne C. Age, neuropathology, and dementia. N Engl J Med. 2009;360:2302–9.
6. Morris JC, Cummings J. Mild cognitive impairment (MCI) represents early-stage Alzheimer's disease. J Alzheimers Dis. 2005;7:235–9. discussion 255–262.
7. Storandt M, Grant EA, Miller JP, Morris JC. Rates of progression in mild cognitive impairment and early Alzheimer's disease. Neurology. 2002;59:1034–41.

8. Jack Jr CR, Knopman DS, Jagust WJ, Shaw LM, Aisen PS, Weiner MW, Petersen RC, Trojanowski JQ. Hypothetical model of dynamic biomarkers of the Alzheimer's pathological cascade. Lancet Neurol. 2010;9:119–28.
9. Petersen RC, Aisen PS, Beckett LA, Donohue MC, Gamst AC, Harvey DJ, Jack Jr CR, Jagust WJ, Shaw LM, Toga AW, et al. Alzheimer's Disease Neuroimaging Initiative (ADNI): clinical characterization. Neurology. 2010;74:201–9.
10. Andreasen N, Hesse C, Davidsson P, Minthon L, Wallin A, Winblad B, Vanderstichele H, Vanmechelen E, Blennow K. Cerebrospinal fluid beta-amyloid(1-42) in Alzheimer disease: differences between early- and late-onset alzheimer disease and stability during the course of disease. Arch Neurol. 1999;56:673–80.
11. Roe CM, Fagan AM, Williams MM, Ghoshal N, Aeschleman M, Grant EA, Marcus DS, Mintun MA, Holtzman DM, Morris JC. Improving CSF biomarker accuracy in predicting prevalent and incident Alzheimer disease. Neurology. 2011;76:501–10.
12. Apostolova LG, Hwang KS, Andrawis JP, Green AE, Babakchanian S, Morra JH, Cummings JL, Toga AW, Trojanowski JQ, Shaw LM, et al. 3D PIB and CSF biomarker associations with hippocampal atrophy in ADNI subjects. Neurobiol Aging. 2010;31:1284–303.
13. Kim S, Swaminathan S, Shen L, Risacher SL, Nho K, Foroud T, Shaw LM, Trojanowski JQ, Potkin SG, Huentelman MJ, et al. Genome-wide association study of CSF biomarkers Abeta1-42, t-tau, and p-tau181p in the ADNI cohort. Neurology. 2011;76:69–79.
14. Albert MS, DeKosky ST, Dickson D, Dubois B, Feldman HH, Fox NC, Gamst A, Holtzman DM, Jagust WJ, Petersen RC, et al. The diagnosis of mild cognitive impairment due to Alzheimer's disease: recommendations from the National Institute on Aging-Alzheimer's Association workgroups on diagnostic guidelines for Alzheimer's disease. Alzheimers Dement. 2011;7:270–9.
15. Yang L, Rieves D, Ganley C. Brain amyloid imaging—FDA approval of florbetapir F18 injection. N Engl J Med. 2012;367:885–7.
16. Landau SM, Lu M, Joshi AD, Pontecorvo M, Mintun MA, Trojanowski JQ, Shaw LM, Jagust WJ. Comparing positron emission tomography imaging and cerebrospinal fluid measurements of beta-amyloid. Ann Neurol. 2013;74:826–36.
17. Hellwig S, Hull M, Meyer PT. Negative amyloid PET imaging in a patient with probable Alzheimer's disease. Clin Nucl Med. 2011;36:1120–1.
18. Weise D, Tiepoltc S, Awissusa C, Hoffmannd K-T, Lobsiend D, Kaisere T, Barthelc H, Sabric O, Gertza HJ. Critical comparison of different biomarkers for Alzheimer's disease in a clinical setting. J Alzheimers Dis. 2015;48:425–32.
19. Ray S, Britschgi M, Herbert C, Takeda-Uchimura Y, Boxer A, Blennow K, Friedman LF, Galasko DR, Jutel M, Karydas A, et al. Classification and prediction of clinical Alzheimer's diagnosis based on plasma signaling proteins. Nat Med. 2007;13:1359–62.
20. Thambisetty M, Simmons A, Velayudhan L, Hye A, Campbell J, Zhang Y, Wahlund LO, Westman E, Kinsey A, Guntert A, et al. Association of plasma clusterin concentration with severity, pathology, and progression in Alzheimer disease. Arch Gen Psychiatry. 2010;67:739–48.
21. O'Bryant SE, Xiao G, Barber R, Reisch J, Doody R, Fairchild T, Adams P, Waring S, Diaz-Arrastia R. A serum protein-based algorithm for the detection of Alzheimer disease. Arch Neurol. 2010;67:1077–81.
22. Fehlbaum-Beurdeley P, Sol O, Desire L, Touchon J, Dantoine T, Vercelletto M, Gabelle A, Jarrige AC, Haddad R, Lemarie JC, et al. Validation of AclarusDx, a blood-based transcriptomic signature for the diagnosis of Alzheimer's disease. J Alzheimers Dis. 2012;32:169–81.
23. Sato Y, Suzuki I, Nakamura T, Bernier F, Aoshima K, Oda Y. Identification of a new plasma biomarker of Alzheimer's disease using metabolomics technology. J Lipid Res. 2012;53:567–76.
24. Kiddle SJ, Sattlecker M, Proitsi P, Simmons A, Westman E, Bazenet C, Nelson SK, Williams S, Hodges A, Johnston C, et al. Candidate blood proteome markers of Alzheimer's disease onset and progression: a systematic review and replication study. J Alzheimers Dis. 2014;38:515–31.

25. Tan L, Yu JT, Liu QY, Tan MS, Zhang W, Hu N, Wang YL, Sun L, Jiang T. Circulating miR-125b as a biomarker of Alzheimer's disease. J Neurol Sci. 2014;336:52–6.
26. Leidinger P, Backes C, Deutscher S, Schmitt K, Mueller SC, Frese K, Haas J, Ruprecht K, Paul F, Stahler C, et al. A blood based 12-miRNA signature of Alzheimer disease patients. Genome Biol. 2013;14:R78.
27. Guidi I, Galimberti D, Lonati S, Novembrino C, Bamonti F, Tiriticco M, Fenoglio C, Venturelli E, Baron P, Bresolin N, et al. Oxidative imbalance in patients with mild cognitive impairment and Alzheimer's disease. Neurobiol Aging. 2006;27:262–9.
28. Puertas MC, Martinez-Martos JM, Cobo MP, Carrera MP, Mayas MD, Ramirez-Exposito MJ. Plasma oxidative stress parameters in men and women with early stage Alzheimer type dementia. Exp Gerontol. 2012;47:625–30.
29. Sultana R, Butterfield DA. Role of oxidative stress in the progression of Alzheimer's disease. J Alzheimers Dis. 2010;19:341–53.
30. Cristalli DO, Arnal N, Marra FA, de Alaniz MJ, Marra CA. Peripheral markers in neurodegenerative patients and their first-degree relatives. J Neurol Sci. 2012;314:48–56.
31. Korolainen MA, Pirttila T. Cerebrospinal fluid, serum and plasma protein oxidation in Alzheimer's disease. Acta Neurol Scand. 2009;119:32–8.
32. Nunomura A, Perry G, Aliev G, Hirai K, Takeda A, Balraj EK, Jones PK, Ghanbari H, Wataya T, Shimohama S, et al. Oxidative damage is the earliest event in Alzheimer disease. J Neuropathol Exp Neurol. 2001;60:759–67.
33. Skoumalova A, Hort J. Blood markers of oxidative stress in Alzheimer's disease. J Cell Mol Med. 2012;16:2291–300.
34. Chang YT, Chang WN, Tsai NW, Huang CC, Kung CT, Su YJ, Lin WC, Cheng BC, Su CM, Chiang YF, et al. The roles of biomarkers of oxidative stress and antioxidant in Alzheimer's disease: a systematic review. Biomed Res Int. 2014;2014:182303.
35. Schrag M, Mueller C, Zabel M, Crofton A, Kirsch WM, Ghribi O, Squitti R, Perry G. Oxidative stress in blood in Alzheimer's disease and mild cognitive impairment: a meta-analysis. Neurobiol Dis. 2013;59:100–10.
36. Glaso M, Nordbo G, Diep L, Bohmer T. Reduced concentrations of several vitamins in normal weight patients with late-onset dementia of the Alzheimer type without vascular disease. J Nutr Health Aging. 2004;8:407–13.
37. Mecocci P, Polidori MC, Cherubini A, Ingegni T, Mattioli P, Catani M, Rinaldi P, Cecchetti R, Stahl W, Senin U, et al. Lymphocyte oxidative DNA damage and plasma antioxidants in Alzheimer disease. Arch Neurol. 2002;59:794–8.
38. Polidori MC, Mattioli P, Aldred S, Cecchetti R, Stahl W, Griffiths H, Senin U, Sies H, Mecocci P. Plasma antioxidant status, immunoglobulin g oxidation and lipid peroxidation in demented patients: relevance to Alzheimer disease and vascular dementia. Dement Geriatr Cogn Disord. 2004;18:265–70.
39. Rinaldi P, Polidori MC, Metastasio A, Mariani E, Mattioli P, Cherubini A, Catani M, Cecchetti R, Senin U, Mecocci P. Plasma antioxidants are similarly depleted in mild cognitive impairment and in Alzheimer's disease. Neurobiol Aging. 2003;24:915–9.
40. Zafrilla P, Mulero J, Xandri JM, Santo E, Caravaca G, Morillas JM. Oxidative stress in Alzheimer patients in different stages of the disease. Curr Med Chem. 2006;13:1075–83.
41. Kim TS, Pae CU, Yoon SJ, Jang WY, Lee NJ, Kim JJ, Lee SJ, Lee C, Paik IH, Lee CU. Decreased plasma antioxidants in patients with Alzheimer's disease. Int J Geriatr Psychiatry. 2006;21:344–8.
42. Maesaka JK, Wolf-Klein G, Piccione JM, Ma CM. Hypouricemia, abnormal renal tubular urate transport, and plasma natriuretic factor(s) in patients with Alzheimer's disease. J Am Geriatr Soc. 1993;41:501–6.
43. Ruggiero C, Cherubini A, Lauretani F, Bandinelli S, Maggio M, Di Iorio A, Zuliani G, Dragonas C, Senin U, Ferrucci L. Uric acid and dementia in community-dwelling older persons. Dement Geriatr Cogn Disord. 2009;27:382–9.
44. Fernandes MA, Proenca MT, Nogueira AJ, Grazina MM, Oliveira LM, Fernandes AI, Santiago B, Santana I, Oliveira CR. Influence of apolipoprotein E genotype on blood redox status of Alzheimer's disease patients. Int J Mol Med. 1999;4:179–86.

45. Irizarry MC, Raman R, Schwarzschild MA, Becerra LM, Thomas RG, Peterson RC, Ascherio A, Aisen PS. Plasma urate and progression of mild cognitive impairment. Neurodegener Dis. 2009;6:23–8.
46. Sinclair AJ, Bayer AJ, Johnston J, Warner C, Maxwell SR. Altered plasma antioxidant status in subjects with Alzheimer's disease and vascular dementia. Int J Geriatr Psychiatry. 1998;13:840–5.
47. Foy CJ, Passmore AP, Vahidassr MD, Young IS, Lawson JT. Plasma chain-breaking antioxidants in Alzheimer's disease, vascular dementia and Parkinson's disease. QJM. 1999;92: 39–45.
48. Bourdel-Marchasson I, Delmas-Beauvieux MC, Peuchant E, Richard-Harston S, Decamps A, Reignier B, Emeriau JP, Rainfray M. Antioxidant defences and oxidative stress markers in erythrocytes and plasma from normally nourished elderly Alzheimer patients. Age Ageing. 2001;30:235–41.
49. Polidori MC, Mecocci P. Plasma susceptibility to free radical-induced antioxidant consumption and lipid peroxidation is increased in very old subjects with Alzheimer disease. J Alzheimers Dis. 2002;4:517–22.
50. Iuliano L, Monticolo R, Straface G, Spoletini I, Gianni W, Caltagirone C, Bossu P, Spalletta G. Vitamin E and enzymatic/oxidative stress-driven oxysterols in amnestic mild cognitive impairment subtypes and Alzheimer's disease. J Alzheimers Dis. 2010;21:1383–92.
51. Zaman Z, Roche S, Fielden P, Frost PG, Niriella DC, Cayley AC. Plasma concentrations of vitamins A and E and carotenoids in Alzheimer's disease. Age Ageing. 1992;21:91–4.
52. Mangialasche F, Westman E, Kivipelto M, Muehlboeck JS, Cecchetti R, Baglioni M, Tarducci R, Gobbi G, Floridi P, Soininen H, et al. Classification and prediction of clinical diagnosis of Alzheimer's disease based on MRI and plasma measures of alpha-/gamma-tocotrienols and gamma-tocopherol. J Intern Med. 2013;273:602–21.
53. Mangialasche F, Xu W, Kivipelto M, Costanzi E, Ercolani S, Pigliautile M, Cecchetti R, Baglioni M, Simmons A, Soininen H, et al. Tocopherols and tocotrienols plasma levels are associated with cognitive impairment. Neurobiol Aging. 2012;33:2282–90.
54. Mangialasche F, Kivipelto M, Mecocci P, Rizzuto D, Palmer K, Winblad B, Fratiglioni L. High plasma levels of vitamin E forms and reduced Alzheimer's disease risk in advanced age. J Alzheimers Dis. 2010;20:1029–37.
55. Sano M, Ernesto C, Thomas RG, Klauber MR, Schafer K, Grundman M, Woodbury P, Growdon J, Cotman CW, Pfeiffer E, et al. A controlled trial of selegiline, alpha-tocopherol, or both as treatment for Alzheimer's disease. The Alzheimer's Disease Cooperative Study. N Engl J Med. 1997;336:1216–22.
56. Arlt S, Muller-Thomsen T, Beisiegel U, Kontush A. Effect of one-year vitamin C- and E-supplementation on cerebrospinal fluid oxidation parameters and clinical course in Alzheimer's disease. Neurochem Res. 2012;37:2706–14.
57. Wayner DD, Burton GW, Ingold KU, Locke S. Quantitative measurement of the total, peroxyl radical-trapping antioxidant capability of human blood plasma by controlled peroxidation. The important contribution made by plasma proteins. FEBS Lett. 1985;187:33–7.
58. Wayner DD, Burton GW, Ingold KU, Barclay LR, Locke SJ. The relative contributions of vitamin E, urate, ascorbate and proteins to the total peroxyl radical-trapping antioxidant activity of human blood plasma. Biochim Biophys Acta. 1987;924:408–19.
59. Li FJ, Shen L, Ji HF. Dietary intakes of vitamin E, vitamin C, and beta-carotene and risk of Alzheimer's disease: a meta-analysis. J Alzheimers Dis. 2012;31:253–8.
60. Devore EE, Kang JH, Stampfer MJ, Grodstein F. Total antioxidant capacity of diet in relation to cognitive function and decline. Am J Clin Nutr. 2010;92:1157–64.
61. Mangialasche F, Solomon A, Kareholt I, Hooshmand B, Cecchetti R, Fratiglioni L, Soininen H, Laatikainen T, Mecocci P, Kivipelto M. Serum levels of vitamin E forms and risk of cognitive impairment in a Finnish cohort of older adults. Exp Gerontol. 2013;48:1428–35.
62. Riviere S, Birlouez-Aragon I, Nourhashemi F, Vellas B. Low plasma vitamin C in Alzheimer patients despite an adequate diet. Int J Geriatr Psychiatry. 1998;13:749–54.

63. Charlton KE, Rabinowitz TL, Geffen LN, Dhansay MA. Lowered plasma vitamin C, but not vitamin E, concentrations in dementia patients. J Nutr Health Aging. 2004;8:99–107.
64. Harrison FE. A critical review of vitamin C for the prevention of age-related cognitive decline and Alzheimer's disease. J Alzheimers Dis. 2012;29:711–26.
65. Engelhart MJ, Ruitenberg A, Meijer J, Kiliaan A, van Swieten JC, Hofman A, Witteman JC, Breteler MM. Plasma levels of antioxidants are not associated with Alzheimer's disease or cognitive decline. Dement Geriatr Cogn Disord. 2005;19:134–9.
66. Serra JA, Dominguez RO, Marschoff ER, Guareschi EM, Famulari AL, Boveris A. Systemic oxidative stress associated with the neurological diseases of aging. Neurochem Res. 2009; 34:2122–32.
67. Aybek H, Ercan F, Aslan D, Sahiner T. Determination of malondialdehyde, reduced glutathione levels and APOE4 allele frequency in late-onset alzheimer's disease in Denizli, Turkey. Clin Biochem. 2007;40:172–6.
68. Gironi M, Bianchi A, Russo A, Alberoni M, Ceresa L, Angelini A, Cursano C, Mariani E, Nemni R, Kullmann C, et al. Oxidative imbalance in different neurodegenerative diseases with memory impairment. Neurodegener Dis. 2011;8:129–37.
69. Bermejo P, Martin-Aragon S, Benedi J, Susin C, Felici E, Gil P, Ribera JM, Villar AM. Peripheral levels of glutathione and protein oxidation as markers in the development of Alzheimer's disease from Mild Cognitive Impairment. Free Radic Res. 2008;42:162–70.
70. Baldeiras I, Santana I, Proenca MT, Garrucho MH, Pascoal R, Rodrigues A, Duro D, Oliveira CR. Peripheral oxidative damage in mild cognitive impairment and mild Alzheimer's disease. J Alzheimers Dis. 2008;15:117–28.
71. Welsh SW, Corrigan FM, Smith WE, Horrobin DF. Plasma thiols, cognitive function and ageing. Clin Chim Acta. 1996;249:183–7.
72. Pincemail J, Degrune F, Voussure S, Malherbe C, Paquot N, Defraigne JO. Effect of a diet rich in fruits and vegetables on the plasmatic antioxidant rates and of the markers of the oxidative damage. Nutrition Clinique Et Metabolisme. 2007;21:66–75.
73. Chico L, Simoncini C, Lo Gerfo A, Rocchi A, Petrozzi L, Carlesi C, Volpi L, Tognoni G, Siciliano G, Bonuccelli U. Oxidative stress and APO E polymorphisms in Alzheimer's disease and in mild cognitive impairment. Free Radic Res. 2013;47:569–76.
74. Tabet N, Mantle D, Walker Z, Orrell M. Vitamins, trace elements, and antioxidant status in dementia disorders. Int Psychogeriatr. 2001;13:265–75.
75. Fernandes MA, Oliveira CR, Oliveira LM, Nogueira AJ, Santiago B, Santana I. Apolipoprotein E epsilon 4 allele is a risk factor for Alzheimer's disease: the central region of Portugal (Coimbra) as a case study. Eur Neurol. 1999;42:183–4.
76. Ceballos-Picot I, Merad-Boudia M, Nicole A, Thevenin M, Hellier G, Legrain S, Berr C. Peripheral antioxidant enzyme activities and selenium in elderly subjects and in dementia. Free Radic Biol Med. 1996;20:579–87.
77. de Lustig ES, Serra JA, Kohan S, Canziani GA, Famulari AL, Dominguez RO. Copper-zinc superoxide dismutase activity in red blood cells and serum in demented patients and in aging. J Neurol Sci. 1993;115:18–25.
78. de Lustig ES, Kohan S, Famulari AL, Dominguez RO, Serra JA. Peripheral markers and diagnostic criteria in Alzheimer's disease: critical evaluations. Rev Neurosci. 1994;5: 213–25.
79. Casado A, Encarnacion Lopez-Fernandez M, Concepcion Casado Mde La Torre R. Lipid peroxidation and antioxidant enzyme activities in vascular and Alzheimer dementias. Neurochem Res. 2008;33:450–8.
80. Snaedal J, Kristinsson J, Gunnarsdottir S, Olafsdottir, Baldvinsson M, Johannesson T. Copper, ceruloplasmin and superoxide dismutase in patients with Alzheimer's disease. A case-control study. Dement Geriatr Cogn Disord. 1998; 9: 239–42.
81. Kharrazi H, Vaisi-Raygani A, Rahimi Z, Tavilani H, Aminian M, Pourmotabbed T. Association between enzymatic and non-enzymatic antioxidant defense mechanism with apolipoprotein E genotypes in Alzheimer disease. Clin Biochem. 2008;41:932–6.

82. Ozcankaya R, Delibas N. Malondialdehyde, superoxide dismutase, melatonin, iron, copper, and zinc blood concentrations in patients with Alzheimer disease: cross-sectional study. Croat Med J. 2002;43:28–32.
83. Rossi L, Squitti R, Pasqualetti P, Marchese E, Cassetta E, Forastiere E, Rotilio G, Rossini PM, Finazzi-Agro A. Red blood cell copper, zinc superoxide dismutase activity is higher in Alzheimer's disease and is decreased by D-penicillamine. Neurosci Lett. 2002;329:137–40.
84. De Leo ME, Borrello S, Passantino M, Palazzotti B, Mordente A, Daniele A, Filippini V, Galeotti T, Masullo C. Oxidative stress and overexpression of manganese superoxide dismutase in patients with Alzheimer's disease. Neurosci Lett. 1998;250:173–6.
85. Ihara Y, Hayabara T, Sasaki K, Fujisawa Y, Kawada R, Yamamoto T, Nakashima Y, Yoshimune S, Kawai M, Kibata M, et al. Free radicals and superoxide dismutase in blood of patients with Alzheimer's disease and vascular dementia. J Neurol Sci. 1997;153:76–81.
86. Padurariu M, Ciobica A, Hritcu L, Stoica B, Bild W, Stefanescu C. Changes of some oxidative stress markers in the serum of patients with mild cognitive impairment and Alzheimer's disease. Neurosci Lett. 2010;469:6–10.
87. Picco A, Polidori MC, Ferrara M, Cecchetti R, Arnaldi D, Baglioni M, Morbelli S, Bastiani P, Bossert I, Fiorucci G, et al. Plasma antioxidants and brain glucose metabolism in elderly subjects with cognitive complaints. Eur J Nucl Med Mol Imaging. 2014;41:764–75.
88. Pincemail J, Cillard J, Neve J, Defraigne JO. [Determination of the plasma global antioxidant capacity: a critical review]. Ann Biol Clin (Paris). 2014;72:413–21.
89. Torres LL, Quaglio NB, de Souza GT, Garcia RT, Dati LM, Moreira WL, Loureiro AP, de Souza-Talarico JN, Smid J, Porto CS, et al. Peripheral oxidative stress biomarkers in mild cognitive impairment and Alzheimer's disease. J Alzheimers Dis. 2011;26:59–68.
90. Martin-Aragon S, Bermejo-Bescos P, Benedi J, Felici E, Gil P, Ribera JM, Villar AM. Metalloproteinase's activity and oxidative stress in mild cognitive impairment and Alzheimer's disease. Neurochem Res. 2009;34:373–8.
91. Krishnan S, Rani P. Evaluation of selenium, redox status and their association with plasma amyloid/tau in Alzheimer's disease. Biol Trace Elem Res. 2014;158:158–65.
92. Bermejo P, Gomez-Serranillos P, Santos J, Pastor E, Gil P, Martin-Aragon S. Determination of malonaldehyde in Alzheimer's disease: a comparative study of high-performance liquid chromatography and thiobarbituric acid test. Gerontology. 1997;43:218–22.
93. Revel F, Gilbert T, Roche S, Drai J, Blond E, Ecochard R, Bonnefoy M. Influence of oxidative stress biomarkers on cognitive decline. J Alzheimers Dis. 2015;45:553–60.
94. Famulari AL, Marschoff ER, Llesuy SF, Kohan S, Serra JA, Dominguez RO, Repetto M, Reides C, Sacerdote de Lustig E. The antioxidant enzymatic blood profile in Alzheimer's and vascular diseases. Their association and a possible assay to differentiate demented subjects and controls. J Neurol Sci. 1996;141:69–78.
95. Vural H, Demirin H, Kara Y, Eren I, Delibas N. Alterations of plasma magnesium, copper, zinc, iron and selenium concentrations and some related erythrocyte antioxidant enzyme activities in patients with Alzheimer's disease. J Trace Elem Med Biol. 2010;24:169–73.
96. Cao G, Prior RL. Measurement of oxygen radical absorbance capacity in biological samples. Methods Enzymol. 1999;299:50–62.
97. Rice-Evans C, Miller NJ. Total antioxidant status in plasma and body fluids. Methods Enzymol. 1994;234:279–93.
98. Sies H. Total antioxidant capacity: appraisal of a concept. J Nutr. 2007;137:1493–5.
99. Butterfield DA, Perluigi M, Sultana R. Oxidative stress in Alzheimer's disease brain: new insights from redox proteomics. Eur J Pharmacol. 2006;545:39–50.
100. Singh M, Nam DT, Arseneault M, Ramassamy C. Role of by-products of lipid oxidation in Alzheimer's disease brain: a focus on acrolein. J Alzheimers Dis. 2010;21(3):741–56.
101. Migliore L, Fontana I, Trippi F, Colognato R, Coppede F, Tognoni G, Nucciarone B, Siciliano G. Oxidative DNA damage in peripheral leukocytes of mild cognitive impairment and AD patients. Neurobiol Aging. 2005;26:567–73.
102. Stadtman ER, Berlett BS. Reactive oxygen-mediated protein oxidation in aging and disease. Chem Res Toxicol. 1997;10:485–94.

103. Vitek MP, Bhattacharya K, Glendening JM, Stopa E, Vlassara H, Bucala R, Manogue K, Cerami A. Advanced glycation end products contribute to amyloidosis in Alzheimer disease. Proc Natl Acad Sci U S A. 1994;91:4766–70.
104. Sultana R, Poon HF, Cai J, Pierce WM, Merchant M, Klein JB, Markesbery WR, Butterfield DA. Identification of nitrated proteins in Alzheimer's disease brain using a redox proteomics approach. Neurobiol Dis. 2006;22:76–87.
105. Conrad CC, Marshall PL, Talent JM, Malakowsky CA, Choi J, Gracy RW. Oxidized proteins in Alzheimer's plasma. Biochem Biophys Res Commun. 2000;275:678–81.
106. Aldred S, Bennett S, Mecocci P. Increased low-density lipoprotein oxidation, but not total plasma protein oxidation, in Alzheimer's disease. Clin Biochem. 2010;43:267–71.
107. Korolainen MA, Nyman TA, Nyyssonen P, Hartikainen ES, Pirttila T. Multiplexed proteomic analysis of oxidation and concentrations of cerebrospinal fluid proteins in Alzheimer disease. Clin Chem. 2007;53:657–65.
108. Sinem F, Dildar K, Gokhan E, Melda B, Orhan Y, Filiz M. The serum protein and lipid oxidation marker levels in Alzheimer's disease and effects of cholinesterase inhibitors and antipsychotic drugs therapy. Curr Alzheimer Res. 2010;7:463–9.
109. Laske C, Leyhe T, Stransky E, Hoffmann N, Fallgatter AJ, Dietzsch J. Identification of a blood-based biomarker panel for classification of Alzheimer's disease. Int J Neuropsychopharmacol. 2011;14:1147–55.
110. Gilbert HF. Molecular and cellular aspects of thiol-disulfide exchange. Adv Enzymol Relat Areas Mol Biol. 1990;63:69–172.
111. Dalle-Donne I, Rossi R, Giustarini D, Colombo R, Milzani A. S-glutathionylation in protein redox regulation. Free Radic Biol Med. 2007;43:883–98.
112. Munch G, Keis R, Wessels A, Riederer P, Bahner U, Heidland A, Niwa T, Lemke HD, Schinzel R. Determination of advanced glycation end products in serum by fluorescence spectroscopy and competitive ELISA. Eur J Clin Chem Clin Biochem. 1997;35:669–77.
113. Tanzi RE, Moir RD, Wagner SL. Clearance of Alzheimer's Abeta peptide: the many roads to perdition. Neuron. 2004;43:605–8.
114. Raucci A, Cugusi S, Antonelli A, Barabino SM, Monti L, Bierhaus A, Reiss K, Saftig P, Bianchi ME. A soluble form of the receptor for advanced glycation endproducts (RAGE) is produced by proteolytic cleavage of the membrane-bound form by the sheddase a disintegrin and metalloprotease 10 (ADAM10). FASEB J. 2008;22:3716–27.
115. Ott C, Jacobs K, Haucke E, Navarrete Santos A, Grune T, Simm A. Role of advanced glycation end products in cellular signaling. Redox Biol. 2014;2:411–29.
116. Liang F, Jia J, Wang S, Qin W, Liu G. Decreased plasma levels of soluble low density lipoprotein receptor-related protein-1 (sLRP) and the soluble form of the receptor for advanced glycation end products (sRAGE) in the clinical diagnosis of Alzheimer's disease. J Clin Neurosci. 2013;20:357–61.
117. Korkmaz GG, Altinoglu E, Civelek S, Sozer V, Erdenen F, Tabak O, Uzun H. The association of oxidative stress markers with conventional risk factors in the metabolic syndrome. Metabolism. 2013;62:828–35.
118. Uzun D, Yanar K, Atukeren P, Cebe T, Mengi M, Ozan T, Kunbaz A, Kuruc AI, Cakatay U, Aydin S. Age-related changes in rat prostate tissue; perspective of protein oxidation. Aging Male. 2015;18:54–9.
119. Dang TN, Arseneault M, Zarkovic N, Waeg G, Ramassamy C. Molecular regulations induced by acrolein in neuroblastoma SK-N-SH cells: relevance to Alzheimer's disease. J Alzheimers Dis. 2010;21:1197–216.
120. Nam DT, Arseneault M, Ramassamy C. Regulation of redox-sensitive signaling pathways in rat primary astrocytes following acrolein exposure. J Alzheimers Dis. 2011;25:263–77.
121. Galbusera C, Facheris M, Magni F, Galimberti G, Sala G, Tremolada L, Isella V, Guerini FR, Appollonio I, Galli-Kienle M, et al. Increased susceptibility to plasma lipid peroxidation in Alzheimer disease patients. Curr Alzheimer Res. 2004;1:103–9.
122. Pincemail J. Contribution à l'étude du rôle de l'oxygène et des radicaux libres chez l'homme et l'animal. Thèse Doctorat en Sciences Biomédicales Expérimentales, Université 1990.

123. Yin H. New techniques to detect oxidative stress markers: mass spectrometry-based methods to detect isoprostanes as the gold standard for oxidative stress in vivo. Biofactors. 2008; 34:109–24.
124. Miller E, Morel A, Saso L, Saluk J. Isoprostanes and neuroprostanes as biomarkers of oxidative stress in neurodegenerative diseases. Oxid Med Cell Longev. 2014;2014:572491.
125. Pratico D, Clark CM, Liun F, Rokach J, Lee VY, Trojanowski JQ. Increase of brain oxidative stress in mild cognitive impairment: a possible predictor of Alzheimer disease. Arch Neurol. 2002;59:972–6.
126. Lauretti E, Di Meco A, Chu J, Pratico D. Modulation of AD neuropathology and memory impairments by the isoprostane F2alpha is mediated by the thromboxane receptor. Neurobiol Aging. 2015;36:812–20.
127. McGrath LT, McGleenon BM, Brennan S, McColl D, McILroy S, Passmore AP. Increased oxidative stress in Alzheimer's disease as assessed with 4-hydroxynonenal but not malondialdehyde. QJM. 2001;94:485–90.
128. Selley ML, Close DR, Stern SE. The effect of increased concentrations of homocysteine on the concentration of (E)-4-hydroxy-2-nonenal in the plasma and cerebrospinal fluid of patients with Alzheimer's disease. Neurobiol Aging. 2002;23:383–8.
129. Klawitter J, Haschke M, Shokati T, Christians U. Quantification of 15-F2t-isoprostane in human plasma and urine: results from enzyme-linked immunoassay and liquid chromatography/tandem mass spectrometry cannot be compared. Rapid Commun Mass Spectrom. 2011; 25:463–8.
130. Janicka M, Kot-Wasik A, Kot J, Namiesnik J. Isoprostanes-biomarkers of lipid peroxidation: their utility in evaluating oxidative stress and analysis. Int J Mol Sci. 2010;11:4631–59.
131. Nam DT, Arseneault M, Murthy V, Ramassamy C. Potential role of acrolein in neurodegeneration and in Alzheimer's disease. Curr Mol Pharmacol. 2010;3:66–78.
132. Huang YJ, Jin MH, Pi RB, Zhang JJ, Ouyang Y, Chao XJ, Chen MH, Liu PQ, Yu JC, Ramassamy C, et al. Acrolein induces Alzheimer's disease-like pathologies in vitro and in vivo. Toxicol Lett. 2013;217:184–91.
133. Marcello A, Wirths O, Schneider-Axmann T, Degerman-Gunnarsson M, Lannfelt L, Bayer TA. Circulating immune complexes of Abeta and IgM in plasma of patients with Alzheimer's disease. J Neural Transm. 2009;116:913–20.
134. Noguchi-Shinohara M, Hamaguchi T, Nozaki I, Sakai K, Yamada M. Serum tau protein as a marker for the diagnosis of Creutzfeldt-Jakob disease. J Neurol. 2011;258:1464–8.
135. Bielewicz J, Kurzepa J, Czekajska-Chehab E, Stelmasiak Z, Bartosik-Psujek H. Does serum Tau protein predict the outcome of patients with ischemic stroke? J Mol Neurosci. 2011;43:241–5.
136. Liliang PC, Liang CL, Weng HC, Lu K, Wang KW, Chen HJ, Chuang JH. Tau proteins in serum predict outcome after severe traumatic brain injury. J Surg Res. 2010;160:302–7.
137. Chalbot S, Zetterberg H, Blennow K, Fladby T, Andreasen N, Grundke-Iqbal I, Iqbal K. Blood-cerebrospinal fluid barrier permeability in Alzheimer's disease. J Alzheimers Dis. 2011;25:505–15.
138. Zipser BD, Johanson CE, Gonzalez L, Berzin TM, Tavares R, Hulette CM, Vitek MP, Hovanesian V, Stopa EG. Microvascular injury and blood-brain barrier leakage in Alzheimer's disease. Neurobiol Aging. 2007;28:977–86.
139. Pincemail J. Oxidative stress or not in healthy older subjects ? In: Oxygen club of California world congress, Valencia, Spain 24–26 June 2015; 2015.
140. Tammen H, Hess R. Collection and handling of blood specimens for peptidomics. Methods Mol Biol. 2013;1023:161–8.
141. Lista S, Faltraco F, Hampel H. Biological and methodical challenges of blood-based proteomics in the field of neurological research. Prog Neurobiol. 2013;101–102:18–34.
142. Henriksen K, O'Bryant SE, Hampel H, Trojanowski JQ, Montine TJ, Jeromin A, Blennow K, Lonneborg A, Wyss-Coray T, Soares H, et al. The future of blood-based biomarkers for Alzheimer's disease. Alzheimers Dement. 2014;10:115–31.

143. O'Bryant SE, Gupta V, Henriksen K, Edwards M, Jeromin A, Lista S, Bazenet C, Soares H, Lovestone S, Hampel H, et al. Guidelines for the standardization of preanalytic variables for blood-based biomarker studies in Alzheimer's disease research. Alzheimers Dement. 2015; 11:549–60.
144. Gubandru M, Margina D, Tsitsimpikou C, Goutzourelas N, Tsarouhas K, Ilie M, Tsatsakis AM, Kouretas D. Alzheimer's disease treated patients showed different patterns for oxidative stress and inflammation markers. Food Chem Toxicol. 2013;61:209–14.

Index

S.C. Bondy, A. Campbell (eds.), *Inflammation, Aging, and Oxidative Stress*, Oxidative Stress in Applied Basic Research and Clinical Practice,
DOI 10.1007/978-3-319-33486-8

MIX
Papier aus verantwortungsvollen Quellen
Paper from responsible sources
FSC® C105338

If you have any concerns about our products,
you can contact us on
ProductSafety@springernature.com

In case Publisher is established outside the EU,
the EU authorized representative is:
Springer Nature Customer Service Center GmbH
Europaplatz 3, 69115 Heidelberg, Germany

Printed by Libri Plureos GmbH
in Hamburg, Germany